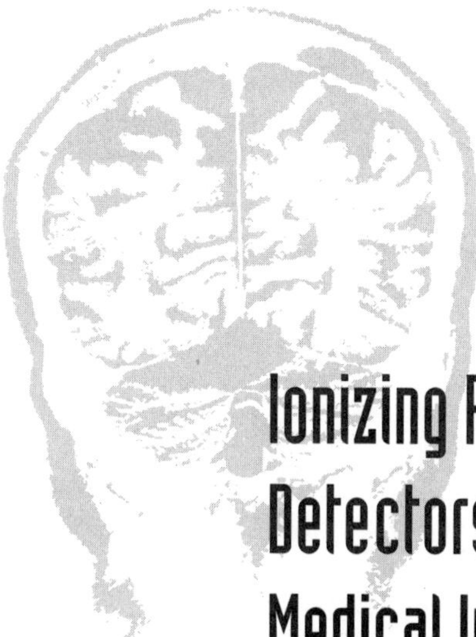

Ionizing Radiation Detectors for Medical Imaging

Ionizing Radiation Detectors for Medical Imaging

Alberto Del Guerra
University of Pisa, Italy

World Scientific

NEW JERSEY • LONDON • SINGAPORE • SHANGHAI • HONG KONG • TAIPEI • CHENNAI

Published by

World Scientific Publishing Co. Pte. Ltd.

5 Toh Tuck Link, Singapore 596224

USA office: 27 Warren Street, Suite 401–402, Hackensack, NJ 07601

UK office: 57 Shelton Street, Covent Garden, London WC2H 9HE

British Library Cataloguing-in-Publication Data
A catalogue record for this book is available from the British Library.

IONIZING RADIATION DETECTORS FOR MEDICAL IMAGING

ISBN 981-238-674-2

Printed by Fulsland Offset Printing (S) Pte Ltd, Singapore

CONTENTS

To Marta, Simone and Nicolò

FOREWORD

The idea of this book originates from a series of lectures on "Detectors Application in Medicine and Biology" that I was asked to give as part of the Academic Training Program at CERN in 1995. In preparing the series of lectures, I realized that I would be talking about detector properties and the medical applications of these detectors to the scientists and engineers who were their inventors. Initially, this realization scared me but, soon after the lectures were delivered, it convinced me about the necessity of writing a book dedicated to detectors for medical imaging, where the properties of the detectors were to be discussed specifically in relation to each medical or clinical applications.

This book is the outcome of this conviction. It took quite a while to become a reality due to the many sub-specialities in Medical Imaging I wanted to be addressed. Intentionally, this book's coverage is limited to Ionizing Radiation Detectors; thus Ultrasound, Magnetic Resonance Imaging and Spectroscopy and other non-Ionizing Radiation Detectors have not been considered.

The book comprises a brief "Introduction" and ten technical chapters, almost 50% of which are dedicated to Radiology and 50% to Nuclear Medicine. The last chapter describes the detectors for Radiotherapy and Portal Imaging. Each chapter completely addresses a specific application. Hence, some properties of one class of detectors may be described or discussed in more than one chapter but I consider this to be a plus. The emphasis is always on detectors and detector properties. When necessary, software and specific applications are described in depth.

The book is intended for students in physics and engineers who want to study Medical Imaging, both at undergraduate and post-graduate level. Scientists, who are working in a specific sub-field of medical imaging, could use this book to acquire an up-to-date description of the state-of-the-art in other related sub-fields, alas within the scope of ionizing radiation detectors. Other Scientists and Physicians could use this book as a reference for Medical Imaging.

Many thanks are due to the various contributors who agreed to write the many chapters, and have been patient with me whilst the book was completed. A special acknowledgement is due to Dr. Deborah Herbert for her careful reading of some chapters.

Pisa, January 31, 2004

Alberto Del Guerra

LIST OF CONTRIBUTORS

Alberto Del Guerra

obtained his degree in Physics at the University of Pisa (Italy) in 1968. He was Visiting Professor and Fulbright Scholar at the Lawrence Berkeley Laboratory, University of California, USA in 1981-82 and became Associate Professor of Physics at the University of Pisa, Italy in 1982. He was then Full Professor of Physics at the University of Napoli "Federico II", Italy (1987-1991), Full Professor of Medical Physics at the University of Ferrara, Italy (1991-1998) and since 1998 he has been Full Professor of Medical Physics at the University of Pisa, Italy. He is Director and Head of the Specialty School in Medical Physics, University of Pisa, Italy. His research activities in the last 25 years have been in the field of Medical Physics, and particularly in medical imaging for radiology and nuclear medicine. PET scanners for functional imaging with small animals has also been a particular focus. He is Editor in Chief of the journal "Physica Medica-European Journal of Medical Physics" and President of the European Federation of Organizations of Medical Physics (EFOMP).

Maria Evelina Fantacci

obtained her degree in Physics in 1992 at the University of Pisa (Italy), where she has held the position of Physics Researcher at the Physics Department since 1997. She has always worked in the field of medical physics. Her present research activities focus on digital mammography. In particular: the use of semiconductor pixel detectors connected via bump-bonding to single photon counting electronics and the development of Computer Aided Detection tool for automated classification of textures and search of micro-calcification clusters and massive lesions.

Andreas Formiconi

holds the position of Physics Researcher at the University of Florence. His research focuses on Single Photon Emission Tomography and on the development of physical and mathematical methods for the solution of inverse problems in the bio-medical field.

Theobald O.J. Fuchs

received his diploma in physics from the University Erlangen Nürnberg, Germany, in 1994. His diploma thesis was partly completed at the European Laboratory for Particle Physics (CERN) in the field of experimental particle physics. He did his Ph.D. research with Willi Kalender at the Institute of Medical Physics (IMP), Erlangen, Germany. Since 1999, he has been working as a fellow researcher at the IMP. His present research activities focus on image quality analysis in CT, artefact corrections, X-ray detectors, and dose optimization.

Willi A. Kalender

received his Masters Degree and Ph.D. in Medical Physics from the University of Wisconsin, Madison, Wisconsin, USA in 1979. In 1988 he completed all postdoctoral lecturing qualifications (Habilitation) for Medical Physics at the University of Tübingen. From 1979 to 1995 he worked in the research laboratories of Siemens Medical Systems in Erlangen. Since 1991 he has been an adjunct Associate Professor of Medical Physics at the University of Wisconsin; from 1993 to 1995 he lectured at the Technical University of Munich. In 1995 he was appointed a full professor and the director of the newly established Institute of Medical Physics at the University Erlangen Nürnberg, Germany. His main research interests lie in the area of diagnostic imaging. The introduction and further development of volumetric spiral CT has been a particular focus. Other fields of research are radiation protection and the development of quantitative diagnostic procedures, e.g. for the assessment of osteoporosis, lung and cardiac diseases.

Ralf Hendrik Menk

obtained the Diplom Physiker at the University of Siegen, Germany, in 1990. He was a Ph.D. student at the University of Siegen, and concluded with the degree "Dr.rer.nat" in 1995. He has been a postdoctoral fellow at the medical beam line at HASYLAB at DESY, Hamburg and a postdoctoral fellow at the medical beam line at the National Synchrotron Light Source at Brookhaven National Lab., Upton, New York, USA, and an Assistant Professor at the University of Siegen, Germany. Since 1999 he has been the head of the Instrumentation and Detector Group at the Sincrotrone Trieste, Italy. His research interest is in the development of instrumentation for synchrotron radiation experiments.

Alfonso Motta

obtained his degree in Physics in 1997 and his PhD in 2000 at the University of Ferrara (Italy). He is working on 3-D algorithms and image reconstruction in Medical Imaging. He now holds a postdoctoral appointment at the Department of Physics "E.Fermi", University of Pisa (Italy).

Roberto Pani

obtained his degree in Physics at the University of Roma "La Sapienza", Italy. He has been appointed an adjunct Associate Professor at the Department of Radiology of Georgetown University in Washington, USA. He holds the position of Associate Professor in Medical Physics at the University of Rome "La Sapienza". He has been working for more than twenty years in the field of advanced detectors for X-ray and gamma-ray spectrometry using scintillators and semiconductors. Since 1990 he has been working on single photon imaging detectors for Nuclear Medicine, with particular emphasis on small animal imaging and scintimammography (SPEM).

Mike Partridge

gained his PhD in 1995 in applied optics and image processing at Cranfield University, UK. He joined The Institute of Cancer Research (Sutton, UK) as a postdoctoral researcher in 1996 and worked as part of a team developing the use of electronic portal imaging for radiotherapy verification. From 2000 until 2002 he worked at the German Cancer Research Centre (DKFZ) in Heidelberg. In 2002 he moved back to the UK and took up a joint position with the Institute of Cancer Research and the Royal Marsden NHS Trust, focusing on the clinical implementation of advanced radiotherapy techniques and research into the use of functional imaging for radiotherapy target definition.

Alessandro Passeri

obtained his degree in Physics in 1985 at the University of Florence, Italy. He initially worked in theoretical nuclear physics before moving to the field of Medical Imaging, where he obtained his PhD in 1990. He holds the position of Physics Researcher at the University of Florence. Since 1996 he has been the director of the research centre on Magnetic Resonance Imaging at the University of Florence. His research focuses on Single Photon Emission Tomography and on the development of physical and mathematical methods for the solution of inverse problems in the bio-medical field.

Paolo Russo

graduated in Physics at the University of Napoli in 1981. He became a Physics Researcher in 1984, an Associate Professor in 1992 and a Full Professor of Medical Physics in 2002 at the Department of Physical Sciences of the University of Napoli “Federico II”, Italy. His research activity has always been in the field of Medical Physics. His scientific interests now focus on the development of semiconductor radiation detectors for medical imaging, for digital radiography and autoradiography. He has been developing silicon microstrip detectors and silicon or GaAs pixel detectors for single photon counting X-ray and beta-ray imaging applications.

Angelo Taibi

obtained his degree in Physics at the University of Ferrara and his PhD in Medical Physics at the same University in 1997. In 1998 he was awarded a Marie Curie Fellowship, to engage his research as a post-doc at the Department of Medical Physics and Bioengineering, University College London, UK. He now holds a post-doc position at the Department of Physics at the University of Ferrara. His research interests concern the physics of X-ray imaging and in particular: optimization of mammography examination, dose reduction by means of quasi-monochromatic X-rays and diffraction enhanced breast imaging.

Guido Zavattini

obtained his degree in physics at the University of Pisa, Italy, working on an experiment to verify the Equivalence Principle with optical techniques, and his PhD in Physics at the University of Bologna, Italy on the development of a fast neutron detector with double pulse shape discrimination. A one-year post-doc fellowship, working for INFN, Trieste, Italy, brought him back to working on optics in a collaboration to measure the magnetic birefringence of vacuum. He now holds the position of Physics Researcher at the University of Ferrara, Italy. In 2002–2003 he spent a sabbatical year at the University of Davis, California, USA. His research interest focuses on position sensitive gamma ray detectors for small animal PET and SPECT imaging.

ACKNOWLEDGMENTS

We would like to explicitly acknowledge the following Publishers, Companies and Authors for the granted permission for the reproduction of the following figures and table in this book:

- the Hamamatsu Photonics K.K., for Figs. 2.4, 9.6, 10.4;
- the Philips' Gloeilampenfabrieken, for Fig. 2.6;
- the International Commission on Radiation Units and Measurements, for Figs. 2.13, 2.14;
- the IEEE Nuclear and Plasma Sciences Society, for Figs. 2.16, 3.21, 3.22, 3.24, 3.27, 3.48, 3.49, 3.50b, 3.51, 3.52, 3.53, 3.54, 7.4, 7.5, 7.6, 10.14, 10.15, 10.16, 10.35, 10.50, 10.53, 10.54, 10.55, 10.58, 10.62, 10.63, 10.64;
- the Radiological Society of North America, for Fig. 2.20;
- the Elsevier Science, for Figs. 3.4, 3.6, 3.8, 3.16, 3.17, 3.30, 3.31, 3.34, 3.39, 3.40, 7.8, 7.9, 8.17, 8.18, 9.1, 9.2, 9.3, 9.5, 9.8, 9.9, 9.10, 9.11, 9.13, 9.14, 9.15, Table 9.1;
- the American Institute of Physics, for Figs. 3.9, 3.15;
- the Nuclear Technology Publishing, for Fig. 3.18;
- the IOP, for Figs. 3.35, 3.41;
- the Publicis KommunikationsAgentur GmbH, for Figs 4.1, 4.2, 4.3, 4.4, 4.5, 4.6, 4.7, 4.8, 4.9, 4.10, 4.11;
- Dr. Bill Ashburn, for Fig. 7.8;
- the American Society of Nuclear Cardiology, for Fig. 7.10;
- the University of Chicago Press, for Fig. 8.5;
- the Nuovo Cimento, for Fig. 8.9;
- the Istituti Editoriali e Poligrafici Internazionali, for Fig. 9.16;
- the Springer-Verlag, for Figs. 10.40, 10.49, 10.61;
- the World Scientific Ltd, for Fig. 11.9.

CHAPTER 1

INTRODUCTION

Alberto Del Guerra
Department of Physics, University of Pisa and INFN,
Sezione di Pisa, Pisa, Italy
(e-mail: alberto.delguerra@df.unipi.it)

1.1 Medical Imaging

The enormous development in detectors for Medical Imaging has been largely due to contributions from the technological innovations in other fields of physics, such as solid state physics, space physics and high energy Physics (HEP). In particular the latter has greatly contributed greatly to the development of new types of detectors, particularly for ionizing radiation. However, it is extremely important to recall that a detector for medical applications is a "*special detector*", the performance specifications being related to patient comfort, diagnosis and eventual therapy. In this respect, some specific concerns will be presented in the next session.

Table 1.1 lists the major fields in medical diagnosis together with the parameters measured and the specific medical applications. The progress in the last twenty years has only been possible with the advent of integrated electronics and fast computers. This has permitted the shift from analog to digital imaging, with more quantitative information being available for diagnosis and prognosis.

X-ray radiology is of course the primary field: radiology has moved from 2-D to 3-D (with the advent of the CT-scanners) and to Digital techniques. It was more than a hundred years ago that the first radiograph was taken by William Roentgen and since then film has been used as "*the detector*" for radiological examination. Only CT-scanners have always made use of digital detectors, and there is still a lot of research in the field of 2-D radiology for developing direct and indirect digital devices instead of film and film-screen systems.

Ultrasound (US) has been a dormant field for many years, but it is now receiving more and more attention particularly because of the absence of the ionizing radiation hazard. The possibility of performing endocavitary US by means of very small probes has a tremendous relevance in clinical practice.

Table 1.1. Bio-Medical Imaging

	(Physical Parameters measured)	*(Medical application)*
X-RAY RADIOLOGY 2-D Film X-ray CT Digital Subtraction Angiography	X-ray absorption Density and average Z Contrast distribution	Anatomy; mineral content; movement of contrast material
ULTRASOUND	Acoustic impedance mismatch; Sound velocity and attenuation	Anatomy; tissue structural characteristics; blood velocity
NUCLEAR MEDICINE Planar scintigraphy, SPECT, PET	Concentration of radionuclides	Metabolism; receptor site Concentration and flow
NUCLEAR MAGNETIC RESONANCE, MRI, MRS	Concentration of nuclides (^{1}H); Relaxation parameters T1 and T2 Frequency shift	Anatomy of tissues; free water content; flow concentration of some molecular species and contrast agents
OTHER Techniques: Biomagnetism, BioImpededance, Optical Imaging,...		

Nuclear Medicine is a specific field where high energy physics has contributed a lot of ideas and innovations; the Anger camera, which started it all, stems in fact from research in new scintillators (NaI) carried out by Hofstadter at Stanford for gamma-ray detection in HEP. Various subspecialities of medicine, such as physiology, neurophysiology, neurology and cardiology, take advantage of this technique, which couples routine clinical applications to medical research. Being digital from the beginning Nuclear Medicine has always been a fertile field for innovation in detectors.

Nuclear magnetic resonance has two major subfields: magnetic resonance imaging (MRI) and magnetic resonance spectroscopy (MRS). MRI has had a major impact in diagnostic radiology in the last 20 years, especially for the clinical study of tissues pathologies and cerebral and heart perfusion.

Finally, other techniques like biomagnetism and bioimpedance are still fighting to enter the clinical arena. It is beyond the purpose of this book to discuss all of these techniques and all of the detectors that are used. I have decided to restrict the topics to the field of ionizing radiation and in particular to the detectors developed and used in radiology and nuclear medicine.

1.2 Ionizing Radiation Detector Development: High Energy Physics versus Medical Physics

Let us consider what happens in high energy physics. There is a fundamental problem of particle physics and we want to make an experiment to solve this problem. We then build the most appropriate detector or series of detectors for that specific experiment. Similarly in medical physics: first of all one has to have a good understanding of the experiment he wants to perform; then one must develop the appropriate detector. It would be quite wrong to say: "I have this very nice detector. Let me find the best experiment for it!"

Let us be more explicit with one example. Let us consider a well known type of detector for experiments in HEP: the microstrip silicon crystal for charged particle tracking. This detector has been specifically developed for high energy physics (see Table 1.2): it has a typical dimension of 5×5 cm^2 and four modules can be assembled together to produce an overall size of 5×20 cm^2; the standard thickness is 300 μm, but can be made thicker. A minimum ionizing particle (MIP) in 300 μm of silicon has an energy loss of about 70 keV; the electronics which has been built is reasonably fast with low noise; integration is available. This electronics needs a trigger and the data acquisition is set up for colliders, with low multiplicity and sparse readout via multiplexer between two bunch crossings. The number of channels is $10^6 – 10^7$, and even greater. The event size is 10^6 bytes (at level #1 trigger). The number of sellable apparatus is one, maybe two, if a confirmation experiment is needed; but of course each apparatus is made of thousands of modules!

Let us now move to Medical Physics: one has the same detector (silicon microstrip crystal) and would like to use it for digital radiography. In Table 1.3 the corresponding list of problems to be solved for this specific application is presented. A typical size for X-ray imaging is 20×20 cm^2. Thus, one should try to put the silicon detector modules in a different geometry with new topological problems. The thickness (300 μm) is probably OK for low energy X-ray, but for the 70 keV (chest radiography), the efficiency is extremely low.

Therefore one must use a thicker counter, at least more than one mm, but a similar result can be achieved by stacking several detectors, because of the necessity to keep a reasonably fast time for the transit of the electrons through each crystal.

Electronics: one does not detect MIP's, i.e. 70 keV energy released. In mammography for instance one has to detect X-rays with an energy as low as 18 keV. Hence the electronics have to be with a very low noise performance but also must be very fast, much faster than for the HEP case. The integration of the electronics for the data taking is mandatory.

Table 1.2. Problems solved for HEP experiments

1.	μ-strip silicon detector for charged particle tracking
2.	Typical dimension: 4× (5×5 cm^2); thickness ≤ 500 μm
3.	Electronics for MIP (in 300 μm ≈ 70 keV energy loss) low noise: 500 – 1000 e$^-$ reasonably fast: 100 – 1000 ns integration on VLSI
4.	External Trigger
5.	DAQ for collider low multiplicity fast acquisition sparse readout
6.	Number of channels: $10^5 - 10^7$
7.	Event size(raw data): 10^6 bytes (level 1 trigger)
8.	Number of sellable apparatus: 1 (maybe two!!) But the apparatus is made of thousands of modules!

Table 1.3. Problems to be tackled for Imaging with X-rays (10-100 keV): Digital Radiology

1.	μ-strip silicon detector for X-rays
2.	Required dimension: 20×20 cm^2; thickness (300 μm – 3 mm)
3.	Electronics for X-rays (down to 10 keV) low noise: 200 e^- fast: 10 – 100 ns integration on VLSI
4.	Self-Triggering
5.	DAQ for Digital Radiology 5×10^4 Hz/mm^2 (on a 20×20 cm^2 2×10^9Hz) 1 s acquisition time (duty cycle 100%)
6.	Number of channels: 10^3 – 10^4
7.	event size: 1 bit – 10 bytes
8.	Number of sellable apparatus: 10^3 – 10^6

There is another difference: there is no trigger (there is no bunch crossing); hence the electronics has to be either self-triggering or free running. In order to have reasonable image contrast, the number of the photons that one has to collect is of the order 5×10^4 photons · mm^{-2} · s^{-1}. If one considers that the dimensions are 20×20 cm^2, and the time to take a radiograph is typically one second then this produces a rate higher than one gigahertz! The standard readout pitch will be one hundred – two hundred μm: this ends up with 10^3 – 10^4 channels. (Channels, not modules.) Event size: this is ridiculously low, since each event consists of one bit: yes or no. One might have 10 bit if some kind of centre of mass positioning is performed. In any case, much less than the 10^6 bits of the HEP case. On the other hand, the number of sellable apparatus could be as large as 10^3 – 10^6. This is evident from looking at the number of films that have been sold and are currently sold.

In summary the experiment for medical physics presents so many different problems and requires so many different solutions that the HEP detector becomes a different and new detector.

1.3 Ionizing radiation detectors for medical imaging

Until recently, the film and film screen systems have had an unrivalled use despite its main limitations, i.e. the limited dynamic range (narrow latitude) and the lack of digital processing. In order to overcome these limitations many indirect and direct digital detectors for X-ray radiology have been proposed and are now in use.

The primary requirements for a digital X-ray detector for radiology are the speed, the area, the dose rate to exposure ratio and the spatial resolution. For any specific application, a trade-off is often necessary between spatial resolution and dynamic range, read-out speed and time resolution, contrast and exposure time. Furthermore, in assembling a detector for clinical applications a large area is often required and the problem of dead area must be solved, in addition to a manageable number of readout channels. The next three chapters of this book are devoted to Conventional Radiology, Digital Radiology and the paramount development of CT-scanners.

The main request from the medical point of view for a new detector is to improve the image quality for the same dose to the patient or to reduce the dose, whilst maintaining the same image quality. It is well known that the X-ray source spectrum plays an important role. Synchrotron radiation has been advocated as a possible optimized source both for improving the image contrast and for performing specific examinations. The topic is fully discussed in Chapter 5.

In the field of medical imaging there is a more and more frequent necessity to work on the biological, cellular, and sub-cellular aspects. In this respect autoradiography is the technique of choice that could act as a link between biological and clinical studies and also as a bridge between morphological and functional imaging. The detector development in this field has been quite remarkable and is illustrated in Chapter 6.

The following three chapters in the book deal with Nuclear Medicine, starting from conventional planar imaging, to Single Photon Emission Computed Tomography (SPECT), to Positron Emission Tomography (PET).

PET in particular has moved from a distinguished research tool in physiology, cardiology and neurology to become a major tool for clinical investigation in oncology, in cardiac applications and in neurological disorders. Many of the PET accomplishments are due to the terrific improvements both in hardware and software aspects, over the last ten years. Nowadays a similar effort is made by many research groups towards the construction of dedicated PET apparatus in new emerging fields such as molecular medicine, gene therapy, breast cancer imaging and combined modalities. Some of the recent developments in specific applications, such as small dedicated camera for functional imaging and small animal scanners are discussed in Chapters 9 and 10, respectively.

Finally the last chapter of the book deals with a discussion of the state-of-the-art detectors in Radiotherapy with particular emphasis to Portal Imaging.

1.4 Conclusions

In recent years new diagnostic and therapeutic methods have attracted more and more dedicated attention from the scientific community. The goal is a better understanding of the anatomy, physiology and pathology of the human being in an effort to find more appropriate medical prevention, diagnosis and therapy. Many of the achievements obtained so far are derived from the use and the optimization of new types of detectors. Such advances have been particularly relevant in the field of ionizing radiation detectors for Medical Imaging.

A further improvement is expected in the near future in Radiology and in Nuclear Medicine with the advent of new digital detectors, such as 1-D, 2-D and pixel solid state detectors, Microstrip and Microgap gas chambers, and new types of scintillator material coupled to advanced Position Sensitive Photodetectors.

CHAPTER 2

CONVENTIONAL RADIOLOGY

Angelo Taibi
Department of Physics, University of Ferrara
via Paradiso 12, I-44100 Ferrara, Italy
(E-mail: taibi@fe.infn.it)

2.1. Introduction

The sensitivity of X-ray film to direct X-ray exposure is low. To avoid large patient doses it is therefore desirable to use a more efficient imaging detector. This is accomplished by converting the X-ray image into light by means of a scintillating screen, also called X-ray intensifying screen, and then recording visible photons on film. The screen absorbs a large fraction of the incident X-rays and also provides signal amplification. The imaging properties of screen-film systems are reviewed in this chapter. Since the understanding of the physics of diagnostic radiology has evolved in parallel with the development of screen-film detectors this chapter also serves to introduce the fundamental physical parameters which define the quality of a radiographic image.

2.2. Physical Properties of X-ray Screens

A scintillating screen is composed of high Z phosphors which emit visible or near-visible light under X-ray irradiation. X-ray-to-light conversion is a multi-step process and a burst of light photons are emitted per X-ray interaction. The scintillation mechanism in inorganic materials is determined by the crystal nature of the phosphor [1]. The photoelectric effect is the dominant type of X-ray interaction within the screen and, due to the transferred energy, a photoelectron has the potential to release many electron-hole pairs, i.e. to move up electrons from the valence band to the conduction band. The return of an electron to the valence band with the emission of a photon is, however, a rather inefficient process because the bandgap energy might be too high to produce a photon in the visible range. If the pure crystal contains small amount of impurities, called *activators*, energy states are created in the forbidden gap. Therefore, as illustrated in Fig. 2.1, de-excitation occurs through these intermediate states and thus enhancing the probability of visible photon emission.

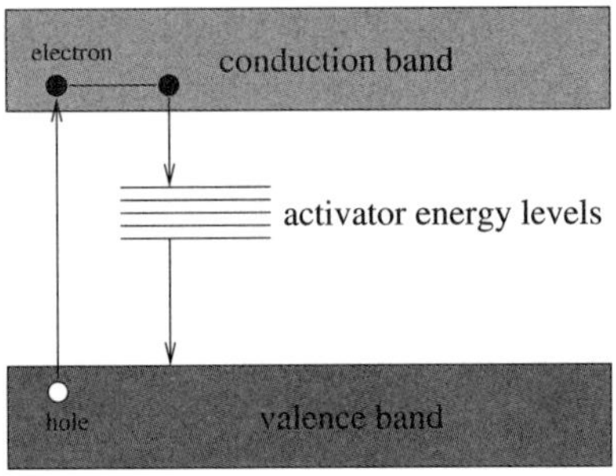

Fig. 2.1. Diagram of energy levels for scintillators used in conventional radiography.

For many years calcium tungstate ($CaWO_4$) was used and more recently terbium-activated rare-earth oxysulphides, such as Gd_2O_2S:Tb and Y_2O_2S:Tb,[a] have been used as phosphor screens because of their high X-ray absorption and high X-ray-to-light conversion efficiency compared to calcium tungstate phosphors [2]. The major advantage of using rare-earth phosphor screens is a reduction of patient exposure due to their high sensitivity.

A screen is generally made up of a plastic support layer, a thin backing layer, a phosphor layer, and a light-transparent protective coating (see Fig. 2.2). Depending upon the application, the backing layer may contain either a reflecting material to increase the light output or an absorbing material to reduce light spread. Phosphor grains (of 5–10 μm in diameter) are embedded in a low-Z transparent binder. The packing factor is generally of the order of 50%. Light diffusion within the screen limits the resolution properties to the order of the thickness of the phosphor layer, as shown in Fig. 2.3. Light spread can be minimized using thinner screens and light absorbing dye in the phosphor layer. Phosphor thickness (normally in the

[a]The name of the activator is generally indicated after the colon.

range 90–360 μm) is a matter of compromise between X-ray absorption and spatial resolution performance. Therefore it has to be optimized depending on the required task.

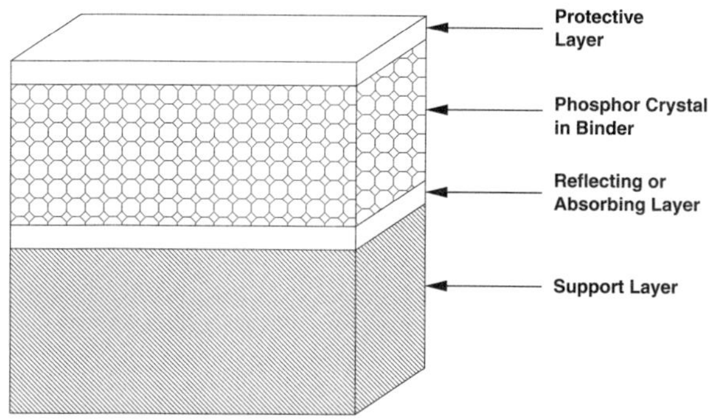

Fig. 2.2. Schematic of X-ray screen structure.

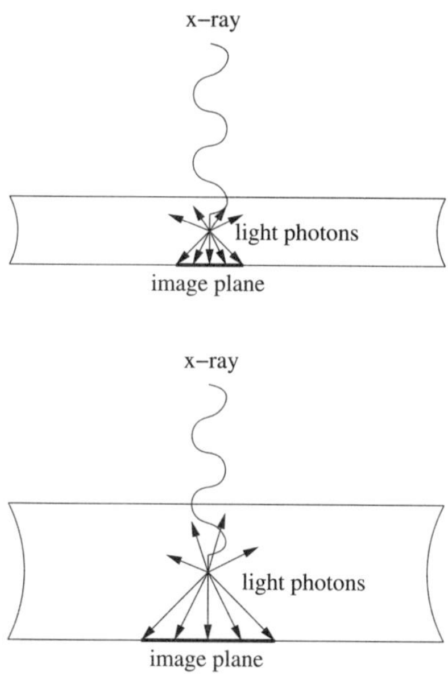

Fig. 2.3. A thick screen (bottom) causes greater light spread than a thin one (top).

Table 2.1 compares physical properties of screens of radiographic interest. It was compiled by adapting the data from various sources [2; 3; 4].

Table 2.1. Characteristics of various X-ray screens. The parameter η_c is the conversion efficiency.

Phosphor	Z (heavy element/s)	Density (g/cm^3)	K-edge energy (keV)	η_c (%)	Emission peak (nm)
$CaWO_4$	74 (W)	6.1	69.5	5.0	425
Gd_2O_2S:Tb	64 (Gd)	7.3	50.2	16	545
Y_2O_2S:Tb	39 (Y)	4.9	17.0	18	420
(Zn, Cd)S:Ag	30/48 (Zn/Cd)	4.8	9.66/26.7	19	530
CsI:Na	55/53 (Cs/I)	4.5	36.0/33.2	13	420
$YTaO_4$:Nb	39/73 (Y/Ta)	7.5	17.0/67.4	8.5	410

Very recently, a new generation of thallium-doped CsI scintillators have become available [5]. Since they are grown in columnar crystals the light emitted within the screen is channeled along the columns thus avoiding light spread as in conventional screens (see Fig. 2.4, from [6]). The advantage is that the traditional compromise between spatial resolution and X-ray absorption can be overcome, the disadvantage is that they are hygroscopic and too fragile for incorporation into screen-film cassettes.

2.2.1. *Screen Efficiency*

The efficiency η of a screen can be expressed as the product of three terms:

$$\eta = \eta_a \cdot \eta_c \cdot \eta_t \, . \tag{2.1}$$

The absorption efficiency η_a is the fraction of incident X-rays detected by the screen. It depends on screen composition and thickness. Figure 2.5 shows mass absorption coefficient μ/ρ as a function of X-ray energy for some phosphors (data were calculated with the program XCOM [7]). The figure also illustrates the effect of photon interaction with K-shell electrons, that largely increases screen efficiency above the K-edge energy.

The conversion efficiency η_c is the fraction of X-ray photon energy converted to light photon energy. The conversion efficiency ranges from 5% to 19% (see Table 2.1). It is worth noting that due to their high light emission, rare-earth screens allow the increase of system sensitivity. This requires, however, the use of a film whose response to the light emission spectrum is optimally matched [2]. The optical emission spectra of some

Fig. 2.4. Needle-shaped CsI scintillator (left) and grain-shaped Gd_2O_2S scintillator (right). A modeling of light dispersion within the phosphor is also shown (reprinted with permission from Hamamatsu Photonics K. K.).

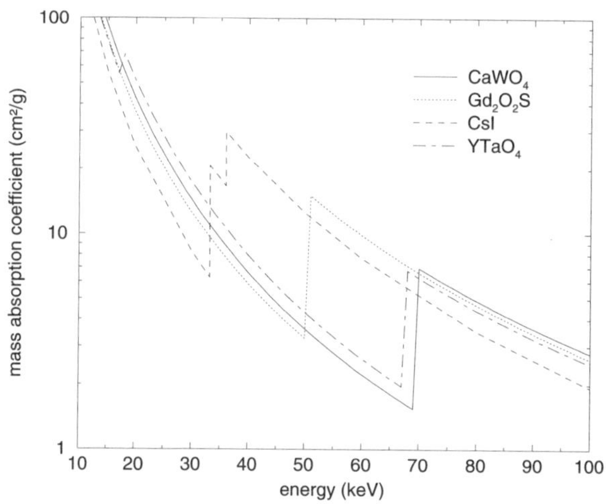

Fig. 2.5. Calculated mass absorption coefficient as a function of X-ray energy for various phosphor screens.

scintillating screens are shown in Fig. 2.6 together with the spectral response of X-ray films (from [8]). Unlike other phosphors that exhibit broad spectral emission curves, rare-earth screens are line emitters.

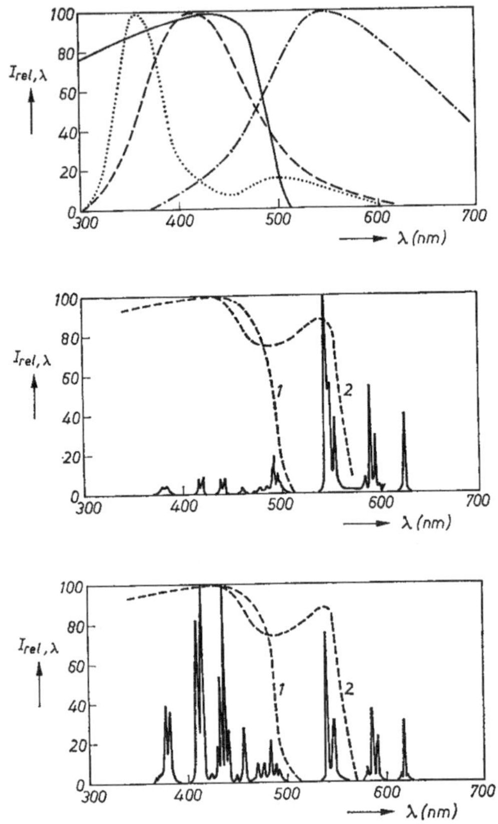

Fig. 2.6. Emission spectra of X-ray intensifying screens. Top: emission spectra of CsI:Na (dashed curve), CsI:Tl (dot-dashed curve), and CsBr:Tl (dotted curve); the continuous curve is the spectral response of a blue-sensitive X-ray film. Middle: emission spectrum of Gd_2O_2S:Tb. Bottom: emission spectrum of Y_2O_2S:Tb. The curves labeled 1 and 2 are the spectral response of a blue-sensitive and a green-sensitive X-ray film, respectively (reprinted with permission from Philips Electronics N. V.).

The transmission efficiency η_t is the fraction of emitted light photons which exit the screen. Light traveling within the phosphor can be absorbed before reaching the screen surface, especially in thick screens. Besides, due to the addition of dyes to increase spatial resolution, a further light absorption occurs in the screens. The *optical attenuation coefficient*, i.e. the rate at which the light is attenuated while it travels within the phosphor, depends on such factors as the relative indices of refraction of phosphor and binder materials, as well as on phosphor grain size and shape. Optical attenuation coefficients for various phosphors have been discussed by [3]. A typical value for the fraction of light escaping the screen is 50%.

As pointed out by [9], the efficiency of a screen can be illustrated by determining the number of light photons per incident X-ray photon which escape the screen to expose the film. Figure 2.7 shows the number of light photons for a Gd_2O_2S screen as a function of X-ray energy in the mammography range, calculated as follows:

(1) The absorption efficiency η_a as a function of the X-ray energy E is given by

$$\eta_a = 1 - e^{-\frac{\mu}{\rho}(E)t}\,. \tag{2.2}$$

A typical value for the coating weight t of a mammographic screen is 0.034 g/cm^2.

(2) When the energy E of an absorbed X-ray is converted with a 100% efficiency into light, the number g of light photons produced is given by the ratio E/E_l where $E_l = hc/\lambda$, and λ being the pertinent wavelength of emission peak. Since for Gd_2O_2S phosphors $\lambda = 545$ nm and the conversion efficiency is 16% (see Table 2.1) then the number of emitted light photons per absorbed X-ray is

$$g = 0.16 \times \frac{E}{2.28} \tag{2.3}$$

(3) Thus, if we know that the transmission efficiency $\eta_t = 0.5$ the calculation is complete.

Measured values of the average photon number per absorbed X-ray, together with other image information properties, can be found in [10] and references therein.

2.2.2. *Swank Noise*

In his classic paper [11] showed that in an X-ray imaging system utilizing a scintillation phosphor, where the scintillation pulses are integrated rather than counted, the output *signal-to-noise ratio* (SNR) is reduced by a factor I

$$SNR_{out} = (N\eta_a I)^{1/2} \tag{2.4}$$

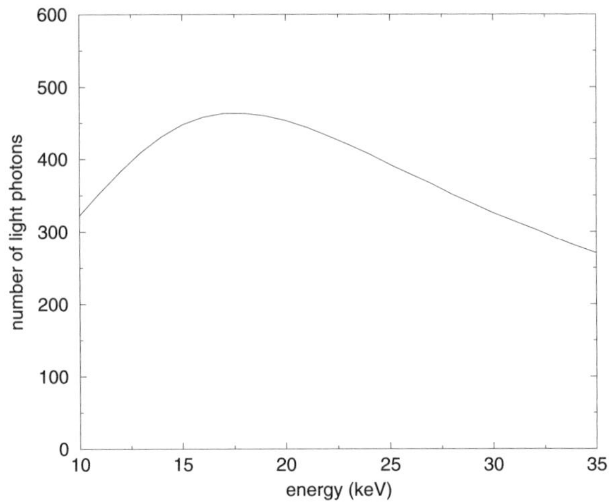

Fig. 2.7. Number of light photons emitted per X-ray impinging on a mammographic screen.

where N is the number of incident X-ray quanta.

From the definition of I as a function of the moments M_i of the probability distribution for photon emission

$$I = \frac{M_1^2}{M_2 M_0}, \tag{2.5}$$

follows the expression ([12])

$$I = (1 + \sigma_g^2/\bar{g}^2)^{-1} \tag{2.6}$$

where $\bar{g}$ is the average number of photons emitted per absorbed X-ray and σ_g^2 is the variance of the scintillation spectrum. Typical I values range from 0.6 to 0.9 for most commercial phosphors and change very little with the energy below the K-edge [13; 14].

The statistics of X-ray-to-light conversion can also be expressed in terms of the parameter ϵ, the so-called "Poisson excess", given by

$$\epsilon = \frac{\sigma_g^2}{\bar{g}} - 1. \tag{2.7}$$

The parameter ϵ is a measure of the excess of the variance in g compared to the Poisson case [15] and will be used later in this chapter.

As pointed out by [16; 17] secondary-quantum noise, i.e. fluctuations in the number of light photons, can be the dominant source of image noise if only few light photons are collected by the optical detector per interacting X-ray. Due to its finite thickness, a phosphor screen is the primary "quantum sink" for the number of detected X-ray quanta is smaller than the incident ones. To ensure that image noise is not further degraded at the subsequent stage in the imaging system, the number of quanta should be considerably larger than the number of X-ray quanta detected otherwise a secondary "quantum sink" would occur. The number of light quanta collected by the film receptor is generally greater than 100 per X-ray absorbed in the screen so this is a problem only in the design of digital systems where fluorescent screens are coupled to a TV or CCD camera [18].

2.3. Physical Properties of Radiographic Films

The film consists of a photographic emulsion deposited on a transparent support, known as the film base, and coated with a thin protective layer (see Fig. 2.8). A double-emulsion film sandwiched between two screens is also used depending on the radiologic examination.[b] As the light exits the screen it is mostly absorbed in the emulsion. However, in double-coated films, some light crosses the film base and exposes the emulsion on the opposite side. This phenomenon causes blurring of the image and is known as *crossover*. Due to high-resolution requirements a single-emulsion/single-screen combination is used for mammography. Since more X-ray photons are absorbed near the surface of the screen a back-screen configuration is used in mammography systems, i.e. X-rays pass through the film first. More advanced systems have been very recently reviewed by [19].

The emulsion is made up of a mixture of silver halide grains and gelatin. The silver halide grains, generally silver bromide, vary in size depending on the application. Image formation in a film is a multi-stage process. On exposure, the light photons emitted by the intensifying screen interact with the silver halide crystals and after electron release, migration and recombination with silver ions, the latent image is eventually built up with sensitivity specks of few silver atoms. To generate a useful image the common photographic processes of development and fixation have to be applied. Development is essentially a detection and am-

[b]Since the front screen absorbs some of the incident radiation such systems are often designed with the back screen thicker than the front screen.

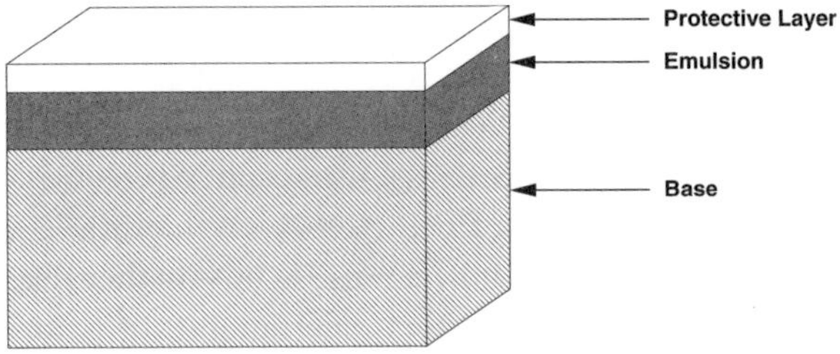

Fig. 2.8. Schematic of a radiographic film structure.

plification process where silver halides crystals are reduced to metallic silver. Latent image centers act as catalyzers in the chemical reaction by absorbing the developer solution which contains electron transfer agents (reduction process) and also agents to regenerate the electron transfer (amplification process). Development of latent image to grains of viewable size results in an amplification factor of about 10^9. Undeveloped crystals are dissolved in the fixing solution. Then, washing and drying complete the film processing stages. The final result strongly depends on variables such as temperature, chemicals, and processing cycle time. These variables must be optimized so as to obtain maximum reduction of exposed crystals and minimum reduction of non-exposed ones and thus obtaining appropriate film contrast. Details on the photographic process are given in [20; 21].

2.3.1. *Film Characteristic Curve*

When the film is viewed in front of a uniformly illuminated light box the blackening depends on the distribution of metallic silver. If the brightness of the incident light is I_0 and the amount of the transmitted light is I, the *optical density* D of the film is defined by

$$D = -\log \frac{I}{I_0} \tag{2.8}$$

The relationship between optical density and X-ray exposure is called the *characteristic curve* of the film, also known as the *sensitometric curve* or *H–D* curve from the name of its proposers, Hurter and Driffield. This is the standard method to represent the response of the photographic film to radiation and is obtained by exposing the screen-film combination to known amounts of radiation and then plotting the resulting D values as a function of the log of the relative exposure E/E_0. The typical characteristic curve shown in Fig. 2.9 provides us with some important film features:

(1) The *baseline* (D of about 0.2) is due to the density of the film base plus the fog level that accounts for the developed grains of an unexposed film.
(2) The so-called *straight-line* is the portion of the curve over which the relationship between optical density and the logarithm of exposure is approximately linear. It defines film *latitude*, i.e. the range of exposures over which the film is usefully sensitive.
(3) The third segment corresponds to large exposures that lead to a saturation of the D signal.

The regions of the curve connecting these three portions are known as the *toe* and the *shoulder* of the curve. A radiograph recorded in the region of the toe or shoulder will result in loss of useful diagnostic information.

2.3.2. *Film Contrast*

Film gradient is defined as the slope of the characteristic curve and in the straight-line portion it is known as the *gamma* (Γ) of the film. The constant Γ is then defined as:

$$\Gamma = \frac{\Delta D}{\Delta \log(E/E_0)}\,. \tag{2.9}$$

In general, since the gradient of the characteristic curve is not constant the instantaneous slope, or "point gamma", is also defined as

$$\dot{\Gamma} = \frac{d\log(I/I_0)}{d\log(E/E_0)} = \frac{dI/I}{dE/E}\,. \tag{2.10}$$

In this form the equation has an important physical significance. In fact, since dE/E is the exposure contrast and dI/I is the brightness contrast, the point gamma is a *contrast gain factor*. The exposure contrast corresponds to what is called the *subject contrast*, defined as the contrast between two areas of the radiologic image, whilst the brightness contrast corresponds to what is called the *radiographic contrast*, defined as the contrast between two areas on a processed film that was exposed to a radiologic image (i.e. the difference between the densities of the two areas). The slope of the characteristic curve expresses quantitatively the *film contrast*. Film contrast represents the ability of the film for translating subject contrast into radiographic contrast [9]. High-gamma films, typically in the range 2–3, are used in

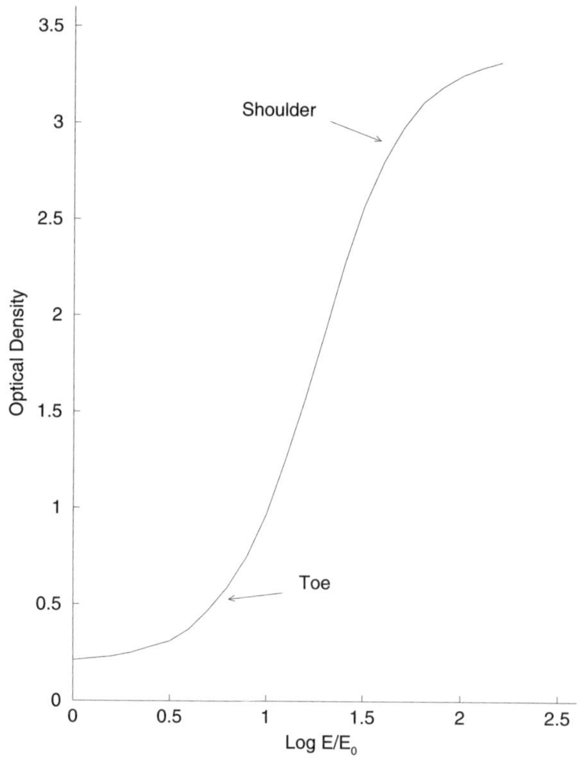

Fig. 2.9. Characteristic curve of a typical screen-film combination.

radiology to take advantage of this contrast gain. There are many factor affecting film contrast and this is the subject of *sensitometry*. A discussion of the sensitometric properties of radiographic films is given in [22; 21].

To obtain the optical density change as a function of the fractional change in exposure Eq. (2.9) is usually written in the more familiar form:

$$\Delta D = 0.434\Gamma \Delta E/E \tag{2.11}$$

i. e. a simple relationship between the radiographic signal ΔD and exposure contrast. Since for X-ray films Γ is usually in the range 2–3 it follows from Eq. (2.11) that a fractional change in exposure will result in about the same density change. This is a useful rule of thumb for many practical applications.

2.3.3. *Contrast vs Latitude*

From the definition of film contrast and film latitude arises one of the main limitations of screen-film radiography. As shown in Fig. 2.10 the high-contrast film has a narrower latitude while the low-contrast film has a wider one. The latitude of the film varies with the reciprocal of film contrast so, due to the characteristic curve, there is no way to get high-contrast film of wide latitude. From the standpoint of radiologic examination this means that contrast and latitude of the film should be closely matched to the application. To achieve consistent film densities independent of tissue characteristics and beam quality an automatic exposure control is nowadays included in all radiographic units.

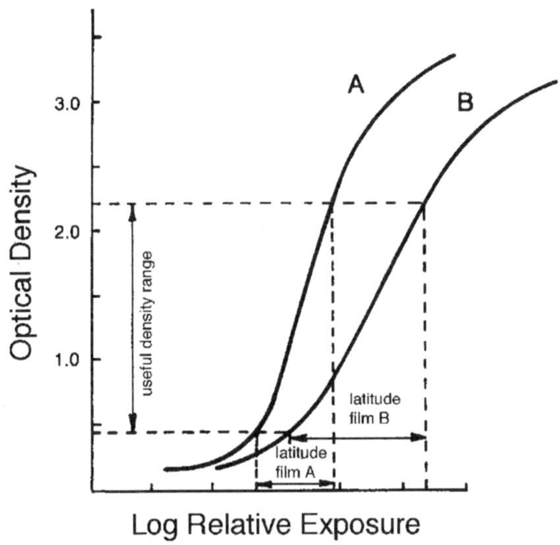

Fig. 2.10. Characteristic curves of two hypothetical radiographic films. Film A has a higher contrast but a narrower latitude than film B.

2.3.4. *Film Speed*

Another important parameter is the sensitivity of a film or *speed*. The speed is inversely related to the exposure required to produce a given optical density, i.e. the speed of a film is high when the required exposure is low.

Its value is defined in terms of the reciprocal of the exposure that yields an optical density of 1.0 above the base plus fog. While the shape of the characteristic curve is related to film gamma, the position of this curve with respect to the E axis is related to film speed (see Fig. 2.11).

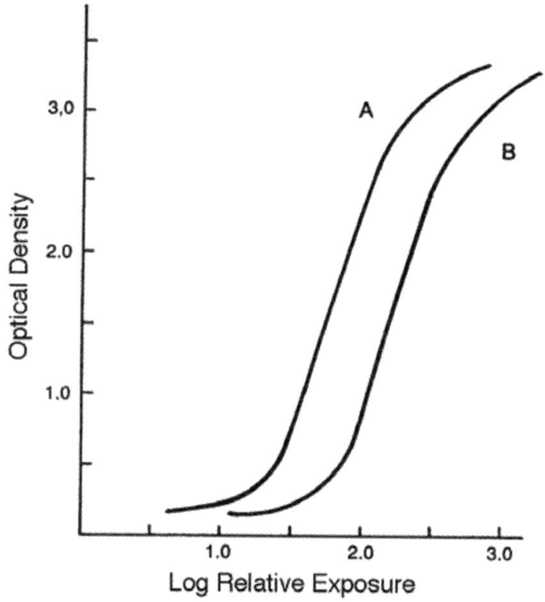

Fig. 2.11. Characteristic curves of two hypothetical radiographic films. The speed of film A is higher than that of film B.

The speed of screen-film systems is influenced by many factors, such as phosphor material and thickness, type of film and screen light, X-ray beam quality and processing conditions [2; 21]. The major advantage of using high-speed systems is the reduction of patient dose. However, faster systems generally result in lower spatial resolution performance. The relationship between sensitivity and other imaging properties has been extensively investigated in recent years [23; 24].

2.3.5. *Reciprocity-Law Failure*

A well-known phenomenon that is important in screen-film systems is the reciprocity-law failure [21]. It is usually assumed that the same D is obtained if the product of X-ray beam intensity and exposure time is constant, provided that the beam quality is unchanged. However, when long exposure times are used the produced density is less and the above condition is no longer satisfied. During the stage of latent image formation, metallic silver specks cannot reach the optimal size for growth and amplification because of the too few light photons emitted over the exposure period. This effect can be important in screen-film mammography when long exposure times are used to compensate for reduced film speed [21]. This phenomenon also occurs for very short exposure times: a greater number of specks results in a less stable latent image, thus producing less density as a function of exposure than would occur normally.

It is worth noting that this effect does not occur in films directly exposed to X-rays for the energy deposited by each interaction is sufficient to create a stable latent image.

2.4. Radiographic Noise

Radiographic noise, or *radiographic mottle* after [25], is the spatial fluctuation of film density obtained by uniformly exposing a screen-film system to X-rays.[c] In his classic paper Rossmann identified three independent random phenomena which contribute to radiographic mottle:

(1) *Quantum mottle*, i.e. spatial fluctuations of X-ray quanta absorbed by the fluorescent screen.
(2) *Structure mottle*, i.e. spatial fluctuations in screen absorption due to structural inhomogeneities in the phosphor coating.
(3) *Film granularity*, i.e. spatial fluctuations of developed silver halide grains in the emulsion.[d]

The standard deviation of the total radiographic mottle is:

$$\sigma(D) = [\sigma_q^2(D) + \sigma_s^2(D) + \sigma_g^2(D)]^{1/2} \tag{2.12}$$

where the subscripts q, s, and g refer to quantum mottle, structure mottle, and film granularity, respectively. As derived by [26] $\sigma_q^2(D)$ and $\sigma_g^2(D)$ can

[c]Due to the granular nature of radiation, the term *uniform exposure* means a distribution of X-ray quanta that follows Poisson statistics.

[d]The original term was *film graininess*. Graininess is actually a subjective impression of density non-uniformity when an observer looks at a radiographic film. Granularity is instead the objective measurement of this non-uniformity expressed in terms of root-mean-square value of mean density.

be written as:

$$\sigma_q^2(D) = \frac{(\log_{10}e)^2\Gamma^2}{\bar{n}A_e}\left(1 + \frac{\langle\Delta E^2\rangle}{\langle E\rangle^2}\right)\left(1 + \frac{\langle\Delta g^2\rangle}{\langle g\rangle^2}\right) \tag{2.13}$$

where $\bar{n}$ is the average number of absorbed X-ray quanta per unit area, A_e the effective sampling aperture (see Sec. 2.6.1) which takes into account both the measuring aperture and the screen blurring, $\langle\Delta E^2\rangle$ the mean square fluctuation in energy E deposited in the screen by an absorption event, and $\langle\Delta g^2\rangle$ the mean square fluctuation in light photons g emitted per unit absorbed energy

$$\sigma_g^2(D) = \frac{(\log_{10}e)D\langle a\rangle}{A_g}\left(1 + \frac{\langle\Delta a^2\rangle}{\langle a\rangle^2}\right) \tag{2.14}$$

where D is the net optical density, $\langle\Delta a^2\rangle$ the mean square fluctuation in grain area a, and A_g the densitometer aperture.

Equation (2.13) intuitively shows that a spread in energy or in light photons increases system noise. Equation (2.14) is known as the Siedentopf relationship [27]. In the special case where all the grains are identical in size, independent of image density level, Eq. (2.14) becomes

$$\sigma_g^2(D) = \frac{(\log_{10}e)Da}{A_g}\,. \tag{2.15}$$

This means that the product of the variance $\sigma_g^2(D)$ and the sampling aperture A_g is constant as noted for the first time by [28] and experimentally measured for a wide range of aperture sizes.

Barnes also showed that the contribution of structure mottle to radiographic mottle is negligible for commercially available screens. Therefore, $\sigma_s^2(D)$ is usually not included in the expression for the radiographic mottle. By knowing the physical parameters of a screen-film system, the standard deviation of the density fluctuations can be calculated and compared with the experimental value measured with a given sampling aperture.

2.5. Definition of Image-Quality

So far we have discussed signal and noise changes over large areas of interest without concerning about their spatial characteristics. The concepts of *gamma* and *standard deviation* are useful to describe large-area transfer characteristics for a radiographic system but say nothing about the spatial structure of signal and noise, respectively. In this section we will introduce some physical quantities whose measurement will provide us with the spatial transfer characteristics of the system.

The assessment of the performance of imaging systems involves the measurement of "working" parameters which define the quality of the recorded image. Image transfer theory has been successfully applied to determine signal and noise transfer characteristics to a known input.[e] The basic theory of a photographic process states that the object intensity distribution $f(x_1, y_1)$, namely the input signal, is related to the image intensity distribution $g(x, y)$, namely the output signal, by a system operator S:

$$g(x, y) = S[f(x_1, y_1)] . \qquad (2.16)$$

If the input to the system is a delta function, i.e. a point source of unity intensity, then

$$g(x, y) = S[\delta(x_1, y_1)] = h(x, y; x_1, y_1) . \qquad (2.17)$$

The response h is called the *point spread function* (PSF). Since any input may be considered to be a linear combination of weighted and displaced delta functions, under the special conditions of *linearity* and *stationarity* the input-output relationship of the system is reduced to a simple convolution formula

$$\begin{aligned} g(x, y) &= \iint_{-\infty}^{+\infty} f(x_1, y_1) h(x - x_1, y - y_1) dx_1 dy_1 \\ &= \iint_{-\infty}^{+\infty} f(x - x_1, y - y_1) h(x_1, y_1) dx_1 dy_1 . \end{aligned} \qquad (2.18)$$

The above relationship states that the output is the convolution of the input with the point spread function. The physical implications of convolution is that the output is blurred due to the spreading of the input-source distribution over the image plane. As shown in Fig. 2.12, a point source on the object plane results in a diffuse blur on the image plane (adapted from [29]).

Convolution in the spatial domain is transformed to multiplication in the frequency domain. Hence, Fourier transform techniques are usually applied for practical measurements. Indeed, by the convolution theorem we obtain

$$G(u, v) = F(u, v) H(u, v) \qquad (2.19)$$

[e]Details of the theory and relative demonstrations can be found in the classic text of [27] and in [30].

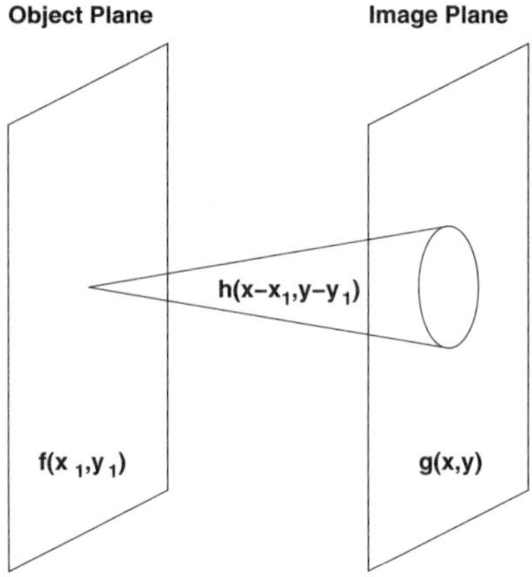

Fig. 2.12. A simplified representation of the image formation process.

where $F(u,v)$, $G(u,v)$, and $H(u,v)$ are the Fourier transforms of $f(x,y)$, $g(x,y)$, and $h(x,y)$, respectively. The above equation shows that the input response function can be easily calculated from the Fourier transform of the output response function by division and inverse transform, provided that the Fourier transform of the PSF is known. The density distribution on a radiographic film can be measured by making use of a microdensitometer so in general, assessment of spatial transfer characteristics is a problem of measuring the PSF of the system. Actually, measurement of PSF is rather difficult due to small aperture sizes to be used which result in very low signals. To overcome these experimental limitations the *line spread function* (LSF) is generally measured. The LSF is defined as the response to a line input and is obtained from the PSF by integrating over one variable

$$l(x) = \int_{-\infty}^{+\infty} h(x, y_1) dy_1 \tag{2.20}$$

Due to the isotropic nature of screen-film system PSFs [23], transfer properties are also completely specified by a section $h(r) \equiv h(x,y)$, where $r^2 = x^2 + y^2$.

2.5.1. *MTF*

To complete signal analysis we now introduce the *modulation transfer function* (MTF), defined as modulus of the normalized Fourier transform of the

LSF:

$$MTF(u) = \left| \frac{\int_{-\infty}^{+\infty} l(x) e^{-2\pi i u x} dx}{\int_{-\infty}^{+\infty} l(x) dx} \right| . \tag{2.21}$$

This physical quantity has an important and simple meaning: it is the ratio of the output amplitude to input amplitude when the incident X-ray pattern is a sinusoidal signal. In other words, since in an imaging process the two-dimensional input function can be expressed in terms of its two-dimensional spatial frequencies, the MTF describes the transmission of such frequencies through the system. It is worth noting that due to Fourier transform properties the total MTF of a cascaded imaging system is the product of the individual MTFs. Needless to say, this is a much easier task compared to complex convolution calculations.

As a practical example, Fig. 2.13 shows in comparison the line spread functions and modulation transfer functions of three screen-film combinations (from [31]). The high-resolution system C exhibits a narrower LSF and thus a broader MTF compared to system A and B. This means that system C is able to transmit high spatial frequency signals better than the two other systems.

2.5.2. *NPS*

It has been long recognized that noise, like signal, has a spatial structure in radiographic imaging systems. Therefore, second order statistics (or joint probability of the density function) has to be studied to investigate the spatial correlation of image noise. The *autocorrelation function* (ACF) lends itself very well to the required task since it is defined as the first joint moment of the second-order probability

$$ACF(x,y) = \lim_{X,Y\to\infty} \frac{1}{4XY} \int_{-X}^{+X} \int_{-Y}^{+Y} \Delta D(x_1,y_1) \Delta D(x_1+x, y_1+y) dx_1 dy_1 \tag{2.22}$$

where $\Delta D(x_1,y_1) = D(x_1,y_1) - \bar{D}$ ($\bar{D}$ is the mean density value). The ACF may be written as the above formula when a single realization of a random process is representative of all the statistics, i.e. the process is called ergodic. By noting that $ACF(0,0)$ reduces to density variance it follows that first and second order statistics are completely specified by the autocorrelation function. However, due to the aperture of the system used

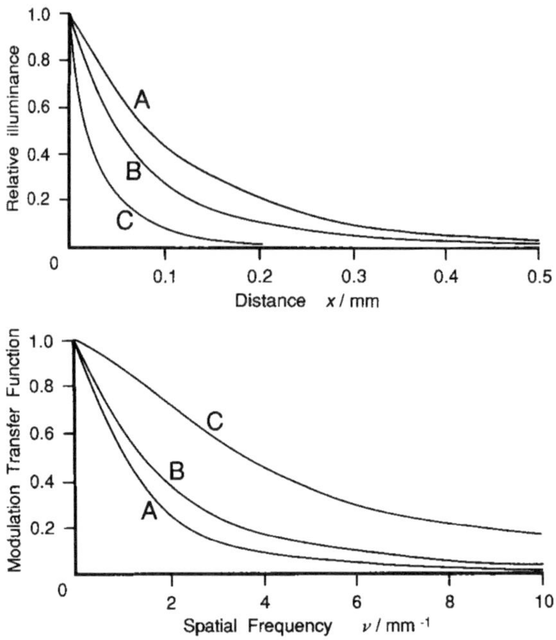

Fig. 2.13. Relationship between line spread functions and modulation transfer functions (reprinted with permission from ICRU, Inc.).

to measure density fluctuations, the measured ACF results in a convolution of the actual ACF with the PSF of the measuring system. Again, to avoid dealing with a convolution procedure the frequency-domain counterpart was introduced. The *noise power spectrum* (NPS) or *Wiener spectrum* of the fluctuations of a stationary ergodic process is defined as

$$NPS(u,v) = \left\langle \left| \lim_{X,Y\to\infty} \frac{1}{4XY} \int_{-X}^{+X} \int_{-Y}^{+Y} \Delta D(x,y) e^{-2\pi i(ux+vy)} dxdy \right|^2 \right\rangle \tag{2.23}$$

where the symbol $\langle\rangle$ denotes an ensemble average, i.e. several realizations of noise are necessary for the spectrum to converge. The Wiener-Khintchin theorem then states that NPS and ACF are Fourier transform pairs

$$NPS(u,v) = \int_{-\infty}^{+\infty} ACF(x,y) e^{-2\pi i(ux+vy)} dxdy\,. \tag{2.24}$$

The NPS characterizes the noise of a system as a function of spatial frequency and that is analogous to the treatment of signal by means of

MTF. However, unlike MTF calculation procedure, current methods for measuring the NPS of screen-film systems directly compute data of the microdensitometric measurements without first finding the ACF [32].

A typical result of a power spectrum of the density fluctuations is illustrated by the curve NPS_{tot} of Fig. 2.14 (from [31]). As discussed in Sec. 2.4, the total noise consists of two independent contributions for practical screen-film systems, quantum mottle and film granularity. The Wiener spectrum of radiographic mottle $NPS_{tot}(u)$ is then given by

$$NPS_{tot}(u) = NPS_q(u) + NPS_g(u) \tag{2.25}$$

where $NPS_q(u)$ and $NPS_g(u)$ are the Wiener spectra for quantum mottle and film granularity, respectively.[f]

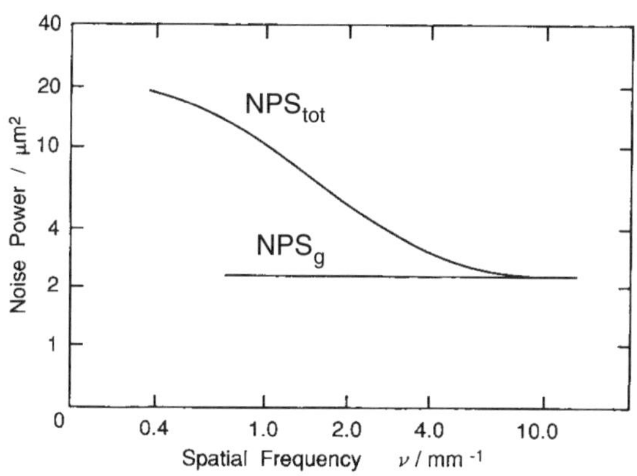

Fig. 2.14. Noise power spectrum for a typical screen-film combination exposed to X-rays (NPS_{tot}) and for the film alone exposed to uniform light (NPS_g) (reprinted with permission from ICRU, Inc.).

It is generally assumed that fluctuations of X-ray quanta are spatially uncorrelated and thus NPS_q is "white" (i.e. all frequencies are present in equal amounts). Unlike X-ray quanta, light photons produced in the screen are spatially correlated and have a variance which exceeds Poisson statistics (see Sec. 2.2.2). Random scattering within the screen will affect the

[f]Although the Wiener spectrum is a two-dimensional quantity, the measured spectrum is usually displayed in one-dimensional form. That is because the Wiener spectrum of radiographic noise is considered as rotationally symmetric [23].

spatial correlation of light photons and they will be distributed according to the point spread function of this scattering process. The Wiener spectrum of quantum mottle is actually related in a complicated way to various parameters (see Eq. (2.13)). However, as a first-order approximation its mathematical representation in terms of optical density fluctuations is [23]

$$NPS_{q,\Delta D}(u) = (\log_{10} e)^2 \frac{\Gamma^2}{\bar{n}} MTF^2(u) \tag{2.26}$$

where $\bar{n}$ is the average number of X-ray quanta absorbed by the screen per unit area. The screen-film combination is then a transducer for white noise that converts X-ray quanta absorbed in the screen to a visible noise pattern (i.e. the density fluctuations in the film). Substituting $NPS_{q,\Delta D}(u)$ in Eq. (2.25):

$$NPS_{tot}(u) = (\log_{10} e)^2 \frac{\Gamma^2}{\bar{n}} MTF^2(u) + NPS_g(u)\,. \tag{2.27}$$

By assuming that the Wiener spectrum of film granularity is "white" too,[g] we are now able to explain the trend of the curves shown in Fig. 2.14. At low frequencies quantum mottle is the dominant source of noise and, due to its imperfect transfer properties, the screen-film system acts like a low-pass spatial frequency filter. At high frequencies NPS_{tot} decreases and film granularity becomes the dominant source of noise.

Although [33] have shown that it is not strictly correct to write the Wiener spectrum as directly proportional to the square of MTF when dealing with screen-film systems, Eq. (2.26) is widely used as an approximation since it has the advantage of relating imaging parameters to the Wiener spectrum in a simple manner.

2.5.3. *DQE*

The *detective quantum efficiency* (DQE) is a fundamental parameter for the evaluation of the imaging characteristics of an X-ray detector. It describes the transfer of signal-to-noise ratio that takes place when recording an image:

$$DQE = \frac{SNR_{out}^2}{SNR_{in}^2}\,. \tag{2.28}$$

[g] That is valid as long as the size of the measuring system aperture is an order of magnitude greater than the largest developed size [34].

As suggested by [27], DQE represents the efficiency of an X-ray imaging system as a photon counter. In fact, if we assume that N incident X-ray quanta per unit area are governed by Poisson statistics, $\sigma_N^2 = N$ and then $SNR_{in}^2 = N$. According to their interpretation it is possible to define a number of detected quanta per unit area N' so that $SNR_{out}^2 = N'$. From Eq. (2.28) follows that

$$DQE = \frac{N'}{N}\,. \tag{2.29}$$

Since the output SNR will always be degraded compared to that of the input, i.e. DQE ≤ 1, it follows that an imaging system acts as an inefficient photon counter, and thus it detects a lesser number of quanta N'. Due to its definition and interpretation SNR_{out}^2 is called *noise-equivalent number of quanta* (NEQ).

To obtain an absolute measure of system performance as a photon detector, the spatial-frequency dependent NEQ was derived by mathematically combining the MTF and NPS. Thus, the effects of signal and noise transfer characteristics of an X-ray detector at a given spatial frequency are expressed by the ratio of NEQ to the input photon fluence, namely the DQE. As reviewed in the paper of [35], the mathematical expression for NEQ(u) of screen-film systems as a function of both sensitometric and information transfer parameters is

$$NEQ(u) = \frac{(\log_{10}e)^2\Gamma^2 MTF^2(u)}{NPS_{\Delta D}(u)} \tag{2.30}$$

where the MTF and NPS enter independently. The subscript in NPS refers to a measurement of Wiener spectrum in terms of density fluctuations. [15] developed a comprehensive model of a screen-film system by including the separate imaging components of screen and film in an overall expression of DQE

$$DQE(N,u) = \frac{\eta_a}{1 + \dfrac{\epsilon}{\bar{g}} + \dfrac{1}{\bar{g}\eta_t DQE_F(q,u) MTF_S^2(u)}} \tag{2.31}$$

where $DQE_F(q,u)$ is the intrinsic detective quantum efficiency of the film, $MTF_S(u)$ is the modulation transfer function of the screen, N is the mean number of incident X-ray quanta per unit area and q is the mean number of light photons which expose the film per unit area. At zero spatial frequency $MTF_S(0) = 1$ and by replacing the expressions of Eqs. (2.6) and (2.7) in

Eq. (2.31), we obtain

$$DQE(N,0) = \frac{\eta_a}{\dfrac{1}{I} - \dfrac{1}{\bar{g}} + \dfrac{1}{\bar{g}\eta_t DQE_F(q,0)}}. \tag{2.32}$$

A similar result was found by [36] who investigated information transfer properties of various X-ray screens. By determining SNR for the screen-film system from the number of grains which are developed after exposure to q light photons they found the following expression of DQE:

$$DQE(0) = \frac{\eta_a}{\dfrac{1}{I} - \dfrac{1}{\bar{g}} + \dfrac{1}{\bar{g}u}} \tag{2.33}$$

where u is a parameter which depends on the film response to light photons emitted by the screen, just like the product $\eta_t DQE_F(q,0)$. If we neglect the information loss due to the film and recall that the average gain $\bar{g}$ is greater than 100 for X-ray screens, both Eqs. (2.32) and (2.33) reduce to

$$DQE(0) = \eta_a I \tag{2.34}$$

which is the expression, first derived by [11], for the DQE of an X-ray screen.

2.6. Image Contrast

For the radiologist the diagnostic task consists in the ability to see the difference in image signal between a region of a suspected abnormality and a normal region. We have seen in Sec. 2.3.2 that such a difference is the radiographic contrast. Since the concepts of contrast, spatial resolution and noise are interdependent we now briefly review how these three fundamental physical parameters are related.

2.6.1. *The Concept of Sampling Aperture*

Any radiographic system correlates the input distribution of X-ray quanta. That results in imaging points as blurs, spreading out the edges, and thereby degrading the contrast of the input distribution. We will now introduce the *effective sampling aperture*, a very useful quantity that represents the average blur size of an imaging component. It was proposed by [37] as a straightforward index of image-quality, utilizing the concept of noise-equivalent passband N_e introduced by [38]. Wagner defined the effective sampling aperture a_e as the reciprocal of N_e where

$$N_e = \iint_{-\infty}^{+\infty} MTF^2(u,v)dudv\,, \tag{2.35}$$

that is to say, in Wagner's words, 'the greater the volume under the squared MTF surface, the smaller the effective sampling aperture, and vice versa'. The properties of Fourier transforms allow us to calculate this figure of merit by knowing the PSF of the system

$$a_e = \left\{ \frac{\iint_{-\infty}^{+\infty} PSF^2(x,y)dxdy}{[\iint_{-\infty}^{+\infty} PSF(x,y)dxdy]^2} \right\}^{-1} . \tag{2.36}$$

The blur function is thus reduced to an equivalent aperture of uniform transmission.

As shown by [39], the effective sampling aperture approach leads to a very important result. In a cascaded imaging system the overall effective aperture is the sum of the apertures of each component[h]

$$a_{tot} = a_1 + a_2 + \cdots + a_n\,. \tag{2.37}$$

Hence, by imaging an object having an equivalent sampling area a_o with a system of equivalent sampling area a_s we obtain $a_i = a_o + a_s$ as a resulting size of the object. This means that the image contrast C_{out} of the object is reduced compared to its input contrast C_{in} by a factor a_o/a_i

$$C_{out} = C_{in}\frac{a_o}{a_i}\,. \tag{2.38}$$

2.6.2. *Noise Contrast*

In the last section we have only considered the effect of the transfer characteristics of the imaging system as a contrast reduction factor without concerning about the fluctuations of the number of X-ray photons which contribute to image formation. It has been long recognized that the fundamental limitation to the image information content is statistical noise [40]. Therefore, the detectability of small or subtle structures in a radiologic image is ultimately dependent on the recorded X-ray quanta. The basic theory of image signal detection is known as the "Rose model", after Albert Rose who first introduced a statistical analysis in the calculation of the minimum detectable contrast [41]. Indeed, an average absorption per unit area of N

[h]This is actually true only for Gaussian distribution of the respective blur functions but it has been found that the above formula can be also used as a general rule.

quanta will have associated with it a fluctuation of quanta characterized by a root-mean-square value $\sqrt{N}$. By supposing that an object of area a introduces a signal difference of $\Delta(Na)$, according to Rose it can be detected only if

$$\Delta(Na) = k\sqrt{Na} \tag{2.39}$$

that is to say, the image signal must be k-times larger than the root-mean-square fluctuations in the background. Rose called $k = \Delta(Na)/\sqrt{Na}$ the *threshold signal-to-noise ratio* and showed experimentally that $k = 5$ is a reasonable value "for not mistaking a noise fluctuation for the real signal" [42]. As pointed out by [39], Eq. (2.39) can be also seen in terms of *threshold contrast* by rewriting it as

$$C_s = kC_n \tag{2.40}$$

where $C_s = \Delta(Na)/Na$ is the signal contrast of an object of area a and $C_n = \sqrt{Na}/Na = 1/\sqrt{Na}$ the noise contrast calculated for the same sampling area a. The motivation for this last relation is that by knowing the object contrast C_s we can easily calculate the number of photons needed to detect it. It is worth noting that the Rose model neglects any additional noise introduced by the imaging system and thereby meaning that X-ray quantum noise is the dominant source of random fluctuations. Such an ideal system is called a *quantum limited system.* We have seen that screen-film systems are not quantum limited at high spatial frequencies (see Fig. 2.14) hence, the Rose model could represent a good approximation only for large-area objects. Also, the human visual system has a threshold of detectability of density differences ΔD on a photographic film. This means that small image signals can be detected by a screen-film system but the image information content is not visible to the radiologist in the image display. Such a *visibility threshold* is of the order of 0.01 and this value was used by [43] to estimate the minimum exposures required for the detection of image signals associated with particular radiological examinations.

2.6.3. *Contrast-Detail Analysis*

The Rose model represents the signal detection theory in its barest essentials. Lesions could be detected if we neglect the contrast degradation due to the spatial correlation of signal and noise, the scatter of X-rays, the variable pattern of normal anatomy which averages out the contrast itself, and the performance of the human eye. Modification to this model were discussed by [41] himself, [44; 45], and more recently by [39]. Wagner attempted to construct a synthesis of the investigation of Rose, Schade and Schnitzler and showed how, with simple modifications of Rose statistical model which include some aspects of vision, an excellent agreement with Blackwell data on the detectability of disks as a function of contrast and size could be obtained. The work of [46] concerned with the determination of the contrast threshold of the normal human observer. He used a single spot of varying size and contrast under a wide range of illumination to report on the ability of the eye to detect these visual stimuli. Test patterns based on arrays of grey disks were also used by Rose to measure the resolving power of a system in terms of size and contrast of single elements. Following the work of Blackwell and Rose, *contrast-detail* (CD) diagrams, i.e. plots of the minimum detectable contrast of an image feature as a function of its diameter, are used nowadays to evaluate the image-quality characteristics of X-ray imaging systems. These diagrams are measured with a test device adapted from the Burger–Rose phantom [47; 42] in which simple objects, such as circular disks, are arranged in a matrix of rows and columns. The objects vary in one direction by size and in the other direction by contrast. An X-ray image of the CD phantom is evaluated by determining the detection threshold for each object-diameter and then averaging the data from the scoring over the observers. A schematic representation of a CD test object specifically designed for image-quality control in mammography is shown in Fig. 2.15 [48]. The typical CD curve of a screen-film combination for mammography in comparison with that of a full-field digital prototype at different entrance doses is shown in Fig. 2.16 [49].

The advantage of CD analysis is that the effects of blurring and noise on visibility can be evaluated in a direct and easily understood way when comparing imaging systems. However, it should be borne in mind that "threshold theory" suffers from the decision-making process of the observer. Statistical decision theory has been developed to provide a more complete description of image-quality. Readers interested in a psychophysical approach to the evaluation of imaging system performance are referred to an excellent report published very recently by the International Commission on Radiation Units and Measurements [31].

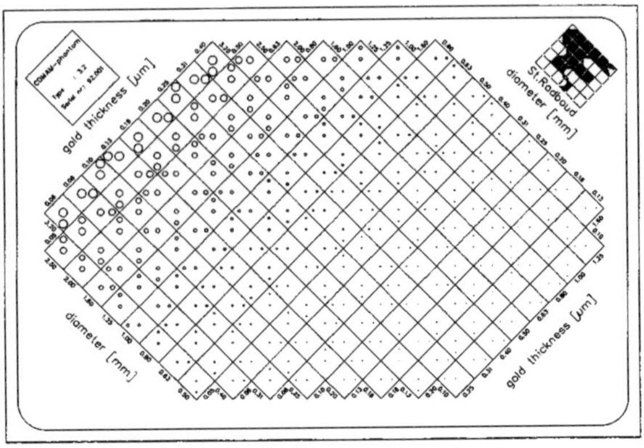

Fig. 2.15. Schematic of a CD phantom specifically developed for mammography.

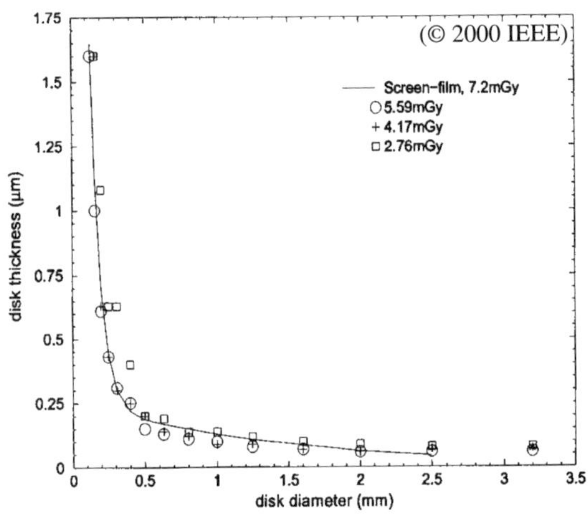

Fig. 2.16. Contrast-Detail curves of mammography systems (reprinted with permission from IEEE, Inc.).

2.7. Image-Quality of Screen-Film Combinations

In this last section performance of screen-film systems commonly used in clinical examinations will be presented. Results will be shown in terms of the physical parameters that have been previously introduced.

2.7.1. *MTF, NPS and DQE Measurement*

A common method of measuring the LSF of radiographic systems is to use a slit of very fine width (10 μm) and then scanning the exposed film by a microdensitometer. Since the screen-film is not a linear system, LSF data are obtained by converting optical density values to X-ray exposure values via the appropriate sensitometric curve. The MTF is then calculated by discrete Fourier transform of LSF data. This subject has been extensively considered in [50] together with other measurements techniques and associated errors. Besides, a theoretical model has been recently developed to compare signal and noise properties of different measurement techniques [51]. MTF curves of screen-film systems available on the market are compared in Fig. 2.17 (adapted from [23]). The screens incorporate rare-earth phosphors and are produced by the same manufacturer for general (screen A) and high-definition (screen B) radiographic procedures. The same film was used in both systems.

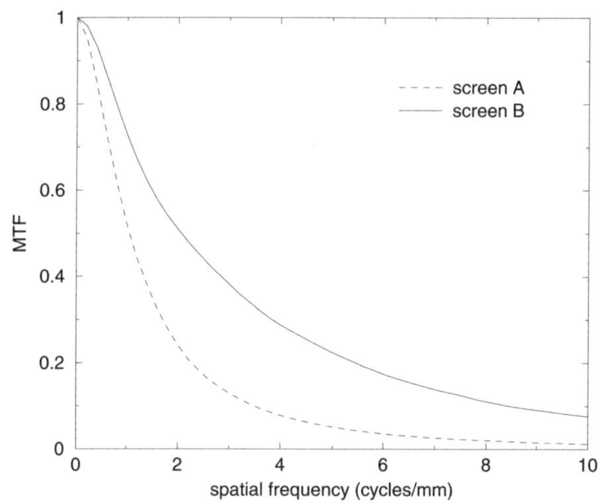

Fig. 2.17. MTF comparison of clinical screen-film combinations.

A widely used procedure to measure Wiener spectra consists in scanning the film sample, which was uniformly exposed to X-rays, by a long and narrow slit. Digital data are Fourier transformed and after corrections for the aperture transfer function one section of the two-dimensional Wiener spectrum is obtained. Details of the calculation and correction methods can be found in [23; 52; 53]. Wiener spectra of systems with different sensitivity and resolution often intersect. This is illustrated in Fig. 2.18 where the same screens (and film) used in the evaluation of MTF are compared in terms of NPS. While spatial resolution properties are governed by the screen, noise properties depend on both screen and film characteristics. Indeed, Fig. 2.19 shows what happens when screen A is coupled to films of different speed and gradient.

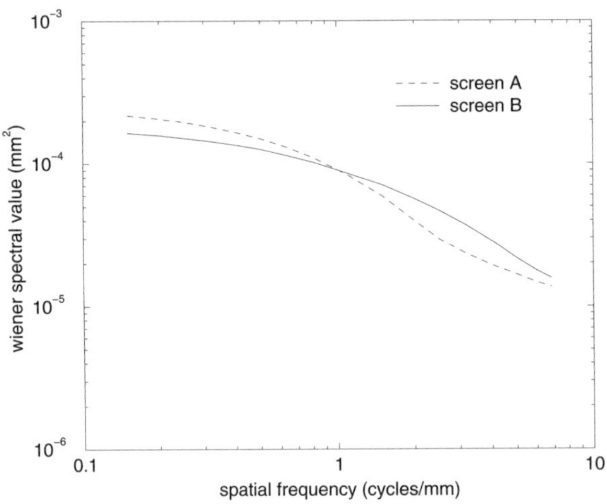

Fig. 2.18. NPS comparison of clinical screen-film combinations.

Once both the MTF and NPS have been measured for that particular screen-film combination the determination of DQE then follows from Eq. (2.30), provided the average number of X-ray quanta used to expose the system is known [35; 54; 55]. Experimental DQE surface for a mammographic screen-film system is shown in Fig. 2.20 (from [21]). At the energies used in mammography the absorption efficiency of the screen is about 0.6 and the reported Swank factor value is in the range 0.6–0.7. This explains the maximum of about 0.3 at low frequencies. Figure 2.20 clearly shows that signal-to-noise ratio capabilities significantly fall off both for spatial frequencies beyond few cycles per millimeter and for exposures outside a relatively narrow range.

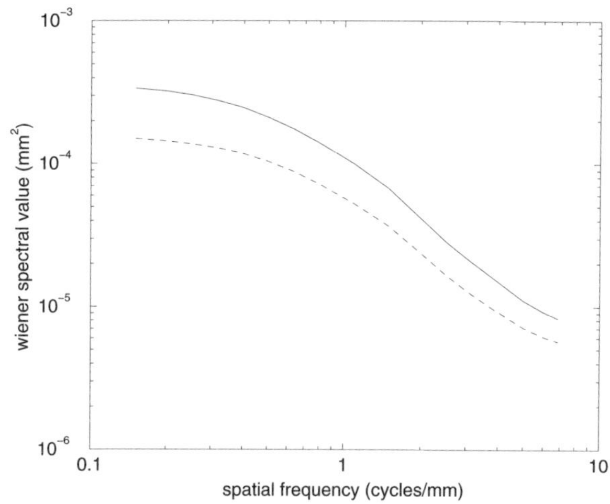

Fig. 2.19. Wiener spectra of two screen-film combinations using the same screen. The combination with the higher-speed, higher-gradient film is represented by the continuos curve.

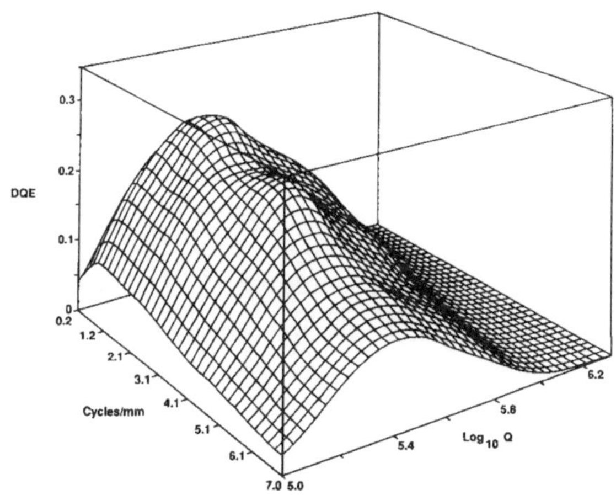

Fig. 2.20. Measured DQE of a mammographic screen-film combination (absolute scale for Q is in photons/mm^2) (reprinted with permission from RSNA, Inc.).

2.7.2. *Quality Indices*

The goal of diagnostic radiology is to produce high-quality images at the lowest radiation dose possible. We have seen that physical parameters which define the quality of the radiograph are related to both image-quality (characteristic curve, spatial resolution, noise) and patient dose (sensitivity). To facilitate a comparison of screen-film systems from different manufacturers and to obtain an overall view of their performance, standardized evaluation methods have been suggested and defined [23; 24; 56; 57; 58].

As an example, basic imaging parameters that can be evaluated with standard instrumentation are compared in Table 2.2 (adapted from [59]). The aim is to show what quality indices are usually taken into account for comparative evaluation of clinical systems. Two manufacturers, here named 1 and 2, were considered in this study and three types of clinical examinations. Systems A, B, and C are suggested by the manufacturer for use in general radiography, chest radiography and mammography, respectively. Depending on the clinical examination different beam qualities are used. Since the speed of a screen-film depends on X-ray energy, a wide variety of speed values may be expected [60]. Details of system evaluation based on such indices can be found in [61].

Table 2.2. Physical parameters and quality indices for different screen-film systems.

Screen-film system	Speed ($\pm 10\%$)	Average gradient	MTF value (2 lp/mm)	Noise Variance ($\times 10^{-3}$)	Base plus Fog
1A	369	2.45	0.27	1.18	0.23
2A	334	2.46	0.45	1.86	0.22
1B	247	2.40	0.21	1.16	0.22
2B	322	1.80	0.29	1.04	0.22
1C	15.5	2.95	0.25*	1.34	0.22
2C	16.4	2.96	0.25*	1.46	0.23

*For mammography systems the MTF value was calculated at 6 linepairs/mm

References

1. Knoll, G. F., *Radiation Detection and Measurement*, 3rd ed. (J. Wiley & Sons, New York) (2000).
2. Arnold, B. E., Physical Characteristics of Screen-Film Combinations, in *The Physics of Medical Imaging: Recording System Measurements and Techniques*, ed. Haus, A. (American Institute of Physics, New York) pp. 30–71 (1979).
3. Zweig, G. and Zweig ,D. A., Radioluminescent imaging: factors affecting total light output, *SPIE Proc.* **419**, pp. 297–304 (1983).

4. MCI Optonix, a division of USR Optonix Inc., NJ, USA. Document available on internet *http://www.mcio.com/phoschar.pdf.*
5. van Eijk, C. W. E., Inorganic scintillators in medical imaging, *Phys. Med. Biol.* **47**, pp. R85–R106 (2002).
6. Hamamatsu Photonics K. K., FOS (Fiber Optic Plate with Scintillator) for Digital X-ray Imaging, Catalogue No. TMCP9003E01 (1996).
7. Berger, M. J. and Hubbell, J. H., XCOM: Photon Cross Sections on a Personal Computer, Center for Radiation Research, National Bureau of Standards, Gaithersburg, MD, USA (1987).
8. Stevels, A. L. N. and Pingault, F., BaFCl:Eu^{2+}, a new phosphor for X-ray intensifying screens *Philips Res. Repts* **30**, pp. 277–290 (1975).
9. Ter-Pogossian, M. M., *The Physical Aspects of Diagnostic Radiology* (Harper & Row, New York) (1969).
10. Ginzburg, A. and Dick, C. E., Image information transfer properties of X-ray intensifying screens in the energy range from 17 to 320 keV, *Med. Phys.* **20**, pp. 1013–1021 (1993).
11. Swank, R. K., Absorption and noise in X-ray phosphors, *J. Appl. Phys.* **44**, pp. 4199–4203 (1973).
12. Rabbani, M., Shaw, R. and Van Metter, R. L., Detective quantum efficiency of imaging systems with amplifying and scattering mechanism, *J. Opt. Soc. Am. A* **4**, pp. 895–901 (1987).
13. Trauernicht, D. P. and Van Metter, R., The measurement of Conversion Noise in X-ray Intensifying Screens, *SPIE Proc.* **914**, pp. 100–116 (1988).
14. Mickish, D. J. and Beutel, J., The determination of X-ray phosphor scintillation spectra, *SPIE Proc.* **1231**, pp. 327–336 (1990).
15. Shaw, R. and Van Metter, R. L., An analysis of the fundamental limitations of screen-film systems for X-ray detection: I. General theory, *SPIE Proc.* **454**, pp. 128–132 (1984).
16. Nishikawa R. M. and Yaffe, M. J., Effect of various noise sources on the detective quantum efficiency of phosphor screens, *Med. Phys.* **17**, pp. 887–893 (1990a).
17. Nishikawa R. M. and Yaffe, M. J., Model of the spatial-frequency-dependent detective quantum efficiency of phosphor screens, *Med. Phys.* **17**, pp. 894–904 (1990b).
18. Maidment, A. D. A. and Yaffe, M. J., Analysis of the spatial-frequency-dependent DQE of optically coupled digital mammography detectors, *Med. Phys.* **21**, pp. 721–729 (1994).
19. Kitts, E. L., Recent Advances in Screen-Film Systems, in *The Expanding Role of Medical Physics in Diagnostic Imaging*, eds. Frey, G .D. and Sprawls, P. (Advanced Medical Publishing, Madison, WI) pp. 153–181 (1997).
20. Wayrynen, R. E., The Photographic Process, in *The Physics of Medical Imaging: Recording System Measurements and Techniques*, ed. Haus A. (American Institute of Physics, New York) pp. 1–15 (1979).
21. Haus, A. G., Screen-Film Image Receptors and Film Processing, in *Syllabus of categorical course on technical aspects of mammography*, eds. Haus, A. and Yaffe M. J. (Radiological Society of North America, Oak brook, IL)

pp. 69–84 (1992).
22. Corney, G. M., Sensitometric properties of radiographic films, in *The Physics of Medical Imaging: Recording System Measurements and Techniques*, ed. Haus A. (American Institute of Physics, New York) pp. 72–82 (1979).
23. Doi, K., Holje, G., Loo, L. N., Chan, H. P., Sandrik, J. M., Jennings, R. J., Wagner, R. F., MTF's and Wiener Spectra of Radiographic Screen-Film Systems, U.S. Department of Health and Human Services Publication FDA 82-8187 (1982).
24. Doi, K., Kodera, Y., Loo, L. N., Chan, H. P., Higashida, Y., Jennings, R. J., MTF's and Wiener Spectra of Radiographic Screen-Film Systems: Volume II, U.S. Department of Health and Human Services Publication FDA 86-8257 (1986).
25. Rossmann, K., Spatial fluctuations of X-ray quanta and the recording of radiographic mottle, *Am. J. Roentgenol.* **90**, pp. 863–869 (1963).
26. Barnes, G. T., Radiographic mottle: A comprehensive theory, *Med. Phys.* **9**, pp. 656–667 (1982).
27. Dainty, J. C. and Shaw, R., *Image Science* (Academic Press, London) (1974).
28. Selwyn, E. W. H., A theory of graininess, *Photogr. J.* **75**, pp. 571–580 (1935).
29. Macovski, A., *Medical Imaging Systems* (Charles C. Thomas Publisher, Springfield) (1983).
30. Evans, A. Li, *The evaluation of medical images* – Medical Physics Handbooks 10 (Adam Hilger Ltd, Bristol) (1981).
31. ICRU Report 54 Medical Imaging–The Assessment of Image Quality (ICRU Publications, Bethesda, USA)(1996).
32. Wagner, R. F., Fast Fourier digital quantum mottle analysis with application to rare earth intensifying systems, *Med. Phys* **4**, pp. 157–162 (1977a).
33. Barrett, H. H. and Swindell, W., *Radiological Imaging*, Vol. 1 (Academic Press, New York)(1981).
34. Barnes, G. T., Noise Analysis of Radiographic Imaging, in *Recent Developments in Digital Imaging*, eds. Doi, K., Lanzl, L. and Lin, P. P.-J. (American Institute of Physics, New York) pp. 16–38 (1985).
35. Sandrik, J. M., Wagner, R. F., Absolute measures of physical image quality: Measurement and application to radiographic magnification, *Med. Phys.* **9**, pp. 540–549(1982).
36. Dick, C. E. and Motz, J. W., Image information transfer properties of X-ray fluorescent screens, *Med. Phys.* **8**, pp. 337–346 (1981).
37. Wagner R. F., Weaver K. E., Denny, E. W. and Bostrom R. G., Toward a unified view of radiological imaging system. Part I: Noiseless Images, *Med. Phys.* **1**, pp. 11–24 (1974).
38. Schade, O. H., Image Gradation, Graininess and Sharpness in Television and Motion Picture Systems, *J. Soc. Motion Pict. Telev. Eng.* **58**, pp. 181–222(1952).
39. Wagner, R. F., Toward a unified view of radiological imaging systems. Part II: Noisy images, *Med. Phys* **4**, pp. 279–296 (1977b).
40. Sturm, R. E. and Morgan, R. H., Screen intensification systems and their limitations, *Am. J. Roentgenol. Radium Ther.* **62**, pp. 617–634 (1949).

41. Rose, A., The Sensitivity Performance of the Human Eye on an Absolute Scale, *J. Opt. Soc. Am.* **38**, pp. 196–208 (1948).
42. Rose, A., *Vision: Human and Electronic* (Plenum, New York) (1973).
43. Motz, J. W. and Danos, M., Image information content and patient exposure, *Med. Phys.* **5**, pp. 8–22 (1978).
44. Schade, O. H., Optical and Photoelectric Analog of the Eye, *J. Opt. Soc. Am.* **46**, pp. 721–739 (1956).
45. Schnitzler, A. D., Image-detector model and parameters of the human visual system, *J. Opt. Soc. Am.* **63**, pp. 1357–1368 (1973).
46. Blackwell, H. R., Contrast Thresholds of the Human Eye, *J. Opt. Soc. Am.* **36**, pp. 624–643 (1946).
47. Burger, G. C. E., Phantom tests with X-rays, *Philips Tech. Rev.* **11**, pp. 291–298 (1950).
48. Bijkerk, K. R., Lindeijer, J. M. and Thijssen, M. A. O., CDMAM-Phantom Manual, Department of Diagnostic Radiology, University Hospital Nijmegen, St. Radbound, The Netherlands (1995).
49. Darambara, D. G., Taibi, A., Speller, R. D. and Gambaccini, M., Contrast-Detail Evaluation of a Full-field Digital Mammography System, *IEEE Trans. Nucl. Sci.* **47**, pp. 870–876 (2000).
50. ICRU Report 41, Modulation Transfer Function of Screen-Film Systems (ICRU Publications, Bethesda, USA) (1986).
51. Cunningham, I. A. and Reid, B. K., Signal and noise in modulation transfer function determinations using the slit, wire, and edge techniques, *Med. Phys.* **19**, pp. 1037-1044 (1992).
52. Vranckx, J., Breesch, P. and De Belder, M., Two-Dimensional Noise Power Spectra of Radiographic Screen-Film Systems, *Photogr. Sci. Eng.* **28**, pp. 134–136 (1984).
53. Koedooder, K., Strackee, J., Venema, H. W., A new method for microdensitometer slit length correction of radiographic noise power spectra, *Med. Phys.* **13**, pp. 469–473 (1986).
54. Nishikawa R. M. and Yaffe, M., J. Signal-to-Noise Properties of Mammographic Film-Screen Systems, *Med. Phys.* **12**, pp. 32–39 (1985).
55. Bunch, P. C., Huff, K. E. and Van Metter, R. L., Analysis of the detective quantum efficiency of a radiographic screen-film combination, *J. Opt. Soc. Am. A* **4**, pp. 902–909 (1987).
56. Moores, B. M., Wall, B. F., Eriskat, H. and Schibilla, H., eds. Technical and physical parameters for quality assurance in medical diagnostic radiology. British Institute of Radiology Report 18, London (1989).
57. Moores, B. M., Wall, B. F., Eriskat, H. and Schibilla, H., eds. Optimization of image quality and patient exposure in diagnostic radiology. British Institute of Radiology Report 20, London (1989).
58. ISO 9236-1, Photography – Sensitometry of screen/film systems for medical radiography – Part 1: Determination of sensitometric curve shape, speed and average gradient, International Organization for Standardization (1996).
59. Gambaccini, M., Evaluation of physical image-quality and sensitometric characteristics of screen-film systems, University of Ferrara (unpublished)

(1997).
60. Buhr, E., Bergmann, D., Hoeschen, D., Ailliet, M., Sirand-Rey, G., Gazzola, C., Uras, S., An interlaboratory measurement of screen-film speed and average gradient according to ISO 9236-1, *Med. Phys.* **27**, pp. 307–311 (2000).
61. Borasi, G., Piccagli, V., Ferretti, P. P., Berardi, P., Evaluation of image quality and sensitivity of radiographic screen-film systems, *Phys. Med.* **V** supp. 1, pp. 287–293 (1989).

CHAPTER 3

DETECTORS FOR DIGITAL RADIOGRAPHY

Paolo Russo

Università di Napoli Federico II, Dipartimento di Scienze Fisiche, and Istituto Nazionale di Fisica Nucleare (INFN), Sezione di Napoli, Via Cinthia, I-80126 Napoli, Italy (e-mail: paolo.russo@na.infn.it)

3.1 Introduction

Digital Radiography (DR) is the science of devising X-ray digital imaging detectors that may substitute the conventional radiographic screen-film combination for use in diagnostic radiology. In projection radiography, a digital imaging detector maps the two-dimensional distribution of X-ray photons transmitted by the irradiated body in discrete spatial units (pixels) and discrete signal units (bits), as opposed to the X-ray film (plus the radiographic intensifying screen, when used in a screen-film combination) which directly provides only an analogue recording of the two-dimensional detected photon distribution. In principle, the digital approach has several significant benefits over the analogue film. First, it gives directly a digital image, able to be processed with computer methods as soon as it is acquired, for image storage, manipulation, archiving and network circulation within the hospital(s). This would avoid the necessity of film digitization with X-ray film scanners. This is an additional processing step after film exposure and development which would equally provide a digital X-ray image from X-ray film, but would add the cost of the film scanner which might represent a significant fraction of the cost of the entire X-ray unit. But the very true advantage of the direct X-ray digital image is related to the linearity and dynamic range of the imaging acquisition process. A digital radiographic detector is expected to provide a linear response to the incident X-ray fluence (number of incident quanta per unit surface) up to several decades, covering the entire range of possible input dose with X-ray examinations. It is well known that the response of (radiographic) film to relative exposure E/E_0 is logarithmic, the optical density *D* of the exposed and developed film being proportional to the log relative exposure only in the (central) linear region of the (sensitometric) characteristic curve $D=D(E/E_0)$. In this linear-response region, $D = \Gamma \log(E/E_0)$, with a constant value of the "gamma" of the film. An ideal linear imaging system (with response *S* for a given exposure *E*) would provide an image contrast $\Delta S/S$ equal to the object contrast $\Delta E/E$ at any exposure level. For the radiographic film with optical transmission *T*, the image contrast $\Delta T/T$ is proportional to object contrast $\Delta E/E$ only in the linear regime (constant gamma). At high exposure levels, saturation of the film response occurs and the slope of the characteristic curve is small; at low exposures, poor film response gives small slope of the characteristic curve and as before, image contrast will be poor. This would prevent the radiographic detection of low contrast objects at very low or very high exposure levels.

The increase in exposure dynamic range permitted by a digital detector in an X-ray imaging system, with respect to the conventional screen-film detector, can be extremely significant. A dynamic range of just 100:1 is 2 to 4 times that of a typical screen-film, but digital X-ray detectors exist that allow a dynamic range as high as 10000:1 or higher. This implies an extremely large range of X-ray attenuation in the body to be registered in the same digital X-ray image. Increased contrast resolution and image post-processing capabilities of digital systems that zoom into the range of exposure levels of interest ("windowing"), then make it possible to detect anatomical details with lower image contrast than with film-based systems. In other terms, the use of digital detectors can improve detail visibility, and then image quality, so improving the diagnostic quality in radiography. With DR systems, lower contrast tissue features could be detected with respect to conventional film radiography—which would require higher exposures to reach the same performance; in addition, image processing can also enhance contrast, so that the technology of DR is also expected to decrease radiation exposure (and dose) to patients. Finally, a digital system would not require chemicals and apparatuses for film development, which means more rapid response and a clear cut to consumer costs, certainly a major voice in the expenses of a film-based Radiology Department.

The possibility of employing Computer Aided Diagnosis (CAD), in addition to teleradiology and digital archiving through currently available hospital information system (HIS), radiology information system (RIS) and Picture Archiving and Communication System (PACS), would then fully disclose the future of filmless, all-digital Radiology Departments, of which a few examples are now in the world, provided that the medical community offer widespread clinical acceptance of the new technologies. Industrial data storage formats and communication protocols exist, e.g. the DICOM standard, that make DR systems a mature reality in many industrialied countries (Fig. 3.1).

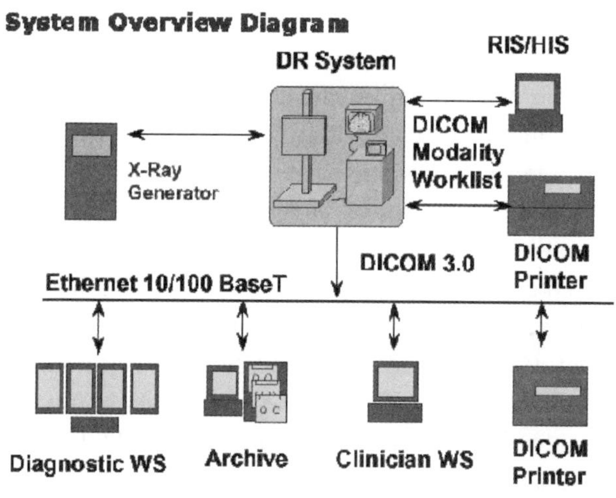

Fig. 3.1. Block diagram of a complete Digital Radiography system (from Canon, 2000).

In principle, the digital detector should perform linearly from zero counts up to the maximum photon counts possible with that system, but in practice the system noise level determines the minimum useful signal level, and count losses are present at high count rates that limit this linear response, normally to 3–4 decades. The dynamic range of present DR systems is in fact between 12 and 16 bits, with most systems featuring 14 bit resolution. However, these figures should be compared to the only 2 signal decades accessible to the X-ray film system, which is well known to be a linear detector only in a restricted range of input levels of logarithmic relative exposures. While the large system-response linearity and large dynamic range are the most relevant imaging parameters to be expected from digital radiographic systems, other important aspects are of concern, like spatial resolution, detection efficiency, system noise, while other less determinant aspects should also be addressed, like temporal response and ease of integration within existing X-ray units.

Given the requested benefits of a DR system and the overall performance and century-long experience with the conventional radiographic film, it can be seen that the realization of a good detector for a digital radiography system is an extremely complex scientific and technological task. The problem of a DR system with a definitely better performance than film-based systems has not been fully solved up to now, although researchers have been working in this field for some decades since the digital medical imaging field was opened by the clinical introduction of Computed Tomography (CT) in 1972. In fact, the first widespread commercial system for DR based on photostimulable phosphor plates, developed by Japanese researchers in the early '80s and still in use, is in fact a mixed analogue-digital system, in which an analogue detector (the imaging plate), after exposure is read-out digitally by optical methods. This system, commercially known as Computed Radiography (CR), provided the first serious way to approach the move to all-digital Radiology Department in hospitals, but its diffusion has been contrasted by the high cost of the dedicated CR units (up to several times that of a conventional system).

Materials science and microelectronics are the research areas most involved with DR, since the detector for DR is essentially a solid state (e.g. semiconductor) large-area X-ray detection system coupled to a fast, mixed (analog + digital) microelectronics read-out system (possibly coupled with a layer of scintillator material). Combined advances in solid state and electronic technologies, first for military and then for large-volume consumer products fields, have given rise in the last two years to a class of commercial DR systems–based on amorphous semiconductor detectors and flat panel display read-out technology–with many features of an ideal X-ray imaging system. These detection systems are being commercialized by several firms both in stand-alone systems and as an integration to existing X-ray units (for general radiography), by substitution of the X-ray film cassette with the new portable digital imaging detector box. A DR system for full field mammography is also on the market.

Digital radiography systems are presently under evaluation in some large Radiology Departments in USA and Europe; unfortunately, as a result also of the large costs for industrial research and development over many years, the cost of a complete DR unit can be between US$400,000 and US$600,000. For comparison, a complete general purpose radiographic X-ray room costs from US$80,000 to US$200,000. It is evident that only large Departments could consider the purchase of such systems, to which one must add also the initial costs for digital workstations and PACS. Also for this cost reason and, in addition, for the changes needed in radiologists university education, related to the necessity of acquiring ability to view images on a digital screen and make diagnosis on them (digital radiographs may show different features than with film radiography, and even the same image may show different features on different viewing stations), and necessary changes in the capacity and retrieval rate of existing image archives, it is confidently foreseen that the radiographic film – this extraordinary image detector – will be exposed and processed for high-quality conventional radiography still for many years all the world over.

In the following sections a brief review of detection systems for digital X-ray radiographic imaging will be presented, with the aim of providing some insight into both well established and emerging technologies, to illustrate experimental systems with potential of industrial exploitation and to indicate possible lines of further research in the digital radiography systems under development.

3.2 Characteristics of X-Ray Imaging Systems

3.2.1 *Figure of merit for image quality: Detective Quantum Efficiency*

The assessment of quality in image science requires the evaluation of many different parameters and image features, like detection efficiency, spatial resolution and image noise. An excellent classic textbook on this subject is available [1]. In comparing the different detector systems for digital X-ray imaging, a main figure of merit will be used, namely, the *detective quantum efficiency*, or *DQE*. Though not the unique parameter for assessing the characteristics of an imaging system, the *DQE* is anyway generally accepted as a quantitative measure of the overall image quality, especially in comparing different digital systems. Furthermore, for its determination, knowledge of other image quality parameters is necessary, related to efficiency and resolution and noise, so that when the *DQE* of an imaging system is known, different image quality aspects are also known for that particular system. At a preliminary stage, the *DQE* of an imaging system can be defined as the squared ratio of the output and input *signal-to-noise ratio* (*SNR*):

$$DQE = \frac{SNR_{out}^2}{SNR_{in}^2} \tag{3.1}$$

In other terms, the *DQE* gives an indication of how the *SNR* of an input image signal is transferred to the output image field. Ideal imaging systems have a *DQE* of 1. Since any real imaging system introduces additional noise from various sources, e.g. electronic, the output *SNR* is always reduced, so that the *DQE* takes values between 0 and 1. Higher *DQE* values mean higher image quality. Let n_0 be the number of incident photons on a pixel of the detector; if the statistics of the input image source is Poissonian, as ultimately in X-ray imaging, then the input noise is $n_0^{1/2}$ and $SNR^2_{\text{in}} = n_0$. This shows that *DQE* as defined above depends on the incident photon fluence. A complete imaging system comprising various parts has a *DQE* given by the product of the *DQE* of the single imaging parts. It can be shown that the *DQE* of an imaging system can be written as

$$DQE = \eta_{QE}\eta_{gen}\,. \tag{3.2}$$

Here, η_{QE} is the quantum efficiency of the detection system (the probability of a single incident photon to generate one electron in the interaction process), and η_{gen} is a corrective parameter, between 0 and 1, that takes into account various inefficiencies in the collection of the input signal and in the generation of the output signal by the complete imaging system. This expression for the *DQE* shows that in order to have a high *DQE*, an (X-ray) imaging system has to show a high quantum efficiency (which depends on the energy of the input photon distribution and on the detector material, through its total linear attenuation coefficient) and low losses in the conversion process of the interacting photon into an electrical signal. For example, in semiconductor radiation detectors, the energy *E* of the interacting photon is converted, with efficiency η_{EA} (energy absorption efficiency), into a number of electron-hole pairs; if electrons or holes are collected at suitable sensing electrodes with efficiency CCE (Charge Collection Efficiency), then η_{gen} includes as factors both terms.

As a matter of fact, both the input and the output image fields contain spatial variations in the number of incident photons, that can be described by the spatial frequency variable *f* (often expressed in line pairs per mm, lp/mm). This implies that the *DQE* as defined in (3.1) corresponds exactly only to the zero spatial frequency detective quantum efficiency *DQE*(0):

$$DQE(0) = \frac{SNR_{out}^2(0)}{SNR_{in}^2(0)}\,. \tag{3.3}$$

Here, $SNR_{out}(0)$ and $SNR_{in}(0)$ represent the zero spatial frequency signal-to-noise ratios of the output and input of the imaging system, respectively. Via Fourier transform and (shift-invariant) linear systems analysis, the description of the image fields can then be operated in the spatial frequency domain, through the optical transfer function of the imaging system, whose modulus is the *modulation transfer function* of the system, *MTF*(*f*). The MTF describes the quality of the imaging system with respect to its spatial resolution properties, i.e. *MTF*(*f*) describes the ability of the system to distinguish variations in the spatial distribution of the incident photon flux. The *MTF*(f) of an imaging system is a decreasing function of spatial frequency, since image blurring degrades the ability of a system to resolve frequent spatial intensity variations in the image. For undersampled systems, like digital radiography systems which imply the 2D spatial sampling of the detected photons distribution in the detector volume, the notion of *presampling* MTF must be introduced. The presampling MTF is the MTF of a digital system before the process introduced by the sampling aperture. This notion allows for evaluating the intrinsic resolution properties of the digital imaging system (which contains both an analogue and a digital part), avoiding the problems introduced by undersampling (aliasing) when MTF is measured directly using a narrow slit or a sharp edge as test objects [2]. The presampling MTF can be evaluated with the method suggested in [3], which requires the acquisition of the X-ray transmission image of a narrow slit slightly tilted with respect to the orthogonal x−y axes of the detector system. The MTF evaluated in this way implies to consider an equivalent detector pixel of a reduced size with respect to the real pixel size, and this causes the presampling MTF to extend to spatial frequencies beyond the Nyquist frequency of the detector. Then, the complete expression for the frequency-dependent *DQE*(f) is:

$$DQE(f) = DQE(0)\frac{MTF^2(f)}{NPS_0(f)} \tag{3.4}$$

where $NPS_0(f)$ is the noise (Wiener) power spectrum of the imaging system normalized to unity as f tends to 0. The *NPS*(f) describes the fluctuations of noise with the spatial frequency: it represents the decomposition in spectral terms of the variance of the image after uniform exposure.

The DQE is a very useful quantitative parameter to estimate the quality of an imaging system, but it is difficult to measure. Determination of input photon fluence, of MTF and of Wiener spectrum is required with the corresponding measurement procedures being possibly complex and time-consuming. In addition to *SNR*(f), *MTF*(f), *DQE*(f) and *NPS*(f), other parameters like the *noise equivalent quanta* (NEQ) and the *contrast transfer function* (CTF) are usually considered when assessing image quality in a radiographic system. The *NEQ*(f) is defined as "the effective number of input X-ray quanta per unit area that would give the same SNR with an ideal imaging system as the real exposure quanta give with the actual imaging system" [4]. The $CTF(f)$ of an imaging system measures the ratio of the output image contrast to the input contrast, as a function of spatial frequency.

It should be noted that the spatial frequency behavior of an imaging system may depend on the direction on the image plane, i.e. on the x-y axes of the detector: this implies that two-dimensional quantities like $MTF(f_x, f_y)$ and $DQE(f_x, f_y)$ should be considered, instead of their one-dimensional forms [5]. For simplicity of exposition, in the following the one-dimensional properties will be used.

3.2.2 *Integrating vs photon counting systems*

With the advent of digital modalities for X-ray imaging, a number of diverse technologies are being used for X-ray detection, which can be separated into two classes. To the first class, termed (*energy* or *charge*) *integrating* systems, belong those detectors which—during the elementary or overall acquisition period—measure the total energy (charge) deposited by all interacting photon in the detector material. This energy, e.g. in the form of electrical charges produced by ionization in the detector, is stored locally until it is read out. For example, in film-based systems, the blackening of the film in any given sensitive area is due to the energy deposited in that area by all photons that interacted with the film (or with the intensifying screen) during the radiography exposure. Similarly, in X-ray imaging systems based on CCD detectors or arrays of photodiodes, the radiation-induced current in each single CCD cell or diode is integrated during the acquisition period, giving a total charge which ultimately is proportional to the total energy deposited in the detector by interacting X-rays. The second class is *single photon counting* acquisition modality: here, each interacting photon depositing an energy (charge) in the detector greater than a detection threshold, is counted as one "hit", independent of the total energy it deposited. While dark current and signal saturation are main problems in integrating mode acquisition, the photon counting modality requires signal discrimination and counting logic in each picture element, a complex microelectronics task for large area, high granularity detectors.

It can be shown that in principle, photon counting imaging systems offer a higher SNR and a better contrast with respect to integrating systems [6]. Experimental data obtained with a single photon counting readout chip and a Si detector confirmed this expectation [7-9]. For the SNR—defined for a uniform illumination as the average counts/pixel divided by the sigma of the counts distribution—on increasing the acquisition time, the SNR increased up to only ~30% of the theoretical (Poisson) limit of a quantum limited device. However, by suitable equalization of the response of single pixels (flat field correction), the SNR can reach 70% of the Poisson limit of $1/\sqrt{N}$ (Fig. 3.2) and, by increasing further the count rate, it can get close to quantum noise limited imaging conditions (Fig. 3.3).

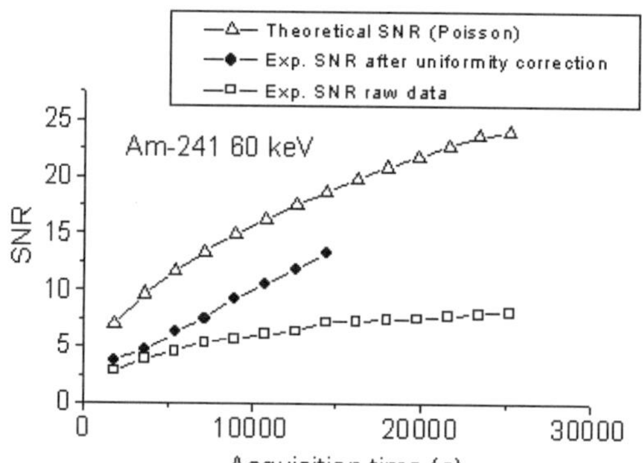

Fig. 3.2. Signal-to-Noise Ratio as a function of acquisition time, obtained with a 60 keV flood illumination from a low-activity Am-241 radioactive source and a single photon counting imaging system (Medipix series). Raw data (middle curve) show a poor SNR, but after correction for detector's non-uniformities the SNR reaches 70% of the Poisson limit for a quantum limited device [9].

This lack of uniformity of response in photon counting systems can be due to detector non uniformity and to slight mismatch in the response of electronic cells in the detector array.

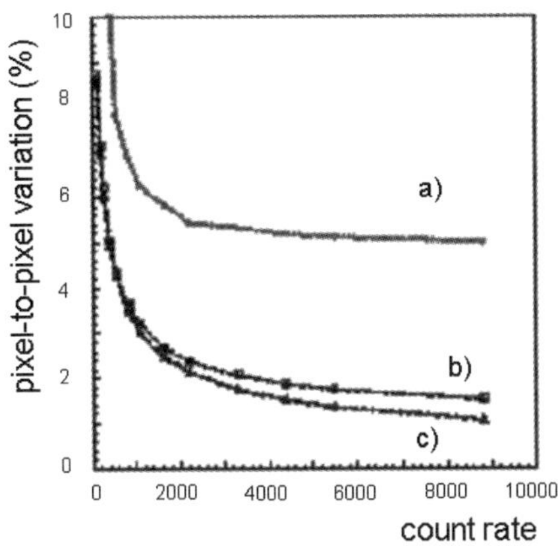

Fig. 3.3. Percentage of noise over pixel matrix as a function of average count rate. (a) Raw data from a flood image; (b) same data multiplied with an efficiency map; (c) Poissonian limit (from [7]).

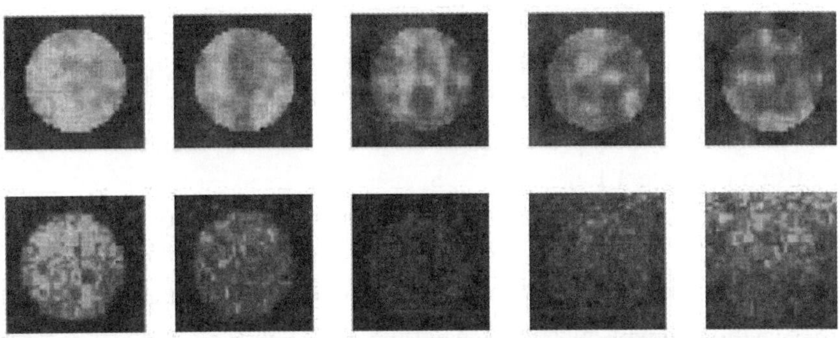

Fig. 3.4. Images of five 6 mm round-shaped low-contrast details of the mammographic phantom TOR(MAS) obtained with a screen-film system (bottom row) and with a 200 μm semi-insulating GaAs pixel detector systemworking in single photon counting (top row). The exposure is 12.5 mAs. The focus-detector distance is 64 cm. Nominal contrast values from left to right are: 8.5%, 5.5%, 3.8%, 2.6%, 2.0% (from [10], reprinted with permission of Elsevier Science B.V.).

As regards the expected improvement in image contrast, Fig. 3.4 shows a graphical comparison between the contrast detail capability of a conventional screen-film combination and that of a single photon counting imaging system equipped with a GaAs detector [10]. These tests, performed with a conventional mammographic unit at normal settings for screening mammography, showed the improvement in detail visibility of this photon counting system with respect to screen-film energy integrating system (minimum contrast reached with the DR system was 2%, reaching about 1% by increasing twice the X-ray dose). This is a result of both high detection efficiency and high charge collection efficiency of the Gas detector (>90%), and of the low noise of the photon counting readout chip.

3.3 Semiconductor Materials for X-Ray Digital Detectors

Both elemental and compound room-temperature semiconductors have been proposed and used as substrate materials for digital radiography. This list includes, among others, Si, CdTe, HgI_2, CdZnTe, GaAs, Se, PbI_2, either in a crystal or amorphous or polycrystal form. However, scientific and technological research in this field is mainly concentrated on the use or improvement of the X-ray detection performance of Si (crystal or amorphous), Se (amorphous), GaAs (bulk and epitaxial), $Cd_{1-x}Zn_xTe$ (with Zn fraction typically of 10% or 20%), especially for digital radiography [11-24].

The main physical properties of these materials for radiation detection are listed in Table 3.1.

Many scientific workshops are dedicated to the study of radiation detectors, and the interested reader could refer to their published proceedings for details on the characteristics and the performance of these semiconductor detector materials (see e.g. [25-27]). Here, it is important to note that among the main features of detectors for digital radiography, there are intrinsic detection efficiency and charge collection efficiency (CCE). The first is a result of detector thickness and material's absorption efficiency in the diagnostic energy range, and it determines the efficiency parameter η_{QE}. The second (CCE) is the ratio of the effective charge collected at the readout electrodes to the expected radiation-induced charge produced by a photon depositing all of its energy into the detector; it should be close to 1 in order to increase the η_{gen} efficiency parameter in Eq. (3.2). So, both parameters are close to 1 in a high-DQE imaging system. As an example, at zero spatial frequency, where

$$DQE(0) = \eta_{gen}\eta_{gen},$$

a conventional screen-film system has a DQE of about 6% and reaches 20% at most, but DR systems are able to reach DQE values greater than 50% [28].

Different problems are related to reaching high detection efficiency and high CCE in semiconductors like Si, epitaxial or semi-insulating (SI) bulk GaAs, or CdZnTe. High-resistivity 300 μm Si substrates are only ~10% efficient for X-ray detection at about 25 keV (Fig. 3.5), but 200 μm thick GaAs detectors have almost 100% detection efficiency for mammographic energies (below 25–30 keV), and 600 μm GaAs detectors are well suited for general radiography in the whole diagnostic energy range (Fig. 3.6).

Table 3.1: Physical properties of semiconductors for X-ray imaging.

Material	**Si**	**a-Si**	**a-Se**	**$Cd_{0.9}Zn_{0.1}Te$**	**GaAs**
Atomic number(s)	14	14	34	48, 30, 52	31, 33
Average atomic number	14	14	34	49.1	32
Density (g/cm^3)	2.33	2.3	4.3	5.78	5.32
Bandgap E_g (eV)	1.12	1.8	2.3	1.572	1.424
Electron mobility μ_e (cm^2/Vs)	1400	1	0.005	1000	8000
Electron lifetime τ_e (s)	$>10^{-3}$	7×10^{-9}	10^{-6}	3×10^{-6}	10^{-8}
Hole mobility μ_h (cm^2/Vs)	480	0.005	0.14	50–80	400
Hole lifetime τ_h (s)	2×10^{-3}	4×10^{-6}	10^{-6}	10^{-6}	10^{-7}
$\mu_e\tau_e$ (cm^2/V)	>1	7×10^{-9}	5×10^{-9}	$3–5\times10^{-3}$	8×10^{-5}
$\mu_h\tau_h$ (cm^2/V)	1	2×10^{-8}	1.4×10^{-7}	5×10^{-5}	4×10^{-6}
Ionization energy W (eV/e-h pairs)	3.62	4	20–60[a]	4.64	4.2
Dielectric constant	11.7	11.7	6.6	10.9	12.8
Resistivity (Ωcm)	$<10^4$	10^{12}	10^{12}	3×10^{10}	10^7

[a] The exact value depends upon the X-ray energy

Both detectors can be pixellated to very small pixel sizes of 50–100 μm with optimum (Si) or acceptable (SI GaAs) spatial uniformity of response over large arrays. CdZnTe detectors are currently produced with thickness of a few mm, they have been pixellated down to 50 μm [29, 30] and have acceptable uniformity of response over about 1 cm^2 sensitive area. Their high detection efficiency (1–2 mm thickness) [31] makes them ideal to extend the energy range to general radiography and to nuclear medicine, in addition to mammography (i.e. a 20–150 keV range).

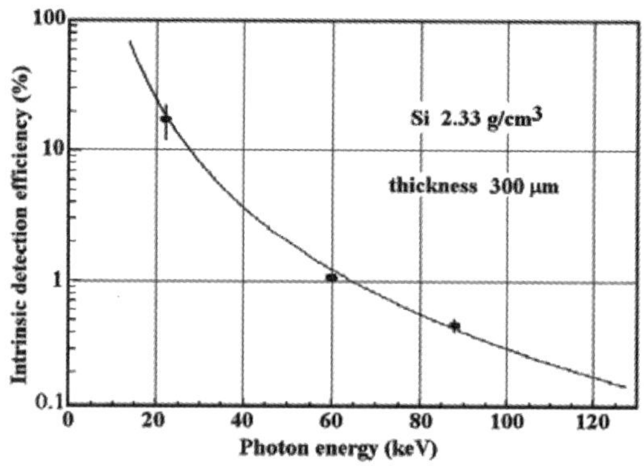

Fig. 3.5. Experimental (points) and computed (line) detection efficiency of a silicon detector, 300 μm thick.

Silicon (as *p-n* junction diodes) and SI GaAs (as Schottky diodes) detectors for imaging arrays [32] are normally operated reverse-biased in the hole-collection modality, and they are normally irradiated from their ohmic side. CdZnTe detectors are operated in photoconductive mode with electron collection, since electrons are better transported than holes in this material (Table 3.1). For full detection efficiency and high CCE, detectors must be biased at full depletion or over-depleted. High-resistivity (about 10^4 Ωcm) Si detectors can be fully depleted at reverse bias voltages below 100 V for 300 μm thickness, and in this condition they can show a CCE close to 100%. In the case of reverse biased high-resistivity (about 10^7 Ωcm) SI GaAs detectors, it is well known that the width of the depletion layer (i.e. the thickness of the region of high electric field that is present in the detector under bias, where drift of the radio-induced charge can take place) increases starting from the junction side to the ohmic side, at a rate of about 1 μm/V [33, 34].

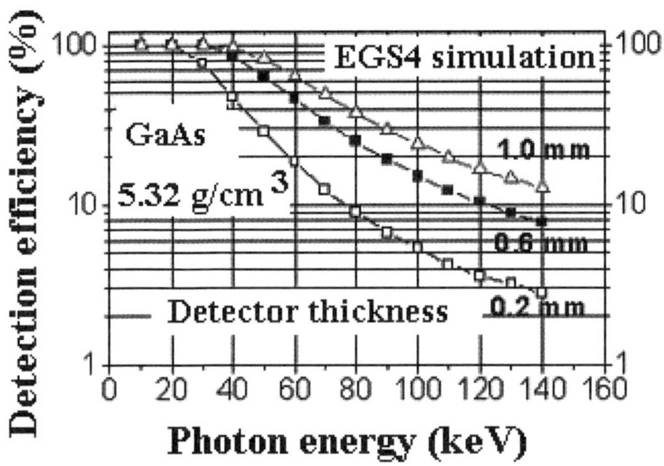

Fig. 3.6. Detection efficiency of GaAs, 0.2 mm or 0.6 mm or 1 mm thick (from [35], reprinted with permission of Elsevier Science B.V.).

Experiments [16] with short-sided single-pad diodes showed that the CCE for SI GaAs detectors is close to 100% only for substrates of 200 μm thickness or thinner, biased at over 300 V (Fig. 3.3). For 600 μm thickness, the CCE of small SI GaAs diodes can reach ~80% at 500 V bias [35]. Slightly lower performance was observed for similar detectors with square pixel sizes greater than 500 μm, even at higher bias voltages [36]. In these conditions of high bias voltages, it can be assumed that the detector volume is almost fully active, namely, that the region of high internal electric field extends almost to the full thickness of the detector, so that a saturation value of CCE versus reverse bias, less than 100%, can be ascribed to intrinsic material defects, e.g. electron trapping centers. On the other hand, for pixellated SI GaAs detectors (200, 300 or 600 μm thick) it has been found impossible to reach high voltages of reverse bias, and hence a fully active detector volume, even with a 200 μm thickness [37-39]. This occurs either because a high bias would produce a too large leakage current, or because high bias voltages could produce electrical breakdown in the detector. In other terms, thick pixellated SI GaAs detectors have shown sub-optimal charge collection efficiency, but also sub-optimal X-ray detection efficiency, since for low bias voltage operation the detector is not fully active all over its thickness (Fig. 3.4). This situation worsens for detectors irradiated from the ohmic side (as in all previous cases referenced to), since in this case the zone of reduced or null electric field is located just on the side of the greater X-ray attenuation. Moreover, in this situation the lower the photon energy, the lower the detection efficiency. These efficiency problems, coupled to a certain spatial non-uniformity in pixel response, due to the lack of detector homogeneity [40], are still to be fully solved for thick SI GaAs pixellated detectors, yet it remains one of the best candidate, with CdZnTe, as a room-temperature semiconductor for DR.

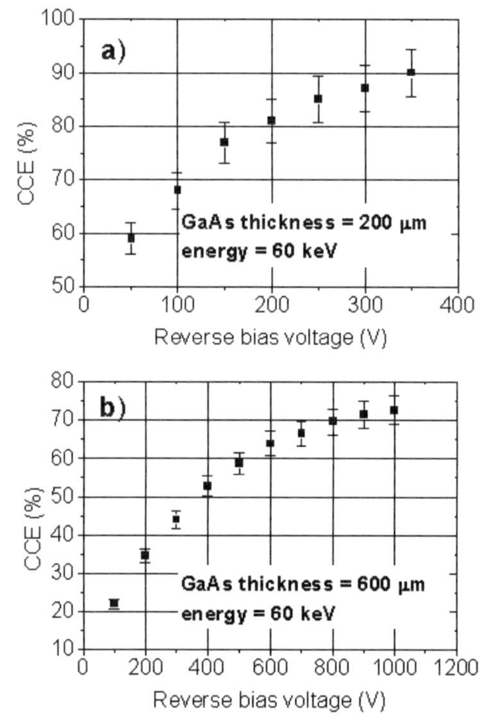

Fig. 3.7. Charge collection efficiency of semi-insulating GaAs detectors (single pad Schottky diodes) as a function of reverse bias voltage, for (a) 200 μm and (b) 600 μm thick substrates, irradiated with 60 keV γ-rays from a ^{241}Am source.

Also an important issue for X-ray radiography is radiation damage to the detectors and its associated readout electronics. A high efficiency detector in digital systems may also shield the downbeam associated electronics from the incident radiation, but even with thin silicon detector, very high radiation damage to the detector and to the underlying readout electronics have been observed only with extremely high flux X-ray sources, as with synchrotron radiation [41].

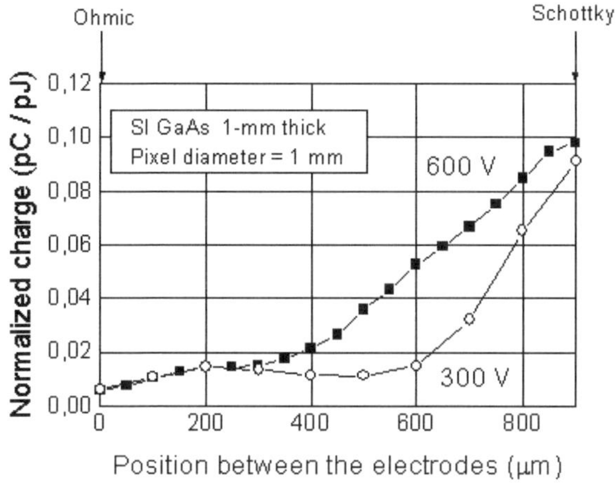

Fig. 3.8. The charge collection efficiency of semi-insulating GaAs detectors (1 mm diameter single pad Schottky diode, 1 mm thick) has been probed by irradiating the detector on a side with an infrared pulsed laser beam, focused to a 50 μm spot at various positions between the ohmic and Schottky contacts. The diode was reverse biased at 300 or 600 V. Once absorbed, the picosecond light pulse generates a charge transient (e–h pairs) that simulates the interaction of an energetic gamma ray. The charge collected at the Schottky electrode as a function of the position of the light spot, indicates that there exist a fraction of the detector volume on the ohmic side (anode) which is not fully active, and that this inactive layer has a thickness which decreases upon increasing the bias voltage. The thickness of the active layer is roughly proportional to the bias voltage (from [34], reprinted with permission of Elsevier Science B.V.).

3.4 X-Ray Imaging Technologies

Research and technological advances in X-ray medical imaging were significant in the last two decades, and the field is still very active mainly in the medical physicists community, but spin-offs from other fields like high energy physics occurred. A (non-exhaustive) list of detector technologies available or under development for radiography systems, includes:

- Fluorescent screen + film
- X-ray image intensifiers with scintillator layers
- Photo-stimulable storage phosphor imaging plate
- Scintillation material (e.g. CsI) + Charge Coupled Devices (CCDs)
- Semiconductor material (e.g. a-Se) + readout matrix array of thin film transistors (TFT)
- Phosphors + semiconductor material (e.g. a-Si:H) + TFT flat panels
- Build-up metal plate + Semiconductor material (e.g. a-Se) + readout matrix array of TFT
- Scintillation material (e.g. CsI) + storage array of thin film diodes (TFD) + readout array of TFT
- 2—D microstrip array on semiconductor crystal (e.g. Si, Cd(Zn)Te) + integrated front-end and readout
- Matrix array of pixels on crystals (e.g. GaAs, CdZnTe) + VLSI integrated front-end and readout
- Multiwire Proportional Counters (MWPCs) [42]
- Gas Microstrip Detectors (GMDs) [43]
- Gas Electrons Muliplier (GEM) [44]
- X-ray-to-light converter plates (AlGaAs).

In the next paragraphs, the above list will be detailed in part in order to describe some technologies already developed or under development for Digital Radiography.

3.4.1 *Photo-stimulable storage phosphor imaging plate*

In conventional screen-film radiography, the (analogue) image sensor is also the image display medium. The first commercially successful digital X-ray imaging system with significant diagnostic performance, to be regarded as a serious digital alternative to general screen-film radiography, is a system in which the X-ray image sensor (still analogue) is separated by the (digital) X-ray image display medium. Introduced in the early ’80s, this system allowed the introduction of digital image processing and digital archiving in Radiography, coupled to a large improvement in image dynamic range resulted eventually in the reduction of patient dose. This kind of digital systems, first produced by

Fuji Photo Film Co. in Japan, is now present in the catalogue of many manufacturers of X-ray diagnostic devices, and is generally known as Computed Radiography (CR). Basic elements of a CR system are: the image sensor, here called the imaging plate (IP), which in a way analogous to film stores in a latent image the detected X-ray image distribution; an image reader, which converts the latent image into an analogue signal and then into a digital signal; an image display like a digital workstation for processing and viewing the radiographs and an image recorder, which allows the users to eventually record the digital image onto a film. The principle of detection is the following ([45] and references therein) (Fig. 3.9) (a thorough review of Computed Radiography technology has been published recently by [46]).

The IP, a flexible plate of about 1 mm thickness, is coated with particular phosphor crystals included in an organic binder. The phosphors, made, e.g. from europium-activated barium-fluorohalide compounds like BaFCl:Eu, BaFBr:Eu, BaFBi:Eu or BaFI:Eu, show quasi-stable energy levels between the valence and conduction bands, which can act as effective long-lifetime traps for electrons and holes excited into higher energy states by the interaction of X-rays (or electrons or UV light, etc.) with the IP (Fig. 3.10).

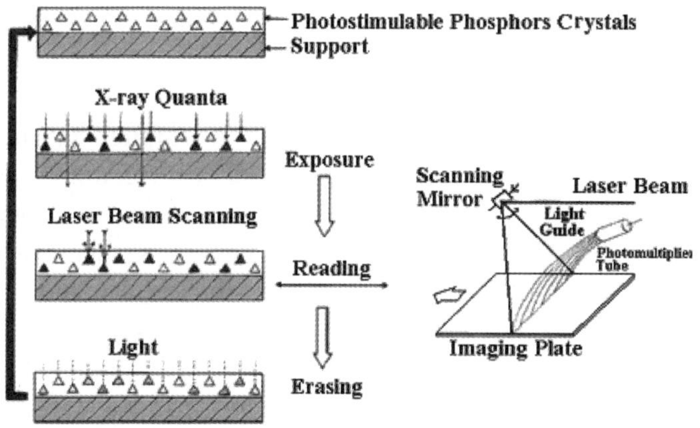

Fig. 3.9. Principle of use of an Imaging Plate for X-ray detection. In the exposure phase, charge trapping occurs in the fluorescence centers (continuous arrrows); in the reading phase, under external excitation with red light, luminescence is produced with relaxation to the ground state (dashed arrows) [45].

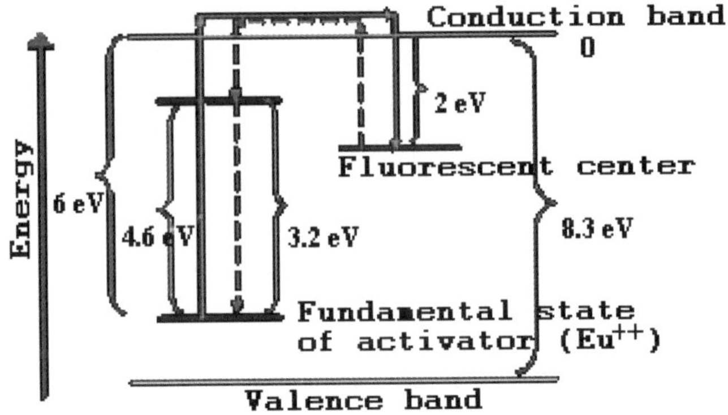

Fig. 3.10. Energy diagram of IP detector material. The dashed path refers to the optical stimulation phase.

As a consequence, when interaction occurs during exposure, charge carriers occupy these traps in an amount locally proportional to the X-ray energy deposited at the incident position in the plate. This produces the spatial distribution of X-ray photon fluence to be stored temporarily in the surface of the IP, forming a sort of "latent" image. In order to "develop" this image, energy is required to produce the detrapping of these charge carriers. This is done optically by generating free carriers in the storage phosphor material, that recombine with the trapping centers and produce emission of visible photons. In fact, the phosphor crystals in the IP, whose dimensions are about 8 μm, are photo-stimulable, in that when stimulated by long-wavelength visible or infrared radiation they emit short-wavelength luminescence radiation (via electronic relaxation to ground energy state), whose intensity corresponds to the amount of X-ray energy absorbed (this effect is known as photo-stimulable luminescence, PSL) (Fig. 3.9). This effect is used in the read-out of the IP: after exposure of the plate to the X-rays, it is raster scanned with a red laser beam (e.g. 633 nm radiation from a low-power helium-neon laser) and the corresponding PSL is collected and directed to a photomultiplier tube (PMT).

The PSL intensity is proportional to absorbed dose of X-rays in the IP: hence, a detector for X-ray imaging can be made (Fig. 3.11). The analogue electric signal from the PMT is then converted to digital (via digitalization at 10 bits resolution) and recorded as a function of the scanning position on the IP, and a digital image matrix is produced, which is processed by the digital image processor and is available for display and for recording on film via optical methods.

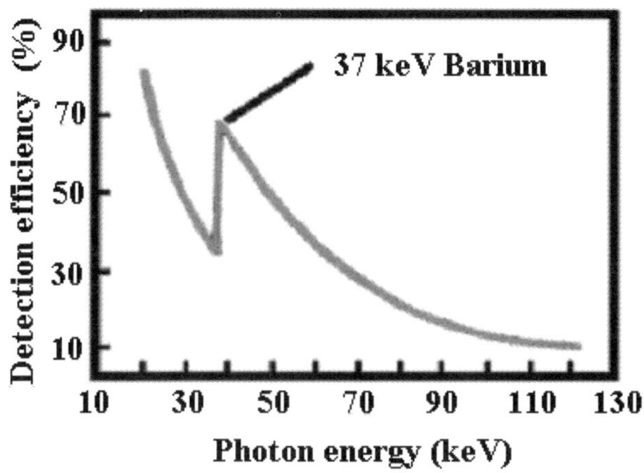

Fig. 3.11. Detection efficiency of IP detector material.

In principle, the read-out process is destructive, i.e. once the latent image has been formed in the IP and then processed by the IP image reader, the information is cancelled and the latent image destroyed. In order to re-use the same IP, the residual energy eventually stored in it must be cancelled: this is done by a flood exposure of the plate with intense light, that "erases" the IP. Erased IPs can be used repeatedly up to hundred times. The IP comes in a cassette unit similar to screen-film combinations, which gives compatibility with existing X-ray equipment.

The phosphor optical stimulation spectrum and the corresponding PSL spectrum depend on the specific barium-fluorohalide compound used in the fabrication of the IP. The optical stimulus for optimum response corresponds to visible wavelengths roughly in the range 500–700 nm, whereas output luminescence spectra are centered at about 400 nm (Fig. 3.12).

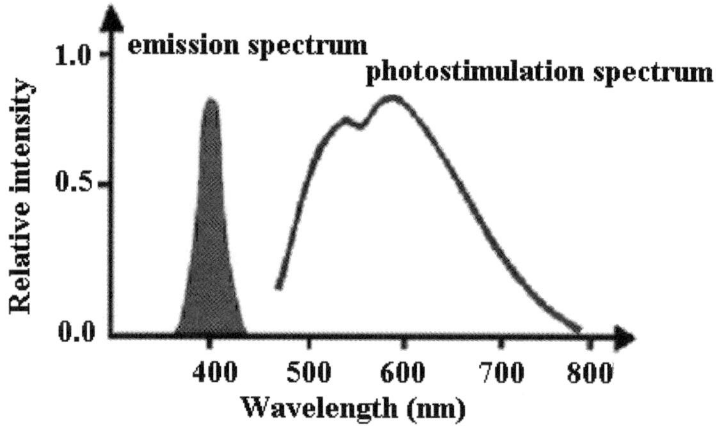

Fig. 3.12. Optical absorption and emission spectra of IP detector material.

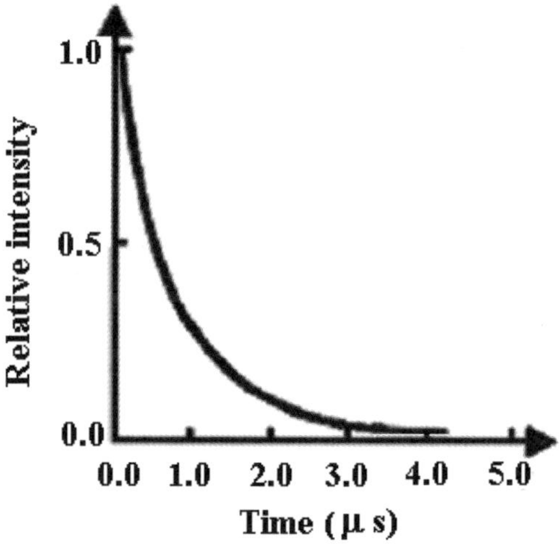

Fig. 3.13. Temporal response of light emission of IP detector material.

Optical response times also vary between 0.6–7.0 μs (Fig. 3.13). This allows for a complete 35 × 43 cm^2 IP cassette for chest radiography to be read in a time as short as 90 s, and about 40 s are necessary to read, process and view a 18 × 24 cm^2 IP in a mammographic CR unit (Fig. 3.14). Finally, multiple-input CR cassette loading system allow increased read-out productivity of the order of 100 IP/h. The spatial resolution of the CR imaging system is determined by the choice of many factors like dimension of the read-out light spot (pixel sampling), read-out velocity, dimension of the image matrix, but ultimately it is limited by the optical diffusion inside the phosphor plate of the probe beam and of the luminescence light.

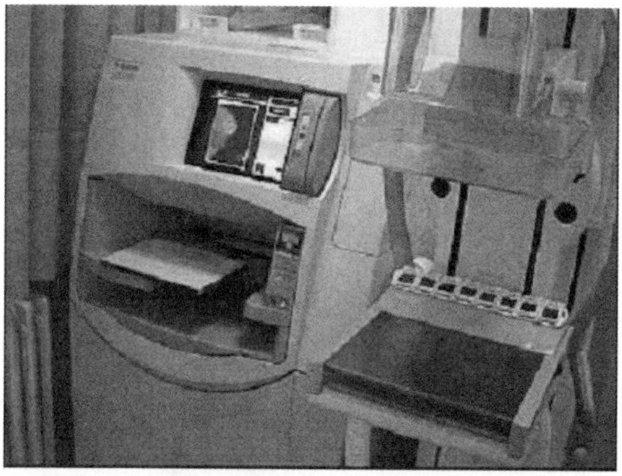

Fig. 3.14. A Fuji FCR5000R-E IP-based mammography system, showing the IP cassette in the foreground and the IP reader on the left.

This diffusion determines a lateral spread of the collected light in the IP reading process, which depends on the thickness of the phosphor plate. Optical read-out velocity and spatial resolution also depend on the power of the read-out laser beam. Improvements in this IP technology points toward new wavelengths of more powerful diode lasers, increased storage time of latent images through changes of phosphor grains chemical composition and increased phosphor grains density for increased detector uniformity and SNR [47]. Research on new photostimulable phosphor materials for digital radiography is still active [48]: here, the main problem is in finding suitable substitutes for the BaFBr:Eu-like phosphors of commercially available systems, which are affected by fading with time of the latent X-ray image. The dynamic range of a CR system is as large as 4 decades, i.e. relative exposure variations as high as 10^4:1 can be accommodated (Fig. 3.15).

Digital image processing is an important phase of image presentation in a CR unit. The most critical algorithms are related to the conversion of the pixel value to a gray level for image viewing. In fact, given the limitations in dynamic range of the display media and of the human eye, the relationship between input pixel value and output gray level is not linear but S-shaped (Fig. 3.16).

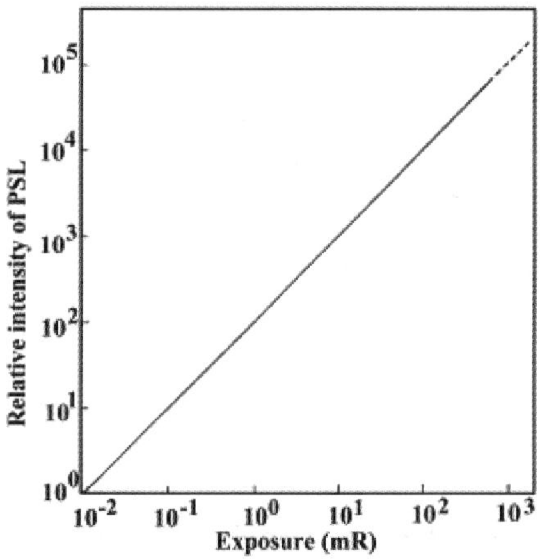

Fig. 3.15. Dynamic range of an IP (from [45]).

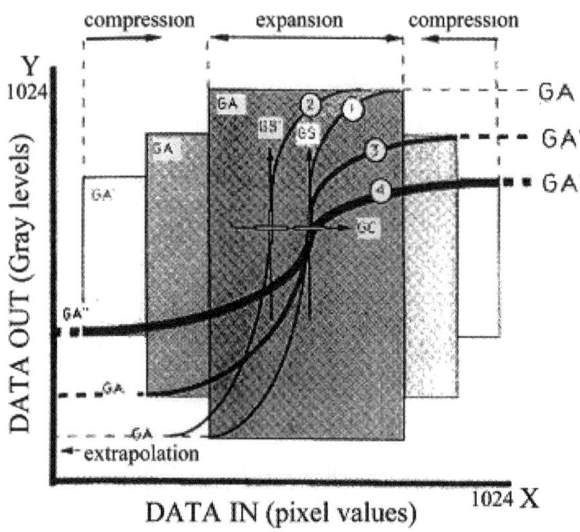

Fig. 3.16. Gradation curves of a CR system (from [47], reprinted with permission of Elsevier Science B.V.).

Image manipulation requires suitable algorithms to control the contrast (gradient) and gray level range (offset) of the display conversion function (so-called gradation curves). Other processing like dynamic range compression and edge enhancement are also available in the digital presentation and interpretation phase at the digital workstation.

The spatial resolution and image matrix size depend on the type of X-ray examination. For mammography, the read-out sampling frequency can be 5–10 pixels/mm, which implies a limiting spatial resolution of 5 lp/mm. Digital image size is then close to 1800×2400 pixels (for a 18×24 cm^2 field) or to 2400×3000 (for a 24×30 field).

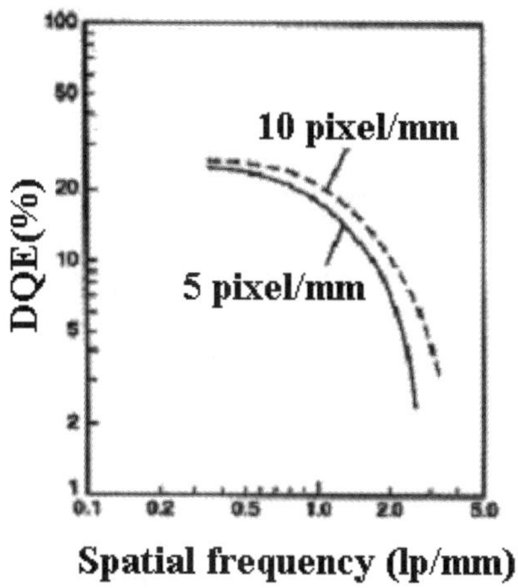

Fig. 3.17. Detective Quantum Efficiency of Imaging Plate in a FCR9000/V mammographic CR unit by Fuji Medical Systems, versus spatial frequency. Reading of the IP was performed at two sampling frequency, 5 and 10 pixel/mm. Exposure was 1 mR (from [47], reprinted with permission of Elsevier Science B.V.).

Fig. 3.17 shows the *DQE(f)* of an IP used in a mammographic CR unit. It can be seen that the maximum DQE is above 20% and that at 2 lp/mm, the DQE can be as high as 10% (10 pixel/mm sampling). While only slight differences exist between the *DQE(f)* response of different types of IP, their DQE values are only slightly higher than that of intensifying screen-film detectors (Fig. 3.18). This translates into a reportedly equal dose efficiency of Computed Radiography and of conventional film-based systems. The great advantage of the extremely high (linear) dynamic range of IP detectors—which eliminates possible under- or over-exposures of radiographic image—is counterbalanced by their lower spatial resolution, so that it is generally judged that the overall clinical image quality of photostimulable storage phosphor screens is essentially equivalent to that reachable with well-exposed, high-quality, film-based systems, for general radiography. Nevertheless, the high throughput, the avoidance of films and chemicals and development procedures, are considered commercial assets of Computed Radiography systems for large Radiology Departments. Nowadays, it is evaluated that about 7500 CR systems are in use worldwide.

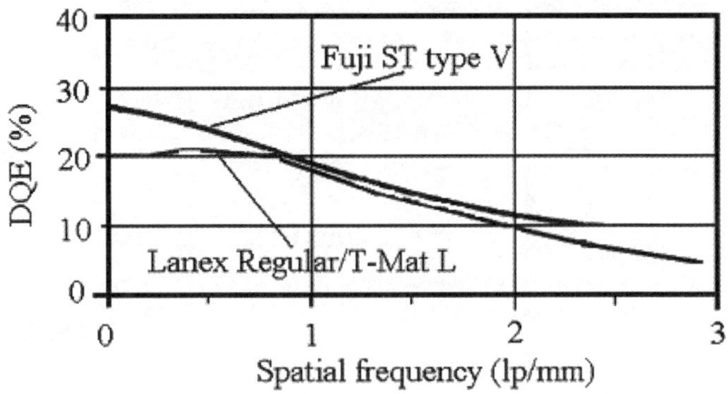

Fig. 3.18. Detective Quantum Efficiency of a Fuji storage phosphor Imaging Plate and of a combination intensifying screen/film, versus spatial frequency (from [49], reprinted with permission of Nuclear Technology Publishing).

3.4.2 *Scintillators/phosphors + semiconductor material (e.g. a-Si:H) + TFT flat panels.*

Hydrogenated amorphous silicon (a-Si:H) has been evaluated, in the past two decades, as a possible semiconductor material for large area radiation detectors in digital radiography. Due to the low efficiency for absorption of diagnostic X-ray photons—as a result of both reduced detector thickness (of the order of microns) and low interaction probability in the low Z semiconductor—amorphous silicon imaging detectors are normally coupled directly to a scintillator or phosphor layer. X-ray photon interaction occurs in the front scintillator layer which covers the detector surface; the underlying a-Si:H detector layer, which is sensitive to the light emitted by the scintillator, is organised in the form of a matrix array of photodiodes which convert this optical signal from the scintillator into an electrical signal. In this way, a digital radiographic system can be produced, in which an *indirect* detection of X-rays in a semiconductor material occurs. The technical distinction between *direct* X-ray imaging systems—where the X-rays convert in the detector—and *indirect* X-ray imaging systems—where the X-rays are first converted to optical photons which are then detected—reflects into peculiar characteristics of either type of imaging systems (Fig. 3.19). In fact, the higher detection efficiency expected for indirect detection systems with respect to direct systems, is counterbalanced by the loss of spatial resolution introduced by lateral spread of the light in the scintillator or phosphor layer due to light scattering.

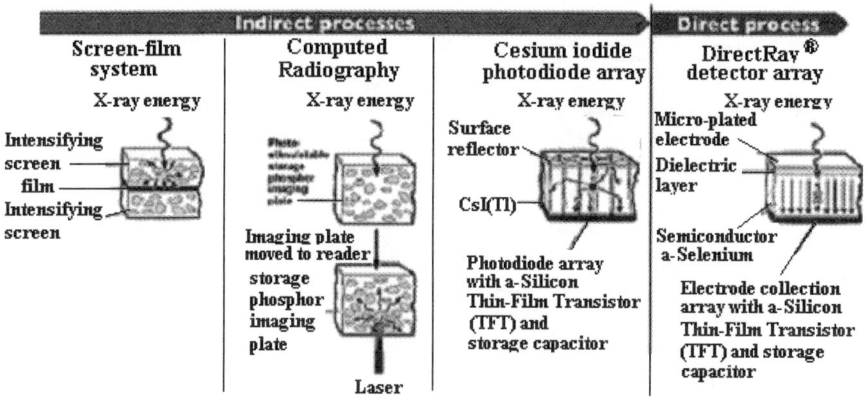

Fig. 3.19. Comparison of different detection technologies in X-ray imaging [50].

Since this spread—measured at the back side of the scintillator, which is in contact with the detection matrix—increases with increasing thickness of the scintillator layer, it is seen that good spatial resolution requires thin converter layers, but a thick layer is required for high X-ray detection efficiency. Both conflicting requirements, anyway, point toward having scintillators with high intrinsic detection efficiency, in addition to high light yield.

Typical optical converters used in combination with a-Si:H based systems, are gadolinium oxysulfides (e.g. $Gd_2O_2S{:}Tb$) phosphors or thallium-doped cesium iodide (CsI(Tl)) scintillator. For a given readout scheme (e.g. flat panel technology), both direct and indirect detection modalities can be implemented (Fig. 3.20). The complete scheme of an amorphous silicon indirect digital radiography system (using Gd_2O_2S) with flat panel readout technology is shown in Fig. 3.21. With amorphous silicon as a semiconductor detector material, large matrices (e.g. 3000×3000 pixels) of single photodiodes can be produced, to be covered with a scintillator layer over a total area as large as 43×43 cm^2, which allows full field DR.

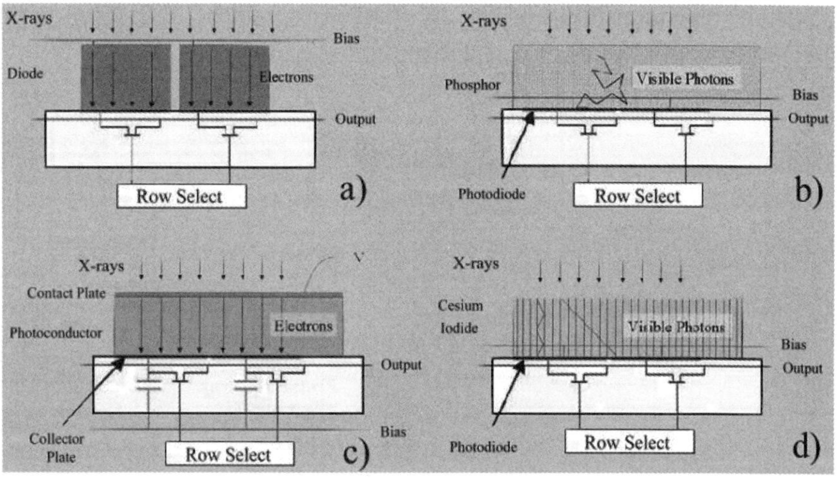

Fig. 3.20. Detector schematics of direct (a, c) and indirect (b, d) digital radiography systems (flat panel radout technology), using: a) single semiconductor diodes or c) photoconductor layer; b) phosphor or d) scintillator converter layer [51].

Each image pixel is represented by the single photodiode layout cell, with a pixel side as small as 140–160 μm. However, electronic read-out of these large matrices at such pixel pitch in a conveniently short time, with detector and readout electronics on the same substrate in a monolithic device, is a major difficulty. Advances in flat panel display technologies for electronic read-out of large area active matrix arrays in liquid crystal displays—promoted by military and consumer industry—

provided the key technology for the read-out of detection arrays in amorphous silicon based digital radiography systems.

Flat panels use arrays of thin film transistors (TFT), each coupled to a thin (≅ μm) a-Si:H sensor, obtained by vapour deposition and suitably structured using standard photolithographic methods. This produces a-Si:H pixel detectors with low efficiency for X-rays but highly sensitive to visible light. The entire matrix is then coated with the thick (≅400–500 μm) phosphor or scintillator layer—whose emission corresponds to green light—in order to increase overall detection efficiency for diagnostic X-rays. This type of detector can be used in a variety of diagnostic imaging procedures like general radiography, mammography, fluoroscopy.

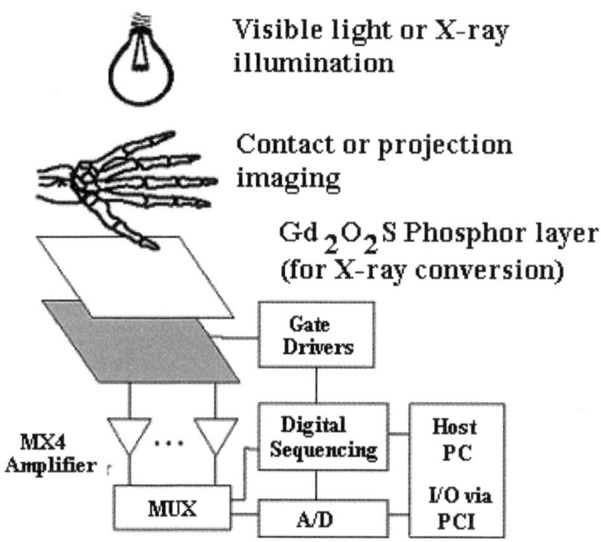

Fig. 3.21. Digital radiography with a-Si:H detector material and flat panel TFT readout technology [52].

Figure 3.22 shows the schematic diagram of the single pixel circuit in the matrix array of an amorphous silicon detector with flat panel readout technology. The sensor comprises a reverse-biased p^+-i-n^+ diode, which produces—during the acquisition time—a photocurrent whose intensity is proportional to the flux of (fluorescence) light emitted by the coupled phosphor layer. This current produces, during the X-ray exposure, the accumulation of a negative charge -δQ on a sensor capacitance, normally fully charged at Q before acquisition via connection to the bias voltage. After exposure, a charge Q-δQ is present on this capacitor, which is then discharged and the output electrical signal +δQ is read (Fig. 3.23).

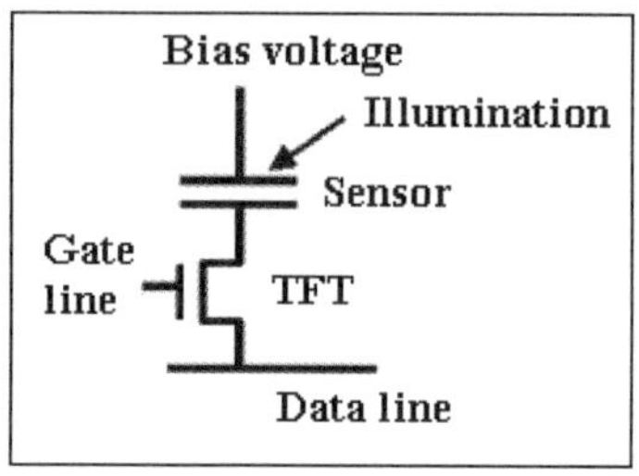

Fig. 3.22. Scheme for the electronic circuit of a single diode cell in an amorphous silicon based detector with Thin Film Transistor (TFT) electronic readout [52].

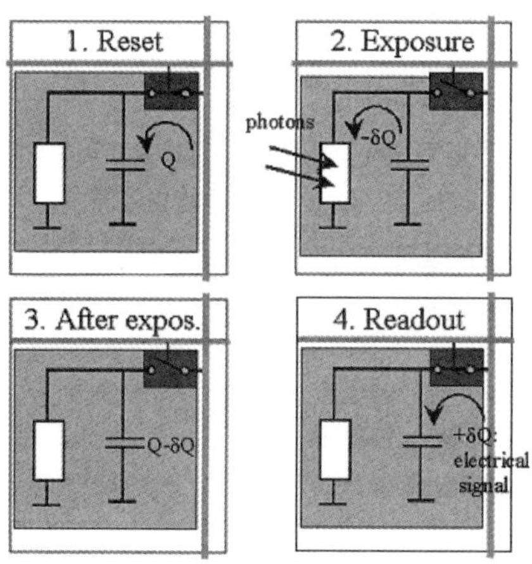

Fig. 3.23. Inizialization, acquisition and readout phases in a a:Si:H diode [53].

When the enable signal is sent to this pixel via the gate line, this electrical signal is then sent to the data line for external readout. The entire array is read line by line in sequence via multiplexer, after the acquisition period, by the readout electronic circuitry. Analog-to-digital converters provide the digitization of the electrical signal that produce the final digital image matrix. The dead area in the pixel cell, due to the presence of the transistor switch, can be reduced with suitable cell layout, producing a pixel *fill factor* close to unity (Fig. 3.24).

The complete structure of a flat panel indirect-radiography digital detector is shown in Fig. 3.25.The TFT technology for readout of pixellated a-Si based devices is still in progress, and custom integrated circuits are being designed for the peculiar task required for the sequential readout of large, high-density matrix arrays [54]. In particular, the row by row readout sequence and the intrinsic parasitic capacitance and high resistance of the transistors, in addition to the requirement of reducing input noise, may limit the time response of flat panel imagers. When coated with a-Si, they have shown a long readout period of a few seconds, which is not a problem in one-shot radiography, but prevents high frame rate acquisitions for cine-radiography.

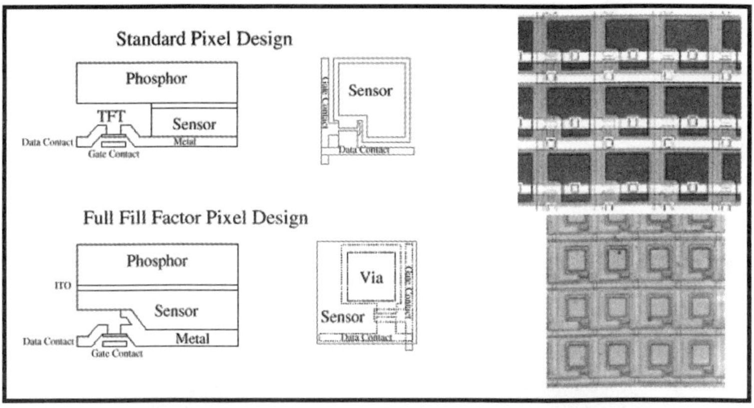

Fig. 3.24. Detailed cell layout of an indirect digital radiography system using TFT readout. In the standard pixel design (upper row) the ratio of the sensor area to the phosphor layer (fill factor) is very limited, thus reducing the total sensitive area. With improved pixel design (bottom row) the fill factor can be increased and reaches values close to 100%. The arrow indicates a single pixel in the matrix array of TFTs [52].

Figure 3.26 shows two commercial types of radiographic "digital cassettes", based on a-Si detector + CsI(Tl) coating. These cassettes (commercially available by many industrial groups) are also intended to replace the analog film cassette in existing radiographic units, which can be upgraded with only digital detector readout technology.

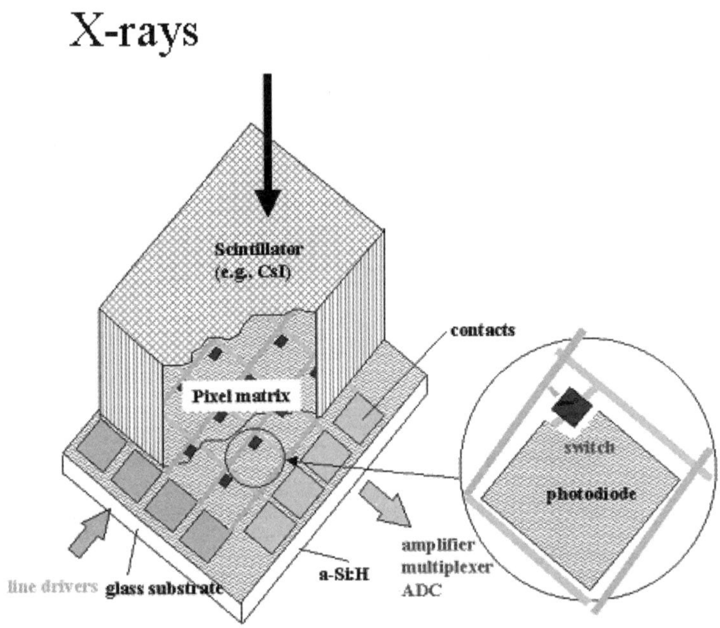

Fig. 3.25. Structure of a flat panel detector based on amorphous silicon technology [53].

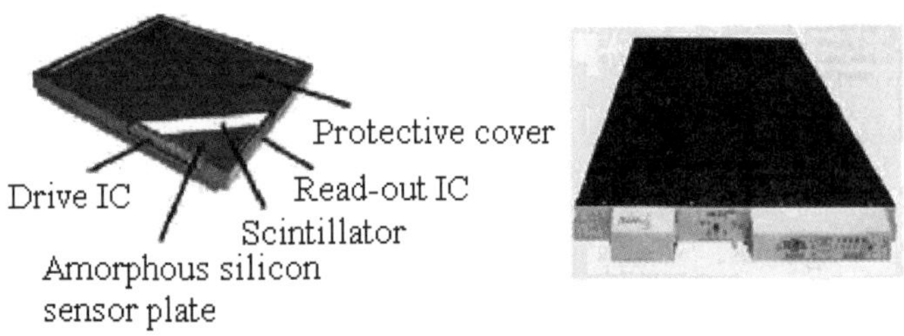

Fig. 3.26. Digital radiography "cassette" containing the amorphous silicon based detector and associated readout electronics. Left: detector assembly structure [55]; Right: a commercial system from Trixell [53].

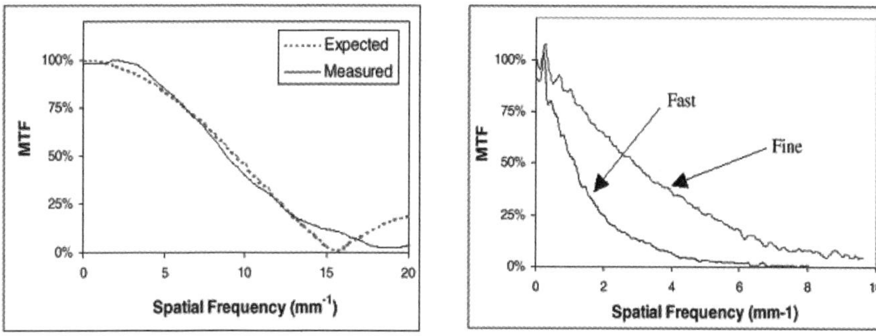

Fig. 3.27. Modulation Transfer Function of 64-μm-pitch amorphous silicon detector array. Left: a-Si detector exposed to green light. Right: a-Si detector array coated with either of two different types of scintillator layers and exposed to X-rays; the "fast" scintillator layer has a greater thickness than the "fine" layer and determines a worser resolution [52].

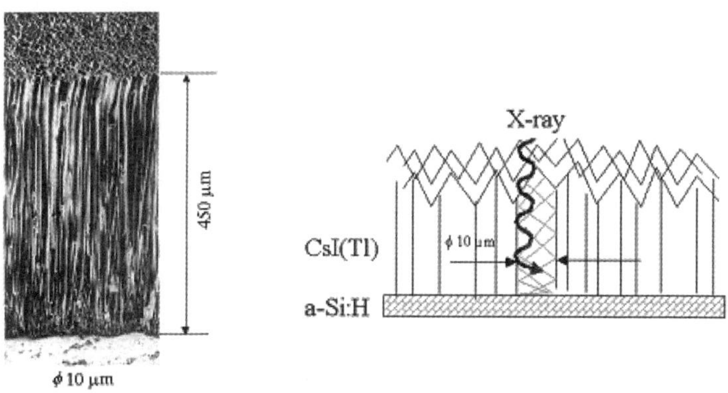

Fig. 3.28. (*left*) "Needle" structure of the scintillator layer in high resolution amorphous silicon radiography. (*right*) Lightguide mechanism in CsI(Tl) needles [53].

The limiting spatial resolution of a scintillator coated detector depends on the spread of the scintillation light in the coating, which is a function of its thickness and of its optical scattering properties (Fig. 3.27). Increasing the phosphor thickness improves detection efficiency but worsens the spatial resolution, with respect to thinner coating layers.

In order to improve the resolving power, in commercial a-Si systems coated with thallium activated CsI scintillator, a particular process for deposition of the coating layer has been adopted, by which CsI appears in the form of thin "needles" of about 10 μm diameter and half a millimeter length (Fig 3-28).
This structure allows for a sort of light guiding in the needles, that contribute to limit the lateral spread of the scintillation light. The resulting spatial resolution increases significantly with the smaller diameter of the needles, as shown in Fig. 3.29.

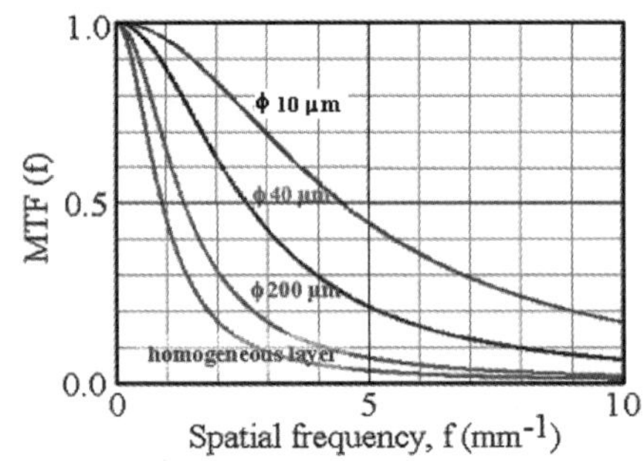

Fig. 3.29. Modulation Transfer Function of CsI(Tl) scintillator for various diameters of the "needles" [53].

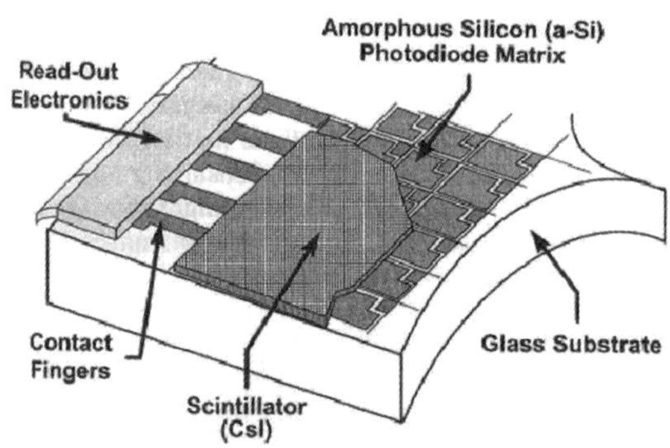

Fig. 3.30. Structure of the flat panel CsI-coated a-Si detector fabricated by General Electric for a full-field digital mammography system. The pixel size is 87 μm arranged at a pitch of 100 μm (from [56], reproduced with permission of Elsevier Science B.V.).

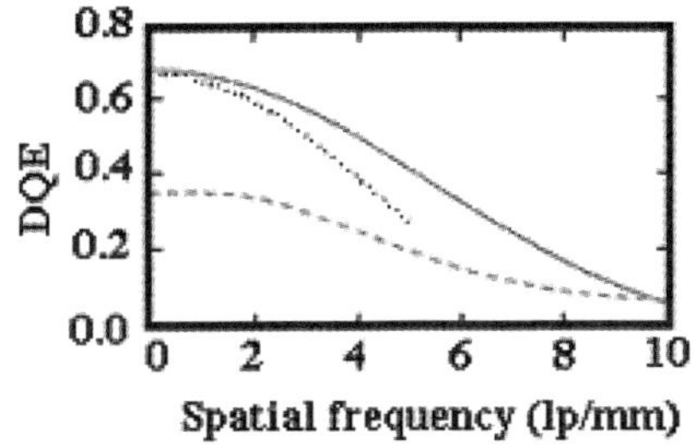

Fig. 3.31. Detective quantum efficiency plot of the General Electric full-field digital mammography system, compared with a screen-film detector (from [56], reproduced with permission of Elsevier Science B.V.).

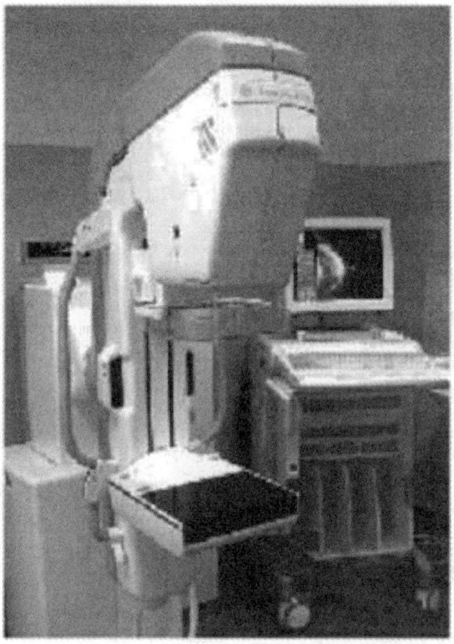

Fig. 3.32. General Electric Medical System's Senograph 2000D digital mammography unit using CsI(Tl) and a-Si:H with flat panel readout technology (100 μm pitch,1929×2304 pixels, 14 bits dynamic range).

This technology has been pushed up to provide needles 5-μm in diameter, for digital mammography, which requires a higher spatial resolution with smaller-size pixels. For example, in the Senograph full field digital mammography system by General Electric, a CsI-coated a-Si detector with 100 μm pitch (and flat panel readout) has been used (Fig. 3.30), which provides high DQE values (Fig. 3.31). This system has a 19×23 cm field of view, at 14 bits/pixel.

Few digital mammography systems based on scintillator coated a-Si flat panels have been developed due to the more stringent requirements of full field mammography (Fig. 3.32), but this type of imagers have shown satisfactory radiographic performance in general radiography (Fig. 3.33) [56, 57].

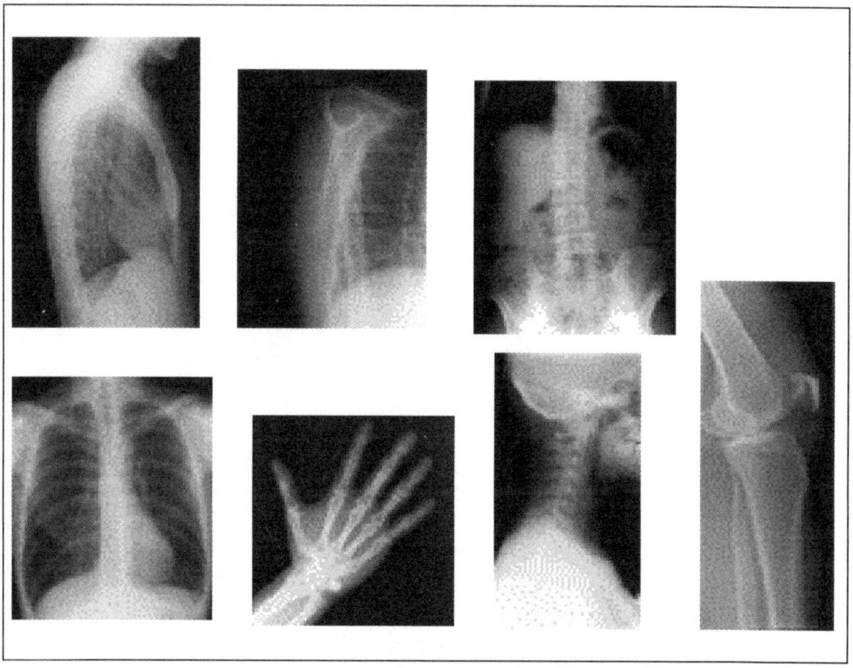

Fig. 3.33. Clinical X-ray images showing the potential of commercial a-Si indirect Digital Radiography system (Images from Evergreen Hospital Medical Center, Kirkland, WA, USA) [55]. Top row: chest side, arm, abdomen; bottom row: chest, hand, head, knee.

This flat panel technology is already in a mature state for replacing screen-film in conventional X-ray unit, however, clinical reports does not seem indicate a definitely better image quality with respect to conventional radiography. It is the author's opinion that true reduction in

X-ray dose and improved contrast expected with digital radiography will be provided only by single photon counting detectors, which are unfortunately only at an experimental stage of development.

Preliminary data regarding this point may come from measurements made by the University of Pisa's medical physics group [28]. By comparing the (pre)sampling MTF and DQE of five radiographic detectors (a radiographic screen-film system, a mammographic screen-film system, a digital image intensifier, an a-Si + CsI + TFT system, a GaAs detector with single photon counting), they found that the photon counting system has always the best response up to the maximum measured frequency of 3.0–3.5 lp/mm (Fig. 3.34). In particular, for the single photon counting detector, it is remarkable to observe a DQE higher than 50%, at spatial frequencies lower than 2.5 lp/mm. These authors also showed that the response of this last digital system (curves E in Fig. 3.34) is practically limited by the pixel aperture function given by sinc($a\nu$), where a = 170 μm is the detector's pixel size in their case. This is an indication of the noise limited performance of this single photon counting system.

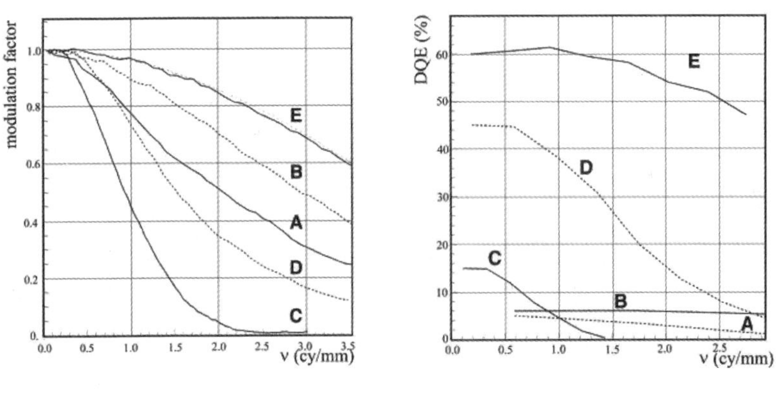

Fig. 3.34. Modulation transfer function (left) and detective quantum efficiency (right) of five different radiographic systems (E: Medipix1 photon counting system bump-bonded to a GaAs detector; A: conventional screen-film system for general radiography; B: conventional mammographic screen-film system; C: X-ray digital image intensifier; D: a scintillator and amorphous silicon flat panel imager) (from [28], reproduced with permission of Elsevier Science B.V.).

3.4.3 *Semiconductor material (e.g. a-Se) + readout matrix array of thin film transistors (TFT)*

The digital imaging systems described in Section 4.2 employ *indirect detection* of radiation, in addition to flat panel readout technology. In digital radiography system based on *direct detection* of radiation, X-ray photons interact directly in a semiconductor substrate which is coated via evaporation on a thin-film-transistor 2D array. Here, incident X-rays generate electron-hole pairs in the substrate layer under a bias voltage applied across the detector structure; charges are collected by the pixel electrodes which are then sequentially read out, one row at a time, by customized electronics within the array, e.g. the TFTs array of an active matrix (Fig. 3.35).

Amorphous selenium (a-Se) has been used for direct conversion of X-rays in digital radiography systems [58], due to its high atomic number and density and its high resistivity (Table 1), and to its capability to be uniformly evaporated over the TFT array in thick layers. Amorphous selenium has been long studied in the past for xeroradiography, and it is finding applications also in portal imaging devices with megavoltage X-rays, where it is bonded directly to a metal plate used as a radiation converter [59].

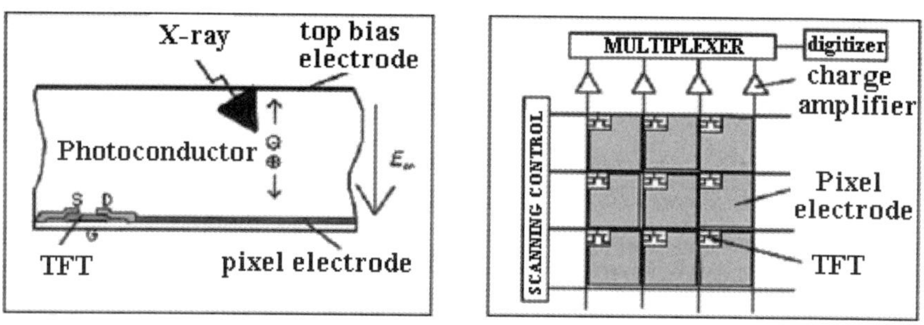

Fig. 3.35. (*left*) Cross-sectional view of one pixel in an direct digital radiography system using a-Se and active matrix readout. (*right*) Schematic view of the TFT array (from [60]; [61], reproduced with permission of IOP Publishing Ltd).

Amorphous selenium can be evaporated up to a thickness of hundreds of micrometers; approximately, 50% detection efficiency for 50 keV beams is obtained with 360 μm of a-Se. When biased at 10 V/μm, its pair creation energy W = 50 eV determines an average of 1000 electron-hole pairs per single 50 keV photon, a charge to be compared with the corresponding value of about 14000 electron-hole pairs for silicon, or 12000 for GaAs. At 20 keV X-ray energy, the reduced amount of about 400–500 charge pairs would be created: a-Se based X-ray imaging systems are less suited for mammography and for single photon counting applications.

The use of indirect-conversion systems, mainly based on amorphous silicon coated with a CsI scintillation layer, seems to be favored by an increasing number of manufacturers of digital radiography units; however, the direct radiography systems may show a number of significant advantages, related to the intrinsically higher spatial resolution determined by the absence of the scintillator layer. The study of thick (20–50 μm) hydrogenated amorphous silicon layers in direct conversion detectors began in the '80s after initial work on 2–15 μm layers [62] for direct detection of charged particles, and research is still active [63] in the field of nuclear radiation detectors. On the industrial side, amorphous selenium technology for direct-conversion digital radiography system seems to be the choice. Such a system was announced by Direct Radiography Corp. (formerly Sterling Diagnostic Imaging, now a subsidiary of Hologic Inc.) in 1999.

Fig. 3.36. The DirectRay® detector from Direct Radiography Corp. (now a subsidiary of Hologic Inc.). Its a 35×43 cm^2 unit employing a-Se and TFT readout technology. Both the (encapsulated) detector panel and the controller logic are visible [50].

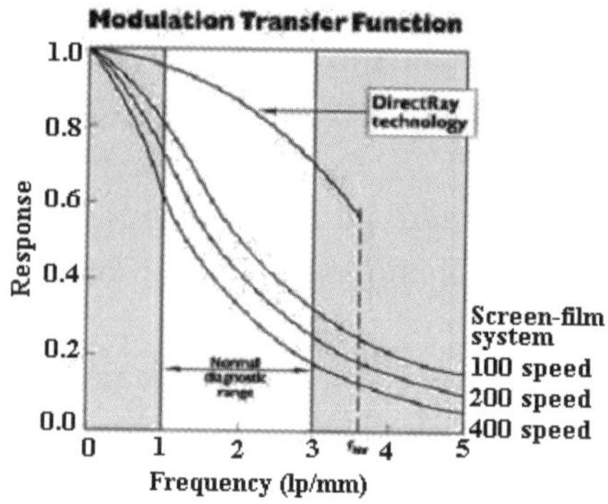

Fig. 3.37. Presampling MTF of the Direct Radiography Corp. a-Se based detector for direct digital radiography, compared with screen-film technology [50].

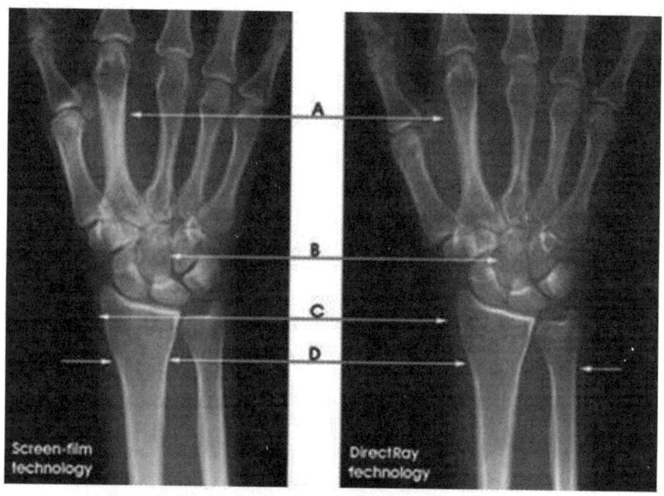

Fig. 3.38. Comparison of radiographies of the same human hand taken at equal dose with screen-film technology (*left*) and Direct Radiography Corp.'s DirectRay® technology (*right*), compared specifically in four points indicated by the arrows, may reveal the comparable or better overall image quality of this a-Se based detector [50].

Figure 3.36 shows the commercial prototype of the Direct Radiography Corp. a-Se + TFT readout cassette for direct digital radiography [50]. The first a-Se flat panel X-ray imager announced commercially has full field capability for general radiography (35×43 cm^2 active area), with a 139 μm detector element pitch (fill factor 86%) and 14-bits dynamic range. With a 3.6 lp/mm detector Nyquist frequency, it features a presampling MTF greater than 0.55 at 2.9 lp/mm (Fig. 3.37). According to the manufacturer, the quality of the radiographic images produced by this detector is comparable or better than the one obtainable with screen-film systems (Fig. 3.38). The industrial production of flat-panel imaging detectors with high uniformity of response over large areas (e.g. 43×43 cm^2), a small pixel pitch and with low background noise is a major technological challenge. As a consequence, the potential for dose reduction of X-ray digital detectors vs. conventional radiography remains partly unexploited with these systems, which require essentially the same X-ray exposure to provide high-quality radiographs as with screen-film systems, yet producing readily-available digital images well-exposed in any condition due to their higher dynamic range.

3.4.4 *Scintillation material (e.g. CsI) + CCD*

Digital radiography requires large sensitive areas, from at least 18×24 cm^2 for full field mammography up to 43×43 cm^2 for chest radiography (with the exception of intraoral dental radiography, with an approximate format of 2×3 cm^2). Large area detectors can be obtained with the a-Si:H or a-Se photoconductors and flat panel readout technology, but small area high resolution detectors based on scintillator coated Charge Coupled Devices (CCD) can also be employed advantageously, in special applications like stereotactic mammography or in full field mammography using a scanning mechanism. CCD devices coated with CsI(Tl) scintillator may offer high spatial resolution (up to 15–20 lp/mm), also due to the availability of pixel sizes as low as 27 μm. The OPDIMA sensor commercialized by Siemens for high-resolution stereotactic digital mammography comprises a 9×5 cm^2 CCD with 30 million pixels, coupled to a CsI layer, for a limiting resolution of 20 lp/mm [64].

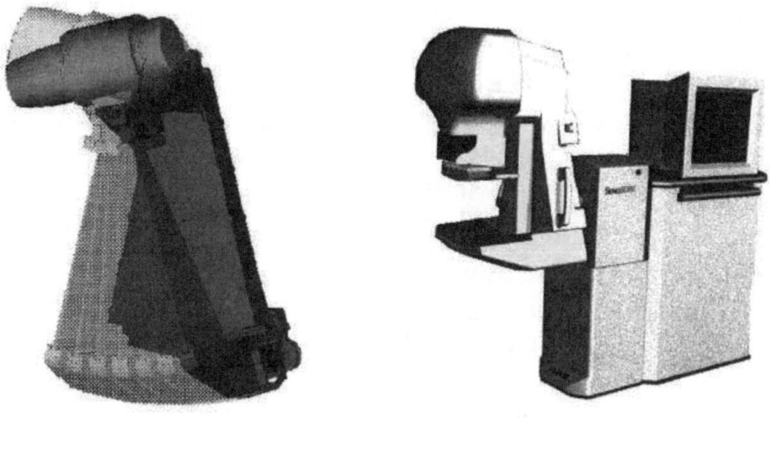

Fig. 3.39. (*left*) Scanning head with X-ray tube, collimator and motor drive, of the SenoScan® slot-scan full field digital mammography system (*right*), employing CsI-coated CCD detector (from [65], reproduced with permission of Elsevier Science B.V.).

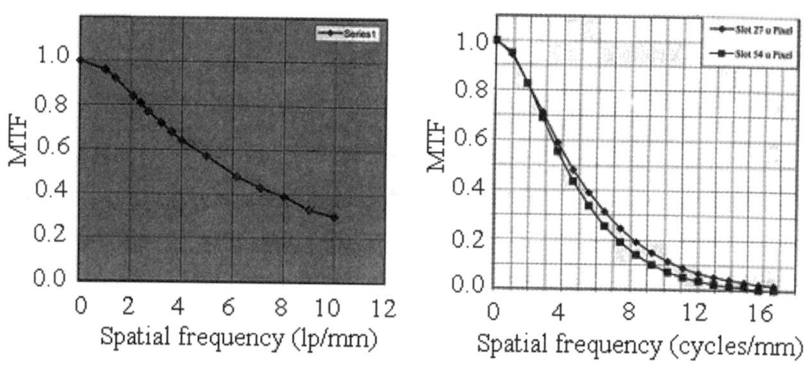

Fig. 3.40. (*left*) MTF of the CsI screen used in the SenoScan® slot-scan digital mammography system. (*right*) MTF in the (transverse) slot direction, for 27 μm pixel size or equivalent 54 μm pixel (from [65], reproduced with permission of Elsevier Science B.V.).

Each CCD module has a sensitive area covering 2048×400 pixels of size $27\times27 \mu m^2$, for a total detector area of about 1×23 cm^2. The dynamic range of the ADC is 12 bit (4096 gray levels). This detector module is scanned continuously sideways to the chest wall; a thin linear beam is used that covers the detector area, obtained by a collimator in front of a transversely mounted X-ray tube (Fig. 3.39). The effective exposure time for the 1×22 cm^2 linear detector is 0.2 s, for a total full-field scan time of 6 s. In order to have a smoothed-operation, in continuously scanning slot detector design, the CCD is read out with the Time-Delayed Integration (TDI) mode, which makes use of a buffering shift register to read the charge accumulated in each detector column just before the detector steps to the next scanning position. As for the resolution properties, the MTFs in the scan and slot (transverse) directions depend on the pixel aperture chosen (27 μm or 54 μm) and the MTF of the scintillator, in addition to the influence of the scanning mechanism. The MTF in the slot direction shows response up to 15 lp/mm (Fig. 3.40).

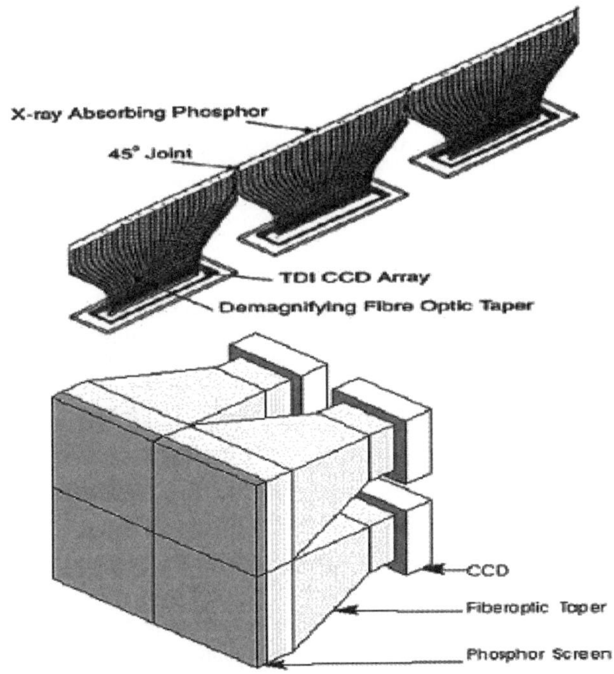

Fig. 3.41. (Top) Linear array of fiber optics connecting the converting phosphor layer to the readout CCD in a scanning X-ray imaging device (from [61], reproduced with permisison of IOP Publishing Ltd.). (Bottom) Mosaic of four tapered fiber optics in a phosphor+CCD 2-D radiography system.

In addition to these linear modules, matrix modules of tapered fiber optics can also be used to connect optically the phosphor layer to the readout CCD in a mosaic of identical X-ray imaging units [61] (Fig. 3.41). An alternative to the use of scintillator converter layers with CCD has been proposed recently [66], in which a GaAs-substrate CCD for low-energy X-ray detection in the so-called Resistive Gate CCD structure provides both X-ray detection efficiency and readout capabilities.

3.4.5 *2D microstrip array on semiconductor crystal + integrated front-end and readout*

Microstrip detectors [67-74] use a semiconductor substrate (e.g. Si, CdTe, CdZnTe, GaAs, HgI_2) on which there are segmented electrodes in the form of thin (e.g, 10 μm) long (e.g. 5–8 cm) metal contacts. These strips act as charge collecting electrodes for the underlying detector structures. These structures can be strip *p-n* junctions or Schottky barrier strip diodes, in the case of high-resistivity Si or GaAs substrates, respectively (Fig. 3.42).

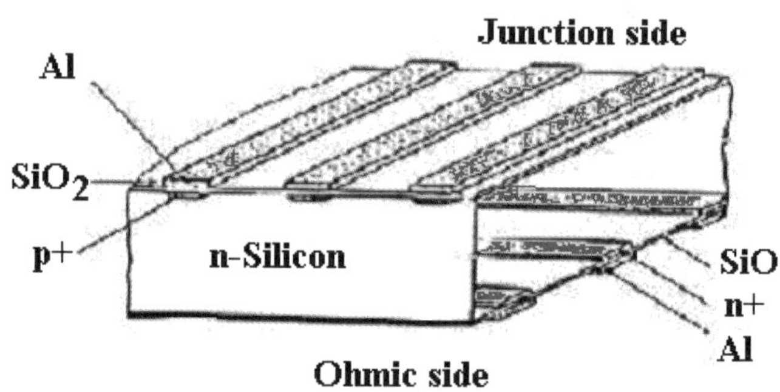

Fig. 3.42. Schematic drawing of a double-side silicon microstrip detectors, showing principle-of-operation of the device for 2D photon counting. On a (Junction) side a high-resistivity *n*-type silicon substrate p^+ strip are implanted to form a series of parallel junctions; n^+ parallel strips are implanted on the back (Ohmic) side, orthogonally to the other set of strips. In addition to the protective oxide layer in the interstrip zone, an aluminium contact is deposited on the strips for electrical connection to the readout circuit. Present technologies for fabrication of this type of detectors may introduce some changes to this layout.

For Cd(Zn)Te, ohmic contacts as well as metal–semiconductor barriers can be realized on either or both detector sides. In very schematic and simple terms, their principle of operation is the following. For the realisation of a 1D detector, only one side of the detector (Junction side) is equipped with *n* parallel microstrips, and the opposite side is covered with a uniform contact. Under external bias, radio-induced charges (electrons and holes) drift toward their respective collecting electrode, and on the Junction side, single-polarity charge is collected only in few adjacent strips. This gives the 1D position information of the coordinate of photon interaction. In 2D (double side) microstrip detectors, the back (Ohmic) contact is also equipped with *m* parallel microstrips, but orthogonal to the *n* microstrips on the junction side. The Ohmic microstrips collect single-polarity charge carriers (electrons, in the case of reverse-biased microstrip detectors) and provide a 1D position information, too. The charge drift motion lasts a few ns, usually. By analysing the temporal coincidence of signals on the Junction and Ohmic side, the coordinate on one side (*X*) can be coupled (electronically or via software) with the orthogonal coordinate (*Y*) on the other side, giving a hit count in the projected (*X*,*Y*) coordinates of the photon interaction points. This allows one to reconstruct the number of interacting photons in each "pixel" of the *n* x *m* image matrix. In the case of charge sharing between adjacent microstrips, which produces above-threshold signals in (a couple of) close strips, the derivation of a *center-of-charge* allows the spatial resolution to increase beyond the limit intrinsically given by the strip pitch.

2D microstrip detectors require only $n + m$ front-end electronics channels, each connected to a single strip, for signal handling, but for many (e.g. $n = 512$) channels, a front-end Application-Specific Integrated Circuit (ASIC) is used for readout. Typically, such ASICs have 64 or 128 channels, and a few of them are enough to cope with the readout needs of large microstrip detectors [75-79]. Normally these same front-end circuits, or their associated digital chips, only allow for the retrieval of the number of photon "hits" per channel, rather than the pulse-height information. In other terms, in order to reduce the complexity of the electronics readout, information is lost on the energy released by the interacting X-ray into the detector. This prevents the possibility to discriminate electronically, at least in part, the difference between primary and (Compton) scattered photons in the beam incident onto the detector.

It is well known that the main problem for this acquisition modality with high count rate applications, like in X-ray radiography—which may require the processing of as many as 10^4–10^5 photons/mm^2—is in the timing performance in the preamplifier and in the coincidence circuitry. If the timing resolution in the coincidence between X and Y strip signals is larger than the minimum time interval between the (X_1,Y_1) and (X_2,Y_2) signals produced by two interacting photons in the detector volume, then a hit could be registered both in the right positions (X_1,Y_1) and (X_2,Y_2), and in the two wrong locations (X_1,Y_2) and (X_2,Y_1) (a "double count"). Since the minimum arrival time of two photons on a given detector area depends on the incident X-ray photon fluence rate on the detector (e.g. 10^4 photons/mm^2/s), and on the total sensitive area, it is seen that this "double count" problem poses a limit on both the maximum sensitive area and on the fast timing requirements of the front-end and coincidence circuits. A number of different approaches have been reported for handling these electronics requirements, both via hardware in *list-mode* or *frame-mode*, and via software [71, 72, 80-88].

Superficially, in list-mode the data streams from the X- and Y-strips are separately time-coded and stored during acquisition, and recovery of coincidence events is carried out after the X-ray exposure. In frame-mode, on-line coincidence between X- and Y-signals is performed and interaction events, in the form of (X,Y) coordinate pairs, are stored and counted.

Microstrip detectors can be realized on a variety of room-temperature semiconductor substrates, but major scientific and industrial efforts have been devoted to imaging systems based on silicon detectors, using a substrate thickness going from 0.3 mm to 0.6 mm. Given the low intrinsic detection efficiency of these Si detectors in the diagnostic energy range (Fig. 3.5), this restricts their possible use only to mammographic energies around 20 keV. Test images of mammographic phantoms or small animals have been shown with prototype 2D microstrip detectors (Fig. 3.43, Fig. 3.44).

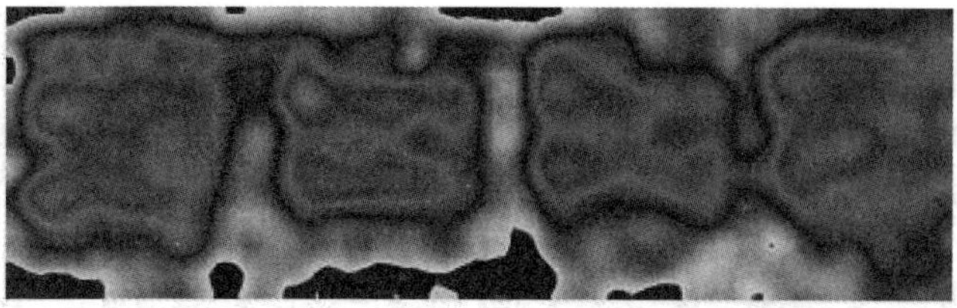

Fig. 3.43. Pseudo-color coded microradiography of part of the spine of a quayle. Four vertebrae are visible. Image obtained with 2D microstrip silicon detector with a strip pitch of 100 μm. Dimensions are about 10×3 mm^2.

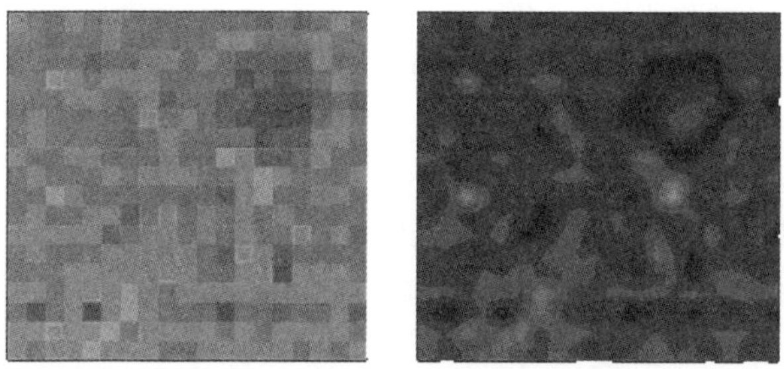

Fig. 3.44. (left) A 3×3 mm^2 raw image at 60 keV of a mammographic phantom containing a 500 μm microcalcification, obtained with a prototype double-sided Si microstrip detector with 100 μm pitch. The color intensity is proportional to the pixel counts of the single photon counting system. (Right) After digital image processing, the detail is clearly visible (top right).

Though a revival of interest has been shown in recent years on this microstrip detector technology for digital radiographic applications—which for all the above-mentioned limitations could likely be limited to small-field mammography (e.g. like in stereotactic mammography during breast biopsy)—it is evident that the co-occurrence of such heavy requirements as high detection efficiency, large sensitive area (e.g. 18×24 cm^2), high spatial resolution (e.g. 0.1 mm in mammography), high count rate ($\approx 10^6$ $cm^{-2}s^{-1}$), imposes such a formidable set of technological challenges that cannot be solved by the state-of-the-art semiconductor microstrip detector technology, for digital radiography applications.

It is worth noting that, by releasing strongly some of the above requirements, useful imaging systems have been assembled with this microstrip technology. For example, in an application like autoradiography in which beta radiomarkers with activity in the order of some 10^{-2}–10^{3} Bq are used to label biological material, small area systems have been first demonstrated [89] and then produced commercially (IDE), with a sensitive area of several cm^2, which show real-time imaging capabilities. Analogously, silicon microstrip detectors can be used in nuclear medicine applications, e.g. as detectors in compact probes for beta and gamma radioguided surgery, a relatively low coun rate application.

3.4.6 *Matrix array of pixels on crystals + VLSI integrated front-end and readout*

The microstrip detectors described in the previous Section are a form of so-called *hybrid detectors*, in which the semiconductor detector substrate can be physically separated from the front-end electronics (usually fabricated on a low-resistivity silicon substrate). Semiconductors used, at room temperature, for such hybrid detectors are Si, CdTe, GaAs, HgI_2, CdZnTe. The structure of a detector hybrid—the radiation detector coupled to its associated readout electronics—is the alternative to the structure of a *monolithic detector*, where, on the same substrate, both the detector and the integrated readout electronics are present. Principal motivation for hybrid detector structures is that the physical separation of the detector and read-out electronics permits the optimization of the detector's response on one side, and of the integrated electronics, on the other side. For example, a silicon CCD detector is a monolithic device with high spatial resolution and low noise, but totally inefficient for direct detection of X-rays; on the other hand, monolithic devices on high Z, efficient semiconductor substrates have been proposed [90, 66]. These microelectronic circuits for hybrid sensors are designed for the analog, or analog and digital, processing of the detector signals. Also in this case, these circuits can implement either a photon counting acquisition modality, or a charge integration scheme for readout. The charge signals are produced by interaction of the incident photons directly in the semiconductor detector. The readout electronics is usually implemented in the silicon CMOS technology.

The *hybrid technology* can be thought of as including both classes of *pixel detectors* and *microstrip* detectors. In a pixel detector, an $n \times m$ matrix array of single pixels is designed as metal contact pads on one detector surface; to these pads correspond either an underlying diode (*p–n* junction or metal–semiconductor Schottky junction) or a pure ohmic contact. On the opposite detector surface, a uniform back contact is realised, usually as a large ohmic contact. In the usual reverse-biased modality, all pixels on the junction side are connected to a lower electrical potential with respect to the ohmic side, and this situation can in principle reconstruct the ideal configuration of $n \times m$ separated diodes, under the same reverse bias (Fig. 3.45).

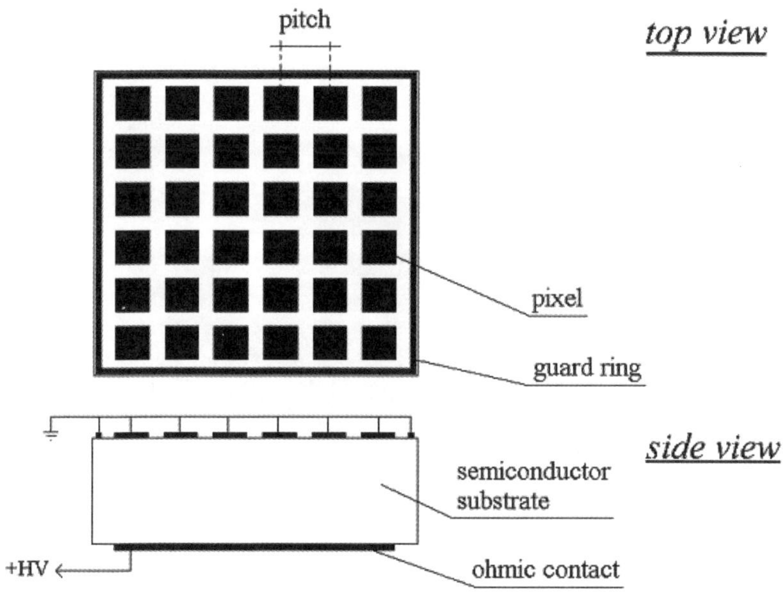

Fig. 3.45. Scheme of a reverse-biased semiconductor pixel detector. When hybridized with a read-out microelectronic circuit, photon irradiation occurs from the ohmic side and charge collection is from the pixel side.

In addition to GaAs, also Si, CdTe and CdZnTe have been used as substrate materials for hybrid pixel detectors, in the form of photoconductors with ohmic response or junction diodes. The structure of a hybrid pixel detector is shown in Fig. 3.46 [91]. Here, a GaAs pixel detector consisting of a matrix array of Shottky diodes and a uniform back ohmic contact, is connected to the underlying readout chip.

In this flip-chip technique, an ad hoc ASIC VLSI circuit must be designed for the readout of each specific pixel detector type. This ASIC readout chip has a matrix structure of single microelectronic cells, dimensionally analogous to the matrix (usually square) array of detector pixel to which it must be connected electrically (a single electronic cell to each single detector pixel), in order to transfer the charge (electrons or holes) produced by the ionizing particle in the detector pixel volume, to the input transistor of the preamplifier circuit of the corresponding ASIC cell.

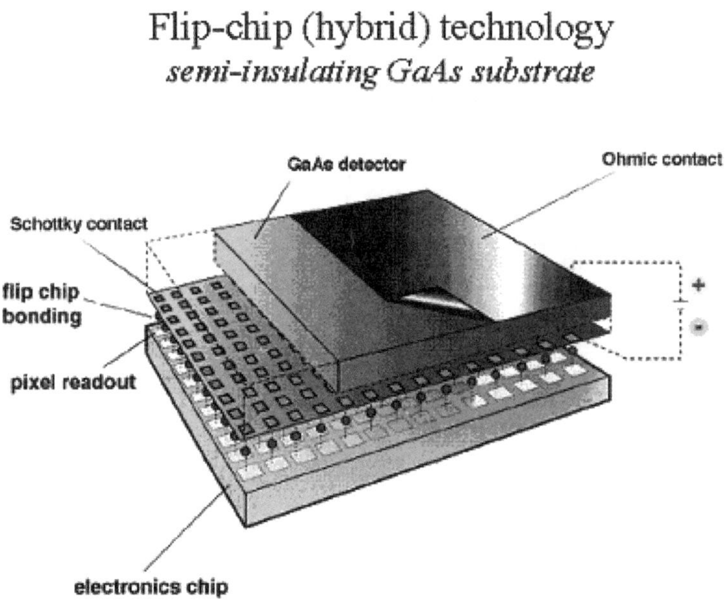

Fig. 3.46. Hybrid technology for pixel detectors (a GaAs detector is indicated here). The flip-chip connection between detector and readout microelectronic chip is obtained via the bump-bonding technique [92].

This connection can be done with the so-called bump bonding technique, which employs small (≈10–20 μm) metal drops (mainly indium, but Pb-

Sn solder or gold can be used), deposited on specific contact micropads of the electronics cell of the matrix, in between the chip and the detector (Fig. 3.47).

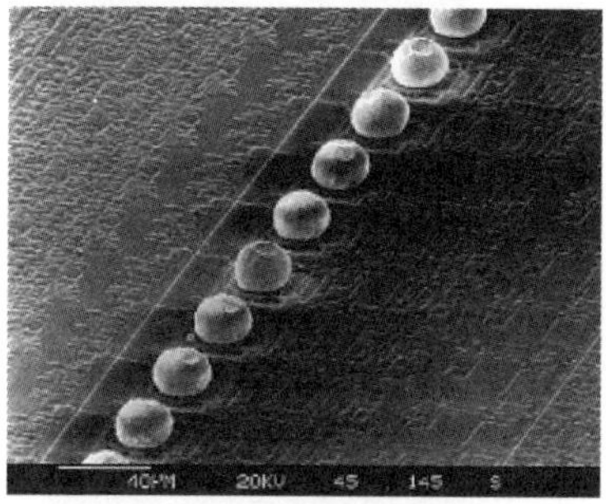

Fig. 3.47. Micrography of solder bumps (about 24 μm in diameter) deposited on pads for the bump-bonding technique.

In the case of indium bump-bonding, a bump is deposited also on the detector pixels, and the two parts of the hybrid are then aligned, pressed and cold-welded together, while hot-welding is used with solder bumps. The pitch of the readout chip cells is the same as the detector's pitch: hence, a limit to this pitch (and to the granularity of the detector) comes from the resolution constraint of the microelectronics technology (e.g. CMOS: 1.0 μm or less) used in the fabrication of the readout ASIC. Use of the bump-bonding technique for hybrid detectors has been reported for Si, GaAs and CdZnTe substrates by several research groups in USA and Europe in the last several years [29, 30, 39, 93, 94, 114], but this remains a critical industrial technology, expensive and with good reliability mainly for silicon-on-silicon substrates. In Europe, the AMS (Alenia Marconi Systems, Rome, Italy) industrial group has bump-bonding capabilities on Si and GaAs detectors; Fraunhofer IZM (Berlin, Germany) provides bump-bonding on Si, as does VTT Electronics in Finland: here, the AJAT company has large experience on bump-bonding on CdTe detectors. In USA, in the past, in collaboration with Universities, some aerospace industrial groups provided bump-bonding for CdZnTe detectors for medical imaging applications, but now only in-house bump-bonding service is provided. MCNC (Durham, NC) provides bump-bonding for silicon detectors, while Advanced Interconnection Technology (Hong Kong) offers now bump-bonding for Si, GaAs, CdTe and CdZnTe detector substrates.

Fig. 3.48. The layout of the Medipix1 photon counting chip [95], containing 170 μm square cells for the readout of 64×64 pixel detectors in a so-called hybrid structure. Readout of the chip is accomplished via the pad connections visible at the bottom side. The chip contains 64 rows of cells for pixel readout, plus one additional row for detector's leakage current measurement and compensation. Chip readout occurs in columns, in a minimum of 384 μs in the 16-bit I/O bus, at 10 MHz clock frequency.

The dimension of the bump connections is at minimum ≈ 10 μm, a ultimate limit for the size of a detector pixel. Since the area of the detector pixel is practically coincident with the area of the cell in the readout chip, independently of the size of the pixel, the *fill factor* for these hybrid detectors is essentially unity. This represents a clear advantage with respect to a-Si or a-Se flat panel imagers, whose fill factor decreases for decreasing pixel size. Various readout ASIC families for hybrid detectors for DR have been designed. Among them, there are the MPEC series, the Medipix series, the MARY series, the ALADIN series. Much of the research activity in hybrid pixel detectors comes from high energy physics, and the CERN microelectronics group has designed various readout ASICs for hybrid (and monolithic) pixel detectors [96, 97]: among the recent ones, the OMEGA series [96] has been used also for biomedical imaging applications [98] and was the forerunner of the Medipix series specially designed for medical imaging.

Medipix1 is a circuit [92] fabricated with the SACMOS 1.0 μm technology, for the readout of a 64×64 array of hybrid pixel detectors using photon counting [95] (Figs. 3.48 and 3.49). This chip provides single photon counting with up to 15 bit per pixel at a maximum rate of 2×10^6 photons/s per pixel. Its cell has a 170×170 μm^2 size and contains about 400 transistors, functionally divided in an analog and a digital part, for a complete electronic chain (charge preamplifier, shaper, adjustable-threshold comparator, pseudo-random counter) (Fig. 3.49).

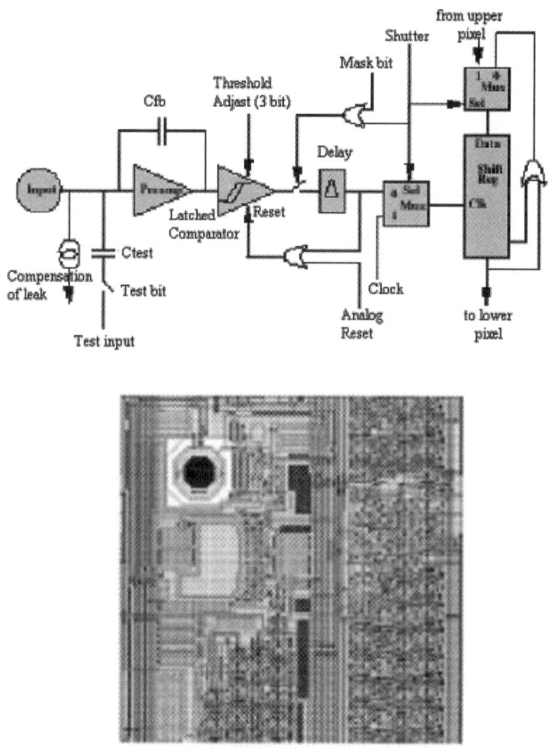

Fig. 3.49. (Top) Schematics of the electronic chain in the single cell of the Medipix1 readout chip. Single photons depositing energy above threshold in the corresponding detector pixel, are counted in the 15-bit (pseudo-random) counter. Maximum counting rate per pixel is 2 MHz. The average Equivalent Noise Charge is 150 electrons rms. Measured minimum detection threshold is 5-6 keV (Si). (Bottom) the layout of one pixel cell (170×170 μm^2). The input signal (ionization charge) is fed through the solder bumps through connection to bonding pad visible in the upper left corner of the cell [95]

In the Medipix1 chip provision is made of a leakage-current compensation circuit. In fact, the leakage current of the reverse-biased detector provides a noise current input to the preamplifier of each cell of the readout array. This circuit allows for the compensation of the very different leakage currents of, e.g. a 300 μm thick silicon or 200–600 μm thick semi-insulating GaAs detectors as used by the Medipix collaboration [10, 32, 37, 99-102].

Input-Output (I/O) operations from such ASICs are performed by chip-specific readout electronics and software. For the Medipix1 chip, the current readout system is the MUROS1 hardware and Medisoft software systems developed by the Medipix collaboration [103]. The Medipix1 "image frame" (64×64 pixels, 15 bit) can be read in as low as 384 μs at 10 MHz clock frequency. The maximum frame rate demonstrated is 135 frame per second. The Medipix photon counting chip has been bump-bonded to Si (300 μm) or GaAs (200 or 600 μm) detectors. For mammographic X-ray applications, GaAs is preferred, since the detection efficiency and the charge collection efficiency of 200 μm GaAs are close to 100%. On the contrary, 300 μm high-resistivity silicon detectors have only <20% detection efficiency for mammographic X-ray energies, but 1-mm thick Si detectors can be produced. This would permit an increase in the detection efficiency without loosing detection performance. From the measurement of the pre-sampling MTF, the spatial resolution of Medipix chip bonded to Si and GaAs detector has been measured to be 4.3 lp/mm [37]. As regards to noise, the available Medipix1 chip has always shown better performance when bonded to a Silicon detectors than to a SI GaAs detector. The minimum detection threshold over the whole array was (6 keV ± 10%) in Si, and (10 keV ± 10%) in GaAs [7, 100]. The total system noise can be as low as 140 e^- in Si [104], and 250 e^- in GaAs [100].

The next version of this readout chip (Medipix2) [105] features a 55 μm pitch, 256×256 cells and capability to readout bump-bonded CdZnTe or CdTe detectors, in addition to Si and GaAs. During the years 2003–2004, using a multi chip module setup (i.e. a large sensor array read-out via a number of smaller ASICs), small field of view X-ray imaging systems with the Medipix2 readout chip and 300 or 600 μm thick Si pixel sensors are expected to be shown by the Medipix2

European collaboration [92]; the sensitive area of such a module is 28×28 mm^2 (512×512 pixels), and modules of 28×56 mm^2 are foreseen. This next-generation single photon counting ASIC will have low noise and window energy discrimination, and after bump-bonding with Si detector is expected to show a noise threshold equivalent to about 5 keV, with an energy resolution better than 0.5 keV. Preliminary data available [106, 107] indicate that low noise operation of Medipix2 chip—assembled with a 300 mm thick silicon detector—can be obtained, at around 6 keV threshold.

Though, in principle, a pixel detector with a pitch as low as 50 μm would give access to a spatial resolution of 100 μm details or 10 lp/mm, required by such demanding applications as mammography (where microcalcifications have less than 500 μm size), a pixel pitch lower than ~100 μm poses problems both of charge sharing between adjacent pixels and of reduced counting statistics. Charge sharing is due mainly to lateral spread by diffusion of the charge created by the interacting radiation and to the shape and strength of the electric potential field in the biased detector close to the pixel collecting electrodes, and varies for different semiconductor substrates. Simulations [108] showed that for SI GaAs and hole collection (as used in Medipix1 chips), with square pixels of 50 μm size separated by 10 μm, in single photon counting, in order to suppress completely the charge sharing effect a detection threshold as high as 90% of the total deposited charge must be set, which implies lowering the detection efficiency to 65%. On the other hand, for 170 μm pitch and SI GaAs detectors, no charge loss has been observed at 60 keV irradiation [109].

The MPEC 2.1 readout chip [94] features 32×32 pixels, each 200 μm square, 18-bit counters, energy windowing and threshold correction, a maximum of 72 e- ENC noise; it has been bonded to Si and GaAs sensors. The MARY chip [29, 30] is an ASIC developed for readout of charge signals from bump-bonded semiconductor detectors like Si or CdZnTe, specifically for digital mammography. It uses a CCD with a Time Delayed Integration technique, an allows the readout of 192×384 pixels arrays with 13 bit dynamic range. The ALADIN chip is the readout ASIC for the LAD1 X-ray imaging system developed by CLRC in U.K. [110-112]. The LAD1 system is one of the largest-area hybrid pixel detector known: however, its predicted use is in time-resolved X-ray diffraction studies rather than in digital radiography. It features a ladder architecture of pixel detectors for a total 300×300 mm^2 sensitive area, pixel detectors each with 64×64 150 μm square pixels, single photon counting at 15 bits, 1 MHz maximum acquisition rate per pixel, up to 1000 fps, 3 keV system noise and 8–25 keV energy sensitivity range when bump-bonded to 300 μm thick Si detectors.

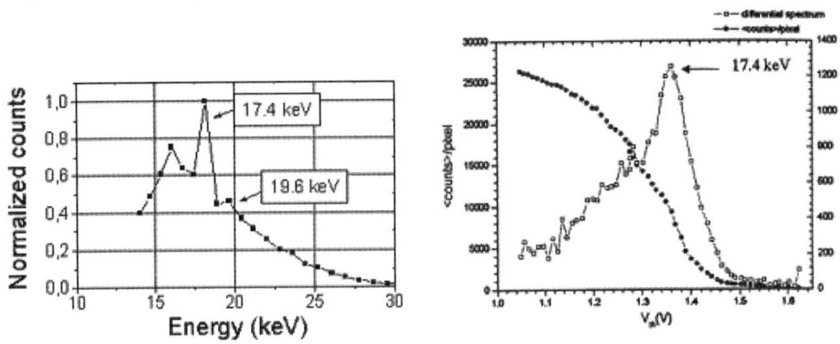

Fig. 3.50. (Left) Energy spectrum of an unfiltered Molybdenum-anode X-ray tube operated at 30 kV, obtained with the variable-threshold technique using a Medipix1 chip bonded to a 300 μm silicon detector. In this case, a sequence of 21 acquisitions (0.3 s each) where required [9]. (Right) The same type of detector was used to acquire the cumulative and the differential spectra of a mammographic beam, Mo anode, 28 kV, 1 mm Be + 0.003 mm Mo, 24 mAs (on the horizontal axis is the voltage threshold, proportional to energy) [39].

Though in principle the single photon counting acquisition modality does not provide any photon energy information, readout circuits like the Medipix or the ALADIN chips have a variable-threshold option that allows the energy spectrum of the X-ray source to be reconstructed (but energy-sensitive hybrid pixel detectors exist [38] or are under study [113]. This is done by taking successive exposures of the source at increasing detection thresholds (cumulative spectrum). Next, a numerical differentiation process of this curve yields the energy spectrum of the source with a resolution dependent on the threshold scan step and on the noise of the imaging system. This feature could be used in the monitoring of the output beam of X-ray units, for beam quality measurements [114] (Fig. 3.50).

Once bump-bonded to a pixellated detector, a photon counting chip requires a chip characterization and optimization phase. The procedures used aim at determining the best chip operating voltages in order to have minimum noise and uniform response, and at knowing the location of malfunctioning or noisy pixels. This is essentially due to defective electronic cells, or dead pixels after the bump-bonding process. A yield of 98-99% "good" pixels in a 64×64 matrix array is considered normal in the Medipix ASIC, which implies up to 40–80 sparse defective pixels in an array. These "bad" pixels are normally shut off (masked) during acquisition. This chip test process comprises chip wafer probing, electronic tests (by sending voltage pulses to each cell in the array, after assembling with the detector) and voltage versus energy calibration of the chip detection threshold (using X-ray laboratory sources) (Fig. 3.51). These semi-automatic test procedures are fundamental in any pixellated hybrid detector systems for radiography.

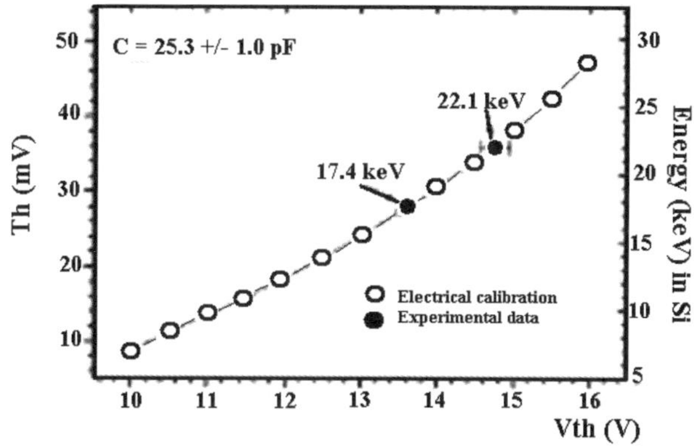

Fig. 3.51. The relationship of the variable detection threshold voltage (Vth, Volts) versus the equivalent energy in a Si detector (*E*, keV) bump-bonded to the Medipix1 readout chip. This calibration curve can be obtained either experimentally by using radioactive gamma sources (points at 17.4 keV and 22.1 keV) or by sending test pulses of amplitude Th (mV) to the test input of the charge preamplifier of the ASIC cells [39].

The Medipix readout chips, bump-bonded to pixellated detectors with Si (300 μm) or semi-insulating GaAs (200 or 600 μm) substrates, are being used for dental radiography, synchrotron radiation X-ray imaging, angiography, mammography, electron microscopy, materials analysis, defectoscopy, nuclear radiation monitoring, neutron radiography. During the year 2000, a 3–year industrial project started in Italy, scientifically supported by INFN, Italy, for producing a prototype detector head for digital mammography using a mosaic of Medipix1 chips bonded to Si or GaAs detectors. Digital mammography systems would also allow one to use Computer Aided Diagnosis (CAD) algorithms for tumour mass and microcalcification identification, directly embedded in the software system for readout of the detector matrix array. In fact, these CAD procedures now require the use of a scanner for digitizing the film mammography [115].

For single photon counting, direct digital mammography, a large series of tests with Si or GaAs detectors bump-bonded to Medipix1 readout chip showed the potential for high contrast detail visibility, reduction in X-ray dose in excess of a factor of 3:1 with respect to conventional screen-film mammography (GaAs detectors) [10, 28, 32, 39, 93, 99, 114] (Fig. 3.52).

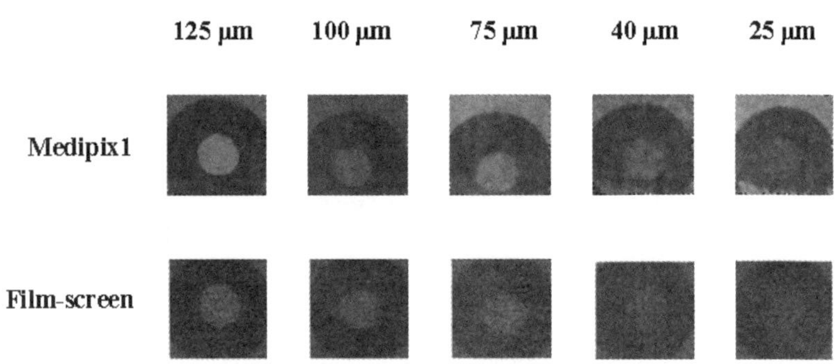

Fig. 3.52. Comparison of the X-ray images of a lucite mammographic phantom acquired with a screen-film system and a 200 μm thick SI GaAs system with Medipix1 read-out. The film radiograph (X-ray tube voltage = 28 kVp, Mo anode, additional 0.025 mm Mo filtration, 32 mAs, 1 s exposure) has been digitized with a 12-bit scanner at an equivalent pixel resolution of 170 μm, equal to the pixel size of the Medipix1 cell. The phantom is 40 mm thick and contains, within 12 mm diameter, 3 mm thick holes filled with wax, a series of cylindrical, 4 mm diameter aluminium details, of various thickness down to 25 μm. Even at the lowest radiographic contrast (25μm thick Al detail), the Medipix1 assembly produces better quality images [93].

The Medipix2 chip bump-bonded to a 300 μm thick Si detector, has shown excellent linearity with X-ray tube output between 0.3 and 80 mAs, and an MTF of 58% at 8.5 lp/mm [116]. Recently, Medipix2 has been bump-bonded also to 1 mm thick CdTe detectors and test images have been presented [117]. Excellent X-ray test images with CdTe and CdZnTe detectors bump-bonded to readout ASICs have been reported [118], this hybrid technology appearing very promising for future DR system.

The great problem for current hybrid pixel detectors is in the realisation of large detector arrays via a mosaic of single chips. In fact, one should avoid the presence of dead spaces between adjacent detectors, a condition not currently possible due to the presence, at least on one side of the chip, of its readout logic circuitry, which could require chip areas a few mm wide. Hence, in a mosaic detector of many single hybrid pixel detectors, the dead areas between chips must be accounted for, e.g. by multiple exposures with the shifted-detector technique. Or, with hybrid detectors featuring no dead space at least on one chip side, a linear array could be fabricated, which could be employed in small-field slot scan imagers. For example, the Medipix2 ASIC quoted above has three-sides buttability, having a dead space of less than one pixel on three sides, the other side showing the presence of the external I/O lines and supply lines connections.

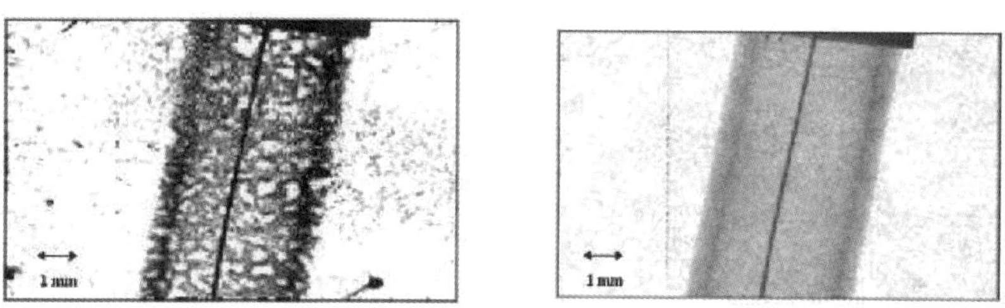

Fig. 3.53. X-ray images of an electrical fuse taken at 60 keV with the ReadOut Integrated Circuit (ROIC) chip developed by IMC (Sweden) and a 200 μm GaAs detector, 3838 μm^2 pixels: (*left*) raw image, (*right*) images corrected for noisy pixels and background non-uniformity [119].

In Fig. 3.53 a 320×240 pixels, 38-μm pitch ReadOut Integrated Circuit (ROIC) chip developed by IMC (Sweden), indium bump-bonded to the semi-insulating GaAs detector, was employed to image an electrical fuse. Some defective response in the ROIC and in the GaAs detector produced a background non-uniformity which was corrected via software. This image correction allowed the SNR to increase from a value of about 4 to about 13 [119]. The MTF of this setup, when compared to a conventional screen-film combination or with a CCD camera coupled to scintillator, showed a significantly better performance up to the maximum characteristic spatial frequencies (Fig. 3.54). It is to be noted that with this hybrid pixel detector, which operates in the charge (energy) integration mode, large spatial and temporal variations in the X-ray sensitivity of the SI GaAs detector were systematically observed [120], that decreased the SNR. These variations in sensitivity were not observed for SI GaAs detectors bonded to the Medipix photon counting chip [99], which proved almost quantum-noise (Poisson) limited imaging [7, 8].

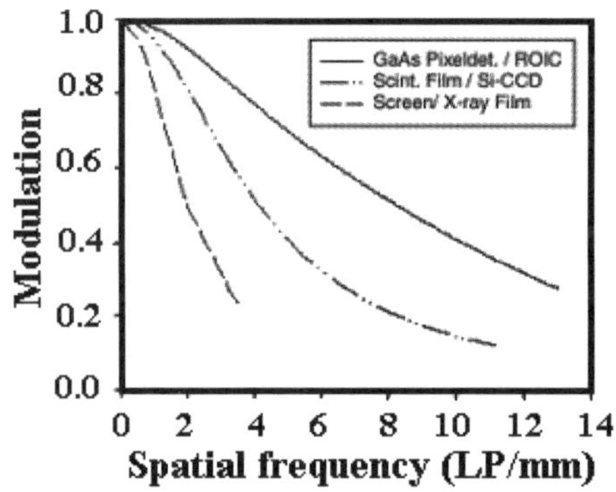

Fig. 3.54. Modulation Transfer Function of a hybrid pixel X-ray detector (200 μm GaAs + ReadOut Integrated Circuit). The response of this single photon counting system is always better than that of conventional screen-film system or of a silicon CCD camera coated with scintillator layer [119].

3.4.7 *X-ray-to-light converter plates (AlGaAs)*

J. Pozela and collaborators at the Semiconductor Physics Institute in Vilnius, Lithuania, showed that a high-energy particle detector can be fabricated by measuring the light intensity emitted in the recombination of electron-hole pairs generated in a semiconductor traversed by those particles [121]. The semiconductor was a graded-gap $Al_xGa_{1-x}As/GaAs$ heterostructure, where the Aluminium fraction, x, varied linearly between 0 and 0.3 in the material thickness. Subsequently, they showed that this detection principle could be employed for acquiring projection X-ray images using a mammographic X-ray tube, hence establishing a new detector technology in X-ray imaging. This new type of detector works without collecting any electric signal and acts—in a manner analogous to fluorescent screens—as a frequency-shifter plate, where high-frequency X-ray photons are converted to lower-energy visible photons and give an optical image that can then be viewed, e.g. with a common TV camera. The physical principle is as follows.

In a semiconductor particle detector, electrons and holes generated in pairs by the interaction of a high-energy particle in the semiconductor material are usually collected electrically and give the output response. During the collection phase, in their process of losing kinetic energy, electrons and holes can also recombine, emitting light (in a time period of the order of 0.1-1 ns) whose intensity can be measured. In the case of a $100\times100\times100\ \mu m^3$ GaAs detector, with about 40 keV total energy loss in the material by an interacting particle, they calculate that the light emitted power after 1 ns can be as high as 2 μW, at a wavelength strictly close to 0.83 μm, corresponding to the GaAs band gap energy. However, since at this wavelength light is also strongly absorbed in GaAs, a graded-gap $Al_xGa_{1-x}As$ material was used in order to collect the emitted light outside of the detector surface.

In Fig. 3.55 the band energy structure is plotted versus the percent concentration x of Al, which increases linearly from the top to the bottom surface of the detector.

The band gap energy E_g increases with increasing concentration x, from its value for pure GaAs up to a few tens of meV, sufficient for decreasing the optical absorption coefficient of the compound by two orders of magnitude (from 10^4 to $10^2\ cm^{-1}$). This means that emitted light can escape with negligible absorption from a ≈100 μm thick $Al_xGa_{1-x}As$ layer, exiting from the side of higher Al concentration. As a consequence of the different energy gap in the graded-gap detector, particles interacting at different depths in the detector produce recombination light with a slightly different wavelength than that corresponding to GaAs bandgap.

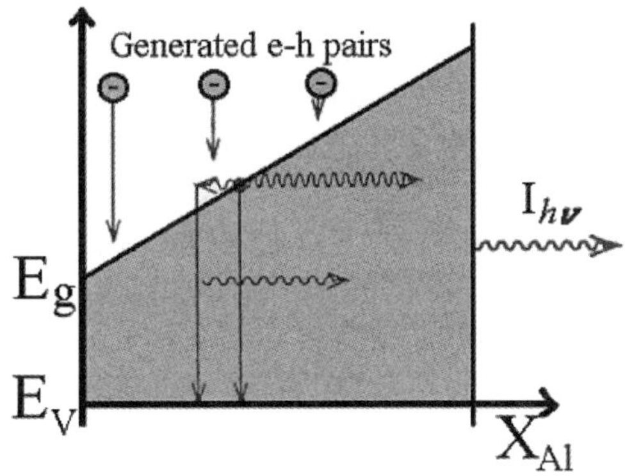

Fig. 3.55. Schematic view of the band structure in a graded-gap $Al_xGa_{1-x}As$ detector (courtesy of Dr. K. Pozela).

Hence, emitted light has a wide spectrum. At this stage of the conversion process, one may add a surface GaAs thin layer that has the function of absorbing this wide-spectrum luminescence and re-emitting a monochromatic light by electron-hole recombination in this GaAs layer, at an energy corresponding to the bandgap, E_g.The enormous advantage of this technology is that it does not require any detector bias or front-end electronic circuits (like charge-sensitive or current-sensitive amplifiers) for processing the signal, apart from the optical readout with a moderately sensitive CCD camera. Fiber-optic coupling between the rear side of the semiconductor plate and the read-out system is also possible, as well as relay-lens optical coupling, both allowing one to introduce magnification in the viewing of the interacting particle beam. In addition, side particle irradiation is feasible, while keeping the surface of optical read-out of the detector, normal to the direction of the incident beam. This group has also reported results on using this new detector technology in X-ray imaging. Irradiation of low-contrast test objects with an X-ray tube at mammographic energies produced clear radiographic projection images when viewed "through" the AlGaAs graded-gap detector [122]. Samples up to 200 μm thick have been produced, over an area of several cm^2. By 20-40 kV X-ray imaging of thin metal wires, Pozela and collaborators reported a limiting spatial resolution of about 30 μm (Fig. 3.56). Improvements in the light yield and photoconversion efficiency of this graded-gap $Al_xGa_{1-x}As$ X-ray detectors has also been shown recently [123].

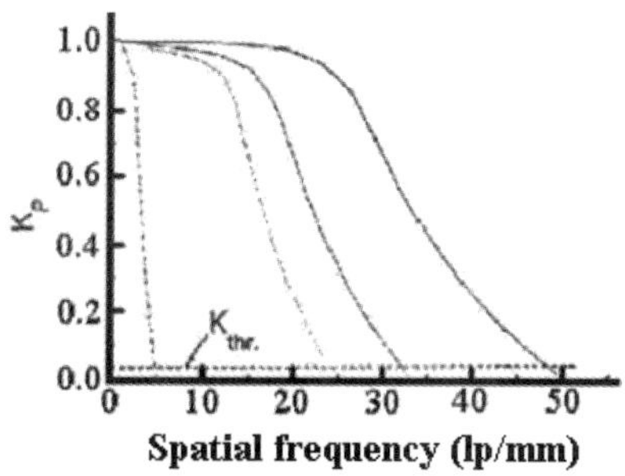

Fig. 3.56. Contrast resolution of the graded-gap $Al_xGa_{1-x}As$ detector under 20-40 kVp X-ray tube irradiation (courtesy of Dr. K. Pozela).

3.5 Conclusions

In the last 10–15 years, scientific and technological research in the design and fabrication of detectors for digital radiography has made a largely significant progress, with the final realization, now gone commercial, of flat panel imagers. These detectors, which are mostly based on amorphous silicon semiconductor coupled to a scintillator layer, can be fabricated with large sensitive areas with a small pixel size. This result is now in the hands of major companies and vendors of radiological equipment, who invested huge funds in the last decades for this and related R&D. Hence, there is little doubt that in the next decade this flat panel imaging will be the technology of choice for general radiography, with further development steps required for realizing similar systems with improved performance for digital mammography. These X-ray imaging systems produce images of good quality, well exposed in all conditions due to their large dynamic range, but they have still to prove clinically, in large trials, their much advocated benefits in terms of improved *diagnostic* quality with respect to film radiography. This phase will likely slow the process of technology acquisitions in hospitals, in addition to the other obstacle represented by their great cost and by the practical problems in maintaining large hospital PACS.

On the other hand, direct DR systems based on hybrid pixel detectors have the potential to surpass the performance of flat panel imagers, in terms of spatial resolution, DQE, frame rate, minimum detectable contrast. University research on this technology is active both in Europe and in USA, but financial sources are still inadequate, and as mentioned, large industrial investments in this field are not likely to come, at least until it will demonstrate large area imaging capabilities with efficient X-ray detectors. This, in fact, is the principal scientific goal in this field, which requires investigation on uniform, thick detectors (CdZnTe, CdTe, GaAs), fabrication of pixel readout ASICs with reduced or no dead space and/or multichip modules for large (e.g. 2k×2k, 50 μm pixels) sensitive areas. On the contrary, for less demanding (as regards large sensitive area) applications other than DR, the small area hybrid detector technology is essentially already available, and could gain scientific, industrial and commercial interest. Areas of interest could be, for example, X-ray imaging in materials science with X-ray tubes or synchrotron radiation, or CT thomosynthesys (all high-count-rate applications, for which high chip readout rates are still to be reached). Additional areas of enormous scientific interest for these detectors would be nuclear medicine (e.g. 140 keV scintigraphy) and astrophysics with space-borne X-ray and gamma-ray telescopes. For the former application, high detection efficiency must be showed; for the latter case, low-threshold operation (about 1 keV equivalent) and Fano-limited energy resolution are also required. These combined industrial and scientific interests could provide additional consensus and relevant financial sources for large research efforts by cooperating Universities and Research Organizations, needed to realize a large-area hybrid detector for digital radiography and mammography, in the next decade.

Acknowledgments

The author is grateful to Prof. Ennio Bertolucci for helpful discussions and suggestions.

References

1. J. C. Dainty, and R. Shaw. Image Science. Academic Press, New York (1974)
2. L.G. Mänsson. Methods for the evaluation of image quality: a review. *Rad. Prot. Dosim.* **90**, 89–99 (2000)
3. H. Fujita, et al. A simple method for determining the modulation transfer function in digital radiography. *IEEE Trans. Med. Imag.* **11**, 34–39 (1992)
4. R. Wagner, et al. Signal detection theory and medical image assessment. In: Recent developments in digital imaging, K. Doi, L. Lanzl, P. P. Lin (eds.), American Association of Physicists in Medicine, Medical Physics Monograph no. 12, American Institute of Physics, New York, pp. 39–59 (1985)
5. M. Bath, et al. Determination of the two-dimensional detective quantum efficiency of a computed radiography system. *Med. Phys.* **30**, 3172–3182 (2003)
6. J. Watt. Photon counting vs. integration. Presented at the 2nd International Workshop on Radiation Imaging Detectors, Freiburg, Germany, July 2-6 (2000)
7. B. Mikulec. Single photon detection with semiconductor pixel arrays for medical imaging applications. PhD Thesis, University of Vienna, Austria, June 2000, CERN-THESIS-2000-021 (available for download from http://documents.cern.ch/archive/electronic/cern/preprints/thesis/thesis-2000-021.pdf)
8. M. Chmeissani, et al. Limits in the detection of low contrast objects in single photon counting X-ray radiography. (2001)
9. A. Sbrizzi. The digital mammography system (MAMMOview) and digi I sistemi per mammografia digitale (MAMtal autoradiography system (BETAview) based on the Medipix VLSI chip series. M.S. Thesis (Physics), University of Naples "Federico II", Naples, Italy (2001, in Italian)
10. S. R.Amendolia, et al. MEDIPIX: a VLSI chip and GaAs pixel detector for digital radiography. *Nucl. Instrum. Meth.* A **422**, 201–205 (1999)
11. M. Conti, et al. Use of EGS4 Monte Carlo code to evaluate the response of HgI2 and CdTe semiconductor detectors in the diagnostic energy range. *Nucl. Instrum. Meth.* A **322**, 591–595 (1992)
12. E. Bertolucci, et al. Use of silicon and GaAs pixel detectors for digital autoradiography. *IEEE Trans. Nucl. Sci.* **NS-44**, 929–933 (1997)
13. E. Bertolucci, et al. Imaging performance of single-element CdZnTe detectors for digital radiography. *IEEE Trans. Nucl. Sci.* **NS-45**, 406–412 (1998)
14. E. Bertolucci, et al. Development of semi-insulating GaAs detectors for digital radiography. *Nucl. Phys.* B (Proc. Suppl.) **61B**, 633–637 (1998)
15. E. Bertolucci, et al. GaAs pixel radiation detectors as an autoradiography tool for genetic studies. *Nucl. Instrum. Meth.* A **422**, 242–246 (1999)

16. E. Bertolucci, et al. Spectroscopic performance of semi-insulating GaAs detectors for digital radiography. *Nucl. Instrum. Meth.* A **422**, 247–251 (1999)
17. E. Bertolucci, et al. Investigation on semi-insulating GaAs detectors using laser-induced current pulses. *Nucl. Instrum. Meth.* A **458**, 158–163 (2001)
18. S. R. Amendolia, et al. A project for digital mammography based on a GaAs pixel detector and on a self-triggering single photon counting acquisition system. *Physica Medica* **XIII**, 157–165 (1997)
19. S. R. Amendolia, et al. Charge collection properties of GaAs detectors for digital radiography. *Physica Medica* **XIV** (suppl. 2), 17–19 (1998)
20. S. R. Amendolia, et al. Performance of a medical imaging system for photons in the 60-140 keV energy range. *Nucl. Instrum. Meth.* A **461**, 422–424 (2001)
21. W. Bencivelli, et al. Evaluation of elemental and compound semiconductors for X-ray digital radiography. *Nucl. Instrum. Meth.* A **310,** 210–214 (1991)
22. W. Bencivelli, et al. Use of EGS4 for the evaluation of the performance of a silicon detector for X-ray digital radiography. *Nucl. Instrum. Meth.* A **305**, 574–580 (1991)
23. G. Batignani, et al. A detailed Monte Carlo study of the performance of a silicon crystal for X-ray detection in the diagnostic energy range. *Physica Medica* **VI**, 39–44 (1990)
24. M. Fiederle, et al. CdTe and (Cd, Zn)Te as detector materials. Presented at the 2nd International Workshop on Radiation Imaging Detectors, Freiburg, Germany, July 2-6 (2000)
25. T. E. Schlesinger. Semiconductors for room temperature nuclear detector applications, T. E. Schlesinger and R. B.James (eds.), vol. 43, Academic Press (1995)
26. R. B. James. Room Temperature Semiconductor Detectors. Proceedings of the 11th International Workshop on Room Temperature Semiconductor for X- and Gamma-Ray detectors and Associated Electronics, Vienna, Austria, October 11–15, 1999 (R. B. James and P. Siffert, eds.). *Nucl. Instr. Meth.* A **458** (2001)
27. NIMA. Proceedings of the 11th Int. Workshop on Room Temperature Semiconductor X- and Gamma-ray Detectors and Associated Electronics, Vienna, Austria, October 11–15, 1999, *Nucl. Instrum. Meth.* A **458** (2001)
28. S. R. Amendolia, et al. Evaluation of imaging properties of digital radiographic systems. *Nucl. Instrum. Meth.* A **461**, 389–392 (2001)
29. S. Yin, et al. Hybrid direct conversion detectors for digital mammography. *IEEE Trans. Nucl. Sci.* **NS-46**, 2093–2097 (1999)
30. S. Yin, et al. Direct conversion Si and CdZnTe detectors for digital mammography. *Nucl. Instrum. Meth.* A **448**, 591–597 (2000)
31. G. C. Giakos, et al. Study of detection efficiency of $Cd_{1-x}Zn_xTe$ detectors for digital radiography. *IEEE Trans. Instrum. Meas.* **47**, 244–251 (1998)
32. S. R. Amendolia, et al. Imaging performance of a GaAs pixel detector. *Nuovo Cimento* **112A**, 167–177 (1999)
33. D. S. McGregor, and J. E. Kammeraad. Gallium arsenide radiation detectors and spectrometers. In: Semiconductors for room temperature nuclear detector

applications, T. E. Schlesinger and R. B. James (eds.), vol. 43, pp. 383–442, Academic Press (1995)
34. P. Russo, et al. Response of SI-GaAs detectors to near-infrared light pulses. *Nucl. Instrum. Meth.* A **466**, 105–114 (2001)
35. P. Russo, and G. Mettivier. Characterization of 600-μm-thick SI-GaAs detectors for medical imaging. *Nucl. Instrum. Meth.* A **466**, 91–96 (2001)
36. E. Bertolucci, et al. Detection performance of SI GaAs detectors for nuclear medicine. *Nucl. Instrum. Meth.* A **460**, 123–126 (2001)
37. C. Schwarz, et al. Measurements with Si and GaAs pixel detectors bonded to photon counting readout chips. *Nucl. Instrum. Meth.* A **466**, 87–94 (2001)
38. M. Lindner, et al. Comparison of hybrid pixel detectors with Si and GaAs sensors. *Nucl. Instrum. Meth.* A **466**, 63–73 (2001)
39. S. R. Amendolia, et al. Imaging and spectroscopic performances for a Si based detection system. *Conference Record* of the IEEE Nuclear Science Symposium and Medical Imaging Conference, Lyon (France), October 15–20, pp. 4–123/126 (2000)
40. J. V. Vaitkus. Instabilities in LEC GaAs Schottky barrier pixel detector imaging arrays. *Nucl. Instrum. Meth.* A **460**, 204–206 (2001)
41. C. Ponchut, et al. Evaluation of a photon-counting hybrid pixel detector with a synchrotron X-ray source. *Nucl. Instrum. Meth.* A **484**, 396–406 (2002)
42. J. E. Bateman. The application of multiwire proportional counter technology to medical diagnosis and research. In: Nuclear Science Applications. Harwood Academic Publishers, U.K., vol. 3, pp. 59–75 (1988)
43. D. G. Gobbi, et al. Photon-counting radiography with the gas microstrip detector. *Phys. Med. Biol.* **44**, 1317–1335 (1999)
44. F. Sauli. GEM: a new concept for electron amplification in gas detectors. *Nucl. Instrum. Meth.* A **386**, 531–534 (1997)
45. H. Kato, J. Miyahara, and M. Takano. Computed radiography with scanning-laser stimulated luminescence. In: Recent developments in digital imaging, K. Doi, L. Lanzl, P. P. Lin (eds.), American Association of Physicists in Medicine, Medical Physics Monograph no. 12, American Institute of Physics, New York, pp. 237–55 (1985)
46. J. A. Rowlands. The Physics of Computed Radiography. *Phys. Med. Biol.* **47**, R123–R166 (2002)
47. B. Jouan. Digital mammography performed with computed radiography technology. *Eur. J. Radiol.* **31**, 18–24 (1999)
48. H. Nanto, et al. New photostimulable phosphor materials for digital radiography. *IEEE Trans. Nucl. Sci.* **NS-47**, 1620–1624 (2000)
49. U. Neitzel. Recent technological developments and their influence. *Rad. Prot. Dosim.* **90**, 15-20 (2000)
50. Direct Radiogarphy Corp. website, www.directradiography.com (2000)
51. Varian. Varian Medical Systems website: www.varian.com/xry/prd003b.html (2003)
52. J. T. Rahn, et al. High resolution X-ray imaging using amorphous silicon flat-panel arrays. *IEEE Trans. Nucl. Sci.* **NS-46**, 457–461 (1999)

53. J. Freudenberger, E. Hell, and W. Knupfer. Perspectives of medical X-ray imaging. Presented at the 2nd International Workshop on Radiation Imaging Detectors, Freiburg, Germany, July 2–6, 2000. *Nucl. Instrum. Meth.* A **466**, 99–104 (2001)
54. R. J. Yarema, et al. A programmable, low noise, multichannel ASIC for readout of pixelated amorphous silicon arrays. *Nucl. Instrum. Meth.* A **439**, 413–417 (2000)
55. Canon. Canon Medical Systems website, www.usa.canon.com (2000)
56. S. Muller. Full-field digital mammography designed as a complete system. *Eur. J. Radiol.* **31**, 25–34 (1997)
57. S. Peer, et al. Direct digital radiography versus storage phosphor radiography in the detection of wrist fractures. *Clin. Radiol.* **57**, 258–262 (2002)
58. W. Zhao, and J. A. Rowlands. X-ray imaging using amorphous selenium: feasibility of a flat panel self-scanned detector for digital radiology. *Med. Phys.* **22**, 1595–1605 (1995)
59. D. Mah , J. A. Rawlinson, and J. A. Rowlands. Detective quantum efficiency of an amorphous selenium detector to mefavoltage radiation. *Phys. Med. Biol.* **44**, 1369–1384 (1999)
60. W. Zhao, et al. Web page: www.sunnybrook.utoronto.ca:8080/~selenium (2001)
61. M. J. Yaffe, and J. A. Rowlands. X-ray detectors for digital radiography. *Phys. Med. Biol.* **42**, 1-39 (1997)
62. V. Perez-Mendez, et al. Detection of charged particles in amorphous silicon layers. *Nucl. Instrum. Meth.* A **252**, 478–482 (1986)
63. C. Hordequin, et al. Nuclear radiation detectors using thick amorphous-silicon MIS devices. *Nucl. Instrum. Meth.* A **456**, 284–289 (2001)
64. T. Mertelmeier. Medical imaging: from data to images. Presented at the 2nd International Workshop on Radiation Imaging Detectors, Freiburg, Germany, July 2-6 (2000)
65. M. M. Tesic, et al. Full field digital mammography scanner. *Eur. J. Radiol.* **31**, 2–17 (1997)
66. J. Ludwig, et al. Development of GaAs-CCDs for X-ray detection. *Nucl. Instrum. Meth.* A **460**, 72–75 (2001)
67. C. Rönnquist, et al. Double-sided silicon microstrip detectors and low noise self-triggering multichannel readout chips for imaging applications. *Nucl. Instrum. Meth.* A **348**, 440–443 (1994)
68. M. Overdick, et al. A Bioscope system using double-sided silicon microstrip detectors and self-triggering read-out chips. *Nucl. Instrum. Meth.* A **392**, 173–177 (1997)
69. E. Beuville, et al. High resolution X-ray imaging using a silicon strip detector. *IEEE Trans. Nucl. Sci.* **NS-45**, 3059–3063 (1998)
70. R. D. Speller, et al. Digital X-ray imaging using silicon microstrip detectors: a design study. *Nucl. Instrum. Meth.* A **457**, 653–664 (2001)
71. R. Beccherle, et al. Development of a digital radiography system based on silicon microstrip detector. *Physica Medica* **XII**, 17–24 (1996)
72. R. Beccherle, et al. Microstrip silicon detectors for digital radiography. *Nucl. Instrum. Meth.* A **409,** 534–536 (1998)

73. M. Lundqvist. Evaluation of a Photon Counting X-ray Imaging System. *Conference Records* of the IEEE Nuclear Science Symposium and Medical Imaging Conference, Lyon (France), October 15–20, 2000, pp. 3–6/12. *IEEE Trans. Nucl. Sci.* **NS-48**, 1530-1536 (2001)
74. A. Papanestis, et al. A Radiographic Imaging System Based Upon a 2-D Silicon Microstrip Sensor. *Conference Records* of the IEEE Nuclear Science Symposium and Medical Imaging Conference, Lyon (France), October 15–20, 2000, pp. 3–1/5 (2000)
75. F. Arfelli, et al. A high efficiency photon counting silicon matrix for position sensitive X-ray detection. *Conference Records* of the IEEE Nuclear Science Symposium, November 9-15, 1997, Albuquerque, NM, USA, vol. 1, pages 304–307 (1997)
76. F. Arfelli, et al. SYRMEP: SR digital imaging of breast tissue samples. *Conference Records* of the IEEE Nuclear Science Symposium, November 9-15, 1997, Albuquerque, NM, USA, vol. 2, pages 1345–1348 (1997)
77. R. Beccherle, et al. PD3, a low-noise precise timing analog front-end integrated circuit. *Conference Records* of the IEEE Nuclear Science Symposium and Medical Imaging Conference, Albuquerque, NM, USA, Vol. 1, November 11-15, 1997, pp. 240–243 (1997)
78. M. Prest, et al. FROST: a low-noise high-rate photon counting ASIC for X-ray applications. *Nucl. Instrum. Meth.* A **461**, 435–439 (2001)
79. L. Sartori, et al. Low-noise front-end amplifier and channel encoder for a 2-D X-ray digital imaging system with single photon counting capability. *Nucl. Instrum. Meth.* A **460**, 213–220 (2001)
80. B. Alfano, et al. First X-ray images with a double-sided microstrips silicon crystal. *Phys. Med. Biol.* **37**, 1167–1170 (1992)
81. B. Alfano, et al. A neural network approach to structured background subtraction for digital radiography. *Physica Medica* **VIII**, 123–129 (1992)
82. B. Alfano, et al. Digital imaging in radiology: preliminary results obtained with a high spatial resolution 2D silicon detector. *IEEE Trans. Nucl. Sci.* **NS-40**, 987–991 (1993)
83. M. Conti, et al. A transputer based "list mode" parallel system for digital radiography with 2D silicon detectors. *IEEE Trans. Nucl. Sci.* **NS-40**, 996–1000 (1993)
84. M. Conti, et al. Modular data acquisition system based on transputer technology for bidimensional time coincidence counting. *Nucl. Instrum. Meth.* A **345**, 120–125 (1994)
85. A. Cisternino, et al. Single photon 2-D imaging system prototype for biopsy digital mammography. *Physica Medica* **XIII**, Suppl. 1, 214–217 (1997)
86. A. Del Guerra, et al. Photons counting of a 2-D silicon μ-strip detector in high X-rays flux environment: double coincidence recovery. *Proceedings* of the IEEE Nuclear Science Symposium and Medical Imaging Conference, San Francisco, October 21–28, 1995, vol. 3, pp. 1642–1646 (1995)
87. A. Del Guerra, et al. A high-rate X-Y coincidence VLSI system for 2-D imaging detectors. *Nucl. Instrum. Meth.* A **394**, 191–198 (1997)

88. W. Bencivelli, et al. X-ray imaging test of a microstrip silicon detector with a transputer DAQ. *IEEE Trans. Nucl. Sci.* **NS-41**, 1522–1525 (1994)
89. E. Bertolucci, et al. Autoradiography with silicon strip detectors. *Nucl. Instrum. Meth.* A **381**, 527–530 (1996)
90. D. A. Boardman, and P. J. Sellin. Design and characterization of high electron mobility transistors for use in a monolithic GaAs X-ray imaging sensor. *Nucl. Instrum. Meth.* A **466**, 226–231 (2001)
91. P. Russo. Hybrid Semiconductor Pixel Detectors for Low- and Medium-Energy X- and Gamma-Ray Single Photon Imaging Using the Medipix Read-Out Chip. In: Encyclopedia of Imaging Science and Technology, J. P. Hornak (Ed.), Wiley Interscience, John Wiley & Sons, Inc. New York (2002)
92. Medipix. The Medipix1 photon counting chip and related hardware and software have been developed by a collaboration between CERN, Geneva (CH), University of Glasgow (UK), University of Freiburg, (D), Universities of Pisa and Napoli Federico II (I). The Medipix2 circuit is being developed by the larger Medipix2 European collaboration. A full description of this partnership and their research activity related to the applications of Medipix1 chip and the development of its successor Medipix2, can be found on the collaboration's web site (http://medipix.web.cern.ch/MEDIPIX)
93. S. R. Amendolia, et al. Low contrast imaging with a GaAs pixel digital detector. *IEEE Trans. Nucl. Sci.* **47**, 1478–1482 (2000)
94. M. Lindner, et al. A hybrid pixel detector for digital X-ray imaging. Presented at the 2nd International Workshop on Radiation Imaging Detectors, Freiburg, Germany, July 2-6 (2000)
95. M. Campbell, et al. Readout for a 64x64 pixel matrix with 15-bit single photon counting. *IEEE Trans. Nucl. Sci.* **NS-45**, 751–753 (1998)
96. E. H. M. Heijne, et al. LHC1: a semiconductor pixel detector readout chip with internal, tunable delay providing a binary pattern of selected events. *Nucl. Instrum. Meth.* A **383**, 55–63 (1996)
97. P. Delpierre, et al. Development of silicon micropattern (pixel) detectors. *Nucl. Instrum. Meth.* A **315**, 133–138 (1992)
98. C. DaVià, et al. Gallium arsenide pixel detectors for medical imaging. *Nucl. Instrum. Meth.* A **395**, 148–151 (1997)
99. S. R. Amendolia, et al. Test of a GaAs-based pixel device for digital mammography. *Nucl. Instrum. Meth.* A **460**, 50–54 (2001)
100. B. Mikulec, et al. Characterization of a single photon counting pixel system for imaging of low contrast objects. *Nucl. Instrum. Meth.* A **458**, 352–359 (2001)
101. C. Schwarz, et al. Dose-dependent X-ray measurements uising a 64x64 hybrid GaAs pixel detector with photon counting. *Nucl. Instrum. Meth.* A **460**, 91–96 (2001)
102. J. Watt. Applications of pixellated GaAs X-ray detectors in a synchrotron radiation beam. *Nucl. Instrum. Meth.* A **460**, 185–190 (2001)
103. G. Bardelloni, et al. A new read-out system for an imaging pixel detector. *Conference Records* of the IEEE Nuclear Science Symposium and Medical Imaging Conference, Lyon (France), October 15–20 (2000)

104. A. Fornaini, D. Calvet, and J. L. Visschers. Soft X-ray sensitivity of a photon-counting hybrid pixel detector with a silicon sensor matrix. *Nucl. Instrum. Meth.* A **466**, 142–145 (2001)
105. X. Llopart, et al. Medipix2, a 64k readout chip with 55 mm square elements working in single photon counting mode. Presented at IEEE Nuclear Science Symposium and Medical Imaging Conference, San Diego, CA, USA, 4-10 November, 2001. *IEEE Trans. Nucl. Sci.* **NS-49**, 2279-2283 (2002)
106. M. Campbell. First Test Measurements of a 64k pixel readout chip working in single photon counting mode. Presented at 4th Int. Workshop on Radiation Imaging Detectors, Amsterdam, The Netherlands, September 8-12, 2002 (available online at www.iworid2002.nl). Llopart, X., and Campbell, M. First Test Measurements of a 64k pixel readout chip working in single photon counting mode. *Nucl. Instrum. Meth.* A **509**, 157–163 (2002)
107. G. Mettivier, M. C. Montesi, and P. Russo. First images of a digital autoradiography system based on a Medipix2 hybrid silicon pixel detector. *Phys. Med. Biol.* **48**, N173–N181 (2003)
108. M. Chmeissani, and B. Mikulec. Performance limits of a single photon counting pixel system. *Nucl. Instrum. Meth.* A **460**, 81–90 (2001)
109. L. Abate, et al. Noise and interpixel dead space studies of GaAs pixellated detectors. *Nucl. Instrum. Meth.* A **458**, 164–168 (2001)
110. P. Seller, et al. Photon counting hybrid pixel detector for X-ray imaging. *Nucl. Instrum. Meth.* A **455**, 715–720 (2000)
111. S. Passmore, et al. Characterisation of a Si single photon counting pixel detector. *Nucl. Instrum. Meth.* A **466,** 202–208 (2001)
112. G. Iles, et al. Large-area pixellated photon counting X-ray imaging system. *Nucl. Instrum. Meth.* A **458**, 427–430 (2001)
113. D. San Segundo Bello, B. Nauta, and J. L. Visschers. Design of analog-to-digital convertes for energy-sensitive hybrid pixel detectors. *Nucl. Instrum. Meth.* A **466**, 218–225 (2001)
114. S. R. Amendolia, et al. Spectroscopic and imaging capabilities of a pixellated photon counting system. *Nucl. Instrum. Meth.* A **466**, 74–78 (2001)
115. S. R. Amendolia, et al. The CALMA project. *Nucl. Instrum. Meth.* A **466**, 428–429 (2001)
116. M. G. Bisogni, et al. A Medipix2-bassed imaging system for digital mammography with silicon pixel detectors. Presented at IEEE Nuclear Science Symposum and Medical Imaging Conference, Portland, OR, USA, October 19-25 (2003)
117. M. Chmeissani, et al. First experimental tests with a CdTe photon counting pixel detector hybridized with a Medipix2 readout chip. Presented at the IEEE Nuclear Science Symposium and Medical Imaging Conference, Portland, OR, USA, October 19–25, 2003, and to be published in *IEEE Trans. Nucl. Sci.* (2003)
118. T. O. Tuemer, et al. High-resolution pixel detectors for second generation digital mammography. *Nucl. Instrum. Meth.* A **497**, 21–29 (2003)
119. R. Irsigler, et al. X-ray imaging using a 320x240 hybrid GaAs pixel detector. *IEEE Trans. Nucl. Sci.* **NS-46**, 507–512 (1999)

120. R. Irsigler, et al. 320x240 GaAs pixel detectors with improved X-ray imaging quality. *Nucl. Instrum. Meth.* A **460**, 67–71 (2001)
121. J. Pozela, V. Jucien, and K. Pozela. Light emitted GaAs particle detectors. *Nucl. Instrum. Meth.* A **410**, 111–114 (1998)
122. J. Pozela, et al. The $Al_xGa_{1-x}As$ X-ray imaging detector. *Nucl. Instrum. Meth.* A **460**, 119–122 (2001).
123. K. Pozela, et al. Optical response of the graded-gap $Al_xGa_{1-x}As$ X-ray detector. *Nucl. Instrum. Meth.* A **466**, 58–62 (2001)
124. AIT. Advance Interconnection Technology web site: www.ait.com.hk
125. IDE. Integrated Detector Electronics AS (www.ideas.com)
126. A. Bandettini, et al. Characterization of the response of a double side silicon detector to X-rays in the diagnostic energy range. *IEEE Trans. Nucl. Sci.* **40**, 983–986 (1993)
127. E. A. Babichev, et al. Photon counting and integrating analog gaseous detectors for digital scanning radiography. *Nucl. Instrum. Meth.* A **419**, 290–294 (1998)
128. T. Mali, V. Cindro, and M. Mikuz. Silicon microstrip detectors for digital mammography—evaluation and spatial resolution studies. *Nucl. Instrum. Meth.* A **460**, 76–80 (2001)
129. S. Manolopoulos, et al. X-ray imaging with photon counting hybrid semiconductor pixel detectors. *Nucl. Instrum. Meth.* A **434**, 38–43 (1999)
130. C. Schwarz, et al. X-ray imaging using a hybrid photon counting GaAs pixel detector. *Nucl. Phys.* B (Proc. Suppl.) **78**, 491–496 (1999)
131. Š Stres, et al. Design optimization of a breast imaging system based on silicon microstrip detectors. *Phys. Med. Biol.* **45**, 2029–2041 (2000)

CHAPTER 4

DETECTORS FOR CT SCANNERS

Willi A. Kalender and Theobald O.J. Fuchs

Institute of Medical Physics, University of Erlangen-Nürnberg, Germany

4.1 Introduction

Within this chapter we will discuss most of the requirements on X-ray detectors which are specific to Computed Tomography (CT) scanners. In order to facilitate a detailed understanding of the special properties a detector should provide for CT measurement, we will give a short introduction into the basics of CT measurement, the standard configuration of modern clinical CT systems and some typical performance parameters of a typical CT scanner. This will include the mechanical and geometrical set-up, special alignment features, the X-ray source, filtration and collimation. In the following, the physical detector principles and properties and their influence on image quality are discussed. Wherever necessary we will take into account modern concepts for multi-row detectors which are state of the art with the latest generation of modern clinical CT scanning units. Special fields of medical CT such as 3-dimensional (3D) angiography or non-destructive material testing with Micro-CT will be addressed in the outlook. Although there is very interesting work going on in these fields, we will not treat these topics in great detail, but rather focus this contribution on modern clinical CT systems for standard radiological applications.

4.2 Basic Principle of CT Measurement and Standard Scanner Configuration

To be able to compute an image in acceptable quality following Radon's theory, a sufficiently high number of attenuation integrals or projection values have to be recorded. It is necessary to carry out measurements in all directions, i.e. at least over an angular range of 180°, and to determine many narrowly spaced data points for each projection. (For general aspects of CT see any textbook, e.g. [1]).

A simple measurement set-up fulfilling this purpose is sketched in Fig. 4.1a. A radiation source with adequate collimation emits a pencil beam and the intensity, attenuated by the object, is registered by the detector placed opposite. For a given angular position, this set-up of radiation source and detector is moved linearly (translation), and the intensity is measured either at single discrete points or continuously. This results in an intensity profile recorded for parallel rays. By determining the ratios of the primary intensity recorded in the periphery and the attenuated intensities recorded behind the object and taking their logarithms, an attenuation profile results which is generally termed a projection. Projections are measured successively for successive angular positions. The complete set of projections, here determined in parallel ray geometry over 180°, is then transferred to the data processing unit.

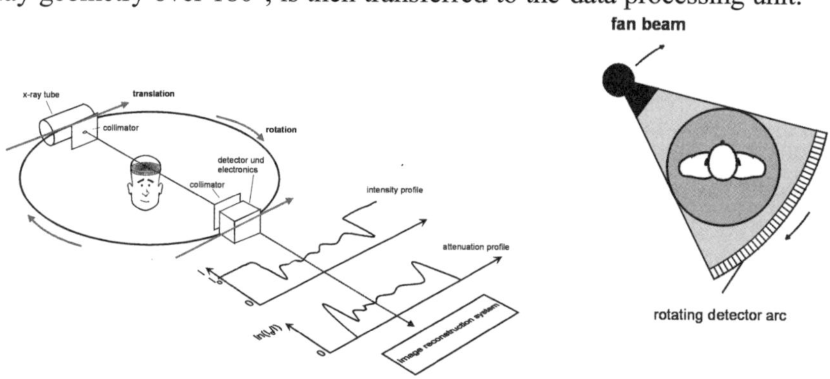

Fig. 4.1. Basic principle of CT measurement. In the simplest case X-ray intensities and therefore object attenuation will be measured with a pencil beam for many different angular positions (a). Standard fan-beam configuration of a so-called 3rd generation CT system where an x-ray tube and a detector arc are both continuously rotating around the patient (b).

CT scanners today measure typically in fan-beam geometry (Fig. 4.1b) over an angular range of 360°. The extension to 360° resulted from several considerations, initially focussing on image quality and better data sampling. By proper geometrical alignment of the detector, the quarter detector offset, overlapping sampling is achieved (see below). The path of a single attenuation integral is determined by the position of the respective detector element within the arc and the current position of the focal spot with respect to the patient. Therefore, high geometrical precision of the whole system in general and the detector assembly in particular are crucial for the quality of the resulting images. Practical considerations also demand 360° scanning; for spiral CT in particular, this is a prerequisite.

Modern CT scanners typically measure 800 to 1,500 projections during one rotation with 600 to 1,200 sampled data points per projection. Accordingly, the number of physical detector channels assembled within the detector arc of a 3rd generation scanner can be up to 1,000 in each row with a spacing of typically 1.0 mm between the centres of adjacent detector elements. Already in the first CT scanner built by Hounsfield in 1972 the scan times were reduced by measuring two slices simultaneously. Nevertheless, over decades a single-slice detector was the standard configuration in medical CT. In 1998 several manufacturers introduced the latest generation of multi-row detector systems, which are characterized by a varying number D of physical detector rows whereas the number of thereby acquired slices is $M = 4$ (see below, Table 4.3). For instance, a multi-row detector system with $D = 16$ rows can thus contain up to 16,000 detector elements. With rotation times down to 0.5 s the data acquisition system has to process over 40 million signals per second; thereby the integration time of the detector signal is in the order of a few hundred microseconds. Another challenging demand on X-ray detectors for clinical CT is the relatively high count rate the detectors have to cope with. Due to the large aperture more than 2×106 photons per mA and s have to be registered by a single channel with tube voltages in the range of 120 to 140 kV. Thus the detectors cannot be operated in a photon counting fashion but work with charge integration.

4.3 Mechanical Design

The two largest single components of a CT unit are the actual examination unit, the so-called gantry, and the patient bed. They are similar in design for most CT units. A typical unit is shown in Fig. 4.2, taking the installation at the Institute of Medical Physics, Erlangen, as an example.

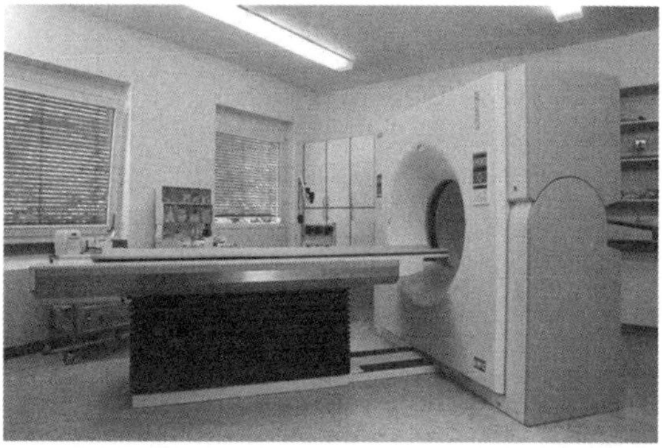

Fig. 4.2. Typical CT examination suite (SOMATOM Volume Zoom in the IMP, University of Erlangen). Photo of the examination room showing the patient bed and the gantry, which typically can be tilted by ±30° and contains the x-ray components and the measuring system.

The actual measuring system, which is hidden in the gantry, is shown schematically in Fig. 4.3a; the geometry and dimensions given here refer to the specific scanner shown in Fig. 4.2, but can be taken as typical values. A field of measurement (FOM) of about 50 cm and a gantry opening of about 70 cm, which allow the examination of patients with larger cross-sections, can also be considered as an 'industry standard' in the medium and upper price segments. The different components and the definition of the fan beam are described in detail below.

The mechanical design of a CT scanner, which may appear to have been a trivial matter for the first CT units, has always imposed high demands on engineering due to the required mechanical accuracy of the sampling process. Among these are the demand for a quarter detector offset to improve sampling, where the distance between the central ray and its direct neighbour measured at the centre of rotation is taken as the sampling distance. If the central ray is offset by one fourth of the sampling distance from the axis of rotation, then this ray will again be shifted by one fourth of the sampling distance in the opposite direction after 180° rotation.

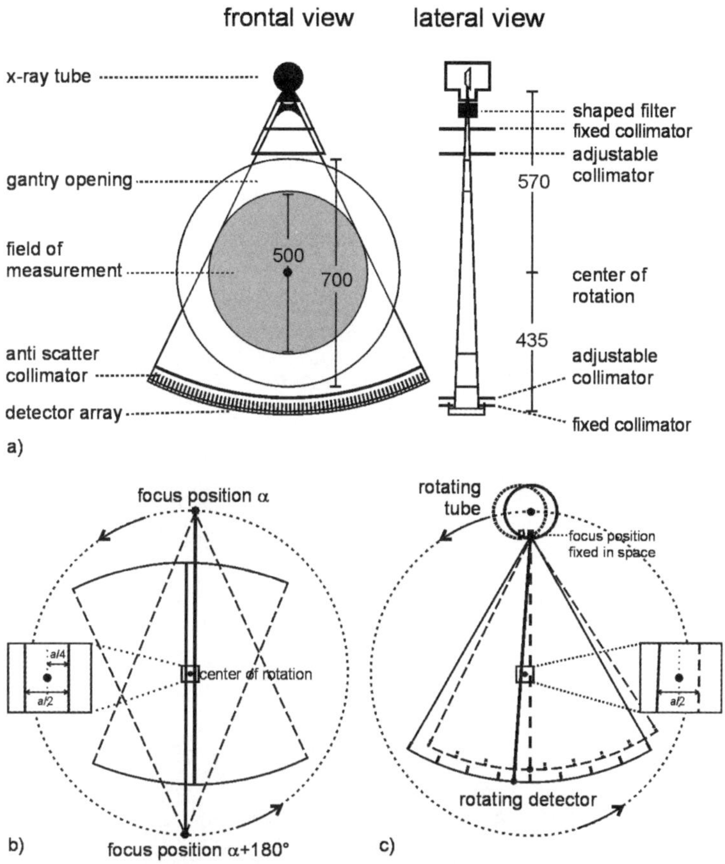

Fig. 4.3. Schematic representation of the scanning geometry and important components of the CT measuring system in frontal view (*x/y* plane) and in lateral view (*y/z* plane) **(a)**. By misaligning the measuring system by one fourth of the sampling distance, a doubling of the sampling frequency is achieved for 360° measurements **(b)**. An additional similar effect can be reached by use of the so-called flying focus **(c)**.

This means that the two measurements with the central ray taken in opposite direction (and in an analogous fashion for all other rays) are offset relative to each other by half the sampling distance (Fig. 4.3b). The sampling frequency is increased by a factor of two compared with a measurement over 180°. The so-called flying or dynamic focal spot serves a similar purpose. This technical performance feature of some CT X-ray tubes allows controlling the focus position electro-magnetically. That is, if the focus is moved on the anode opposite the direction of movement of the X-ray tube, it is then kept fixed in space for two successive measurements. Thereafter the focus 'flies' back to its starting position on the anode, and the process repeats itself. For each focus position in space two measured interlaced projections result, since the detector continues to move continuously. The sampling frequency is thereby doubled (Fig. 4.3c), thus enhancing the spatial resolution.

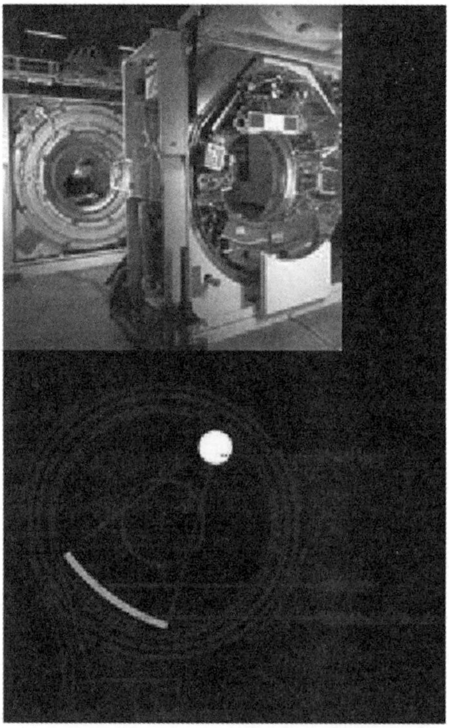

Fig. 4.4. Modern CT systems employ slip ring technology to allow for continuous data acquisition. Photograph of a measuring system during assembly with slip rings (left) and components (right) visible (**a**). Schematic representation of slip rings, which allow the transfer of electrical energy to the x-ray components and data from the detector system to the image reconstruction unit (**b**).

The demands on mechanical design have risen continuously with the rotation speeds offered today. The resulting centrifugal forces pose further technological challenges. To appreciate this, one has to take into account that the rotating part of the gantry, i.e. the components shown in Fig. 4.3, represent a mass of typically 400 - 1000 kg. To accelerate these masses to two rotations per second, linear drives integrated into the rotary bearing are preferably used which also guarantee sufficiently uniform motion.

The transfer of electrical energy to the X-ray tube and to the other components on the rotating part of the gantry and data transfer back to the stationary part is mostly achieved by slip rings (Fig. 4.4). The data from the data acquisition system can alternatively be transferred optically or by high frequency transmission systems. The choice of the respective components will be made according to cost and reliability.

Further functions which the CT system must offer can easily be provided with today's technical means. Among these are the capability of gantry tilt; most CT scanners allow a tilt of the scan plane up to ±30° with respect to the axis of rotation. This allows the selection of scan planes directly through the structures of interest, which is important for critical regions, such as the base of the skull or the lumbar spine.

4.4 X-Ray Components

CT has always demanded relatively high X-ray power. This demand has risen dramatically with the increase in scan speed and the introduction of spiral CT. Typical values for the maximum power are 20 to 60 kW, with the high voltage ranging from 80 to 140 kVp. The maximum ratings provided by a given system constitute a limit which cannot be sustained over a longer period of time, such as the typical scan times of 30 to 60 s for spiral CT. In particular the heat storage capacity of anode discs available today would be exceeded, but also the generator capacities, which are mostly adapted in their specifications to the respective X-ray tube. Longer pauses for tube cooling were therefore required for examining larger volumes by sequential single slice scanning. In much the same way, in spiral CT the tube current and thus the power have to be reduced, which often leads to the choice of slice widths thicker than what can be considered optimal. A typical sketch of power versus time is shown in Fig. 4.5. This documents the dependence of the maximal selectable tube current on the desired spiral scan duration.

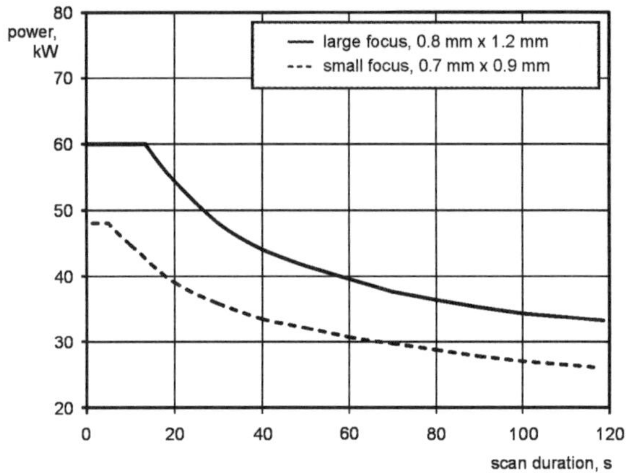

Fig. 4.5. Maximal x-ray power depends on scan duration and prior tube load. For spiral scans of long duration power has to be reduced to avoid excessive loading. The reduction of total scan time offered by multi-row detectors allows reduction of tube loading or, alternatively, the selection of higher power values and/or the choice of the thin slice settings necessary for high resolution in all three dimensions.

These technical limitations, which will remain for the foreseeable future, explain the particular attractiveness of multi-row detector systems. The available X-ray power is used much more efficiently, since M slices offer a solid angle which is M times larger than for a single one. To a first approximation, the scan duration will be reduced by this factor M and allow for a selection of higher power settings. Alternatively, multi-row detectors may allow the utilization of X-ray components with correspondingly reduced power ratings at lower cost. This does not appear generally desirable, since high power ratings are obligatory for many special applications and for single slice scans as will be seen.

X-ray tubes used in CT today offer focal spot sizes of typically 0.5 to 2.0 mm. Often, a selection of foci is available. The SOMATOM Volume Zoom, for example, offers two focus sizes of 0.8 mm × 1.2 mm and 0.7 mm × 0.9 mm. Smaller foci are preferable for scans with thin slices and high resolution; for scans of large volumes with the primary aim of high soft tissue or contrast resolution, larger focus size with higher power ratings are desirable (Fig. 4.5). The optional performance feature of a 'flying' focus, which allows controlling the focus position on the anode during the scan, has already been alluded to in the previous section.

With respect to spatial resolution, the limiting factors are the focal spot size and the spacing between successively sampled rays, i.e. the physical spacing of the detector channels in connection with the data acquisition mode. For example, in the case of a detector quarter shift (interlaced sampling) a physical spacing of the detector elements of 1.0 mm corresponds to a sampling distance of 0.284 mm at the center of rotation assuming a geometrical set-up as shown in Fig. 4.3 (magnification factor 570/1005). In comparison the focal spot size of 0.7 mm for the same set-up corresponds to 0.303 mm (magnification factor 435/1005 if projected from a point on the detector array). Thus the focal spot size is more critical in this case, but detector dimensions are critical just the same, of course.

4.5 Collimators and Filtration

CT systems feature various collimators, filters and shielding designs which provide filtration of the X-ray spectra, definition of the measured slices, guarding the detector against scattered radiation and general radiation protection (Fig. 4.3). They may vary from scanner type to scanner type, but in principle they always offer the same functions. A first collimation is provided very close to the focus to reduce the generated radiation roughly to the maximally anticipated beam for the given detector and geometry. This first reduction of the radiation cone is provided by the lead housing of the X-ray tube itself, which provides an aperture for rough definition of the fan beam. In a second step the maximum permitted fan is defined exactly by a fixed collimator. An additional adjustable collimator allows variable collimation to the desired slice width S and $M \times S$ for single slice and multi-slice systems, respectively. The latter collimator is placed as far from the focus as possible, i.e. close to the gantry housing, to minimize penumbral regions caused by the finite focus size.

The influence of collimator geometry and focus size on the definition of the slice is illustrated in Fig. 4.6. For an ideal point focus a perfect rectangular slice profile results. For any real extended focus penumbral regions and, to a first approximation, a trapezoidal profile result. The deviation from a rectangle can be reduced by a movable collimator in front of the detector; such means are used in particular for defining very thin slices, since here the highest relative deviations from the ideal rectangular form have to be expected. It must be accepted, however, that the dose profile will be wider than the sensitivity profile. This simply reflects the situation that the exposed section is thicker than the section measured by the detector (Fig. 4.6 right).

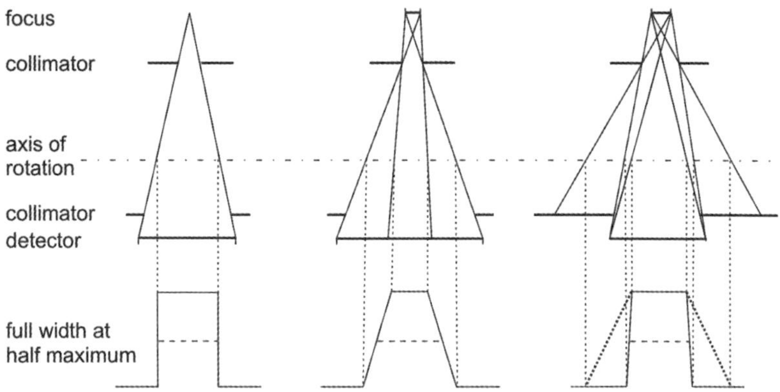

Fig. 4.6. Definition of the scanned slice. The width and profile of the exposed slice are determined by the focus size, collimator geometry and width. The use of a detector collimator (right) will have a positive influence on the slice profile (solid), whereas the dose profile (dotted) will remain wide and therefore exceed the slice profile.

Collimators in front of the detector—in addition to the optional movable collimator there is always a fixed collimator with a width corresponding to the maximum collimation—serve to minimize signal contributions from scattered radiation. Optional anti-scatter collimators serve the same function, reducing contributions from scattered radiation; these are mostly implemented as a system of thin lamellae made of strongly absorbing material, e.g. 100 μm thick tantalum sheets, which are positioned between the single detector elements and aligned exactly in the direction of the X-ray focus. This alignment, although technically demanding, is necessary to avoid a reduction of signal-carrying primary radiation and can only be implemented in scanners with a rotating detector. Scanners with stationary detector rings (so-called 4th generation) in consequence suffer a severe disadvantage with respect to the suppression of scattered radiation. This can be considered as one reason why, apart from a dual-row detector system for electron beam CT, multi-row detector systems are not offered with stationary detectors.

Several components add to the filtration of the X-ray spectrum. In addition to the inherent filtration of the X-ray tube, which typically amounts to 3 mm aluminium-equivalent thickness, flat or shaped filters are also used. Flat filters, for example copper sheets of 0.1 to 0.4 mm thickness, shift the spectrum to higher energies. Low energy portions, which contribute strongly to the dose, but less to the signal, are reduced to a high degree. The use of additional filtration is therefore welcomed, but also demands higher X-ray power.

So-called shaped or 'bow-tie' filters attenuate radiation hardly at all in the centre, but strongly in the periphery (Fig. 4.3a frontal view). They reduce the demands on the dynamic range of the detector system and scattered radiation intensities arising from peripheral object zones, and also the patient dose. The material for shaped filters should be of low atomic number in order to keep the spectral and beam hardening differences between the centre and the periphery of the fan beam as small as possible. Teflon, for example, is an efficient filter material for this purpose due to its high density combined with a relatively low effective atomic number.

One further component of the collimator system, which is employed close to the detector in some systems, is the so-called 'high-resolution comb'. The teeth of the comb overlap with the gaps between detector elements and, depending on their width, reduce the detector aperture to improve spatial resolution. It is clear that this will also reduce the geometric dose efficiency of the total system. Such devices are therefore only used for high contrast scanning, where increased noise does not pose a particular problem.

With multi-row detectors there is a technical problem associated in respect to dose. To ensure proper normalization of all measured intensity values it is desirable that all detector rows are offered the same primary intensity value I0. This implies that penumbra regions of the fan beam which are of no or little concern in single-slice CT should be avoided in multi-slice CT. This situation is sketched in an exaggerated fashion in Fig. 4.7. While penumbral radiation intensities are fully utilized in single-slice acquisition, they are to be excluded in practical MCT implementations to exclude errors in attenuation measurements. This implies a reduction of geometric efficiency and a corresponding increase in dose values. Typical increases of 20% to 30% have to be expected depending on the design and the slice width used [2].

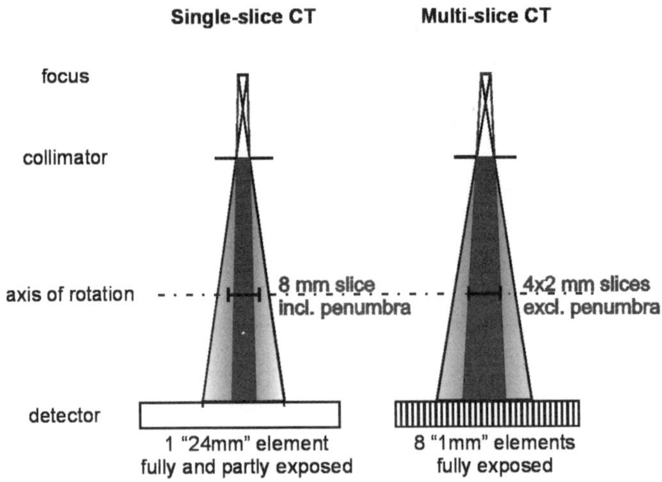

Fig. 4.7. While radiation intensity in the penumbral region is fully utilized in single-slice acquisition it is generally excluded in multi-slice CT implying a loss in geometric efficiency and a relative increase in dose.

Since real systems are inevitably susceptible to thermal, mechanical or other disturbances, manufacturers try to build in a safety margin by keeping the dose profile wider. Proper control of all system parameters, in particular of the focus position, is the alternative. It will remain a particular challenge for MSCT designers to limit dose increases due to lacking usage of penumbral radiation. The effect's relative importance will decrease with increasing values of $M \times S$, i.e. it will be of less significance when wider arrays become available, since the relative proportion of the penumbral region decreases. It will necessarily remain significant for submillimeter slices and small M, e.g. for special collimations such as 2×0.5 mm, which should therefore be reserved for high-resolution high-contrast imaging.

4.6 Detector Systems

The detector, the system for quantitative recording of the incident ionizing radiation, constitutes one of the most important and technologically most critical components of the entire CT system. It has to transform the incident X-ray intensity into a corresponding electrical signal, to amplify this signal and to convert it from analog to digital form. The decisive components are the X-ray sensitive detector elements and their geometrical configuration, the preamplifiers and the analog-to-digital converters.

The demands on the electronics, i.e. preamplifiers and analog-to-digital converters (ADC), can be specified relatively easily. The electronic noise level generated by these components should be significantly lower than the statistical fluctuations in X-ray intensities due to quantum noise. A frequently stated requirement is that the electronic noise must not be higher than half the magnitude of the maximally expected quantum noise, i.e. $\sigma_E < 0.5\ \sigma_Q$. The demands on the dynamic range are high, too. Since the rotation time can typically vary between 0.5 and 2.0 s, the integration time of each detector channel can vary by a factor of 4; the tube current is typically 50 to 500 mA, a factor of 10 in intensity; slice width S too can be selected within a wide range, typically from 0.5 mm to 5.0 mm. Thus the intensity incident to a single detector channel can vary by a factor of 400 alone due to technique factors which are determined by the operator. On the other hand, with realistic patients attenuation values as high as 12 can occur, which means that the dynamic range of the ratio of intensities is larger than 216. Usually, 16-bit ADCs are employed, which alone would not be sufficient. Additionally, the dynamic range is increased by on-line gain-switching of the preamplifiers and block-wise scaling the digital data of complete projections depending on the maximum X-ray intensity reaching the central parts of the detector array. Typical gain selections are 1, 16 and 64, thus providing additional 6 bits, resulting in effective 22 bits, which is sufficient to avoid digitization noise to become visible in the reconstructed images.

Several other characteristics of the detector system and their evaluation are more difficult to specify, however. Some of the essential demands are summarized in Table 4.1.

Table 4.1. Demands on CT detector systems

Demands on CT detector systems	**Acceptable values**
High dynamic range	10^5 - 10^6
High linearity over the complete dynamic range	
High quantum absorption efficiency	> 90 % (ideally 100 %)
High luminescence efficiency* for scintillation detectors	> 5 % (ideally 100 %)
High geometric efficiency	80 % - 90 % (ideally 100 %)
Fast temporal response and decay	Decay constant < 10 μs
Low afterglow	< 0.01 %, 100 ms after end of irradiation
Low radiation drift	≤ 0.5 % for longest scan duration
Electronic noise low compared with quantum noise for all scan modes	$\sigma_E \leq 0.5 \cdot \sigma_Q$
Low cross-talk between detector elements	< 3 %
High homogeneity of the material of each detector element for good artefact behaviour	i.e. purity of the material > 99.99 %
Equal response of all detector elements within a detector array	< 0.1 % difference (after optional correction)
Detector material machinable in a simple manner with high precision	± 10 μm tolerance
Feasibility for configuration as single or multi-row array	$D \geq 4$ rows
Environmental acceptability of the detector material	Low toxicity, low disposal cost
Chemical stability	e.g. resistant to moisture
Stable against environmental influences	e.g. thermal expansion coefficient < 10^{-5} per °C
Feasibility of collimation against scattered radiation	
Low cost and easy to service	

* Ratio of the energy of visible light to be detected to the energy of the absorbed X-ray intensity

For the implementation of detectors, many approaches have been discussed and pursued. In CT two conversion principles and detector types have predominantly been set to use:

- Ionization chambers, mostly filled with the noble gas xenon under high pressure
- Scintillation detectors (Fig. 4.8a) in the form of crystals, such as cesium iodide or cadmium tungstate, and ceramic materials, such as gadolinium oxysulfide. Specific details about various ceramic scintillators can be found, e.g. in [3, 4].

A sketch of these two detector types and their operating principles is shown in Fig. 4.8. Semiconductor materials, solid state detectors which provide an electrical signal directly and are therefore often called direct converters, have not been used in clinical scanners until now. Such developments are not expected in the short term.

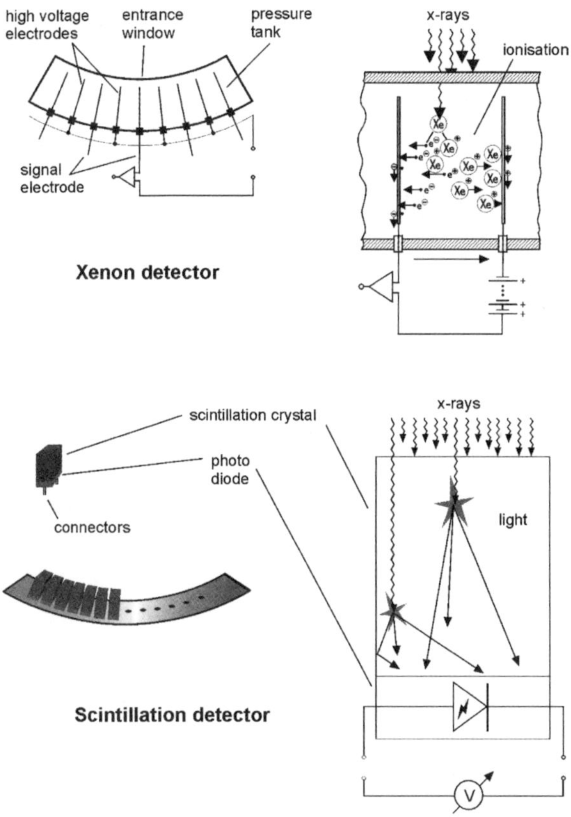

Fig. 4.8. Two detector systems dominate CT: xenon ionization chambers (**a**) and scintillation detectors (**b**).

Xenon ionization chambers (Fig. 4.8a) offer several advantages: their construction is relatively simple in principle, and the sensitivity of the individual detector channels is exactly the same since constant gas pressure exists for the complete detector pressure chamber. The temporal response of xenon, with fast decay and low afterglow, has frequently been cited as a further decisive advantage. Temporal characteristics will be explained and discussed below in comparison to solid state systems.

Table 4.2. Detector efficiency for typical detector systems

Detector / Object	20 cm H_2O	20 cm H_2O and 2 cm bone	40 cm H_2O and 4 cm bone
120 kVp			
Xenon[1] (10 bar, 3 cm)	42.8 %	39.2 %	32.9 %
Xenon[1] (25 bar, 6 cm)	73.8 %	74.0 %	72.7 %
Ceramic scintillator[2]	89.9 %	88.1 %	84.5 %
140 kVp			
Xenon[1] (10 bar, 3 cm)	38.4 %	34.3 %	27.1 %
Xenon[1] (25 bar, 6 cm)	71.0 %	70.3 %	67.0 %
Ceramic scintillator[2]	85.3 %	83.0 %	78.2 %

[1] 1.3 mm aluminum entrance window, chamber pressure and depth given in parentheses
[2] Gadolinium oxysulfide with 1.4 mm depth

Lower quantum efficiency has to be listed as a disadvantage of xenon detectors in comparison with solid state detector materials. This does not mean in general, however, that xenon systems are necessarily inferior to scintillator systems. All factors of influence must be taken into account. The total efficiency of the detector system is not given by the quantum absorption efficiency of the detector material and depth alone, since a number of additional factors also contribute. Geometric efficiency is also essential and mostly determined by the dead spaces between the individual elements. These are typically of the order of 0.1 to 0.2 mm in the fan direction for detector element widths of 1 to 2 mm. Xenon detectors may vary strongly in their performance depending on the design: their efficiency may vary drastically, depending on the pressure, chamber depth, thickness of the entrance window and other construction details (Table 4.2).

Image quality in general, in particular detector-dependent artefact behaviour, influences low-contrast detectability and thus also directly or indirectly the dose efficiency of the complete system. Xenon detectors offer advantages in this respect due to their homogeneous gas distribution and the resultant uniform response. This explains why they were considered the system of choice for a long time. In combination with their temporal response, this also explains why some manufacturers changed their systems from solid state detectors to xenon in the late eighties when faster systems were demanded.

The demand for very short decay times has gained in importance with the introduction of sub-second scan times, as was mentioned before. To illustrate this point, Fig. 4.9a shows decay curves for some typical detector materials. The temporal decay of a signal after a short radiation pulse is determined in essence by two phenomena: (1) by its decay, the rapid exponential falloff of the largest signal components, and (2) by its afterglow, a second much slower decay phase with smaller signal contributions which can also be approximated by a multi-exponential function. The superior characteristics of UFC (Ultra Fast Ceramic), a sintered ceramic gadolinium oxysulfide (Gd_2O_2S) material with a decay time of 10^{-6} s, are evident. Their effect on spatial resolution and image quality can be demonstrated by means of simulation. For this purpose measured data of a SOMATOM PLUS 4 with UFC detector were manipulated in the following way: to each of the measured data obtained in intervals of 0.5 ms, an afterglow contribution was added according to the function $\exp(-t/\tau)$. Even for a decay constant $\tau = 1$ ms a significant smoothing influence on anatomic structures becomes visible (Fig. 4.9b). This influence of the decay characteristics on spatial resolution has also been verified in a direct comparison of a UFC detector and a xenon detector in an otherwise unchanged CT system; a significant improvement in spatial resolution was found to result for the UFC detector [5]. Any advantage with respect to image quality amounts directly to an advantage with respect to dose efficiency; accordingly, temporal response has to be considered an important detector characteristic with respect to dose in addition to quantum efficiency. In any case, for the high scan speeds reached today, xenon detectors are no longer considered to be the gold standard of today's technology.

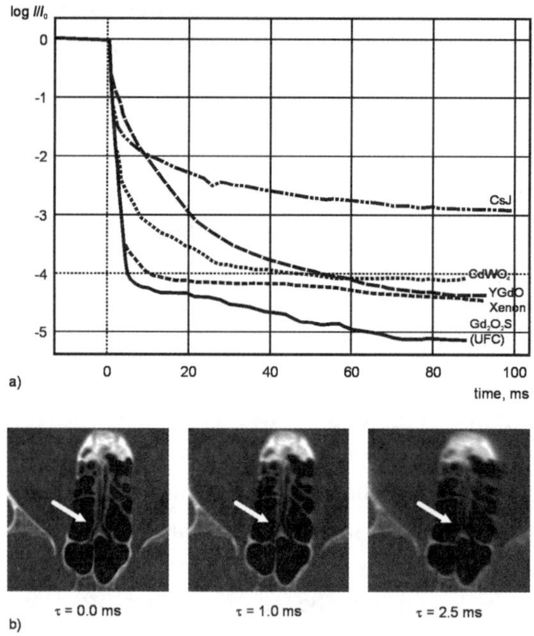

Fig. 4.9. (a) Temporal response of detectors: Decay characteristics of different detector materials after a short x-ray pulse. (b) Slow signal decay may influence image quality, in particular spatial resolution, in a negative way. Simulations of this effect demonstrate that this is already the case for a decay constant $\tau = 1$ ms.

A further disadvantage of xenon detectors – and this may possibly be the decisive disadvantage with respect to future developments – is the fact that linear arrays for single-slice scanning can be built very easily, but multi-row designs or complete arrays are extremely difficult to manufacture. Accordingly, all multi-row detectors in the past and all new detector systems have been built with ceramics or scintillation crystals (Table 4.3).

High geometric efficiency is a requirement for detectors which has particular importance for multi-row detectors. This implies that dead spaces should be as small as possible. The septa and, whenever in use, the anti-scatter collimators between the individual detector elements which are oriented towards the focus, have a width of typically 0.1 to 0.2 mm, while the separation of elements in the z-direction is approximately 0.1 mm. Values of geometric efficiency of 80 to 90% are therefore the best available today. For arrays with a finer separation of detector elements in the z-direction a stronger reduction of geometric efficiency must be accepted.

Table 4.3. CT scanners with multi-row detector systems. D represents the number of detector rows, M the number of simultaneously scanned slices.

Manufacturer	Scanner type	No. of rows D	No. of slices M	Year
EMI	Mark I	2	2	1972
Siemens	SIRETOM 2000	2	2	1974
Siemens	SOMATOM SD	2	2	1977
Imatron	C-100	2	2	1983
Elscint	Twin	2	2	1994
GE	LightSpeed	16	4	1998
Siemens	SOMATOM Volume Zoom	8	4	1998
Marconi	Mx8000	8	4	1998
Toshiba	Aquilion	34	4	1998

4.7 Concepts for Multi-Row Detectors

Examples for multi-row detector arrays are shown in Fig. 4.10. Due to cost considerations complete separate electronic channels are not provided for each individual detector element. Whenever slice thickness larger than the minimum thickness given by the array pitch is selected, the signal from several elements is combined in the z-direction, amplified and A/D-converted. This implies the capability to define the slice width by an electronic combination of signals, whereas the X-ray beam is collimated close to the X-ray source in the usual way.

Figure 4.10a shows the technical solution promoted by GE Medical Systems. The 16 individual rows of the detector array which offer a minimum slice thickness of 1.25 mm at the centre of rotation can be combined to single slices of 1.25 mm, 2.5 mm, 3.75 mm or 5 mm, respectively. Toshiba Medical Systems offers a similar solution of a more or less isotropically structured matrix array in a technically more demanding way with a 34-row array design for its Aquilion. The innermost four rows define a 0.5 mm slice each, followed on each side by 15 rows of 1 mm. Here also, only four slices can be measured simultaneously, i.e. in this case four slices of 0.5 mm, 1 mm, 2 mm, 4 mm or 8 mm, respectively.

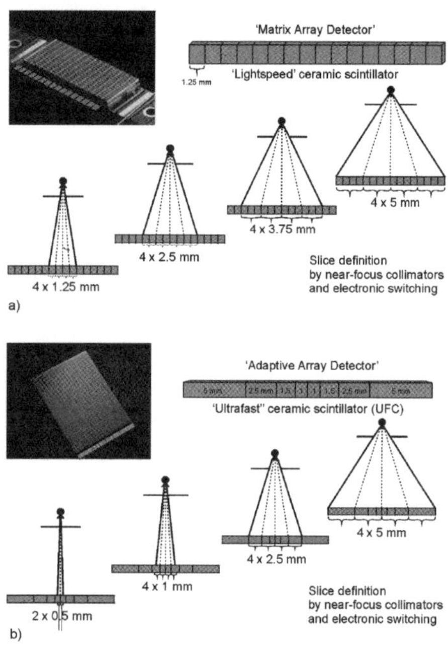

Fig. 4.10. Multi-row detector systems for spiral CT. The number of the slices scanned simultaneously and their width are defined by pre-patient collimation and post-patient electronic combination of signals from the detector array. (**a**) Isotropically structured detector array (GE LightSpeed). (**b**) Anisotropically structured detector array Siemens SOMATOM Volume Zoom and Marconi (Mx8000).

An alternative solution which in principle represents a similar approach, but attempts to minimize dead spaces, was developed by Siemens in cooperation with Elscint and is provided in the SOMATOM Volume Zoom and in the Marconi Mx8000. The design of such a so-called adaptive array is shown in Fig. 4.10b, where only the two innermost detector rows define thin 1 mm slices directly, while offering increasing slice widths off centre. Here also, slices are defined by pre-patient collimation and the post-patient combination of measured signals. Of course additional post-patient collimation is always possible also. An important advantage of an adaptive array of this type is that the outermost thicker slices, in this case 5 mm slices, do not contain septa and do not cause a reduction in geometric efficiency. A definition of very thin slices is possible by narrowing the pre-patient collimation and limitation to only the two innermost detector rows.

Both detector concepts presented here, regularly structured isotropic and "adaptive" anisotropic arrays, represent decisive technological advances which will characterize CT in the years ahead. An expansion to higher numbers of rows and slices are to be expected for the near future. They require greater efforts and investment with respect to detector electronics. Arrays expanded in the z-direction constitute a continuous transition from fan beam to cone beam. Consequently they require further efforts with respect to image reconstruction and data corrections to reduce or eliminate a negative influence on image quality due to the increasing cone beam angles and contributions from scattered radiation which will increase drastically.

4.8 Outlook

The next significant improvement of CT systems can be expected to be the utilization of large area semiconductor detection devices, which are currently being introduced into projection radiography.

The transition from scanning one or only a few slices to data acquisition for an entire field means the transition from fan beam to cone beam geometry. The underlying considerations represent a decisive future-oriented development for computed tomography. Various approaches to this goal are under investigation today; nevertheless, a solution which would lend itself to clinical CT imaging with its high demands on image quality and, in particular, low-contrast detectability has not yet been established.

There is no exact definition in the literature for when a fan beam, assumed to be planar, becomes a cone beam. For the clinical systems in use today, for example, strict planar geometry is always assumed, even for four fans although they cover up to 4×8 mm in z-direction. It may therefore be appropriate not to base the definition on the detector or the type of scanning, but instead on the type of reconstruction. For all multi-slice CT scanners in use today an ideal fan beam geometry is assumed; there are no additional assumptions with respect to geometry or corrections for divergence of the fan. A 2-dimensional reconstruction is simply applied in all cases. Whenever the non-planar geometry is taken into account and a 3D reconstruction is carried out this should be referred to as cone-beam CT.

The data sets recorded by any single row in cone beam geometry will become increasingly inconsistent with increasing distance of the specific row from the central plane. If this is ignored, artefacts will result whenever the cone angle exceeds a few degrees. For this reason, true cone-beam CT with area detectors is employed in practice today only when compromises with respect to image quality are permitted. This is mostly the case for high contrast displays, which require high spatial resolution, but do not require the degree of freedom from artefacts necessary for low-contrast detail detectability. Examples are the imaging of vessels after administering contrast medium, preferably by an intra-arterial injection, or applications in non-destructive material testing in medicine or industry.

In most cases conventional area detectors were used. As an example, 3D CT angiography scans can be acquired with a rotating C-arm and standard image intensifier after intra-arterial injection [6]. This application only allows the assessment of vascular anatomy and pathology, but not of soft tissues, which exhibit low contrast. Micro-CT systems are in widespread use today and already commercially available, e.g. for examinations of biopsies and small animals, or for non-destructive material testing.

Most systems make use of CCD arrays (typically 1024×1024) with cesium-iodide or gadolinium-oxysulfide scintillators [7, 8], but in the meanwhile the applicability of several prototype flat-panel detectors for Micro-CT is under examination (see e.g. [9-13]). Usually, an approximate 3D image reconstruction of the Feldkamp algorithm type is employed. Spatial resolution is extremely high, consistent with the given task and the implementation using about 100 μm and 10 μm effective detector element dimensions, respectively. The requirements for low-contrast detectability, which would have to be provided by a clinical CT scanner, have not yet been fulfilled. New reconstruction procedures and corrections for the increasing contributions from scattered radiation must still be developed and evaluated to achieve this. Proposals are available by now [15, 16]. Both the development of high-quality detector arrays and cone-beam reconstruction algorithms constitute the greatest challenge to physicists and engineers in CT today.

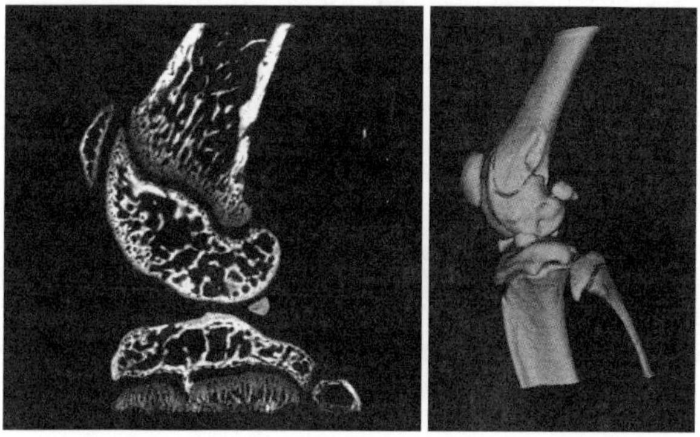

Fig. 4.11. The knee joint of a mouse scanned with a micro-CT unit that employs a 1024^2 CCD camera as detector. A sagittal slice (left) and a corresponing shaded surface display of the whole specimen (right) are shown. The scan was performed with the following parameters: tube voltage 70 kV, tube power 3 W, focus size 5 μm; size of the reconstructed voxels $(6\ \mu m)^3$ [14].

Acknowledgement

Parts of the text and the figures are taken from [1] with friendly permission of the publisher.

References

1. W. A. Kalender. Computed Tomography. Wiley & Sons, New York (2000)
2. C. H. Collough and F. E. Zink. Performance evaluation of a multi-slice CT system. *Med. Phys.* **26***(11)*: 2223–2230 (1999)

3. W. Rossner and B.C. Grabmaier. Phosphors for xray detectors in computed tomography. *Journal of Luminescence* **48–49**: 29–36 (1991)
4. C. Greskovich and S. Duclos. Ceramic scintillators. *Annu. Rev. Mater. Sci.* **27**: 69–88 (1997)
5. T. O. J. Fuchs, M. Kachelrieß, and W. Kalender. Direct comparison of a Xenon and a solid-state CT detector system: Measurements under working conditions. *IEEE Transactions on Medical Imaging* **19***(9)*: 941–948 (2000)
6. R. Fahrig, S. Fox, S. Lownie, and D. W. Holdsworth. Use of a C-arm system to generate true 3-D computed rotational angiograms. *Am. J. Neuroradiology* **18:** 1507–1514 (1997)
7. W. A. Kalender, K. Engelke, and S. Schaller. Spiral CT: Medical Use and Potential Industrial Applications. In: Developments in xray tomography. Bonse, U. (ed). Proceedings of SPIE **3149**: 188–202 (1997)
8. K. Engelke, M. Karolczak, A. Lutz, U. Seibert, S. Schaller, and W. A. Kalender. Mikro-CT. Technologie und Applikationen zur Erfassung von Knochenarchitektur. *Der Radiologe* **39***(3)*: 203–212 (1999)
9. J. H. Siewerdsen, and D. A. Jaffray. Cone-beam computed tomography with a flat-panel imager: Effects of image lag. *Med. Phys.* **26***(12)*: 2635–2647 (1999)
10. J. H. Siewerdsen and D. A. Jaffray. Optimization of x-ray imaging geometry (with specific application to flat-panel cone-beam computed tomography). *Med. Phys.* **27***(8)*: 1903–1914 (2000)
11. J. H. Siewerdsen and D.A. Jaffray. Cone-beam computed tomography with a flat-panel imager: Magnitude and effects of x-ray scatter. *Med. Phys.* **28***(2)*: 220–231 (2001)
12. D. A. Jaffray and J.H. Siewerdsen. Cone-beam computed tomography with a flat-panel imager: Initial performance characterization. *Med. Phys.* **27***(7)*: 1311–1323 (2000)
13. R. E. Colbeth, S. Boyce, R. Fong, K. Gray, R. Harris, I. D. Job, I. Mollov, B. Nepo, J. Pavkovich, et al. 40 × 30 cm flat panel imager for angiography, R&F, and cone-beam CT applications. *SPIE Medical Imaging*: 4320–4354 (2001)
14. K. Engelke, L. Wachsmuth, U. Taubenreuther, A. Lutz, M. Karolczak, R. Raiss, and W. A. Kalender. High resolution in vitro μCT of osteoarthritis in a mouse model. Presented at High Resolution Imaging in Small Animals Conference, Rockville (Maryland) (2001)
15. M. Kachelriess, S. Schaller, and W. A. Kalender. Advanced single-slice rebinning in cone-beam spiral CT. *Med. Phys.* **27***(4)*: 754–772 (2000)
16. M. Kachelriess, T. O. J. Fuchs, S. Schaller, and W. A. Kalender. Advanced single-slice rebinning for tilted spiral cone-beam CT. *Med. Phys.* **28***(6)*: 1033–1041 (2001)

CHAPTER 5

SPECIAL APPLICATIONS IN RADIOLOGY

Ralf-Hendrik Menk

Sincrotrone Trieste S.C.p.A. di interesse nazionale
Strada Statale 14 - km 163,5 in AREA Science Park
34012 Basovizza, Trieste ITALY

5.1 Introduction

Worldwide synchrotron radiation facilities [1] provide an extraordinary kind of X-ray source, which offers a large variety of application in several fields such as material science, crystallography, micro-spectroscopy, X-ray diffraction and more recently a variety of medical applications as for example are described in [2].

The main feature of these sources is the wide and continuous energy spectrum providing a very high photon flux over an energy range up to some 50 keV or even higher. Moreover, the beam comprises a high natural collimation at least in the vertical direction and a high degree of coherence in both, space and time. In the horizontal direction one can produce a large area, fan-like photon beam shape.

All the unique features mentioned in combination with sophisticated optics make synchrotron sources well suited instruments, e.g. for medical applications. Thus during the past years several synchrotron radiation (SR) facilities have developed dedicated medical beamlines. Those make not only use of the excellent source characteristics but also from fact that the high intensity SR spectrum allows one to select and to tune monochromatic photon beams with a very narrow energy bandwidth. Besides, the laminar geometry reduces scattered radiation and thus avoids image blurring. As a consequence an enhancement of the image quality is observed while the dose is conserved or is even reduced at the same time. This is simply due to the fact that for each application the right energy window can be chosen. Moreover, the monochromaticity can be utilized to implement multiple energy techniques.

A dedicated imaging system is required for each specific medical application and various approaches have been pursued with the development of different X-ray digital detectors with appropriate features like the geometry, the X-ray detection mode and the electronics. In general a high detective quantum efficiency (*DQE*) is the essential parameter for medical radiology (see appendix). During the last two decades several systems for medical imaging based on synchrotron radiation have been developed. Of note here are synchrotron-based full field digital mammography, transvenous coronary angiography, bronchography and multi energy tomography. Moreover, new imaging modalities have been recently developed aiming a further improvement in the image quality based on the detection of phase shift information, important effect especially for low absorbing biological materials.

5.2 Special Applications

5.2.1 Mammography

Mammography is a screening technique with the highest sensitivity for detecting early breast cancer, which statistically lead to a significant improvement in breast cancer survival [3]. The demands on the image performance in mammography, in terms of contrast and resolution are beyond those of any other medical imaging modality. Subsequently the requirements on both, the X-ray source and the detector are very high. Of note are here the detection of low contrast details and small size details such as low contrast masses and clusters of micro-calcifications, which are possible indicators of early breast cancer. At the same time special care has to be taken for an efficient use of the radiation dose to the patient: since the breast is one of the most radiosensitive organs the risk of cancer induced by X-ray exposure has to be minimized. The visibility, e.g. of micro-calcifications, which are embedded in a tissue background, can be described by the signal to noise ratio ($SNR_{\Delta S}$). Taking into account the Poisson statistics of the photons the $SNR_{\Delta S}$ it is defined as:

$$SNR_{\Delta S} = \frac{N_1 - N_2}{\sqrt{N_1}} \tag{5.1}$$

where N_1 is the average fluence of photons on the image in the background region and N_2 is the average fluence in the area of the detail. The SNR is higher at lower energies because the difference of the attenuation coefficients decreases with the energy. On the other hand lower energy means higher absorbed dose in the organ. A parameter which describe the image quality versus the energy is the Figure of Merit (*FoM*) which is defined as the signal-to-noise ratio normalized to the mean glandular dose (MDG):

$$FoM = \frac{SNR_{\Delta S}}{\sqrt{MGD}} \tag{5.2}$$

The *FoM* curve shows (Fig. 5.1) a maximum corresponding to a certain energy and consequently at that energy the image quality is maximized for the same radiation dose. Since the *FoM* depends on the tissue thickness and the tissue composition, a selectable monochromatic energy should be chosen according to the breast thickness and variation in the composition between more transparent adipose and denser glandular tissue. The example in Fig. 5.1 shows how the maximum moves to higher energies as the thickness of the sample increases. Utilizing now monoenergetic X-rays generated by a synchrotron radiation beam line it is possible to tune the photon energy always on the maximum of the *FoM*.

Since the first mammographic units became available in the 1970s several technical improvements have been reported on the X-ray beam quality, adequate breast compression and automatic exposure control. Unfortunately, there are still several limitations and approximately 10% of clinical obvious breast cancers are not visible with mammography [3].

Even today screen film systems are dominating as imaging detectors in mammographic units. Even though film screen systems feature excellent spatial resolution they have several inherent limitations [4]. One of the main drawbacks of film systems is their non-linear response. The contrast properties of a radiographic film are described by its characteristic curve, which has typically a sigmoidal shape (Fig. 5.1). That means that the range of the X-ray exposure with linear correspondence with the optical density is limited to a factor of about 25.

The noise in this system is due to the structure of the fluorescent screen and, mainly, to the granularity of the film emulsion that prevents the system to be "quantum limited" decreasing the contrast resolution.

An important challenge in film screen mammography is the capability to diagnose early stages of breast cancer in radiological dense breast (or breast types that contain a large amount of fibroglandular tissue). Subtle attenuation variation in breast tissue requires a high contrast sensitivity that means a steep slope in the characteristic curve of the film. On the other hand a trade-off between the contrast sensitivity and the film dynamic range (latitude) has to be considered.

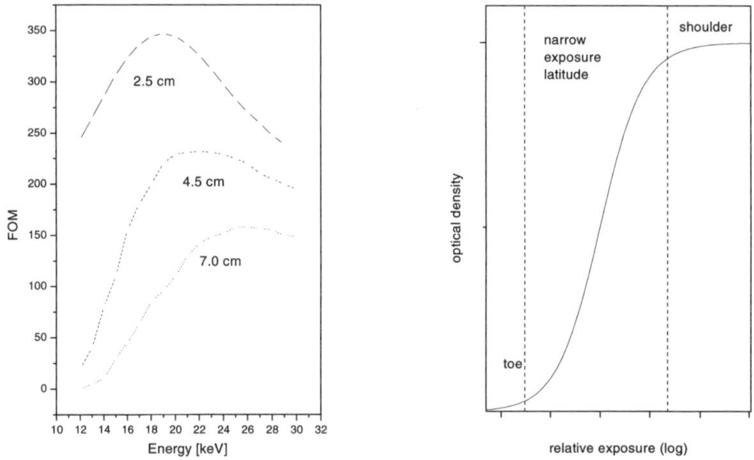

Fig. 5.1. (Left): Figure of merit (*FoM*) versus photon energy. The *FoM* is shown for three different tissue thickness (2.5 cm, 4.5 cm and 7.0 cm, respectively). (Right): Typical exposure curve of a film screen system.

High contrast sensitivity in not suitable for a large range of transmitted exposure due to the presence of dense (radiographically opaque) and adipose (transparent) part in breast composition. As a consequence dense parts can be underexposed or transparent parts can be overexposed corresponding to the toe and to the shoulder of the characteristic curve. Another limitation of film screen mammography is the compromise between spatial resolution and X-ray interaction efficiency.

To perform a radiograph at a reasonable dose the film is used in contact with a fluorescent screen that converts the X-rays in visible light. The absorption efficiency depends on the phosphor thickness, but as the thickness increases also the screen blur increases reducing the spatial resolution. This is due to the diffusion of light from the point of absorption of the X-ray to the film. In mammography high spatial resolution (18–20 LP/mm (Line Pairs per mm)) is essential thus the screen must be relatively thin with a consequent reduction of the detection efficiency.

The *DQE* of screen-film system reaches a maximum of 0.25–0.30 (Fig. 5.2) at low spatial frequencies and drops to less than 1% at higher frequencies [5]. The loss of *DQE* is attributable primarily to film granularity noise.

For a diagnostic interpretation, e.g. of micro-calcifications, a signal to noise ratio of at least 3 is required. Utilizing a detection system with matched pixel size and monoenergetic photons the skin dose D_{skin} is given according to [6] by

$$D_{\text{skin}} = \frac{2e^{\bar{\mu}} L\, SNR_{\Delta S}^2}{DQE(0)\bar{\mu}^2 w^4 C_\mu^2} E_\gamma \left(\frac{\bar{\mu}}{\rho}\right) \tag{5.3}$$

Here L is the length of tissue traversed by the beam, $\bar{\mu}$ and ρ are the absorption coefficient and the density of tissue, respectively. $C\mu=\Delta\mu/\bar{\mu}$ is the object contrast, i.e. the relative change of the absorption coefficient in the object with respect to its neighborhood. Note that D_{skin} increases with the inverse fourth power for a cubic object of the size w and increases with the square of the signal to noise ratio $SNR_{\Delta S}$ required. If one consider a micro-calcification of w = 0.025 cm embedded in breast tissue of L = 4.5 cm (length of compressed breast) and $SNR_{\Delta S}$ =5 then the following values for contrast, count rate in a 75·75 μm^2 detector pixel and skin dose can be found (Table 5.1).

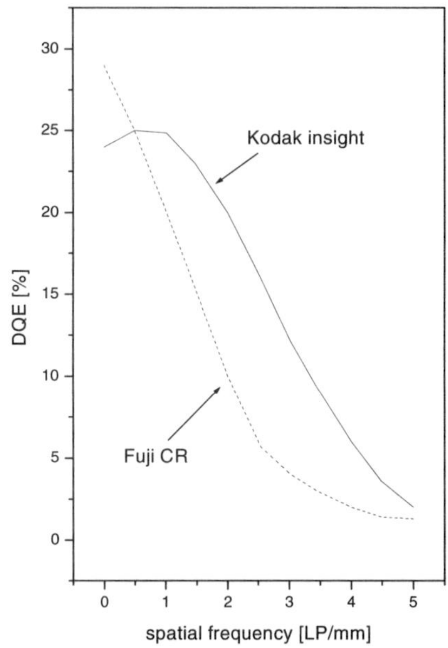

Fig. 5.2. DQE curves as function of the spatial frequencies for a Kodak (Kodak Insight) and a Fuji (Fuji CR) film screen system [7].

Table 5.1. Contrast as function of the (monochromatic) photon energy and the associated count rate per pixel and the skin dose per *DQE(0)* of the imaging detector.

Energy [keV]	C_μ [%]	Fluence [photons/pixel]	D_{skin} [J/kg]
20	5.9	1600/DQE(0)	0.00045/DQE(0)
30	1.9	9070/DQE(0)	0.0004/DQE(0)

5.2.2. Digital Mammography with Synchrotron Radiation

To overcome the present limitations in mammography both the detection device and the source have to be optimized.

In order to improve the source quality SR-based mammography experiments have been carried out. Comparing the energy spectrum of SR source and of a mammographic X-ray tube it turns out that the SR intensity is orders of magnitude higher than that of a clinical unit. A monochromator can be utilized in order to obtain a beam with high flux and with a narrow energy bandwidth. The appropriate energy can be tuned according to the density and thickness of the breast to the best *FoM* value. In addition the highly vertical collimated SR geometry inherently reduces the scattering in the image thus avoiding the use of anti-scatter grids. Approximately a decade ago first pioneering work was done at ADONE (Frascati, Italy) using a channel-cut Si(111) crystal as monochromator and a conventional screen-film system [8].

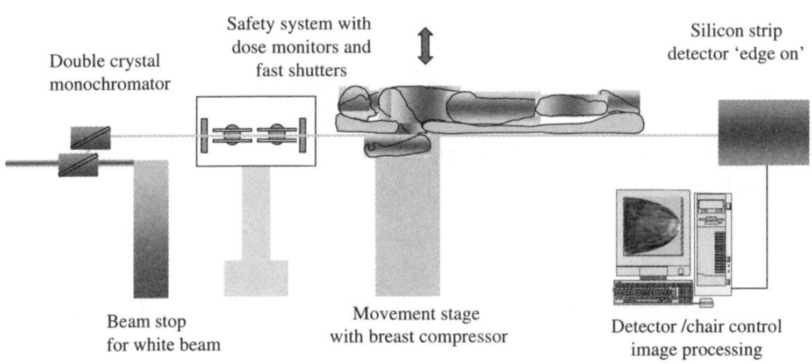

Fig. 5.3. Principal set up for synchrotron based mammography with SR and digital detector in line scan geometry.

The conclusion of these experiments was that the images of test object as well as breast tissue specimens taken with the monochromatic SR have higher contrast resolution at similar or less radiation dose. Over the past years other preliminary investigations have been carried out by the medical research group at NSLS (Brookhaven, USA) with screen-film and imaging plate detectors [9] and at SRS facility (Daresbury, UK) with screen-film system [10] using in both cases flat Si(111) crystal monochromators.

Utilizing digital detectors for mammography several limitations can be overcome [11, 12]. It should be pointed out that in digital mammography the image acquisition, display and storage are performed independently allowing optimization of each procedure. The response of detector is linear over a much wider dynamic range (12 bits or more), therefore more ideal for radiological dense breast types with improved lesion detectability.

To fulfill the severe requirements of mammography the digital detectors must feature low intrinsic noise and a very high detection efficiency in order to keep the dose to minimum, hence, a high DQE system is necessary. The spatial resolution should be also very high and reasonable pixel size is supposed to be about 100 – 50 μm^2. Moreover, a digital image presents all the advantages of the computer manipulations to improve the detection of subtle and small suspicious lesions with off-line contrast enhancement procedures.

In order to fully validate and exploit the method of digital mammography with synchrotron radiation a dedicated beam-line called SYRMEP was designed and built at the SR facility ELETTRA, Trieste (Italy) and it is under operation since 1996 [13, 14]. A Si (111) double crystal monochromator in Bragg mode provides a 150 mm wide laminar fan beam at the experimental hall (Fig. 5.3). The energy of the beam can be tuned in the energy range from 10 to 35 keV. As depicted in Fig. 5.3 two-dimensional radiographs are obtained by a vertical scan of the sample/patient through the laminar beam.

From the *FoM* curve in Fig. 5.1 it turns out that for a tissue thickness of 4.5 cm the highest value can be expected at energies of about 20 keV. Such a tissue thickness is obtained by the compression of the breast. Since this is a painful procedure higher compressions are normally not suitable. Hence the advised imaging energy for synchrotron radiation based mammography is around 20 keV. According to Table 5.1 for this energy the fluence per detector pixel ($75 \cdot 75 \mu m^2$) of 1600 [photons/pixel] for a detector featuring a $DQE(f=0)=1$ and 3200 for a $DQE(f=0)=0.5$ is required in order to resolve a microcalcification of 0.25 mm with a $SNR_{\Delta S}=5$. These values seem to be suitable for single photon counting. Thus a counting detector should be a better solution than an integrating one if the dead time can be kept in the order of some hundred ns (see figure in the appendix). In the scanning approach, however, not the fluence but the fluence rate [photons/pixel /s] is crucial and thus the requirements on the detector become fairly challenging as shown in the following. Taking into account, that the patient can withstand the compression some ten seconds and that longer scan time induce image blurring due to the patient movement (like breathing) an entire image has to be taken within 10 s. If the field of view is 7.5 cm in the vertical direction and the spatial resolution is 75 μm then the fluence rate is $1.6 \cdot 10^6$ photons/pixel/s for a detector with $DQE(f=0)=1$ and for $SNR_{\Delta S}=5$. If $SNR_{\Delta S}=3$ is sufficient then the fluence rate is around 600 kHz. In both cases a counting device can be employed for this application however, a commercial device featuring these rates is not available. Thus included in the SYRMEP project study is the development of a novel digital detector based on a linear silicon array of pixels that is able to digest these fluence rates. The prototype detector consists of a micro-strip silicon detector which is irradiated from the side in an "edge on" geometry [15]. In this configuration it acts as a linear pixel detector with a horizontal pixel size of 200 μm defined by the strip pitch and a vertical pixel size of 300 μm due to the wafer thickness. Hence the absorption efficiency is of about 80% – 90% in the energy range of interest since the beam impinges parallel to the longitudinal dimension (10 mm) of the strips which becomes the effective detector thickness. If dead time effects are negligible the $DQE(f=0)$ is equal to the absorption efficiency. Higher efficiency values cannot be obtained for this kind of detector because of the presence of an insensitive volume. This is due to the fact that the strips are implanted 250 μm from the edge [16]. Thus, photons absorbed in this region are not detected.

The electronics is realized in a custom made VLSI and works in single photon counting mode [17]. Each channel features an analogue part, which consists of a charge sensitive preamplifier, a shaping amplifier, an analogue buffer, a threshold discriminator, and a 16 bit pulse counter. The read-out of all counters in the circuit is done in serial up to a rate of 20 MHz. The counting rate of the first prototype electronics using periodic signals is close to 100 kHz per pixel which is in fact one tenth of the required flux. At present a counting electronics up to 1 MHz per channel is under production.

Joining the SR features with the digital mammography approach promising experimental results have been obtained with both, phantoms and in vitro full breast tissue samples. It turned out that for the images taken with the digital detector the applied dose to the sample was smaller or comparable to that of the clinical unit. However in terms of contrast resolution the digital ones are superior (Fig. 5.4)

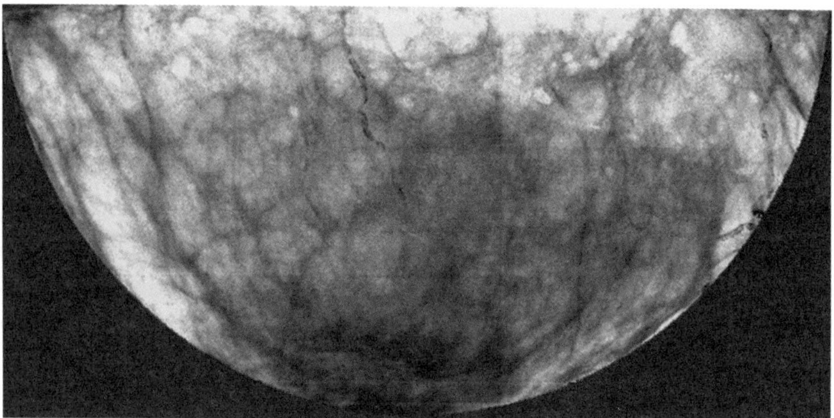

Fig. 5.4. Full breast tissue recorded at the Syrmep beam line at the Sincrotrone Trieste, Italy with the "edge on" silicon strip detector [18].

5.2.3. Subtraction Techniques at the k-Edge of Contrast Agents

Applying formulae (5.3) it can be shown that dose-contrast considerations prevent direct imaging of low absorbing details such as the arterial system in the heart or the bronchi. Diagnostically both are from great interest since heart/lung diseases are the leading causes of dead in most regions of the industrialized world [19].

In order to increase the image contrast, for instance of the coronary arteries in clinical routine iodine based contrast agents are utilized, which are power injected for example into the lumen of each of the coronary arteries. For this procedure it is necessary to introduce a catheter into a peripheral artery and threading it into the ostium of one of the two coronary arteries. During and after the power injection transmission radiographs of the upper torso are recorded either in a static fashion or in a time sequence (so-called cines). In this fashion the coronary arteries are imaged with excellent signal to noise ratio. This imaging modality represents the most reliable evidence that can be obtained in vivo about the status of the coronary artery circulation [20]. However, hazards associated with the catheter placement are numerous spanning from arterial perforation to embolic obstructions in various distal arteries including those of heart and brain.

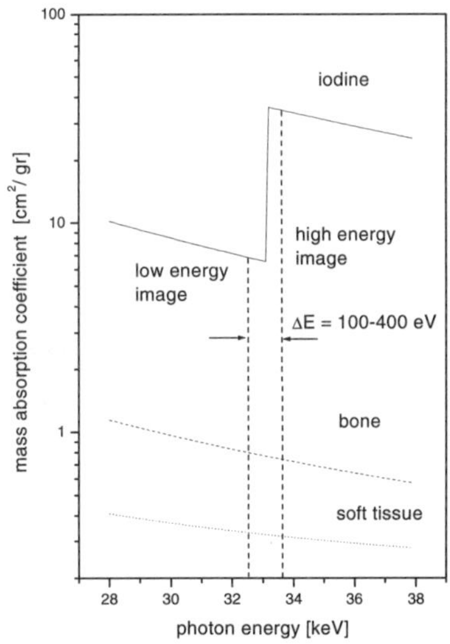

Fig. 5.5. Mass absorption coefficients for iodine, bone and soft tissue versus energy. Depicted are also the positions of the high and low energy band, which are used for dichromography.

Already in 1953 [21] efforts were reported to avoid the arterial catheter and image the coronary arteries after an intravenous injection of the contrast agent. Since in this modality the iodine bolus is substantially diluted before it arrives in the coronary arteries (factor 20 or more) the iodine contrast is in the order of 1% and is hooded by higher absorbing structures like bones and muscles. The idea of Jacobson was to use dichromatic absorption radiography that is to say two images are recorded one after the other with two different energies bracketing somehow the iodine absorption edge (Fig. 5.5). The image which is taken with energy beyond the *k*-edge of the iodine (33.18 keV) — in the following called "high image" comprises absorption effects from iodine, bones, fat and muscles while the image taken at lower energy ("low image") has negligible iodine contrast and serves as a mask. In order to increase the iodine contrast one simply has to perform a logarithm subtraction of the high image and the low image. In the image attained in this fashion only iodine should contribute to the signal.

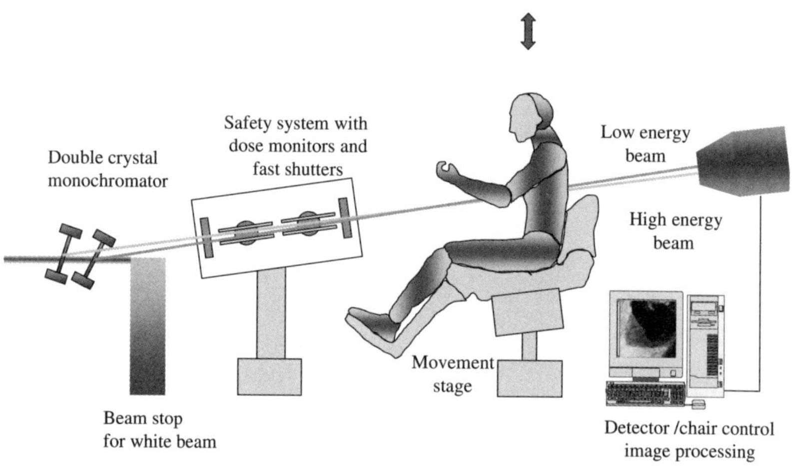

Fig. 5.6. Sketch of the angiography beam line at HASYLAB at DESY.

Even though the idea was advanced that time this modality actually was never really utilized since the sources available at that time would not deliver sufficient flux. Neither appropriate detection systems were available. Moreover, high and low images were taken in a time sequence that is to say one after the other. Since the heart is a rapidly moving organ and the patient could move in the mean time the subtraction never really worked out and moving artifacts destroyed the image contrast. This problem could be overcome thirty years later and was due to the availability of synchrotron sources [19] and more advanced detection systems. Utilizing perfect crystals it is possible to generate two monochromatic X-ray fan beams out of the polychromatic synchrotron radiation whose energy closely bracket the iodine absorption k- edge (100 eV – 400 eV). Both beams are simultaneously available and are spatially separated at the source and at the detector but intercept in the patient (Fig. 5.6).

By translating the patient in vertical direction through those beams two images are taken simultaneously by a dual line digital detector. In order to avoid movement artifact of the heart the scan should be completed within 250 ms.

Under the (simplified) assumption that the patient consists only of water and iodine (see appendix) one can calculated an iodine image and a "water image" by a simple matrix inversion. If S denote the logarithmic and normalized images taken at the two energies, ρ, μ are the densities and absorption coefficients for water (tissue) and iodine and x describes the thickness of the specific material, then one obtains an equation system of two unknown. These are the water image ($\rho_{water}\, x_{water}$) and the iodine image ($\rho_{iodine}\, x_{iodine}$) which can be obtained by a matrix inversion on a pixel basis as shown in the appendix of the following matrix:

$$-\begin{pmatrix} S_{low} \\ S_{high} \end{pmatrix} = \begin{pmatrix} \dfrac{\mu_{iodine}^{low}}{\rho_{iodine}} & \dfrac{\mu_{water}^{low}}{\rho_{water}} \\ \dfrac{\mu_{iodine}^{high}}{\rho_{iodine}} & \dfrac{\mu_{water}^{high}}{\rho_{water}} \end{pmatrix} \cdot \begin{pmatrix} \rho_{iodine} & x_{iodine} \\ \rho_{water} & x_{water} \end{pmatrix} \qquad (5.4)$$

Here the subscripts low and high, respectively, refer to the energy below and above the iodine *k*-edge. Depicted in Fig. 5.7 are the two simultaneous taken energy images and the resulting water and iodine image, demonstrating how the right coronary artery (RCA) emerges from the background by the subtraction method. Visible in the proximate part of the RCA is a stenosis indicated by the arrow RCA, which was treated by a bypass (CABG-RCA).A metal loop indicates the attachment of the bypass to the aorta.

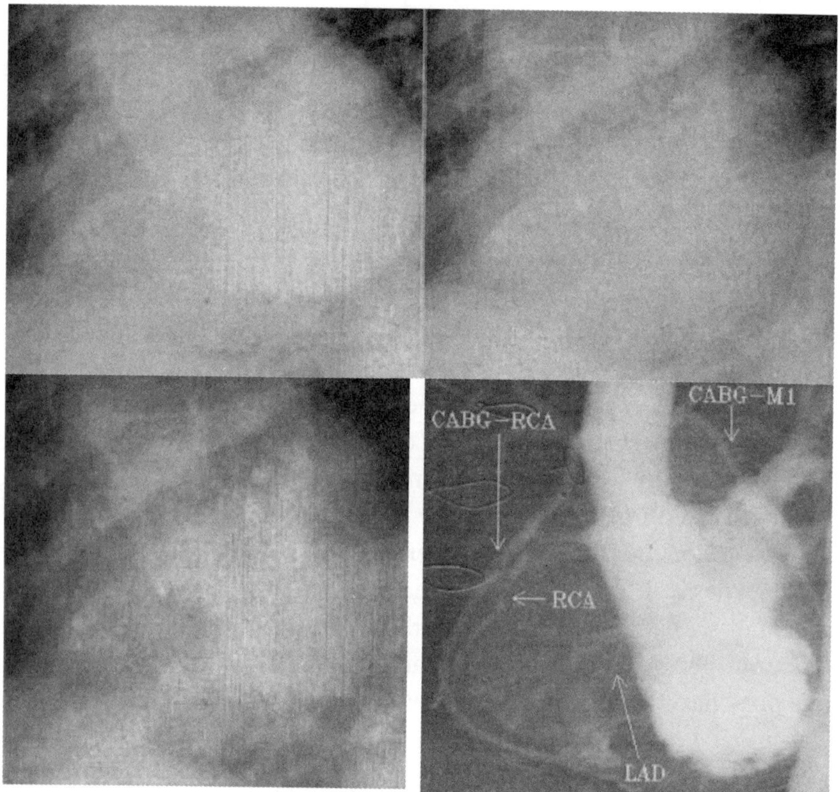

Fig. 5.7. Upper row: Energy image above (left) and below (right) the iodine *k*-edge recorded with the NIKOS 4 system at DESY, Germany. The imaging detector is a highly segmented 2-line ionization chamber with 19 bit dynamic range. Depicted also are the associated water (lower row left) and iodine images (lower row right) obtained by the inversion of the matrix of the mass absorption coefficients.

The skin dose per iodine image is about 40 mSv and the total dose for three scans in two different projections including some low dose positioning scans amounts approximately 300 mSv, which is somewhat smaller than the dose of a standard invasive coronary angiography. It should be noted that the RCA but also the left coronary artery (LCA) are relatively easy to image since both can be projected without being obscured by other (bigger) iodine filled structures such as the ventricle, the aorta or the atrium. Typically the fluence in the vicinity of these vessels is at least 10^9 photons/s/mm^2. Hence sufficient flux is available and according to the graph in the appendix the *DQE* even for a "noisy" integrating detector can be reasonable high. Counting detectors, however, featuring such a count rate and subsequently an accordingly short deadtime – that is to say high *DQE* values- are difficult to realize and very expensive.

Therefore integrating detectors are more suitable for imaging free projected arteries in this special application Obviously the described imaging modality can be easily modified for imaging for instance pulmonary structures. Instead of iodine, here a xenon oxygen gas mixture (80% and 20%) is used which is inhaled while the image sequence is taken.

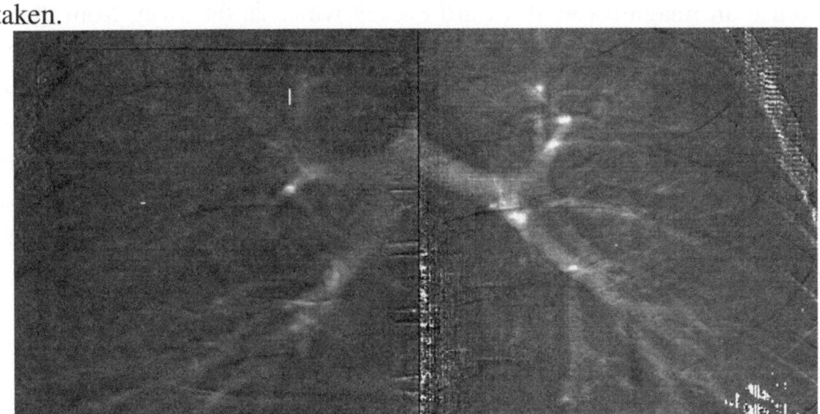

Fig. 5.8. Dichromography bronchograms in anteroposterior projection of the right and left lung, respectively with a xenon concentration at the level of the first bronchial bifurcation of 75%. Third and fourth order bronchial branches are visualized. The study was approved by the institutional committees on human research at both the Stanford University medical center and the Brookhaven National Laboratory and was carried out at the medical beam line at the NSLS/BNL.

Since the k – edge of Xenon is with 34.56 keV slightly higher than that of iodine the values in the mass absorption coefficient matrix in (5.4) have to be adjusted accordingly.

Dichromography can provide clinically significant diagnosis information and appears to be well suited for the detection of intraluminal lesions or of parenchymal lesions that distort the adjacent airways [22].

Initial results on synchrotron radiation based bronchography in humans were obtained at the medical beam line at the National Synchrotron Light (NSLS) at Brookhaven National Laboratory (BNL) [22]. The spatial resolution in the images acquired with this modality is at least three times higher than of those recorded with novel MRI approaches based on ^{129}Xe or ^{3}He, [23-25]. The approximation done in (5.4) that the patient consists of water and iodine/xenon only is clearly a rough approach. And in fact the formula produces artifact in those parts of the image, where the mass absorption coefficient is substantially different from that of water, for instance in the bones, such as the rips or the spine (Fig. 5.7 or Fig. 5.8).

Besides an iterative software approach to overcome this problem one can also use an image taken at a third energy which is far away from those energies that are bracketing the k-edge of the contrast agent. For instance one can use the third harmonics of the wiggler spectrum which is also transmitted by the monochromator system. Normally the contamination of the third harmonics is in the order of a fraction of a per cent. But blocking the ground wave utilizing an appropriate filter one can obtain a pure third harmonic beam. At this point formulae 5.4 has to be expended to the third energy and one gets

$$-\begin{pmatrix} S_1 \\ S_2 \\ S_3 \end{pmatrix} = \begin{pmatrix} \frac{\mu^1_{\text{iodine}}}{\rho_{\text{iodine}}} & \frac{\mu^1_{\text{water}}}{\rho_{\text{water}}} & \frac{\mu^1_{\text{bone}}}{\rho_{\text{bone}}} \\ \frac{\mu^2_{\text{iodine}}}{\rho_{\text{iodine}}} & \frac{\mu^2_{\text{water}}}{\rho_{\text{water}}} & \frac{\mu^2_{\text{bone}}}{\rho_{\text{bone}}} \\ \frac{\mu^3_{\text{iodine}}}{\rho_{\text{iodine}}} & \frac{\mu^3_{\text{water}}}{\rho_{\text{water}}} & \frac{\mu^3_{\text{bone}}}{\rho_{\text{bone}}} \end{pmatrix} \begin{pmatrix} \rho_{\text{iodine}} \cdot t_{\text{iodine}} \\ \rho_{\text{water}} \cdot t_{\text{water}} \\ \rho_{\text{bone}} \cdot t_{\text{bone}} \end{pmatrix} \quad (5.5)$$

Once three images S_i $(i = 1,2,3)$ are measured at the three different energies the inversion of the mass absorption matrix leads to the density images of water, iodine and bone. Some feasibility studies on three-energy or multi energy tomography (MECT) were carried out at the NSLS/BNL [26, 27] to verify this approach (Fig. 5.9). It turned out that this method allows a direct segmentation of tomographic reconstruction, a common problem in diagnostic computer tomography. It should be noted, that the benefit of image processing in noise limited images should not be overestimated. Any improvement in intensity resolution is at the expense of the spatial resolution and vice versa. Clearly, the subtraction of large structures can help for the interpretation, as does the *k*-edge subtraction method in spite of the factor of $2^{1/2}$ loss in the intensity resolution for constant skin dose [28].

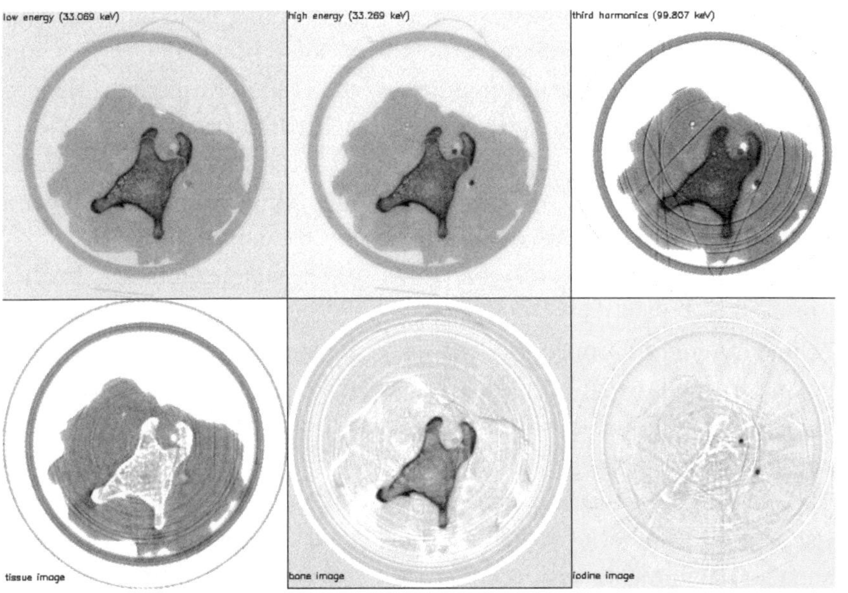

Fig. 5.9. Feasibility study of three-energy tomography on a cow tail sample. Shown are the three energy images recorded at the NSLS/BNL, USA at three energies (33.069 keV, 33.269 keV and 99.807 keV) and the associated water, bone and iodine image (two 1 mm tubes filled with 37 mg/ml iodine). The imaging detector was a Li drifted Si detector operated in integrating mode.

5.2.3.1 *Detectors and detector requirements for dichromography*

It becomes obvious from the discussion above and in the appendix, that the requirements on the detector for transvenous coronary angiography are exceptional high. To summarize: A signal of a coronary artery of a diameter of 1 mm filled with 20 mg/ml iodine lies in the order of 1% or less. This weak signal is superimposed by a strong varying background with variations of at least three orders of magnitude and should be imaged in at least 250 ms (or less than 1 ms per line) in order to avoid motion artifacts. The field of view is $10 \cdot 10$ cm^2 and the required spatial resolution is around 1 LP/mm. Moreover, dose considerations suggest a *DQE* close to one.

Of note is that the left circumflex branch (LCX) is always partially obscured by the left ventricle. The fluence rate here can drop down to 3.2 10^5 photons / mm^2/s (see appendix). On the other hand the system has to digest the maximum photon fluence of some hundred GHz per mm^2 which is the typical fluence rate behind the lungs of a patient. Hence it turns out that the noise equivalent of the detector system has to be 10^5 photons per mm^2 and 1 s integration time while the maximum photon fluence is around 10^{11} photons/mm^2/s. Therefore the required dynamic defined as the maximum signal divided by the noise is 10^6. Feasibility studies on single photon counting devices and pulse integration detectors [29] showed that the $DQE(f=0)$ in high fluence regions such as behind the lungs dropped tremendously due to dead time effects. A high-count rate photo multiplier tube system was developed in Novosibirsk [30]. For this device a maximum count rate up the 6 MHz per cell was reported.

One possibility to increase the count rate of a counting device is to increase the number of segments in the beam direction however, making such a system technically very difficult and expensive. Therefore almost all systems developed so far for transvenous coronary angiography in human studies are integrating devices using electronics with a dynamic range of at least 16 bits.

As a representative example of detectors used in this specific application a low noise ionization detector developed by the University of Siegen, Germany in collaboration with HASYLAB, DESY, Germany will be briefly discussed in the following. More details can be found in [31, 32]. This detector consists of two segmented anodes realized on printed circuit boards and a shared drift cathode defining the two detector lines. For the discussed application a strip pitch of 0.4 mm was chosen, since a spatial resolution of 1 LP/mm is sufficient to image an artery of a diameter of 1 mm. In order to cover a field of view of 10 cm in the horizontal direction each line comprises 336 anode segments. In the direction of the beams the strips measure 56 mm. As indicated in Fig. 5.10 each line comprises also a grid made of gold coated tungsten wires mounted at a distance of three mm above and below the drift cathode. All these components are accommodated in a pressure vessel capable of holding noble gases such as Xe or Kr under medium high pressure up to 15 bar. In the discussed energy range the X-ray photons are mainly absorbed via the photoelectric effect releasing a certain amount of electrons and ions as discussed in the appendix. The formers are collected on the drift cathode, while the latter form the signal collected on the anode strips, once they passed the grids. These are almost transparent for electrons, but provide a sufficient electrostatic shielding for the ions. Without the grids the ions would induce slow signal changes on the strips due to their small drift velocities and thus produce cross talk between adjacent image lines.

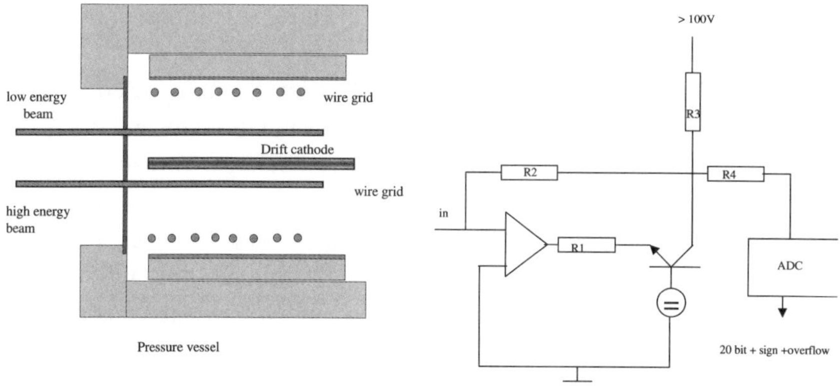

Fig. 5.10. (Left): lateral view of the multi-channel ionization chamber. (Right): read out electronics for one channel (Patent No. 19758363.6)

Table 5.2. List of groups that are working on human studies in transvenous coronary angiography

Facility	Detector	Monochromator	Operation
NSLS Brookhaven (USA)	Dual line Si(Li), integrating mode, 5 mm thick, 250 μm pixel, 4 ms integration time /line, 16 bit dynamic $DQE_{max}(f=0) = 0.5$	1 bent Laue Si(111)	Started in 1990 Human studies 30 patients
Hasylab Hamburg (Germany)	Dual line ionization chamber, integrating mode, 400μm pixel, integration time 0.8 ms per line 19 bit dynamic, $DQE_{max}(f=0) = 0.8$	2 bent Laue Si(111)	Started in 1981 Dedicated beamline for a large program on human studies since 1997, 400 patients
ESRF Grenoble (France	HPGe dual line integrating mode, 2 mm thick, 350 μm pixel, integration time 0.8 ms per line, 16 bit dynamic, $DQE_{max}(f=0) = 0.8$	1 bent Laue Si(111)	Started in 1998 first animals. Human studies beginning of 2000, 6 patients

As a consequence a decrease in the spatial resolution in the vertical direction would be unavoidable.

Each of the anode strips is connected via a gas tight connection to an electronics channel that comprises a current to current converter and a 20 bit ADC[a] (Fig. 5.10) outside the pressure vessel. All 672 channels can be read out via a fast fiber link within 0.2ms. With a noise equivalent of 3.1 photons / pixel and integration time a dynamic of almost 19 bit was achieved.

A detailed comparison of two detection systems used in this application (NIKOS at the HASYLAB at DESY, Germany and the image detector at the NSLS at BNL, USA)) has been performed using the methods described in the appendix [33]. Worldwide several groups are working on transvenous coronary angiography with humans. Those are summarized in Table 5.2. A complete summary of the groups' approaches in using synchrotron for intravenous coronary angiography with differences in X-ray optics and the types of digital detectors can be found in [18].

[a] Type DDC 101 from Burr Brown

5.2.4. Phase Effects

As seen in the previous chapters these special applications rely not only on the tunability of the SR but also on the natural collimation and hence on the natural scatter rejection. A careful analyzes of the images obtained with SR however, shows that even here a substantial scatter contribution mainly due to refraction and/or scattering in the μrad range contributes to the images.

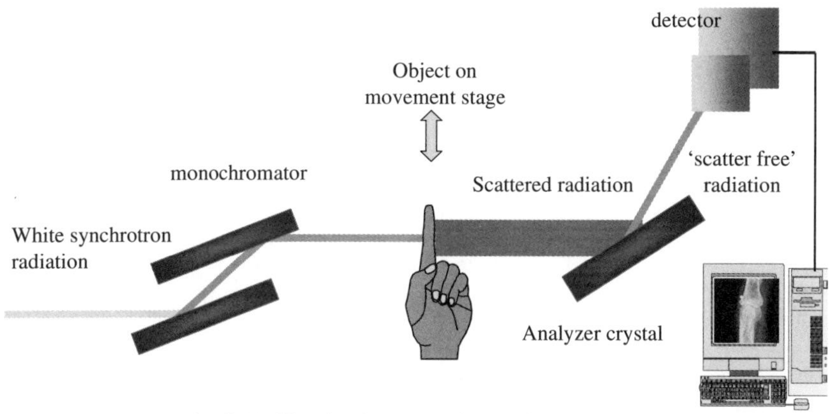

Fig. 5.11. Sketch of a setup with analyzer crystal.

The scatter angle can be expressed as the gradient of the phase shift given in the lowest order by

$$\varphi(x, y) = -r_e\lambda \int \rho_e(x, y, z)dz \tag{5.6}$$

where r_e is the classical electron radius, λ the wavelength of the electromagnetic wave (X-ray) traveling in z-direction through an object with the electron density ρ_e .

If the phase shift depends on one of the transverse variables, for instance x, the wave is refracted by an angle

$$\Delta\alpha \approx \frac{\lambda}{2\pi}\frac{\partial\varphi(x)}{\partial x}. \tag{5.7}$$

Note, that the phase shift and subsequently the scattering angle is proportional to ρ_e and hence to the nuclear charge Z. This dependency is similar to that of the Compton cross-section. The latter, however, is well known to produce little contrast in images when compared to absorption imaging providing much higher contrast since here the cross section is proportional to Z^5.

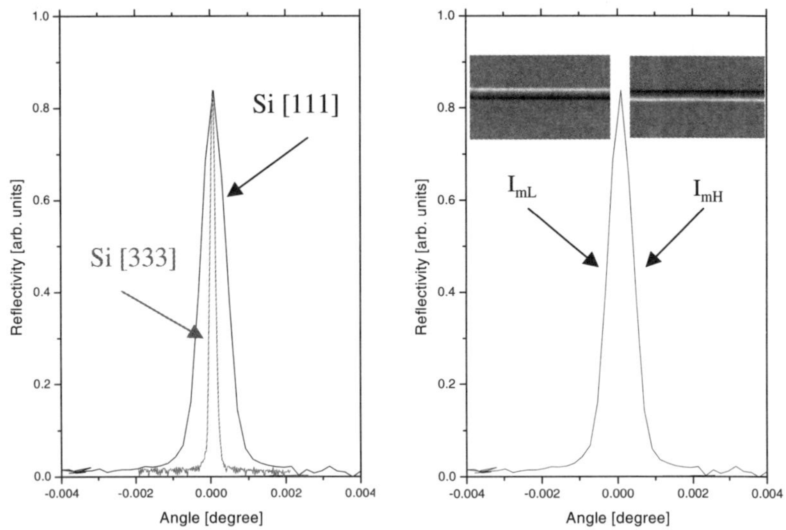

Fig. 5.12. (Left): comparison between the reflectivity curves of the Si [111] and the Si[333] reflection. (Right): Radiographs obtained of a 700 μm Nylon wire with a detuned analyzer crystal. Arrows indicate the angular positions on which the images were recorded.

But the expectation of faint images in phase imaging is not justified for three reasons:

- both Compton and photo effect cross-sections are proportional to r^2_e instead of the much larger factor $r_e\lambda$ appearing in Eq. (5.6),
- the phase shift at a given energy is only weakly coupled to absorption through the dispersion relations, and
- the interference with the primary wave gives a strong enhancement [28].

Both the phase shift as well as the scattering angle can be exploited as sources of radiographic contrast if the X-ray source is sufficiently coherent such as synchrotron radiation sources. The former is used in the so-called phase contrast imaging modality, where the detector is simply placed at a substantial distance downstream of the object. Here the image contrast arises by interference effects between the undisturbed wave and the refracted wave, which occurs on the boundary of objects with different refractive index [34-36]. In order to appreciate these interference fringes detectors with high spatial resolution (such as high-resolution films direct coupled CCDs) are necessary.

Even if interference is not visible the scattering in angular range mentioned before can be exploited as shown first by Foerster [37], Produrets [38], Somenkov [39] and Beliavsky [40]. By means of an analyzer crystal placed between the object and the detection device (Fig. 5.11) it is possible to convert the angular distribution of the refracted beams into intensity changes onto the detector. This is due to the fact that a perfect crystal has a very narrow angular acceptance (around μ radian) for X-rays of a certain energy/wavelength. For monoenergetic radiation, only the beams reflected are those, whose angles with respect to the surface of the crystal are within a certain range around the Bragg angle. The reflectivity as function of the Bragg angle — also known as rocking curve — is shown in Fig. 5.12 for a Si [111] and Si [333] crystal and X-ray of energy of 25 keV. If the analyzer crystal is perfectly aligned with respect to the direction of the incident wave (and its energy) then only those rays are reflected on the detector and subsequently form the image there that are within the FWHM of the rocking curve. Beams refracted under higher angles however are rejected. In this configuration the analyzer crystal serves as a perfect slit with almost no contributions from refracted X-rays.

Beyond simple scatter rejection a crystal analyzer reveals another more sophisticated feature when detuned (e.g. moved to the slope of the rocking curve). In this fashion well-defined amounts of the scattered/refracted radiation contribute to the measured image I_m recorded by the detector. I_m can be now considered as a mixture of two components only — namely a pure absorption (I_R) and a pure refracted part ($\Delta\Theta_z$). Initially the weight of both components is unknown. If, however, two images I_{mL} and I_{mH} are recorded — one with the analyzer crystal detuned to smaller angles (Θ_L) and the other with the analyzer detuned to higher angles (Θ_H) as indicated in Fig. 5.12 — then the

equation with two unknowns can be solved by a simple matrix inversion on a pixel basis as follows: At first the images are recorded at the steepest slope of the rocking curve *R(Θ)*.
Here *R(Θ)* is almost linear and can be expanded in a first order approximation in a Taylor expansion and one obtains

$$\begin{aligned} I_{\mathrm{mL}} &= I_{\mathrm{R}}\left(R(\Theta_{\mathrm{L}}) + \frac{\partial R}{\partial \Theta}(\Theta_{\mathrm{L}})\Delta\Theta_{\mathrm{z}} \right) \\ I_{\mathrm{mH}} &= I_{\mathrm{R}}\left(R(\Theta_{\mathrm{H}}) + \frac{\partial R}{\partial \Theta}(\Theta_{\mathrm{H}})\Delta\Theta_{\mathrm{z}} \right) \end{aligned} \tag{5.8}$$

where $R(\Theta_L)$, $R(\Theta_H)$ are the values of the rocking curve at the angle Θ_L and Θ_H , respectively and $\partial R/\partial\Theta\ (\Theta_L)$ and $\partial R/\partial\Theta\ (\Theta_H)$ are the values of the derivation at those angles. Since the rocking curve and its derivation are known (5.8) can be solved for the unknown I_R and $\Delta\Theta_z$. One gets

$$\begin{aligned} I_{\mathrm{R}} &= \frac{I_{\mathrm{mL}} \left.\frac{dR}{d\Theta}\right|_{\Theta_H} - I_{\mathrm{mH}} \left.\frac{dR}{d\Theta}\right|_{\Theta_L}}{R(\Theta_{\mathrm{L}}) \left.\frac{dR}{d\Theta}\right|_{\Theta_H} - R(\Theta_{\mathrm{H}}) \left.\frac{dR}{d\Theta}\right|_{\Theta_L}} \, . \\ \Delta\Theta_{\mathrm{Z}} &= \frac{I_{\mathrm{mH}} R(\Theta_{\mathrm{L}}) - I_{\mathrm{mL}} R(\Theta_H)}{I_{\mathrm{mL}} \left.\frac{dR}{d\Theta}\right|_{\Theta_H} - I_{\mathrm{mH}} \left.\frac{dR}{d\Theta}\right|_{\Theta_L}} \end{aligned} \tag{5.9}$$

I_R and $\Delta\Theta_z$ are called 'apparent absorption Image' and 'refraction image' while the technique itself is called 'Diffraction Enhance Imaging (DEI) and was suggested and proven first by Chapman [41, 42]. The name implies that the diffraction of a perfect analyzer crystal is utilized to enhance the contrast of the images.

Sources of contrast in the apparent absorption image are absorption and extinction/scatter rejection. Extinction here means missing intensities of those X-rays that are refracted under angles substantially higher than the acceptance of the analyzer (Fig. 5.13). In contrast the refraction image gives for each pixel the angle of refraction in the object plane. Thus the white horizontal line in the refraction image (Fig. 5.13) indicates the upper boundary of the wire where the X-rays are refracted to positive angles. The black line represents the lower boundary of the wire. Here X-rays are refracted to negative angles. Note that the steeper the slope of the rocking curve the higher is the contrast enhancement. Therefore the use of higher reflections such as Si [333] instead of the Si [111] (see Fig. 5.12) is of substantial advantage.

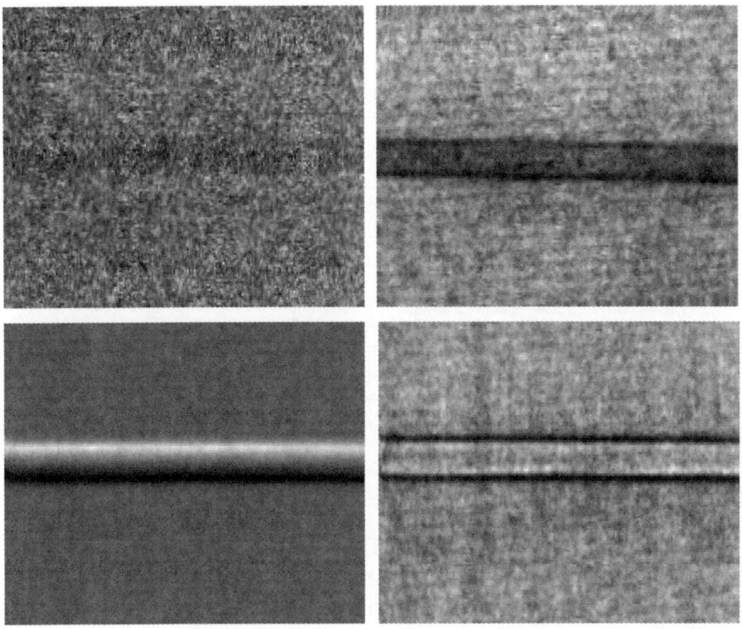

Fig. 5.13. Images of a 700μm Nylon wire collected at Elettra at 25 keV. The upper right image was recorded in transmission mode like a standard radiography at 25 keV. Depicted in the upper right radiograph is the same wire but this time with the analyzer in place and tuned on the peak of the rocking curve of the Si [111] crystal. The images in the lower row are the refraction and the apparent absorption image calculated from the images shown in figure 5.12. The high contrast of the apparent absorption image is due to extinction of X-rays.

In order to understand the improvements of this technique with respect to normal transmission radiography a simple phantom made of wires with different diameters and for instance a hollow cylinder can be used [43]. Depicted in Fig. 5.13 is a small part of this phantom.

For the wires shown, a contrast improvement by an average factor of about 1.5 has been observed moving from the standard transmission image to the DEI images using a Si [111] analyzer. A further improvement by an average factor of around 5 is obtained when instead of the Si [111] reflection the Si [333] reflection is used. The contrast C here is defined as

$$C = \frac{N_1 - N_2}{N_1} \tag{5.10}$$

where N_1 and N_2 are the average counts per pixel measured respectively on the background and on the detail of the image (Fig. 5.14).

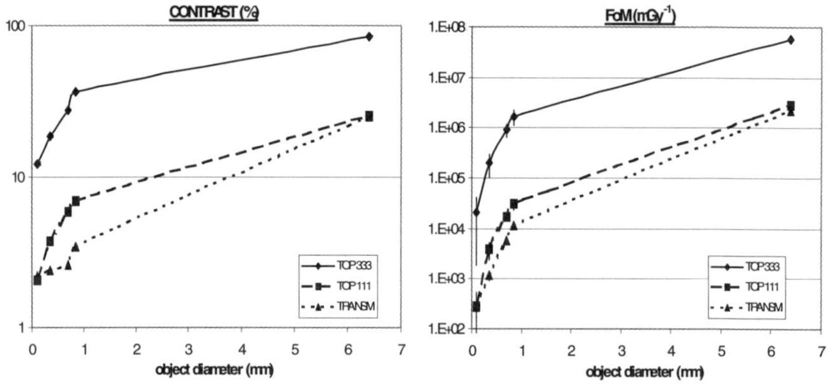

Fig. 5.14. C and FoM for the images of the nylon wires and the Plexiglas cylinder acquired at Elettra at 25 keV. The transmission, the DEI [111] (peak) and the DEI [333] (peak) images are compared.

These improvements are also apparent in the *FoM* plot in Fig. 5.14: the average gain factor for the wires when moving from transmission to [111] is approximately 2.5, and from [111] to [333] is as large as 60. Overall, the visibility of the wires (as expressed by the *FoM* as defined in Eq. (5.2) is improved by more than two orders of magnitude in the [333] DEI with respect to the transmission image. As an example for the tremendously improved image quality a DEI radiography of skull of a bird is shown in figure 5.15. This image was taken at photon energy of 20 keV. Without the contributions of scattering even small details such as the feathers can be appreciated as demonstrated in this post mortem image taken at the medical beam line at Elettra. The apparent absorption image is more pronounced than a normal transmission radiograph. Additional information gives the refraction image. Here the overall attenuation differences between bone, soft tissue, and feathers are not seen in the conventional fashion. Moreover, edges are enhanced. Since the physical effects upon which these images are based are sensitive to boundaries, one can also see inclusions in the bones that are not visible in the associated apparent absorption image.

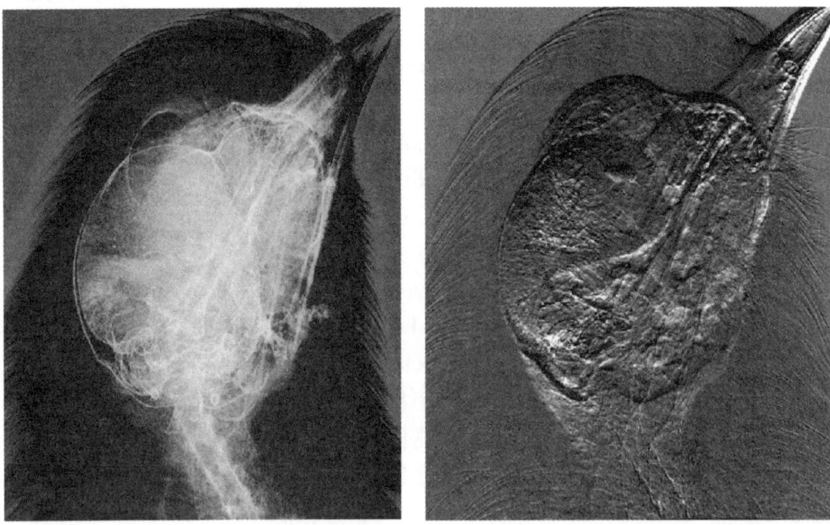

Fig. 5.15. Apparent absorption image (left) and refraction image (right) of a skull of a robin.

5.2.4.1 *Detectors for Phase Imaging*

Up to now no specific detectors are build for this new imaging modality. All groups working on this subject use rather "off the shelf" detectors such as image plates, film screen systems, high-resolution films and CCD cameras equipped with phosphors. None of these feature high DQE and good spatial (better than 10 μm) resolution at a time. In other words the spatial frequency depending DQE is too small for all of these devices. Good candidates at least in the energy range below 30 keV could be silicon strip detectors. Also suitable even for higher energies are multi-segmented, high-pressure gaseous detectors filled for instance with xenon. The latter can feature a high dynamic range (of 19 bit or more) as in the case of the imaging detector for non invasive coronary angiography and a good spatial resolution (< 50μm). For the DEI formalism as discussed above this is in principle sufficient. However, the analyzer can be used also in another fashion. Since the scattering angle is inverse proportional to the object size one can produce scattering from special spatial arrangements like muscle fibers, collagen fibers etc. In principle this is done in so-called USAXS experiments using a Bonse Hart interferometer [44]. However, the scatter probability several FWHM far away from the peak of the rocking curve is fairly low (10^{-4}–10^{-10}).

Thus the dynamic of the system has to be in this order of magnitude. This is crucial in order to record a diffraction pattern or refracted rays very close to the direct beam (e.g. typical flux around 10^{12} photons /($mm^2 \cdot s$) at a wiggler beam line at Elettra), which represents large structures (> 2500 Angstrom) as well as the diffraction signal for weak scattering samples that has to be recorded with a precision of a single photon. At present no detector exists that fulfills such high requirements on the dynamic range. However, recent developments in gas amplification structures such as MicroCATs [45] or CATs [46], but also GEMs [47], and MicroMegas [48] enable several new modes of operation for gaseous detectors in the domain of X-rays. Due to their rate capability these new Micro-pattern devices are well suited for combining gas gain operation and integration at the same time.

It was demonstrated [49] on a 128 channel prototype detector that the gas gain mechanism can be utilized to adjust the total charge in the detector according to the incoming photon flux. In this fashion for each single photon a signal is generated which is significantly higher than the noise background of the integrating electronics. Hence, for almost all photon fluxes single photon precision can be obtained (without using ultra-low noise electronics) and thus the maximum of the DQE curve (figure in appendix) can be shifted to whatever lower photon fluxes.

Such a MicroCAT detector (see Fig. 5.16) is operated either with argon–based or xenon–based gas mixtures for converting the energy of incoming photons into primary charge. By adjusting the gas mixtures and pressures (typically 1–5 bar are used) detection efficiencies larger that 70% can be maintained for photon energies between up to 30 keV. A constant drift field within the conversion region transports the primary electrons towards the gas gain and encoding stage. The system is equipped with a novel gas gain structure, namely the Micro-CAT, which performs a charge amplification of the primary electrons by factors between 1 and 10^4. The working principle of this structure is based on a strong increase of the electrical field in the vicinity of micro-holes and is very similar to that of other gas gain structures mentioned above.

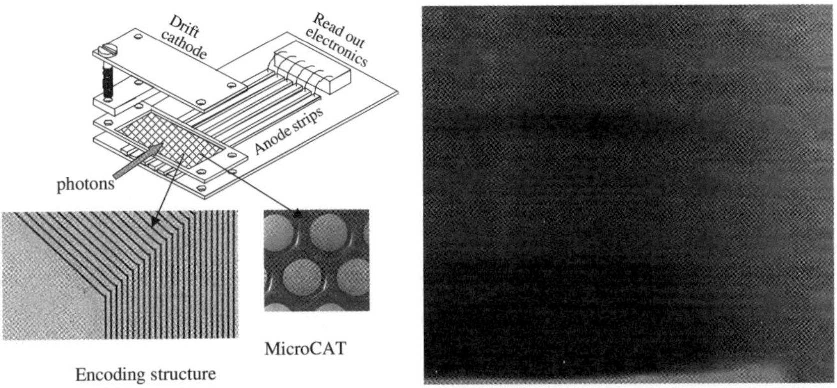

Fig. 5.16. (Left): Sketch of a prototype MicroCAT detector. (Right): Direct beam and diffraction pattern of a rat-tail tendon collagen recorded with a MicroCAT detector.

The Micro-CAT is supplied with negative high voltage (cathode) and is mounted at a short distance above the grounded read out structure (anode). The detector records the deposited charge by charge integration. In this case the gas gain mechanism is used to adjust the total charge in the detector according to the incoming photon flux, so that for each single photon a signal is always generated which is significantly higher than the noise background of the integrating electronics. Hence, for almost all photon fluxes single photon precision can be obtained.

Position resolution is obtained by collection the deposited and optionally amplified charge on a fine structure with gold-plated parallel strips with a pitch of 150 μm and a length of 20 mm. For reading out the 128 existing strips two VLSI chips are used. One chip has 64 input channels with low noise integrating stages (integration times down to 100 μs can be adjusted) which are read out via one output channel using an analogue multiplexer (64:1). The two output voltages of the VLSI chips are digitized with a 1 MHz ADC with a precision of 16 bits realized as a PC card (ANALOGIC). During read out the integrating stages are fully sensitive, that is no dead time is introduced.

Due to the unique features of the Micro-CAT structure it is now possible to combine the advantages of the integrating detector with that of the single photon counter and adjust the DQE curve with respect to the photon flux available to the application of interest. If high flux measurements are required the Micro-CAT detector is basically working as an ionization chamber while for single photon resolution the gas gain characteristics are utilized. Combining the two modes it was possible to image the direct beam and the diffraction pattern of a rat-tail tendon collagen was recorded (Fig. 5.16) featuring more than 8 orders of magnitude.

Another interesting candidate could be a thin Si–strip detector build on a Si [111] or Si [333] silicon waver. In this fashion the strip detector could be used for both, the analyzer crystal (but this time in transmission mode and not as described above in reflection mode) and as imaging detector. Similar to dark field microscopy the detector would measure only scattered radiation, when tuned to the peak of its rocking curve. The under–the–Bragg–angle deflected beam could then be recorded for instance by a high DQE MicroCAT detector.

5.3. Conclusion and Outlook

The unique features of synchrotron radiation enables a variety of medical applications among them those that were briefly discussed here. The success of these applications however depends strongly on the performance of the imaging detectors. For instance non-invasive coronary angiography would not have been possible without custom-made detectors. Further improvements in image quality and image contrast at reasonable dose will be only possible if the DQE of the detector systems is optimized for each specific application. For some future applications a new generation of ionizing radiation detectors is required. The effort in terms of human resources as well as the financial effort will be high but will pay off at the end.

Acknowledgement

Parts of the work presented here were supported by the European commission through the contracts CT961694, FMGE-CT-980104 and HPRI –CT- 1999-50008. I would like to thank F. Arfelli for numerous discussions and the critical remarks on the manuscript.

Appendix

A. Image Formation and Detector Characterization

A comparison of the performance of different detectors is not straightforward. In order to do so appropriate methods have to be used and a common measure has to be found. In the following such a common measure, namely the detective quantum efficiency (*DQE*), will be developed for both, integrating and counting X-ray detectors with spatial resolution.

In X-ray radiographs image formation is a statistical process and the intrinsic limitations are given by Poisson noise only. A common measure of the image quality is the signal (S) to noise (σ) ratio (SNR_{out}) in the entire image or parts of the image.

Consider a homogeneous area A illuminated for a certain time T (a flat field image). At each single point of A, a flux density (in the following denoted $\phi(x,y,t)$) can be assumed to be superimposed with noise due to the Poisson statistics. The signal to noise ratio within this illuminated area is then given by

$$SNR_{\text{in}} = \frac{\int\limits_{A}\int\limits_{T} \phi(x,y,t)\cdot dt\cdot da}{\sqrt{\int\limits_{A}\int\limits_{T} \phi(x,y,t)\cdot dt\cdot da}} = \frac{N}{\sqrt{N}} = \sqrt{N}\,. \qquad \text{(A.1)}$$

N denotes the integral number of quanta in the entire area A. Note that this expression describes the intrinsic limitation for all kinds of photon imaging systems. Photon integrating or, more precisely, energy integrating devices such as CCDs ionization chambers, Silicon strip detectors in integrating mode etc. contribute additional noise to the image. Thus, in general, the image information is degraded at each step of the detection. An image of the flat field recorded with a real, and hence noisy detector, delivers the following degraded SNR_{out}:

$$SNR_{out} = \sqrt{DQE} \cdot SNR_{in} . \qquad \text{(A.2)}$$

This relation might be considered as a definition of the *DQE*. Obviously, a reduced *DQE* value requires an increased number of quanta (or dose) in order to achieve the same signal to noise ratio in the recorded images. To avoid unnecessary dose applied to patients in medical application, the *DQE* values of detector systems should be close to 1. With the knowledge of both the SNR_{out} and the SNR_{in} it is possible to evaluate the *DQE* for a given detector system.

Images are sampled at discrete positions in the space domain with the period defined by the pitch of the segmentation. Obviously this kind of image recording underlies the sampling theorem and the smallest detail transmitted is defined by the Nyquist frequency in the reciprocal domain of the space. In the following this domain is denoted as the domain of spatial frequencies *f* and *g*f* is the frequency in a direction parallel to the segmentation of the detector. Formula (A.2) can be rewritten in terms of the spatial frequencies as

$$DQE(f,g) = \left(\frac{SNR_{out}(f,g)}{SNR_{in}(f,g)} \right)^2 = \left(\frac{S_{out}(f,g)\sigma_{in}(f,g)}{\sigma_{out}(f,g)S_{in}(f,g)} \right)^2 \qquad \text{(A.3)}$$

where $SNR_{out}(f,g)$ and $SNR_{in}(f,g)$ are, respectively, the output and input signal to noise ratios at the spatial frequencies *f* and *g*. The transformation in the domain of the spatial frequencies allows the convenient evaluation of the *DQE(f,g)* in terms of experimental measurements as shown in the following sections. It will turn out that the unsampled input signal S_{in} as well as σ_{in} is independent of spatial frequencies since the latter is considered to be white noise. Thus the SNR_{in} in the domain of the spatial frequencies is constant for all spatial frequencies. This can be shown by utilizing a pin hole aperture at the position $x = 0$, $y = 0$ which covers the flat field introduced earlier. The input signal in the spatial domain is then a convolution of the flux density with the pin hole which can be written in terms of a two dimensional Dirac distribution. One gets

$$S_{\text{in}}(x, y) = \phi_0 \delta(x, y) \tag{A.4}$$

and in the domain of the spatial frequencies

$$S_{\text{in}}(f, g) = \int_{-\infty}^{\infty}\int_{-\infty}^{\infty} (\phi_0) \delta(x, y) e^{-2\pi i f x} e^{-2\pi i g y} dxdy = \phi_0 . \tag{A.5}$$

The input noise in the Poisson case is

$$\sigma_{\text{in}}^2(f, g) = \int_{-\infty}^{\infty}\int_{-\infty}^{\infty} \phi_0 \delta(x, y) e^{-2\pi i f x} e^{-2\pi i g y} dxdy = \phi_0 \tag{A.6}$$

which results in $SNR_{\text{in}}(f, g) = \sqrt{\phi_0}$.

The detector system, however, responds to the input impulse at the object coordinates $x = y = 0$ with a characteristic transfer function. Due to the effects of charge generation within the detector the ionization which consists of the image information does not necessarily remain at this specific position but rather spreads over adjacent pixels. This causes a blurring or enlargement of the signal at the image coordinates x', y'. This effect is described by the point-spread-function (*PSF*). Mathematically the output signal S_{out} in the space domain of the image is a convolution of the input signal with the PSF of the detector and with the quantum efficiency ε of the detector. With the assumption of ε = *constant* one gets

$$\begin{aligned} S_{\text{out}}(x, y) &= \varepsilon \int_{-\infty}^{\infty}\int_{-\infty}^{\infty} \phi_0 \delta(x', y') PSF(x - x', y - y') dx' dy' \\ &= \varepsilon \phi_0 PSF(x, y) \end{aligned} \tag{A.7}$$

Formula (A.7) is suitable for all kinds of photon integrating detectors but has to be extended, when dealing with energy integrating detectors, such as the discussed. The total charge released in a particular detector pixel with a pixel width of B, a beam or pixel height of A and during an integration time T is

$$Q = \frac{FE_\gamma}{W_{ion}} \phi_0 ABTe , \tag{A.8}$$

if we consider a constant flux ϕ_0 of photons with the energy E_γ. Here FE_γ denotes the total absorbed energy in a detector element, W_{ion} is the ionization energy required producing an ion – electron pair (or one electron – hole pair) and e denotes the electron charge magnitude. F is a factor that takes into account that not all the energy carried by a photon is really deposited in the detector and hence F is always less than one. For example a fluorescence photon generated after a photoelectric effect can escape out of the active detector volume. With (A.8) a new expression for S_{out} in (A.7) can be found

$$\begin{aligned} S_{\text{out}}^{\text{Q}}(x,y) &= \frac{FE_\gamma}{W_{ion}} \phi_0 ABT\chi\varepsilon ePSF(x,y) \\ &= K\varepsilon\phi_0 E_\gamma PSF(x,y) \end{aligned} \tag{A.9}$$

where χ is a property of the acquisition electronics that describes the conversion of charge to gray values in the image. Note that $K\phi_0 E_\gamma$ is without dimension.

A Fourier transform delivers the output signal in terms of its spatial frequencies and one obtains

$$\begin{aligned} S_{\text{out}}(f,g) &= KE_\gamma\,\varepsilon \int_{-\infty}^{\infty}\int_{-\infty}^{\infty} \phi_0 PSF(x,y) e^{-2\pi ifx} e^{-2\pi igy} dxdy \\ &= KE_\gamma\,\phi_0 \varepsilon OTF(f,g) \end{aligned} \tag{A.10}$$

where *OTF* is the optical transfer function. The modulus of the *OTF* is the modulation transfer function (*MTF*). In practice the *MTF* is a measure of how well a detail of a certain size is recorded with the detector system. Note that for an ideal detector $MTF(f\,,\,g) = 1$ at all spatial frequencies. For such a detector the noise would be independent of spatial frequencies, similar to the input noise in formula (A.6).

The low pass characteristics of a real detector, however, degrade the high spatial frequency signals as well as the high frequency noise. In general the image noise can be subdivided into three major constituents: one part due to the Poisson statistics, a second part that describes the fluctuations in the generated charges or the absorbed energy and a third additional noise source in the following denoted with σ_{add}. The latter consists of inherent detector noise, electronics noise and ADC noise, which add in quadrature since they are statistically independent. In the model developed so far the total image noise in the space domain utilizing error propagation would be

$$\sigma_{\text{tot}}^2(x,y) = \left(\frac{\partial S_{\text{out}}(x,y)}{\partial(\varepsilon\phi_0)}\right)^2 \left(\Delta(\varepsilon\phi_0)\right)^2 + \left(\frac{\partial S_{\text{out}}(x,y)}{\partial(E_\gamma)}\right)^2 \left(\Delta(E_\gamma)\right)^2 + \sigma_{\text{add}}^2 \tag{A.11}$$

In this notation σ_{add} has the dimension of a photon fluence at the input of the detector. With respect to formula (A.8) and with $\Delta(\varepsilon\phi_0) = \frac{1}{ABT}\sqrt{\varepsilon\phi_0}$ equation (A.11) can be rewritten to

$$\sigma_{\text{tot}}^2(x,y) = \frac{K^2 E_\gamma^2 PSF^2(x,y)\varepsilon\phi_0}{(ABT)^2} + \left(K^2 PSF^2(x,y)\varepsilon^2\phi_0^2\right)\frac{\int_0^\infty n(E)E^2 dE - \left(\int_0^\infty n(E)E dE\right)^2}{E_\gamma^2} + \sigma_{\text{add}}^2 . \tag{A.12}$$

The first term is due to Poisson statistics, the second term describes the contribution of the single event energy resolution

$$\phi_0 = \int_0^\infty n(E)dE \quad and \quad E_\gamma = \int_0^\infty n(E)EdE . \tag{A.13}$$

The Fourier transform of (A.12) delivers directly the noise in the domain of spatial frequencies $\sigma^2(f,g)$ when utilizing the Parseval Theorem:

$$\sigma^2(f,g) = \int_{-\infty}^{+\infty}\int_{-\infty}^{+\infty} \phi_0\varepsilon \, \frac{K^2 E_\gamma^2 PSF^2(x,y)}{(ABT)^2} e^{-2\pi i f x} e^{-2\pi i g y}\, dxdy$$

$$+\int_{-\infty}^{+\infty}\int_{-\infty}^{+\infty} \phi_0\varepsilon K^2 PSF^2(x,y)\, e^{-2\pi i f x} e^{-2\pi i g y} \frac{\int_0^\infty n(E)E^2 dE - \left(\int_0^\infty n(E)E dE\right)^2}{\left(\int_0^\infty n(E)E dE\right)^2} dxdy$$

$$+\int_{-\infty}^{+\infty}\int_{-\infty}^{+\infty} \sigma_{\mathrm{add}}^2\, e^{-2\pi i f x} e^{-2\pi i g y} dxdy$$

$$\begin{aligned}\sigma^2(f,g) &= \phi_0\varepsilon\, K^2 E_\gamma^2 \left(\frac{|MTF(f,g)|^2}{(ABT)^2} \right.\\ &\quad + \varepsilon |MTF(f,g)|^2 \frac{\int_0^\infty n(E)E^2 dE - E_\gamma^2}{E_\gamma^2}\\ &\quad \left. + \frac{W_{add}}{E_\gamma^2 \phi_0 \varepsilon\, K^2} \right)\\ &= K^2\phi_0\varepsilon E_\gamma^2 (W_P + W_d) = K^2 \phi_0 \varepsilon E_\gamma^2 NPS(f,g)\end{aligned} \tag{A.14}$$

W_p and W_d are the so called Wiener spectrum of the Poisson noise and the additional noise, respectively. With equations (A.5), (A.6), (A.10) and (A.14) the following expression for the *DQE* is found

$$DQE(f,g) = \varepsilon \frac{|MTF(f,g)|^2}{NPS(f,g)} \tag{A.15}$$

or for the one dimensional case

$$DQE(f) = \varepsilon \frac{|MTF(f)|^2}{NPS(f)} \qquad \text{(A.16)}$$

According to general conventions $|MTF(f,g)|^2$ is normalized in such a fashion that $|MTF(0,0)|^2 = 1$. From special interest is a detector property, which relates variations in detector exposure at zero spatial frequency to variations at the output. This is often denoted with Zero Spatial Frequency *DQE (f=0)*

$$DQE(f=0) = \frac{\varepsilon}{1 + \frac{\varepsilon\left(\int_0^\infty n(E)E^2 dE - E_\gamma^2\right)}{E_\gamma^2} + \frac{W_{\text{add}}}{\varepsilon K^2 E_\gamma^2 \phi_0}}. \qquad \text{(A.17)}$$

Formula (A.17) reveals three characteristics of the *DQE(f=0)* for integrating detectors:

- For high photon fluence the expression is dominated by the quantum efficiency ε of the detector. Obviously *DQE(f=0)* values close to 1 are achievable by detectors with a high absorption efficiency only.
- In the energy range considered here the second term in the denominator becomes negligible if the fluorescence and Compton contributions are small.
- Since the additional noise term shifts the *DQE(f=0)* rather than changing its shape, W_{add} should be as low as possible in order to achieve high *DQE(f=0)* values even for small photon fluence.

At least for the frequency depending part the *DQE* for a multi channel single photon counting detector is the same as discussed before. Changes occur in the *DQE(f=0)*, however. A rigorous treatment of the DQE problem taking into account the real case that the counting statistics is modified by the dead time would go beyond the scope of appendix. Therefore in the following discussion the simplified assumption that the dead time of a counting detectors does not alter the Poisson statistics is applied.

For single photon counting detectors one has to distinguish between the non-paralyzing (detector does accept another photon while the actual photon is processed) and the paralyzing case (the detector does not accept any other photon while the actual photon is processed). If n is the real rate in the detector, τ the deadtime of the system then the measured rate m is in the non-paralyzing case is given by

$$m = \frac{\varepsilon n}{1 + n\tau} \tag{A.18}$$

The associated Poisson error is then given as

$$\Delta m = \sqrt{\frac{\varepsilon n}{1 + n\tau}} \tag{A.19}$$

and thus the SNR^2_{out}

$$SNR^2_{\text{out}} = \frac{\varepsilon n}{1 + n\tau}. \tag{A.20}$$

Since the real rate in the detector is n photons per time unit the $SNR^2_{\text{in}} = n$ and thus the $DQE(f=0)$ according to (A.2)

$$DQE(f = 0) = \frac{\varepsilon}{1 + n\tau} \tag{A.21}$$

for the non paralyzing case.

For the paralyzing case the measured rate m is given by

$$m = \varepsilon n e^{-n\tau} \tag{A.22}$$

with an associated error of

$$\Delta m = \sqrt{\varepsilon n e^{-n\tau}}, \tag{A.23}$$

which leads accordingly to

$$DQE(f = 0) = \varepsilon e^{-n \cdot \tau} \tag{A.24}$$

It is obvious that the noise properties of an imaging system determine the radiation dose which a patient has to accept for a given signal to noise ratio. The fraction of the radiation dose a patient has to bare due to imperfections of the imaging system is *1-DQE.* Hence, for medical applications the *DQE* should be close to one. Any dose quoted without proper consideration of the spatial and intensity resolution is therefore meaningless. As a consequence there is only one reliable method to compare the performance of different detectors, namely to compare the DQE as a function of the spatial frequencies at the most critical intensity [28].

B. Digital Subtraction Technique

The measured transmission ϕ_1 of the high energy beam through a patient, considered to consists of water and iodine only, is

$$\phi_1 = \phi_{01} e^{-\frac{\mu_i^>}{\rho_i}\rho_i \Delta x_i} e^{-\frac{\mu_w^>}{\rho_w}\rho_w \Delta x_w} \tag{A.25}$$

and that of the low energy image ϕ_2

$$\phi_2 = \phi_{02} e^{-\frac{\mu_i^<}{\rho_i}\rho_i \Delta x_i} e^{-\frac{\mu_w^<}{\rho_w}\rho_w \Delta x_w} \tag{A.26}$$

where ϕ_{01} and ϕ_{02} are the fluences for the high and the low energy beam in front of the patient and μ/ρ are the mass absorption coefficients for iodine and water below (<) and above (>) the iodine *k*-edge. ρ_w and ρ_i are the densities and Δx_w and Δx_i are the effective thickness of water and iodine.

Normalized to the incident flux (A.25) and (A.26) can be rewritten as

$$-\begin{pmatrix} \ln(\frac{\phi_1}{\phi_{01}}) \\ \ln(\frac{\phi_2}{\phi_{02}}) \end{pmatrix} = \begin{pmatrix} \mu_w^> & \mu_i^> \\ \mu_w^< & \mu_i^< \end{pmatrix} \begin{pmatrix} \rho_w \Delta x_w \\ \rho_i \Delta x_i \end{pmatrix}. \tag{A.27}$$

The inversion of the absorption coefficient matrix gives the water and the iodine images

$$\begin{pmatrix} S_{\text{water}} \\ S_{\text{iod}} \end{pmatrix} = \begin{pmatrix} \rho_w \Delta x_w \\ \rho_i \Delta x_i \end{pmatrix} = \begin{pmatrix} \frac{1}{\mu_w^>}\left(1-\frac{\mu_i^>}{\mu_i^> - \mu_i^<}\right) & \frac{\mu_i^>}{\mu_w^<} \\ \frac{1}{\mu_i^> - \mu_i^<} & -\frac{1}{\mu_w^> - \mu_w^<} \end{pmatrix} \begin{pmatrix} -\ln\left(\frac{\phi_1}{\phi_{01}}\right) \\ -\ln\left(\frac{\phi_2}{\phi_{02}}\right) \end{pmatrix} \quad \text{(A.28)}$$

For the iodine image one gets

$$S_{\text{iod}} = \rho_i \Delta x_i = \frac{1}{\mu_i^> - \mu_i^<}\left[\ln\left(\frac{\phi_2}{\phi_{02}}\right) - \ln\left(\frac{\phi_1}{\phi_{01}}\right)\right] = \frac{1}{\Delta\mu_i}\ln\left(\frac{\phi_2}{\phi_{01}}\right) \quad \text{(A.29)}$$

for $\phi_{01} = \phi_{01}$. Since the mass absorption coefficients are well known (A.28.) might be written as

$$\begin{pmatrix} S_{\text{water}} \\ S_{\text{iod}} \end{pmatrix} = \begin{pmatrix} 0.0231926 & 0.0239174 \\ 2.56786 & -0.476898 \end{pmatrix} \begin{pmatrix} -\ln\left(\frac{\phi_1}{\phi_{01}}\right) \\ -\ln\left(\frac{\phi_2}{\phi_{02}}\right) \end{pmatrix} \quad \text{(A.30)}$$

With the assumption that only Poisson noise of the two energy images contributes to the total image noise the minimum flux required for a given *SNR* can be calculated:

$$\phi_{\min} = \frac{SNR^2 2}{(\Delta\mu_i \rho_i \Delta x_i)^2 \Delta T} \quad \text{(A.31)}$$

For example: In order to image a vessel (Ø 1 mm and filled with 10 mg/ml iodine) with a $SNR = 3$, a flux density of at least 20.000 photons /(ms pixel) are required, when a pixel covers 0.5 mm^2. The electronics noise should be at least three times smaller than the Poisson noise of this minimum flux, which is 144 photons/pixel and 1 ms integration time. For a 2 mm vessel, however, a flux of 800 photons/(ms pixel) is required with a noise of 30 photons/pixel for 1 ms integration time.

References

1. H. Winick and S. Doniach, (eds.), *Synchrotron Radiation Research,* (1980)
2. W. C. Thomlinson, Medical applications of synchrotron radiation at the National synchrotron Light Source in Synchrotron Radiaition in Biosciences, *Oxford Science Publications* (1994)
3. NCRP report no.85, National Council on Radiation protection and Measurements, Washington DC (1986)
4. S. A. Feig and M.J. Yaffe, Digital Mammography. *Radiographics.* **18**, 893 (1998)
5. M. B. Williams and L.L. Fajardo, *Acad. Radiol.* **3**, 429 (1996)
6. K. Engelke, PhD thesis, Hamburg, DESY internal report, **F41/89-12**, (1989)
7. Van Metter and Dickerson. Objective performance characteristics of a new asymmetric screen-film system. *Medical Physics* **21:9**, 1483 (1994)
8. E. Burattini, E. Cossu, C. Di Maggio, M. Gambaccini, P. Indovina.,M. Marziani, M. Porek, S. Simeoni and G. Simonetti, *Radiology* **125**, 239 (1994)
9. R.E. Johnston, D. Washburn, E. Pisano, C. Burns, W.C. Thomlinson, L.D. Chapman, F. Arfelli., N. Gmür., Z. Zhong and D. Sayer, *Radiology* **200**, 659 (1996)
10. R.A. Lewis, A.P. Hufton, C.J. Hall, W.I. Helsby, E. Towns-Andrews, S. Slawson and C.R.M Boggis, *Synch. Rad. News* **12**, 7 (1999)
11. M.B. Williams and L.L. Fajardo, *Acad. Radiol.* **3**, 429 (1996)
12. M.J. Yaffe and J.A. Rowlands, *Phys Med Biol* **42**, 1 (1997)
13. F. Arfelli, A. Bravin, G. Barbiellini, G. Cantatore, E. Castelli, M. Di Michiel, P. Poropat, R. Rosei, M. Sessa, A. Vacchi, L. Dalla Palma, R. Longo, S. Bernstorff, A. Savoia and G. Tromba, *Rev. Sci. Instrum.* **66** (1995)
14. F. Arfelli, G. Barbiellini, V. Bonvicini, A. Bravin., G. Cantatore, E. Castelli, L. Dalla Palma, M. Di Michiel, R. Longo, A. Olivo, S. Pani., D. Pontoni, P. Poropat, M. Prest, G. Richter, R. Rosei, M. Sessa, G. Tromba, R. Turchetta and A. Vacchi, *Physica Medica,* **XIII, Suppl. 1** (1997)
15. F. Arfelli, V. Bonvicini, A. Bravin, G. Cantatore, E. Castelli, L. Dalla Palma, M. Di Michiel, R. Longo, A. Olivo, S. Pani, D. Pontoni, P. Poropat, M. Prest, A. Rashevsky, G. Tromba and A.Vacchi, *Radiology,* **208,** 709 (1998)
16. F. Arfelli, G. Barbiellini, V. Bonvicini, A. Bravin, G. Cantatore, E. Castelli, L. Dalla Palma, M. Di Michiel, R. Longo, A .Olivo, S. Pani, D. Pontoni, P. Poropat, M. Prest, A. Rashevsky, G. Tromba, A. Vacchi and N. Zampa, *IEEE Trans. Nucl. Sci.* **44**, 874 (1997)
17. G. Comes, F. Loddo, Y. Hu, A. Kaplon, F. Ly, R. Tuchetta., V. Bonvicini and A. Vacchi, *Nucl. Instr. and Meth.* **A377,** 440 (1996)
18. F. Arfelli, *Nucl. Instr. and Meth.* **A462**, 607 (2001)
19. E. Rubenstein, *Proc. Natl. Acad. Sci., USA 83,* **9724** (1986)
20. A.C. Thompson, W.M. Lavender, L.D. Chapman, N. Gmür, W. Thomlinson, V. Rosso,C. Schulze, E. Rubenstein, J.C. Giacomini and H.J. Gordon, *Nucl. Instr. and Meth.,* **A347**, 545 (1994)
21. B. Jacobson, *Acta Radiologica*, **39**, 437 (1953)

22. J.C. Giacomini, H.J. Gordon, R. O'Neil, A. Van Kessel, B. Carson, L.D. Chapman, W. Lavendar, N. Gmür, R.H. Menk, W. Thomlinson, Z. Zhong and E. Rubenstein *Nucl. Instr. and Meth.* **A406,** 473 (1998)
23. M.S. Albert, G.D. Cates, B. Driehuys, W. Happer, B. Saam, C.S. Springer Jr, and A. Wishnia, *Nature* **370,** 199 (1994)
24. M. Ebert, T. Grossmann, W. Heil, W.E. Otten, R. Surkau, M. Leduc, P. Bachert , M.V. Knopp, L.R. Schad and M. Thelen, *Lancet* **347**, 1297 (1996)
25. P.Bachert, L.R. Schad, M. Bock, M.V. Knopp, M. Ebert, T. Grossmann, W. Heil D. Hofmann, R. Surkau and E.W. Otten, *Magnetic Resonance Med.* **36**, 192 (1996)
26. R.H Menk., Charvet A.M., Arfelli F., Chapman L., Thomlinson W., Zhong Z *PHYSICA MEDICA* **13**, 26 (1997)
27. F.A. Dilmanian, X.Y. Wu, E.C. Parsons, B. Ren., T.M. Button, L.D. Chapman, X Huang, S. Marcovici, R. H. Menk, E.L. Nickoloff, M.J. Petersen, C.T. Roque, W.C. Thomlinson and Z. Zhong, *PHYSICA MEDICA,***13**, 13 (1997)
28. H.J. Besch, *Nucl. Instr. and Meth.* **A419**, 202 (1998)
29. H.J. Besch, E.J. Bode, R.H. Menk, H.W. Schenk, U. Tafelmeier, A.H. Walenta and H.Z. Xu, *Nucl. Instr. and Meth.* **A310**, 446 (1991)
30. E.N. Dementyev et al, *Rev. Sci. Instrum.* **60**, (1989)
31. R.H. Menk, W.R. Dix, W. Graeff, G. Illing, B. Reime, L. Schildwaechter, U. Tafelmeier, H.J. Besch, U. Grossmann, R. Langer, M. Lohmann, H.W. Schenk, M. Wagner, A.H. Walenta, W. Kupper, C. Hamm and C. Rust, *Rev. Sci. Instrum.* **2235** (1994)
32. M. Lohmann, H.J. Besch, W.R.Dix, O.Dunger, M. Jung, R.H. Menk, B. Reime and L. Schildwachter, *Nucl. Instr. and Meth.* **A419**, 276 (1998)
33. R.H. Menk et al, *Nucl. Instr. and Meth.* **A398,** 351 (1997)
34. S.W. Wilkins, T.E. Gureyev, D. Gao, A. Pogany and A.W. Stevenson, *Nature* **384** (1996)
35. Snigirev, I. Snigireva, V. Kohn, S. Kuznetsov and I. Schelokov, *Rev. Sci. Instrum.* **66**, 5486 (1995)
36. P. Cloetens, R. Barrett, J. Baruchel, J. Guigay and M .Schlenker, *J. Phys D: Appl. Phys.* **29**, 133 (1996)
37. E. Foerster, K. Goetz and P. Zaumseil, *Kristall und Technik* **15**, (1980).
38. K.M. Produrets, V.A. Somenkov and S.Sh. Shil'stein, *Sov.Phys.-Tech.Phys,* **34**, 654-657 (1989)
39. V.A. Somenkov, A.K. Tkalich and S.Sh. Shil'stein, *Sov. Phys.-Tech.Phys,* **36**, (1991)
40. E.A. Beliaevsky, V.P. Epfanov and V.N. Ingal, *Soviet Patent 4934958*
41. L.D. Chapman, W. Thomlinson, F. Arfelli, N. Gmür, Z. Zhong, R.H. Menk, R.E. Johnson, D. Washburn, E. Pisano and D. Sayers, *Rev.Sci.Instrum.* **67**, 9 (1996)
42. L.D. Chapman, W. Thomlinson, R.E. Johnson., D. Washburn, E. Pisano, N. Gmür, Z. Zhong, R.H. Menk, F. Arfelli and D. Sayers, *Phys.Med. Biol.* **42** (1997)
43. L. Rigon, Z. Zhong, F. Arfelli, R.H. Menk, A. Pillon *Proceeding to the SPIE Medical Imaging Conference, San Diego* (2002)
44. U. Bonse and M. Hart, *Appl.Phys.Lett* **.7**, 238 (1965)

45. Sarvestani, H.J. Besch, M. Junk, W. Meißner, N. Sauer, R. Stiehler, A.H. Walenta and R.H. Menk, *Nucl. Instr. and Meth.* **A410,** (1998)
46. P. Bartol, M. Bordessoule, G. Chaplier, M. Lemonnier and S. Megtert, *J.Phys. III France* **6,** (1996)
47. F. Sauli, *Nucl. Instr. and Meth.* **A386,** 531 (1997)
48. Y. Giomataris, P.H. Rebourgeard, J.P. Robert and G. Charpak, *Nucl. Instr. and Meth.* **A376**, 29 (1996)
49. R.H.Menk et al, *Nucl. Instr. and Meth.* **A440** (2000)

CHAPTER 6

AUTORADIOGRAPHY

Maria Evelina Fantacci

Physics Department of Pisa University, Pisa, Italy

Autoradiography is a well suited technique used as a standard and powerful analysis tool in biological, biochemical, medical and biophysical fields. Applications of autoradiography ranges from morphological studies on biological samples and biomolecular studies (as analysis of dot blots, electrophoresis, protein/amminoacid uptake and localization) to dynamic studies of biological mechanisms. It consists in mapping the distribution of a radioactive marker (the most common used are ^{3}H, ^{14}C, ^{35}S, ^{32}P) in a radiolabelled sample, and takes advantage of the fact that animals and plants cannot distinguish between stable and radioactive isotopes of the same elements in their physiological reactions.

6.1 Autoradiographic Methods

"Blots" [1] are membranes such as nitrocellulose or coated nylon to which nucleic acids have been permanently bound. Blot hybridizations with specific nucleic acid probes provide critical information regarding gene expression and genome structure. The most common blot applications used in modern laboratories are Northern blots, Southern blots and dot/slot blots. Regardless of the type of blot, the principles of probe synthesis, hybridization, washing and detection are the same.

Southern blotting [2] involves separating DNA fragments on agarose gels then denaturing them in situ. Fragments are then transferred to a nitrocellulose or nylon membrane where they are immobilized. After prehybridization to reduce non-specific hybridization with the probe, the membrane is hybridized with the desired labeled nucleic acid probe. The membrane is subsequently washed to remove unbound and weakly binding probe, and is then exposed to film.

Northern blotting [3] is in principle the same as Southern blotting except that RNA instead of DNA is isolated, size fractionated on an agarose gel, transferred and linked to a membrane, and then probed. The specifics of Northern blotting, however, differ substantially from Southern blotting for several reasons. The first major consideration is that RNA is both chemically and biologically far more labile than DNA. From a practical viewpoint the extreme sensitivity of RNA to RNases and the wide prevalence and stability of these enzymes means that in order to prepare intact molecules of RNA (particularly large RNA), one must pay careful attention to eliminating RNases. Secondly, only a few percent of the total RNA of a cell is m-RNA, so that blotting against total RNA is not a very sensitive way of detecting rare mRNA. Therefore, although one can often detect a target mRNA in a preparation of total RNA, one can attain greater sensitivity by isolating first the poly A+ mRNA fraction and carrying out Northern analysis.

Western blotting [4] is a method for identifying a specific protein in a complex mixture and simultaneously determining its molecular weight. The name came about this way: DNA blotting was first described by E. M. Southern and, hence, became Southern blotting. RNA blotting then was dubbed Northern blotting. Thus, it was inevitable that protein blotting would become Western blotting. The procedure can be broken down into a series of steps: size separation of the denatured proteins in the mixture by Polyacrylamide Gel Electrophoresis (PAGE); transfer of the separated proteins to a membrane while retaining their relative position; detecting the protein under investigation by its specific reaction with an antibody and determination of its size relative to standard proteins of known size.

The major steps in determining a DNA sequence are: preparation of the template DNA: annealing of the primer; carrying out the sequencing reaction; separating the products on an acrylamide-urea gel; reading the sequence. Primer end labeling during manual sequencing is usually done using ^{32}P-ATP to label the 5' end of the primer with polynucleotide kinase. ^{32}P labeling has the advantage that it is a relatively high energy beta emitter which exposes the film in a short period of time allowing faster acquisition of sequence information. Unfortunately, due to its high energy, ^{32}P beta particles scatter over a relatively large area which causes poorer band resolution than when the DNA is labeled internally with ^{35}S. This resolution problem can be avoided by carrying out end labeling with the more expensive ^{33}P-ATP which has beta particle emissions which are only about one fifth the strength of ^{32}P. This provides less sensitivity than ^{32}P, but provides excellent resolution.

6.1.1 *Traditional autoradiography: methods*

In traditional autoradiography the radiolabelled sample is put as close as possible to a photographic film in order to obtain the autoradiographic image. A photographic film consists of a support covered by a photographic emulsion [5], which is essentially a dispersion of silver halide (AgBr) crystals in a gelatin matrix. The emulsions used for nuclear measurements, as in this case, have several distinguishing features: the silver halide crystals are very uniform in size and sensitivity; there are very few crystals that may be developed without exposure to a charged particle (very low chemical fog); the silver to gelatin ratio is much higher than in a conventional emulsion. When such an emulsion is exposed to the ionizing radiation emitted by the isotopes used as markers, clusters of silver atoms are produced. These are known as latent image centres. Then the emulsion is developed and the latent image is made visible by the reduction of silver ions in the silver halide crystal to metallic silver. Thus, after the development of the film, a negative is obtained which precisely displays the distribution of radioactivity within the sample. The negative can then be read by a microdensitometer. The spatial resolution is determined by various factors: dimensions of the AgBr crystals, distance sample-emulsion, type and energy of radiation. The isotopes most used in autoradiography are β^- emitters. Their mean energies and their ranges in photographic emulsion are reported in Table 6.1.

Table 6.1 Mean energies and ranges for the nuclides most used in autoradiography.

Isotope	β^- mean energy (keV)	Range in photographic emulsion (μm)
^{3}H	5.7	1
^{14}C	50	10
^{35}S	49.2	10
^{32}P	694	800

With film autoradiography a spatial resolution of about 30 μm can be reached. An improvement can be obtained by the use of photographic emulsions in liquid or gel form, without any support. In these cases the coating of the specimens ensures a very close contact between source and emulsion to achieve maximum spatial resolution. The reading of the image in this case is made with light microscopy and a value of 0.5 μm can be reached. The use of monomolecular layers of photographic emulsion read by electron microscopy is also possible. In this other case a spatial resolution of 0.1 μm can be obtained.

Recently [6,7] the atomic force microscopy (AFM) has been also applied for measuring very small etch pits on CR-39 solid state track detectors. In this technique, the CR-39 is etched for a short time and then the etched surface is scanned with the AFM. The resolution is determined by the size of the etch pits which can be observed with the AFM. After irradiation with 1 MeV Helium ions, the position resolution of imaging for the charged particle track was 30 nm (FWHM). The ultimate resolution of this method was thought to be of the order of several nanometers; thus it can be considered as a new technique for subcellular scale imaging.

An other important parameter is the detection efficiency, which critically influences the exposure time. Moreover, a high and precise detection efficiency is necessary if a quantitative measurement is required. The intrinsic efficiency of the film is very low, thus requiring very long exposure times. It can be increased by the use of phosphor intensifying screens [8], which work by generating photons through the interaction of the radiation within the phosphors. In this case the spatial resolution is worsened. The β particles interact with the phosphors to generate a large number of photons which are detected by the film. The optimal sensitivity in such a system is reached when the photons (generated by the screen) wavelength match the spectral response of the film. Conventional intensifying screens work by placing the radiolabelled sample on a sheet of autoradiography film with the intensifying screen laying under the film (i.e. the film is sandwiched between the screen and the sample). To make use of the intensifying screen, the radioactive particles must have enough energy to pass through the film and to come in contact with the intensifying screen. ^{32}P and ^{125}I have sufficient energies to penetrate through the film, but there is no benefit for weak and medium energy radioisotopes such as ^{3}H, ^{14}C, and ^{35}S.

One method to improve the efficiency of autoradiographic detection of weak beta emitting radionuclides [9] is fluorography, which involves the use of a solvent containing the primary scintillator PPO (2,5-diphenyloxazole); following sample chromatographic separation, the paper is immersed briefly in this fluorographic cocktail, and dried leaving the PPO impregnated throughout the sample and then exposed to the photographic film Thus, there is a very close contact between the radionuclide and the scintillator. Within the matrix of the paper the light undergoes attenuation which can be avoided by the use of EA-Wax, a new medium melted at low temperature (55^0) directly on the sample matrix. This method is useful in particular for tritiated samples.

6.1.2 *Traditional autoradiography: limits*

Although traditional film-based autoradiography can provide images of radiolabelled biological specimens with excellent spatial resolution, it is a time consuming process, with very long exposure times (from several days to several months). The main limits of this method, more evident in quantitative measurements, are the nonlinearity of response, the dynamics range (maximum 10^2) and the background noise level. Moreover, this method does not allow for energy and time resolution and a chemical development process is still necessary.

The challenge is the development of a realistic and feasible alternative to film detectors, that has to be digital, faster in acquisition, quantitative in the determination of radioactive isotope, competitive or, at least, comparable in spatial resolution performance and able to perform real time acquisitions for dynamic studies. An other application for such a method is the double-label measurement [10], which can greatly increase the information obtained from a radiolabelling experiment. The limited dynamic range and nonlinear response of the X-ray film make it difficult to determine the contribution of different isotopes to the autoradiographic image. The use of a digital detector has been suggested to eliminate these problems and to make quantitative double-label autoradiography possible and practical.

6.1.3 *New detectors for autoradiography*

Many attempts have been made in the last decades to design systems that would detect and localize the ionizing radiation for autoradiographic applications. We show here the main research fields and describe some commercial systems now available.

6.2 Imaging Plates

6.2.1 *Principles*

The Imaging Plate [11] is a radiation image sensor made of specifically designed phosphors that trap and store the radiation energy. The stored energy is stable until the plate is scanned with a laser beam, which allows the release of the energy as luminescence. In particular, an Imaging Plate is a flexible image sensor in which bunches of very small crystals (grain size about 5 μm) of photo-stimulable phosphors are uniformly coated on a polyester support film. The photo-stimulable phosphors used are BaFX:Eu, with X=Cl, Br or I, and SrS:CeSm. The most used is the barium fluorobromide containing a trace amount of bivalent europium as luminescence center. Exposure of samples to the Imaging Plate is performed in a similar manner to the film case. The exposed Imaging Plate is inserted in a phosphor reader and is scanned with a focused laser beam. The photostimulated luminescence released upon the laser is collected by a photomultiplier tube through a light guide and is converted to electric signals. Depending on the purpose on the collected image, reading density, reading sensitivity and sensitivity range may be selected and image analysis and data processing can be made. The Imaging Plate is reusable after erasing the residual latent image with uniformly irradiated visible light.

6.2.2 *Commercial systems and performance*

The various systems commercially available (for example PhosphorImager of Molecular Dynamics [10], Molecular Imager System of BioRad [12]) comprise: Phosphor Imaging Plates of various dimensions (up to about 600 cm^2) and efficiencies (each optimized for a specific energy range) from various companies (Fuji, Kodak, ecc.), a scanning/reading system and the dedicated software.

Many factors make the Imaging Plate method superior to the conventional film-based sensors [11]: the ultrahigh sensitivity (for ^{14}C marked samples it ranges from 0.1 to 2.1 cpm/mm^2), the wider dynamic range, from 10^4 to 10^5 (to be compared with 10^2 of the photographic method), and the superior linearity, as the fluorescence emission is proportional to the dose in the entire range. These differences are illustrated in Fig. 6.1 [11], where on the abscissa axis there is the radiation dose of a standard sample of ^{32}P used for exposure, measured by liquid scintillation counter, on the left ordinate there is the amount of luminescence from the Imaging Plate and on the right ordinate the blackened density of photo-film. Furthermore, digital electric signals are directly available and this method produces less detection counting errors (about 300 $events/mm^2$) even at high flux density and the accumulated background can be erased before an other use. The spatial resolution attainable is 100-400 μm and the time reduction with respect to the film for each application is reported in Table 6.2 [13]. In Figs. 6.2 [12], 6.3 [12], 6.4 [10] are shown some examples of autoradiographic images made with these systems.

Table 6.2 Comparison between exposure times required by film and imaging plates for different autoradiographic applications. From Ref. [13].

Application	Isotope	Exposure time (film)	Exposure time (image plate)	Reduction factor
Colony hybridization	^{32}P	2 days	20 minutes	1/150
Dot Blot hybridization	^{32}P	Not possible	13 hours	
Southern Blotting	^{32}P	2 days	14 minutes	1/200
Sequencing gel	^{32}P	20 hours	14 minutes	1/90
Chromatography	^{14}C	10 days	4 hours	1/60
Whole body Autoradiography	^{14}C	4 weeks	4 hours	1/170

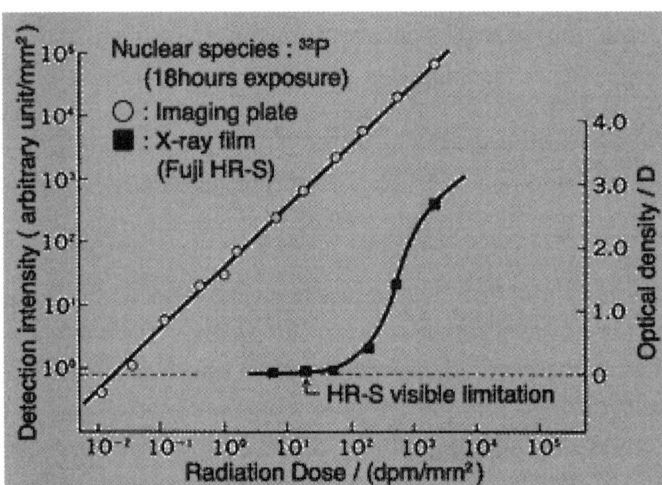

Fig. 6.1. Comparison of characteristics between Imaging Plate and photographic methods for autoradiography. From Ref. [11].

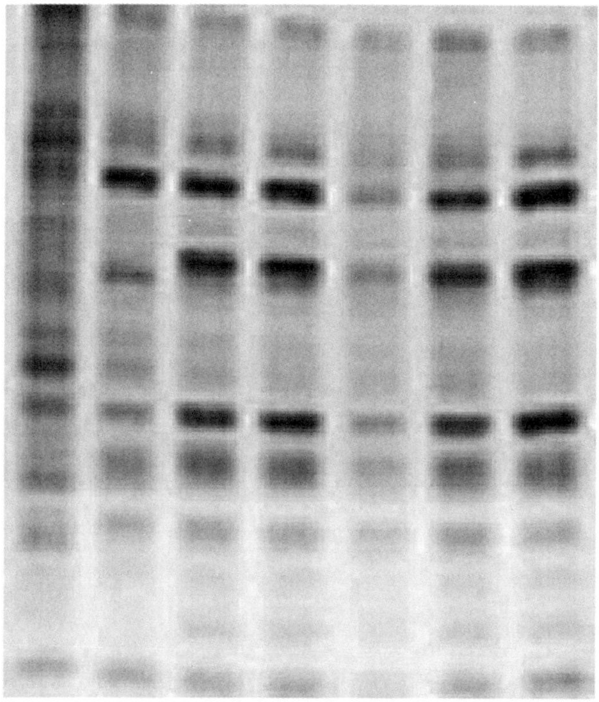

Fig. 6.2. ^{35}S labelled proteins separated on a polyacrylamide gel (by BioRad Molecular Imager System). From Ref. [12].

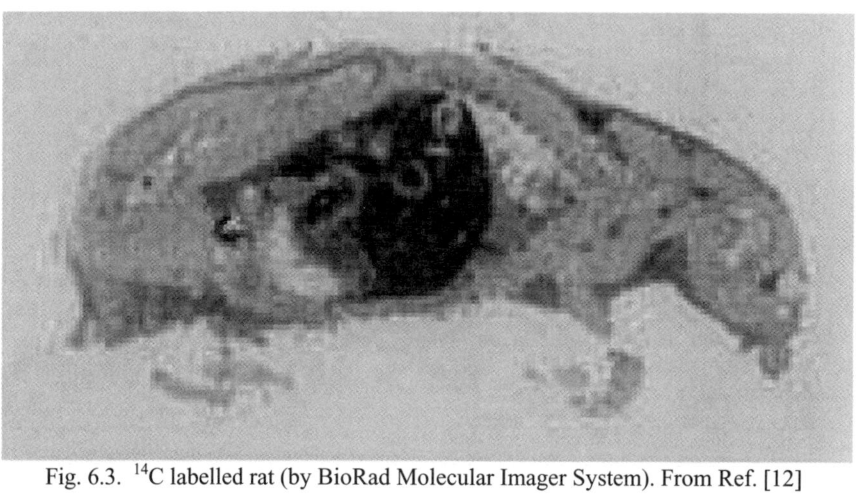

Fig. 6.3. ^{14}C labelled rat (by BioRad Molecular Imager System). From Ref. [12]

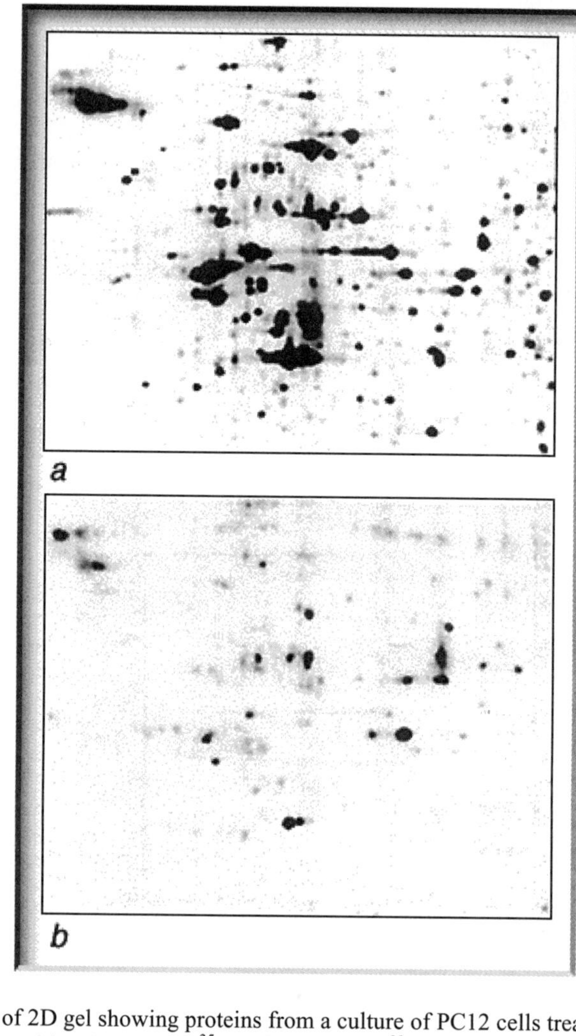

Fig. 6.4. Part of 2D gel showing proteins from a culture of PC12 cells treated with nerve growth factor and incubated with ^{35}S-methionine and ^{32}P-orthophosphate (by Molecular Dynamics PhosphorImager). From Ref. [10].
a)Direct exposure showing both ^{35}S and ^{32}P labelling.
b)Exposure of same region of the gel through copper foil, showing primarily ^{32}P labelling.

6.3 Gaseous Detectors

6.3.1 *Principles*

The simplest example of gas detector [14] consists of a container filled with a gas mixture, in which an electric field is established between two metallic electrodes (anode and cathode) by a bias voltage. When a ionizing radiation interacts with the atoms of the gas, it produces free electrons and positive ions which can produce further ionization and are collected at the electrodes producing an electric signal. The number of electron-ion pairs collected depends on the bias voltage [15]: up to a given voltage V_{min} the field strength is too weak to collect the ion-electron pair before they recombine. At voltages larger than a given V_{max} the output pulse is independent both from the energy lost by the ionizing radiation and the voltage (Geiger–Muller regime). Between V_{min} and V_{max} there lies the proportional region where the applied voltage is high enough to accelerate the electrons to the anode so that they collide with other atoms and cause ionization. In this regime the amplitude of the resulting signal is proportional to the energy lost by ionizing radiation and to the bias voltage.

Depending on the anodes and cathodes geometry and the readout electronics such a detector can give also spatial information about the impact point of the ionizing radiation.

6.3.2 *Research fields*

Since their introduction as particle detectors for high energy physics experiments by G. Charpak and F. Sauli [16,17], gaseous detectors have received great attention for their possible application in biomedical field.

One of the first attempts in using gaseous detectors in autoradiography has been a one-dimensional position sensitive Multi Wire Proportional Chamber (MWPC), designed by Van Eijk et al. [18], for the registration of six columns on a ^{14}C-β radiochromatogram. This MWPC has six anode wires of Au plated W, 30 mm apart (20 μm diameter, 200 mm length). Their positions correspond to those of the six columns on the chromatogram. The entrance window, that acts as the cathode plane, is aluminized on both sides with aluminized mylar foils (12 μm thickness). The other cathode plane is a printed circuit board with strips (2.5 mm width) perpendicular to the anode wires. The chamber is operated with a 90% Argon/10% CH_4 gas mixture. The chromatogram has been recorded with a spatial resolution of 3 mm FWHM.

A 2D MWPC is an area detector which has [19] an accurate bidimensional localization capability in detecting ionizing radiation. One of the main difficulties encountered in applying gaseous detectors to autoradiography is the long range of the β-rays in gas at atmospheric pressure. One way to overcome this limit has been the use of the multistep avalanche chamber. In this device [20] a wire chamber localizes a cloud of electrons from an avalanche produced in the uniform electric field created between two parallel electrodes. The sample with the radioactive compound is placed against the cathode, either outside the gap or so that its surface is the cathode plane. The anode of the gap is a grid transparent to electrons. This introduces a significant improvement in positional accuracy because, when a long-range β-particle penetrates such a gap, it produces ionization electrons along the track, but the multiplication of the electrons is not constant and the ionizing electrons produced upon the entrance of the β-particle give rise to the largest amplification.

One of the first attempt of using MWPC for 2D imaging in autoradiography has been made in the eighties at Pisa University [19,21], where one step of amplification with parallel electrodes has been used. This MWPC has an active area of 25x25 cm^2 with an anode-cathode gap of 4 mm. A cathode coupled delay line readout system is used for both X and Y coordinates. For imaging with ^{14}C β^- rays the MWPC was constructed with a very thin (6 μm) mylar window and a minimized inactive gas volume. A spatial resolution of 4.5 mm (FWHM) has been obtained with a detection efficiency of 20% and a sensitivity of about 1 Bq/cm^2. For imaging with tritium two windowless MWPCs have been built. The sample can be positioned inside the chamber 200 μm far from the cathode plane. The first chamber has an anode plane with a 2 mm pitch and operates at atmospheric pressure and in gas flow conditions. A spatial resolution of 1.5 mm (FWHM) has been measured along the wire direction, with an efficiency of 10% and a sensitivity of about 10 Bq/cm^2. The anode plane of the second windowless chamber has a wire spacing of 1 mm, at 45° with the cathode planes. The MWPC is positioned inside a steel box pressurized up to 4 atm. A spatial resolution of 800 μm has been measured along both directions for ^3H sources. Superimposing two MWPCs allowed also the simultaneous reconstruction of the two dimensional distribution of two tracers, with a spatial resolution of 800 μm for ^3H and 1.5 mm for ^{14}C. These systems have been applied to the study of variations in the ability of cell clones to incorporate a radioactive precursor of DNA biosynthesis, to study the relationship between myocardial regional supply and demand in normal and ischemic segments of the heart wall and to reconstruct a functional innervation mapping of the heart.

Bateman et al. [22,23] have developed a digital autoradiography system based on a multistep avalanche/MWPC consisting of an avalanche gap (4 mm thick in a first version, 1 mm thick in the last version, defined by two electroformed nickel meshes stretched and glued onto glass reinforced plastic frames) immediately against the sample plate, a 10 mm thick transfer gap and a conventional MWPC with delay line readout. The MWPC has an anode plane with 2 mm wire pitch, 20 μm diameter wires, separated by 6 mm from two mutually orthogonal cathode planes of 75 μm wires spaced at 1 mm. This system has demonstrated its ability to image tritium labelled electrophoretograms with a spatial resolution of 0.4 mm (FWHM) and ^{14}C and ^{35}S labelled samples with 0.8 mm (FWHM). The sensitivity levels achieved allow to reduce exposure times (relative to silver emulsions) from months to hours in the case of tritium and from days to hours in the case of higher energy labels.

At CERN [20] an other method has been developed which exploits the advantages introduced by the electron amplification due to avalanches in a uniform electric field in the gas between parallel electrodes, but with a readout method based on the optical imaging of the atoms excited in an avalanche, made visible by the addition of an efficient light-emitting vapour to noble gases such as Argon or Xenon. The basic principle is the following: when avalanches are reproduced by multiplication of electrons in gases under the effect of strong electric fields, a great variety of processes lead to the emission of photons with a spectrum ranging from vacuum ultraviolet (VUV) to visible. Several years of research permitted researchers to find the optimum conditions for the emission of light in a wavelength range convenient for optical imaging with the maximum amount of photons per charge produced in an electron avalanche at the largest charge gains. The optimum structure utilizes parallel-grid electrodes, with addition to the noble gases of TEA or TMAE (not easy to handle) and a thin plastic wavelength shifter. For the construction of the charge-amplifying structures, crossed-wire (50 μm diameter and 500 μm pitch) meshes have been used, which are transparent to light and also to electrons, if multiplication in successive structures is required. As the amount of light emitted by the avalanches is not sufficient to be directly detected by the CCD, it is equipped with an image intensification system. The use of multistep structures permits the gating of the image intensifier and the triggering of the data-acquisition system by the information from the first preamplification stage. This system permits accuracies of the order of 50 μm with ^{3}H labelled compounds and of the order of 200 μm with ^{35}S labelled compounds. The gain in time is two orders of magnitudine compared with photographic film.

6.3.3 Commercial systems

A commercial system which employs this method is the Beta Imager by Biospace Instruments [24]. It consists of a parallel plate avalanche chamber filled with gas mixture (Argon + TEA) and an intensified CCD camera (581 × 756 pixels). As an option, a zoom system (3× magnification) is also available for dedicated sample holders.

This system shows a linear response over a dynamic range of 10^4 and the detection efficiency ranges from 50% to 100% depending on the isotope. The smallest activity detectable is 0.007 cpm/mm^2 for ^{3}H, 0.01 cpm/mm^2 for ^{35}S, ^{14}C, ^{33}P and 0.1 cpm/mm^2 for ^{32}P. In the case of full field of view (20 cm × 25 cm) the spatial resolution is 200 μm for ^{3}H, 350 μm for ^{35}S, ^{14}C, ^{33}P and 500 μm for ^{32}P. In the case of maximum zoom (25 mm × 33 mm area) it is 60 μm for ^{3}H, 120 μm for ^{35}S, ^{14}C, ^{33}P and 160 μm for ^{32}P. The available software for processing allows to differentiate isotopes, and therefore to obtain the spatial distribution of each isotope after discrimination in a single acquisition.

An other commercial system based on gaseous detectors is the InstantImager by Packard [25]. The InstantImager [26] is a fully automated system which analyses and quantifies the distribution of radioactivity on flat samples like blots and gels, TLC plates or tissue sections. It is sensitive to ^{14}C, ^{32}P, ^{33}P, ^{35}S and ^{125}I and many other isotopes (not ^{3}H). The imaging area is 20 cm × 24 cm and the pixel size 500 μm × 500 μm (250 μm × 250 μm with interpolation). It uses a newly patented detector composed of three main components: a protective window, the MIcroChannel Array Detector (MICAD) and the multi-wire proportional chamber (MWPC). The sample is loaded into the sample drawer and covered with a reusable mylar sample comver. The sample holder is elevated to the detector window. The MICAD detector is composed of eight alternating layers of conductive and non conductive material (vrass) which are fused together forming a plate. About 210420 holes or microchannels, which are 0.4 mm in diameter, are drilled into the plate. Above the MICAD detector plate there is a multi-wire proportional chamber. It is composed of 200 ultra-thin gold anode wires flanked by two cathode planes running in the X and Y directions. This configuration of wires and cathode tracks is used to localize electron avalanches as they emerge from the microchannel of the MICAD detector plate. A gas mixture of 95% argon, 2.5% CO_2 and 1% isobutane fills the closed detector system. β particles emitted from the sample enter the nearest microchannel where gas ionization occurs. The voltage potential across the microchannel array plate accelerates the electrons in each channel and causes an electron avalanche which is detected in the anode/cathode above. Each avalanche will cause electrical signals in the tracks of the X and Y cathode planes. A fast digital signal processor is then used to calculate the centroid of the X and Y signals to accurately determine the position of the sample activity. The linear dynamic range is greater than 10^6, the average background per pixel is < 0.005 cpm (corresponding to 0.015 cpm/mm^2 and 750 cpm over the entire 20 cm x 24 cm counting surface area) and the maximum counts in one pixel is 65535. The counting efficiencies are approximately 1.8% for ^{14}C, 2.0% for ^{35}S, 5.0% for ^{33}P, 25.0% for ^{32}P 0.31% for ^{125}I. The microchannels of the MICAD detector prevent spreading of the radioactive emissions. The InstantImager is capable of resolving two bands of ^{14}C, ^{32}P and ^{125}I which are separated by 500 μm, 1.5 mm and 1 mm, respectively. InstantImager has been used for a wide variety of autoradiographic applications, including [25] quantitative analysis of dry and wet polyacrylamide and agarose gels; southern, northern and western blots, dot blots and high density and colony hybridizations; ^{14}C metabolism; ^{32}P DNA adduct assays on thin layer chromatography samples.

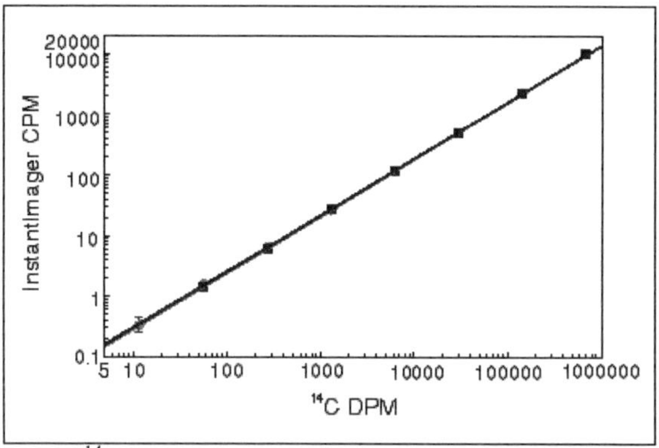

Fig. 6.5 Response to ^{14}C spots applied to silica TLC plate (by Packard InstantImager). Acquisition times were 30 minutes (solid squares) and 16 hours (open circles). Each point represents the mean of four spots for each dilution. From Ref. [25].

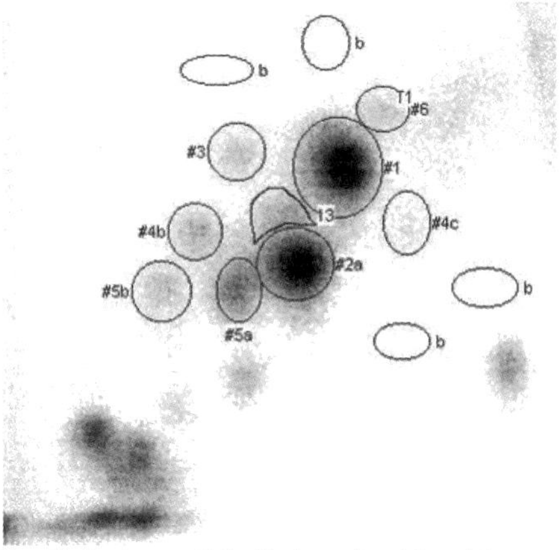

Fig. 6.6. Image of 32P activity in postlabelled nucleotide adducts on a 2D TLC plate (by Packard InstantImager). Regions of interest are overlaid with labels identifying the adducts. Regions used to determine background are labelled "b". From Ref. [25].

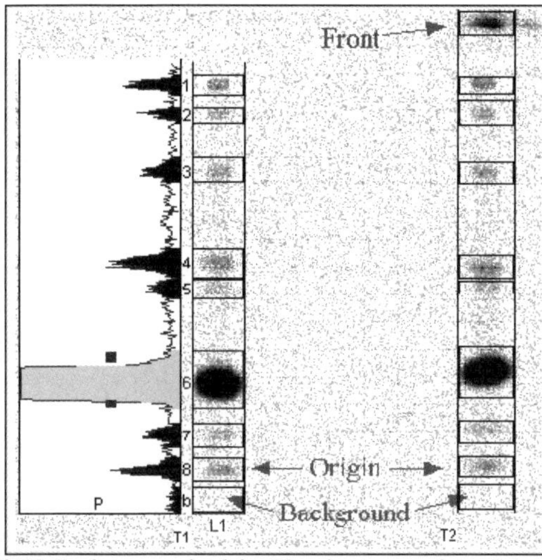

Fig. 6.7. Image of TLC plate acquired in 30 minutes used to determine radiochemical purity of ^{14}C labelled substance (by Packard InstantImager). From Ref. [25].

In Fig. 6.5 [25] the linearity curve of the response of InstantImager to ^{14}C is reported. In Figs. 6.6 [25] and 6.7 [25] we show some examples of autoradiographic images made with these systems.

6.4 Semiconductor Detectors

6.4.1 *Principles*

Semiconductor detectors [27] are based on crystalline materials whose outer shell atomic levels exhibit an energy band structure consisting of a valence band, an energy gap of a few eV and a conduction band. The valence and conduction bands are regions of many discrete levels which are so closely spaced that may be considered as a continuum, while the energy gap is a region in which there are no energy levels at all; electrons in the conduction band are detached from their parent atoms and are free to roam about the entire crystal, while electrons in the valence band are more tightly bound and remain associated to their respective lattice atoms. A few electrons can be excited into the conduction band by thermal energy, thus leaving an equal number of holes in their original positions. The passage of ionizing radiation creates more electron-hole pairs which can be collected applying an external electric field.

Different semiconductor detectors as noncrystalline materials, charge coupled devices and avalanche photodiodes will be treated separately in the following sections.

6.4.2 *Silicon strip detectors*

6.4.2.1 *Strip architecture*

Spatial information can be obtained using separate readout strips instead of surface electrodes. The possible use of a silicon strip detector system in autoradiography has been indagated, for applications as thin layer β-radiocromatography and electroforesis [28], since the time in which only 1D detection systems were available. Such a prototype, consisting of 128 strips (250 μm pitch) and associated electronics, has shown that simultaneously a detection efficiency of 50% and a spatial resolution better than 500 μm were possible with ^{14}C.

More recently, two dimensional microstrip detectors have been developed [29] in order to improve the recognition of tracks in experiments with colliding beams in high energy physics. The excellent spatial resolution for track reconstruction (1 mm) and the good noise performance (< 100 e^-) reached [30] by the last generation of high energy silicon bidimensional detectors are promising features also for autoradiographic applications. In particular, position sensitive microstrip silicon detectors are very well suited [31] for an intrinsically digital autoradiography since they are able to detect both electrons and low-energy photons, have a very good spatial resolution, determined by the microstrip pitch, and are capable of a real time display of the measured activity distribution.

6.4.2.2 *Research fields*

Double-sided silicon microstrip detectors and CMOS low noise self-triggering 16-channels chips have been tested with low energy X-rays and β-emitting sources for imaging of simple patterns by C. Ronnqvist et al [29]. These detectors, of n-type high resistivity bulk material, have strips on both sides, the n^+ strips perpendicular to the p^+ strips. Biasing resistors and coupling capacitors are integrated on the detector. The detectors used in this work are 280 μm thick and have an area of 64 × 64 mm^2 with 128 readout strips. The noise behaviour of the readout chip was measured on a four-channel prototype version to ENC = 155 e^- + 16.4 e^-/pF. With two 16-channels readout chips bonded on the two sides of the detector the image of a probe needle irradiated with 60 keV photons has been acquired with an active detector area of 1.6 x 1.6 mm^2. A VLSI readout chip, dedicated to high resolution β ray imaging with silicon microstrip detector has been realized by R. Bauer et al. [30]. Typical values obtained with a 32-channels prototype version are: gain of 92 mV/MIP; output linearity better than 10% for an input signal range of 10 MIP; peaking time of 600 ns; noise ENC = 110 e^- + 20 e^-/pF for a power consumption of 2.5 mW/channel.

E. Bertolucci et al [31] have developed a digital autoradiography system based on a double sided 300 μm thick silicon strip detector with a pitch of 100μm. Each strip is read out by a charge preamplifier, amplified, and fed into a discriminator; discriminated signals are fed into TDCs to time-stamp the arrival of a particle in that strip. An architecture using a network of transputers [32] compares the signal arrival times on all the *X* and *Y* channels, searching for time coincidence events. By a parallel algorithm the transputers reconstruct the radioactivity distribution. In the first prototype 16*X* and 16*Y* channels have been read and two TDCs and two transputers have been used, so the active surface area was 2.56 mm^2. Larger areas have been obtained with a high precision microscanning of the sample in the plane. This system was able to perform imaging of organic material with specific sensitivity as small as 0.002 nCi/mm^2 and to record activity measurements with good linearity in the range 0.002–20 nCi/mm^2 (specific activity) and have been also obtained autoradiographic images of mammary cells marked with ortho-(^{32}P)phosphate.

6.4.2.3 *Commercial systems*

Recently has been produced by IDE (Integrated Detectors and Electronic) AS (Oslo, Norway) a commercial system called Bioscope. The Bioscope system [33,34] is a 640 × 640 channel radiation detection system based on silicon microstrip detectors and self-triggering read-out chips. The main components are a double-sided silicon microstrip sensor, 10 low noise preamplifier chips, Analog Transceiver Cards, a Digital Transceiver Card, a Digital Signal Processor and a PC for storage, data and image processing.

The detector, manufactured by Hamamatsu, is 300 μm thick and is equipped with 640 strips with a pitch of 50 μm on both the p and n sides so that the sensitive area extends over 32 × 32 mm^2. The ohmic separation on the *n* side is achieved by implanting 16μm wide p-stop strips between the *n* strips. Each strip is read out by a built-in capacitor (AC coupling) consisting of the strip implants, insulating layers of oxide and nitride and a metallization layer. The strips on both sides are biased through individual poly-silicon resistors from 50 to 60 MΩ. The read-out electronics is a 128 channel self-triggering chip in CMOS VLSI technology developed by IDE AS. Its low noise preamplifiers and CR-RC shapers have an ENC = 60 e^- + 15 e^-/pF at 2 μs peaking time. The chip performs a fully data-driven readout cycle and only the relevant data need to be processed by the electronics outside (sparse readout). The Analog Transceiver Cards supply voltages and control currents to the chips and amplify their signals, while the Digital Transceiver Card serves as a coincidence unit, generating a trigger signal only if exactly one channel from each side has fired within an adjustable coincidence interval. The data acquisition system consists of a 16-channel ADC card controlled by a Digital Signal Processor and the data are then transferred into the PC for display, storage and further processing. Until now the Bioscope system has been tested by acquiring images of β^- emitting sources, showing encouraging results with respect to stability, noise and number of insensitive strips. As next step the spatial resolution (now related with the pitch) will be improved by charge interpolation and the energy resolution (now about 5 keV) will be enhanced by implementing an optimized calibration procedure. In Fig. 6.8 [33] is reported the block diagram showing the main components of the Bioscope system and in Figs. 6.9 [33] and 6.10 [33] are reported autoradiograms made with this system.

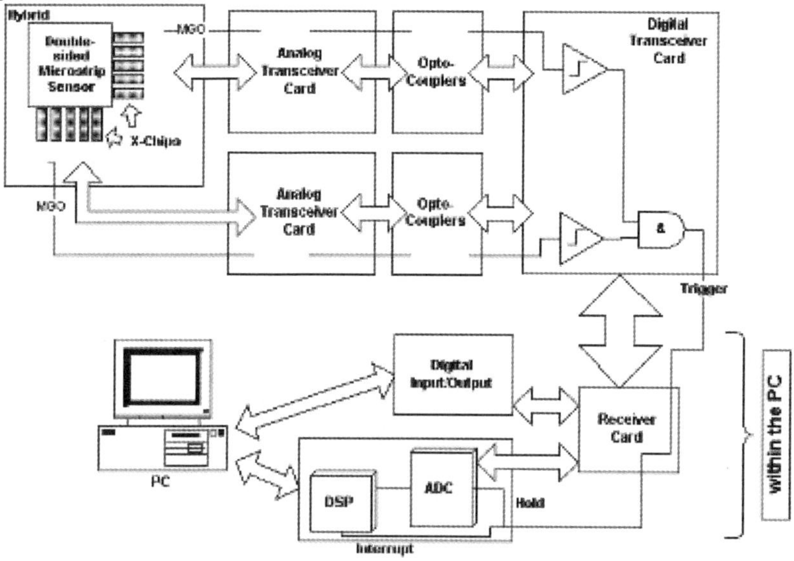

Fig. 6.8 Block diagram showing the main components of the Bioscope system. From Ref. [33].

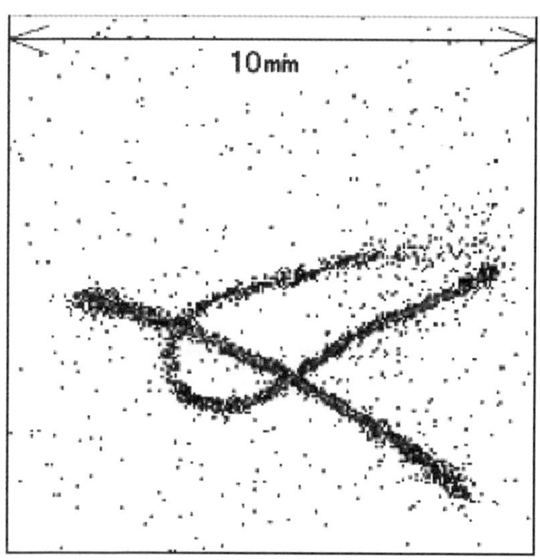

Fig. 6.9 Autoradiogram of two thin hairs labelled with ^{35}S (by IDE AS Bioscope). From Ref. [33].

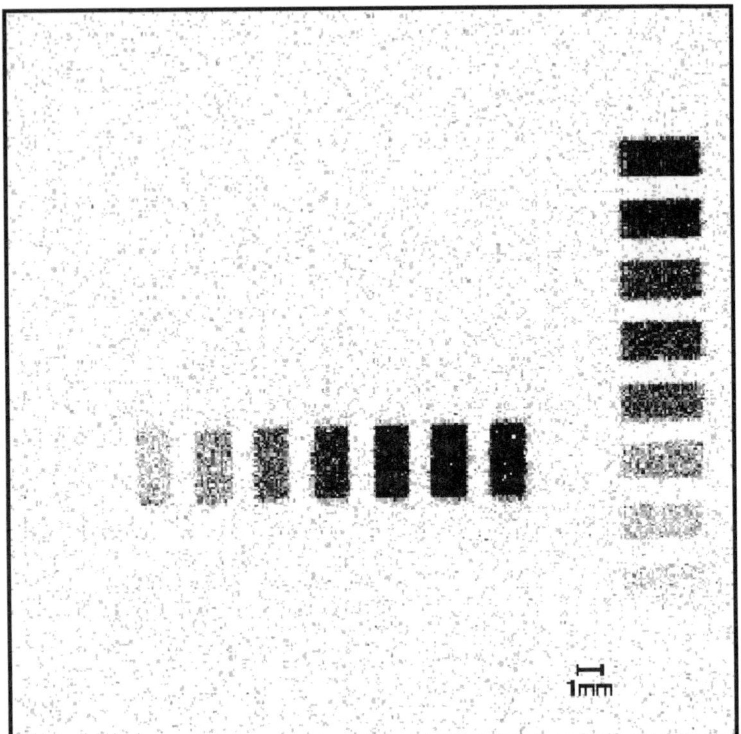

Fig. 6.10 Autoradiogram of two ^{14}C microscales (by IDE AS Bioscope). From Ref. [33].

6.4.3 *Pixel detectors*

6.4.3.1 *Pixel architecture*

With semiconductor detectors a two-dimensional spatial information can be obtained also by pixel architecture, in which one of the two electrodes is segmented in small area cells (pixels). Up to now, however, the use of pixel detectors have suffered [35] from a limit in the available sensitive area due to electronics fan-out. Since the bump bonding technique has been successfully used to connect pixel detectors to their electronics chip, larger sensitive areas have become possible.

6.4.3.2 *Research systems*

A new type of beta camera based on an Imaging Silicon Pixel Array (ISPA) tube combined with planar plastic scintillators or with $SiY_2O_5(Ce)$ scintillating powder has been developed at CERN [36].

The ISPA-tube consists of a multialkali photocathode evaporated on the optic fiber entrance window of a vacuum-sealed tube. Behind the photocathode at 30 mm distance there is a 8 mm × 4.8 mm SACMOS silicon chip (Omega2 chip [37]) containing 1024 pixels of 75 μm × 500 μm edges. Each pixel is bonded via Pb–Sn solder bumps to its proper front-end electronics including a charge amplifier, a comparator with adjustable threshold, a delay line, a coincidence logic and a flip-flop. The 64 columns of 16 pixels each are binary readout at 5–10 MHz in a line-parallel mode, resulting in readout times of 6 to 12 μs. In the prototype the active surface is 7 mm × 4 mm, limited by a diaphragm. As planar beta detectors have been tested several disks of polystyrene scintillators doped with additional wavelength shifters, with thickness ranging from 30 to 120 μm and in some cases aluminized at their entrance faces (forming a reflecting mirror in order to increase the light fraction reaching the photocathode). In addition, 2.6 mg/cm^2 SiY_2O_5 scintillating powder has been compared with the polystyrene disks. To determine the spatial resolution of the beta camera people have used brass templates 1.5 mm (^{90}Sr–^{90}Y) and 0.15 mm (^{14}C, ^{63}Ni) thick with slits 60 μm wide, obtaining values ranging from 105 μm FWHM (^{63}Ni source and 30 μm thick plastic scintillator) and 240 μm FWHM (^{90}Sr–^{90}Y source and 120 μm thick plastic scintillator). The number of photoelectrons produced by the energy losses (photoelectron sensitivity) of β particles along their paths in the planar detector discs ranges between 0.2 and 0.7 photoelectrons/keV. The detection sensitivities for ^{3}H and ^{14}C sources have been evaluated with activity-calibrated autoradiographic microscales, consisting of methacrylate polymers of 21 × 35 mm^2 area and 50 μm (^{3}H) or 120 μm (^{14}C) thickness sandwiched between inactive polymer strip, and amount to 0.1 Bq (^{3}H) with 150 min counting time and to 0.0125 Bq (^{14}C) with 180 min counting time. With this beta camera people have also been performing measurements of "in situ" hybridization of ^{35}S.

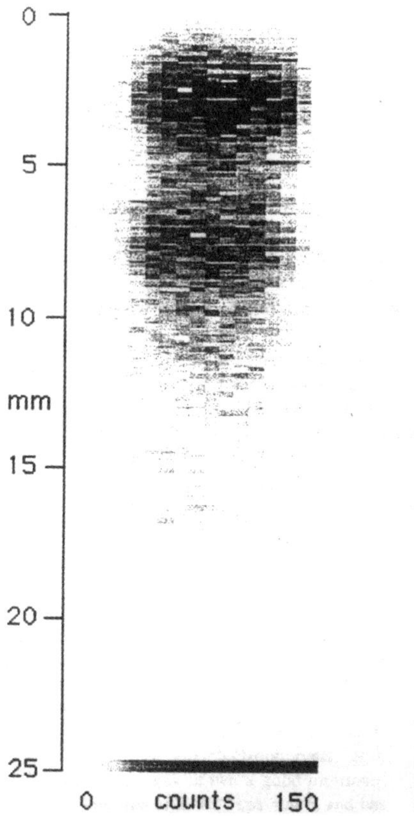

Fig. 6.11 Image of a DNA sequence obtained with a Si pixel detector after 10h exposure. Pixel size is 75x500μm². From Ref. [38].

The same Omega2 electronics has also been used by S.R. Amendolia et al. [35] in conjunction with a Silicon detector to obtain autoradiographic images of clusters of human epithelial cells and DNA fragments separated via electrophoresis, both labelled with ^{32}P. The Si detector is a 300 μm thick Silicon crystal on which a pixel matrix made by 64 rows and 16 columns of 75 × 500 μm^2 pixels has been defined and connected via bump bonding to the corresponding chip. The read-out is triggered using the amplified and discriminated signal from the back side of the detector (common for all pixels), so there is no need for an external trigger, and a threshold value of about 11500 electrons (40 keV in Si) was set for all pixels. This system has shown a good minimum detectable activity per unit area of 2×10^{-4} cps/mm^2, has proved linear for count rates in the typical autoradiographic range 0.2–20 cps, and was able to clearly localize clusters of cells which have incorporated the radioactive tracer and DNA fragments on an electrophoretic gel on paper (blots). In Fig. 6.11 [38] is reported the image of a DNA sequence obtained after 10h exposure.

Together with these measurements, a test with a GaAs detector equipped with the more recent Omega3 electronics has been successfully done also. The detector is a 200 μm thick semi-insulating undoped Gallium Arsenide wafer on which a matrix of 128 rows and 16 columns of 50×500 μm^2 was defined, for a total sensitive area of 6.4×8 mm^2.

Omega3 is a an improvement of the Omega2 system, different in geometry, with similar features from the point of view of the measurements above described, except for the additional feature of the "self-triggered" mode (the read-out strobe is generated on-chip by the fast-OR of each column of pixels).

The evolution of the above mentioned systems is BETA*view*, developed by E. Bertolucci et al. [39,40,41]. It is based on a GaAs detector and a low threshold, single particle counting electronics developed by the Medipix collaboration for medical applications. In general, now the detector can be a 200 μm thick GaAs or a 300 μm thick silicon crystal equipped with 64×64 square pixels (150 μm pixel size, 20 μm separation between adjacent pixels). Each pixel is bump-bonded to a corresponding channel of the Photon Counting Chip [42] which contains an amplifier, a discriminator and a 15-bit counter. Each channel independently records the number of detected particles over the energy threshold with a maximum count rate of 2 MHz (set via a common threshold voltage level, and programmable via software by a 3-bit adjustment value), and after the acquisition the counters are sequentially read, 16 columns at the time. Linearity and sensitivity of this system have been investigated both by simulation [39] and experimental studies [41] with β sources of medical and biological interest such as ^{14}C, ^{35}S and ^{32}P. With threshold values ranging from 6000 to 8500 electrons has been obtained the result of 0.25 cps/cm^2 background count rate.

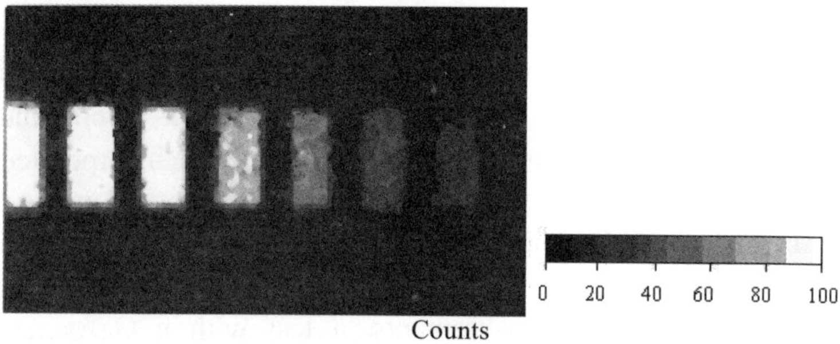

Fig. 6.12. BETA*view* image of [^{14}C] Microscale Autoradiography Standard (1.13÷32.86 kBq/g); the image consists of a 64×128 pixels matrix imaged in two steps, each made of 7h acquisition time over a 64×64 pixel area (1.1×1.1 cm^2). From Ref.[38].

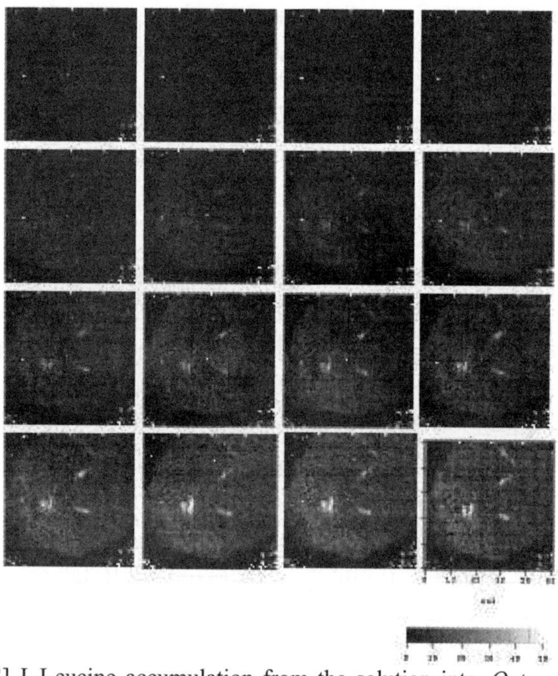

Fig. 6.13. [^{14}C] L-Leucine accumulation from the solution into *Octopus vulgaris* eggs over time. From Ref. [38]. Each frame, in the sequence of 16, shows the total accumulation of counts up to the time interval incremented of 5 min for each frame. The 16th (the frame number increase proceeding from left to right and from top to bottom) frame corresponds to an acquisition time of 16×5 min = 80 min. The large circular area corresponds to the labelled solution contained in the cylindrical sample holder. The five eggs are seen as "hot spots" with increasing radioactivity concentration during the experiment. Some noisy pixels at the chip border can also be noted.

Quantitative measurements of radioactive spots have been done with drops of solutions of radio-labelled proteins (for ^{14}C and ^{35}S) or ^{32}P-phosphates deposited on paper filters; quantitative images of a northern blotting filter (^{32}P marker) have been obtained and the results have been compared with a commercial MWPC (Canberra Packard InstantImager, already described in this chapter). Moreover, thanks to the key feature that an energy threshold can be easily set via software, images of two radioactive markers (^{32}P and ^{35}S) have been obtained. More recently [40] this system showed also the capability of monitoring in real time a dynamic process consisting in the kinetics of the uptake of an amino acid, its synthesis in a biological system and the subsequent migration of the radioactivity into the inner part of the cells. In particular, there has been success with the [^{14}C]L-Leucine uptake by eggs of mollusk cephalopod *Octopus vulgaris* and the consequent migration of marker into the ooplasm, with an extimated activity in each egg less than 1Bq. Figures 6.12 [38] and 6.13 [38] show some results obtained with of this system [40].

6.5 Amorphous Materials

6.5.1 *Principles*

In amorphous materials, semiconductor properties [43] are affected by the addition of dopants in the reaction vessel. For example, in an idealized picture of the resulting material in the case of silicon, the short-range structure is similar to that of c-Si (crystalline silicon) but the long-range is of a more random nature; the addition of hydrogen, to obtain a-Si:H, serves to reduce the density of unpaired valence electrons which create electron and hole traps. The electron and hole mobilities for a-Si:H (2 and 10^{-2} $cm^2/Vsec$, respectively) are less than those for c-Si (1300 and 500 $cm^2/Vsec$, respectively) and decrease with decreasing temperature. Selenium can exist in crystalline form but it is the amorphous form [44] (a-Se) which can be formed in large uniform sheets. a-Se is believed as an inorganic polymer consisting of covalently bonded chains and although not a crystalline structure an effective band gap of 2.2 eV can be ascribed to it. It is a photoconductor, which convert radiation energy directly into electric charge and the intermediate step of conversion into visible light is unnecessary.

The inherent large-area nature of this technology have made amorphous materials based devices very interesting for many applications including medical radiation physics.

6.5.2 *Research and commercial systems*

Antonuk et al. [43] in the nineties have developed two-dimensional arrays [45,46] consisting of a regular arrangement of a-Si:H photodiode sensors coupled to thin-film transistors used in conjunction with scintillating materials.

The photodiodes have a *p–i–n* structure, in which the i-layer constitutes most of the thickness (about 1 μm, compared to the considerably thinner doped layers of 10 – 100 nm) and the FETs have a standard structure (source, drain and gate contacts); there is also a polyimide (nearly transparent plastic material) layer to provide a planarization and passivation layer. Bias and switching voltages to the arrays and conversion of analog signals to digital form are provided by external electronics. Incident radiation is converted by the scintillating material in light photons which generate electron-hole pairs in the depletion layer. During acquisition, with the FETs nonconducting, the electron-hole pairs result in a discharging of the sensor capacitance, which is initially charged to V_{bias}, and the degree of this discharging constitutes the integrated imaging information which is sampled during readout, when the corresponding sensors are reinitialized by rendering the FETs conducting. After a row is read out, the associated FETs are made nonconducting and the process is repeated for the next row until the entire array has been sampled to give a single image. With this kind of device some images have been acquired with arrays of 270, 450 and 900 μm pitch.

a-Si image detection has recently reached a stage of maturity that finally allows the technology to be integrated into commercial products. For example, dpiX [47] supplies image sensors arrays with pixel sizes from 127 to 392 μm and image formats from 20 × 25 to 30 × 40 cm^2

which can be supplied with a coating of cesium iodide as scintillating layer for X detection. Such arrays are presents in Varian and Thomson Tubes Electroniques products. a-Se employs a direct conversion process which eliminates intermediate conversion steps to achieve image quality which exceeds indirect digital methods and is comparable to fine grain radiographic film. The commercially available [48,49] systems based on a-Se technology employ an a-Se photoconductor coated over an a-Si thin-film transistor (TFT) array. P.K. Soltani et al. [50] have described the device used in Agfa products. This device employs a uniform a-Se vapor deposited over field-effect transistor (FET) array forming a full active area of 35 x 43 cm^2. The Se layer has a dielectric and top electrode layers applied to form a capacitor structure where charges generated during exposure are collected under an applied 10 V/μm field which additionally prevents lateral spreading of the generated charges. The charge collection and integration is accomplished with a square charge collection electrode deposited over the FET array. The pixel pitch is 139 μm, with a geometric fill factor of 86%. The charge collection electrode is attached to a signal storage capacitor, which is in turn attached to the drain side of the FET. The readout is similar to that described for a-Si and, after each exposure, the device also performs background corrections by a look up table (LUT) which stores a reference image of the detector array response to a flat exposure and by the acquisition of dark images before and after each exposure to evaluate the amount of dark current during exposure. This system has been characterized also [50] with X-rays in terms of exposure, energy response and imaging capabilities showing significant performance improvements over indirect digital methods, making it useful in film replacement applications.

6.6 CCD Based Systems

6.6.1 *Principles*

Charge Coupled Devices (CCDs) are [51] made of silicon and consist of two dimensional arrays of tiny potential wells. The electrons released after the interaction with the ionizing radiation are trapped in these wells.

This charge information is then readout by successively shifting the charge from one well to the next until it reaches the output electronics.

The main problems in the use of CCD for radioisotope imaging are the small sensitive area and the low detection efficiency for photons with energy from 30keV and charged particles. So their use in association with scintillators has been indagated.

6.6.2 *Systems description and performance*

In 1988 Y.Charon et al. [52] have developed a detector which associates a charge coupled device to a light amplifier applied to the localization of β^-emitters, included in thin slices, with a high resolution and a short acquisition time. The β source is placed above a thin scintillating plastic. When an electron crosses this sheet, it produces photons which are emitted in a small area. Their number depends on the nature and thickness of the plastic, the initial energy of the electrons and their angular distribution. The photons are conducted through an optical fiber entry plate to the light amplifier photocathode. This light is intensified and transferred to the window of the CCD by a pipe of optical fibers plates. A gateable electronic camera drives the CCD and digitalizes the analog value of each pixel. This system is able to analyze a surface of $60mm^2$ with a spatial resolution of 15μm and a detection efficiency of 100% (relative to the measured standard source) for ^{35}S detection.

Karellas et al. [53] investigated the adaptation of a cooled Charge-Coupled Device for some autoradiographic applications. The proposed approach uses a scientific-grade CCD which is optically coupled via a lens to a scintillator without intermediate amplification. The performance of the CCD is investigated with the option of a fiberoptic coupling between the scintillator and the CCD. In this system the area coverage is 9 cm × 9 cm and the sensitive area of the CCD is 31 mm × 31 mm with a square pixel size of 0.015 mm on a side. The experimental prototype uses a lens coupling (*f*/1.2) without intermediate intensification between a very thin Gd_2O_2S:Tb scintillator (< 100 μm thick) and the CCD. In order to reduce the dark current, the CCD is cooled to -55°C using liquid nirogen. This approach was used for radionuclide imaging in the integrating mode for detection of ^{99m}Tc emission. In addition to experiments in the integrating mode, calculations were carried out for the potential adaptation of the device for pulse-height analysis in the counting mode.

More recently, a prototype of digital autoradiographic system, incorporating a cooled scientific CCD, has been developed by J.H. MacDonald et al. at the Physics Department of the Institute of Cancer Research (ICR), Sutton, UK [54]. The ICR system is designed to image β^- particles and low energy X-rays from radiolabelled biological tissue samples, using direct irradiation of the CCD without any optical coupling between the sample and the CCD. The CCD (Table 6.3 [54]), specifically designed for scientific applications, is read out in slow-scan mode (from 50 kHz up to 1 MHz) and can be cooled to any temperature between 140 K and 300 K in order to prevent the build up of dark current. The sensitive area of the device is covered by a 2 mm thick polyimide layer, which protects the electrode structure on the front surface when samples are placed against it. The sensitive volume can be split in two regions: the depletion layer, 7 μm thick, and the field-free layer, 13 μm thick.

Table 6.3 Charge Coupled Device parameters

Parameter	Value
Active area	17.3 x 25.9 mm^2
Pixel format	1152 x 770
Pixel pitch	22.5 x 22.5 μm^2
Total sensitive thickness	20 μm
Depletion region thickness	7 μm
Operating temperature	73 K – 233 K
Peak signal (e-/pixel)	500000
Charge transfer efficiency	99.9995%
Readout noise (e-/pixel)	12

To generate charge in the CCD, photons or particles have to reach the epitaxial layer, or sensitive volume. To do this, they must pass through the polyimide, electrode and oxide layers, which are typically 3 μm thick in total. Photons of energy up to few keV and β^- particles of energy less than about 10 keV will be generally absorbed in these layers and will not generate any appreciable signal. The charge liberated in the depletion layer is collected with virtually 100% efficiency and little radial displacement, while the charge liberated in the field-free layer diffuses radially until it either reaches the depletion layer and is collected, or reaches the potential barrier between the epitaxial silicon layer and the p^+ type silicon substrate. Around 90% of the charge reaching this barrier is reflected towards the depletion layer and collected, but it degrades the spatial resolution.

This prototype system, tested using line phantoms (three lengths of 10 μm diameter gold-plated tungsten wire coated with ^{35}S, ^{125}I and ^{32}P respectively) over an area of about 4 cm^2, shows position resolution of 30 μm, 30 μm and 40 μm respectively, i.e. resolution degrades with higher energy emitting isotopes. System response is linear over a dynamic range of at least 10^4, with a signal to noise ratio approaching 300.

6.7 Avalanche Photodiodes

6.7.1 *Principles*

An avalanche photodiode detector (APD) [55] is a *p–n* junction realized in a silicon wafer structured in such a way that it may be operated near breakdown voltage under reverse bias. The electrons generated by interactions with ionizing radiation and accelerated by the high electric field gain sufficient velocity to generate free carriers by impact ionization, resulting in internal gain which can improve the signal to noise ratio relative to standard silicon sensors. A typical APD consists of several regions including a 20 μm thick drift region and a 120 μm thick active junction which contains a less than 10 μm thick multiplication region. Free charge carriers generated in the drift region undergo transport to the active region. The gradation in dopant concentrations causes a small electric field to impart a net movement of electrons towards the multiplication region. Electrons entering this region quickly attain velocities large enough to cause knock-on collisions with bound electrons in the lattice, releasing additional electron-hole pairs which can also undergo new collisions, so the multiplication process occurs many times producing a significant net gain in the electrical signal. This signal multiplication can be useful in detection of low energy particles as the beta emission of tritium (maximum 18 keV, mean energy 6 keV). Tritium is widely used in neurological and biological research involving proteins, because it does not interfere with their normal conformal structure. Usually detection of these low energy betas requires the destructive and time consuming manipulation of the samples and placing them into a scintillating fluid, while in this case the tritiated samples can be positioned directly under specially processed (with a very thin front window) APDs.

6.7.2 *Systems description and performance*

J.S. Gordon et al. [55] have developed arrays of 4×4 pixels fabricated with 1 mm^2 elements with 0.3 mm dead space between them. The standard processed APDs have a 0.3 μm dead layer consisting of a 0.1 μm native oxide (SiO_2) layer grown on the APD face for passivation and an about 0.2 μm thick diffused boron contact layer. They have succeeded in making arrays with good tritium detection efficiency (27%). The main areas of improvements in detection efficiency were due to fabrication of ultra-thin dead layers (< 0.15 μm) along with the low noise associated to the low capacitance of the small area, well isolated pixels.

6.8 Microchannel Plates

6.8.1 *Principles*

Microchannel Plates (MCPs) are flat glass discs containing millions of small diameter pores [56]. MCPs were developed in the late 1960s [57] as compact imaging electron multipliers, each channel acting as a continuous dynode multiplier. They have been used in image intensifiers and in detectors for photons, X-rays, neutrons and charged particles. MCPs are manufactured by combining individual glass fibres to create the channels and glass matrix which comprise the plate. These fibres are prepared by inserting a bar of etchable "core" glass into a tube of "cladding" glass. These fibres are then stacked into an array which is drawn out to produce a hexagonal "multifibre". The multifibres are stacked to form a hexagonal "boule" and then sliced to produce individual MCPs which are polished and then have the core glass etched out to open up the channels. At this point MCP destined to become electron multipliers are fired in a hydrogen oven to render the glass surface semi-conducting and allow replacement of photoelectrons via the electrodes which are deposited by evaporation on the faces of the plate.

MCP detectors were originally developed [58] at Leicester University (UK) to study the X-ray emissions from active galactic nuclei. The same scientists have applied successfully these detectors in biological research, obtaining very good results in many autoradiographic applications. Imaging of radiolabelled biological samples [59] began in the Physics Department in 1995, with the use of high resolution, low background, large area microchannel plate (MCP) detectors (developed for the NASA Chandra X-ray observatory). In particular, this device seems able in efficiently detecting tritium, for which most digital autoradiography methods show a low sensitivity due to the low energy of the emitted β.

6.8.2 *System description and performance*

The detector developed by J.E. Lees et al. [60] consists of a pair of MCPs in a chevron arrangement with a resistive anode readout. The inter-plate gap is 200 μm wide while the rear MCP to anode distance is 2 mm. The active area of the detector, defined by the rear electrode (93 mm on a side) is 86.49 cm^2. A carbon coated polypropylene window in front of the detector acts as an ion and UV shield. The MCPs, purchased from Philips Photonique [61], were manufactored from their "standard" MCP and propriety "low noise" glass. The anode consists [60] of a square resistive sheet 110 mm on a side on a ceramic substrate with charge collecting electrode at the vertices [62] manufactured by Hymec

[63]. Minimising the anode size keeps the capacitance, and hence the anode thermal noise, as low as possible. Each electrode signal is amplified by a charge sensitive preamplifier and then by a doubly differentiating, doubly integrating filter amplifier. Distorsions associated with resistive anode readout are removed by means of a "look up" table generated by imaging a well known array of pinholes (600 μm diameter holes, 4 mm pitch). In Table 6.4 are described the MCP parameters.

The behaviour of MCP detectors at low count rates is related to background noise. The main source of detector background noise from "standard" plates is from the potassium in the MCP glass (β emission from ^{40}K) [64, 65]. Low noise MCPs, however, are manufactured from material containing no radioactive isotopes (proprietary ^{40}K and ^{87}Rb free glass) therefore all detectors body parts are made from PCTFE (Poly-Chloro-Tetra-Fluoro-Ethylene) [66] rather than materials known to contain radioactive elements [67].

Table 6.4. MCP parameters

Parameter	Value
Thickness (L)	1.5mm
Channel diameter (D)	12.5μm
Channel pitch (p)	15.0μm
L:D	120:1
Channel bias angle	6^0

In this way dark noise rates are 0.12 counts/cm^{2}sec above a discrimination level of 0.05 times the peak detector gain [68]. For these low noise MCPs, most of the backgound is due to 1.46 MeV ^{40}K γ-rays from concrete in the laboratory walls [60]. The detector background can be reduced to cosmic ray levels (0.015 counts/cm^{2}sec) by surrounding the detector with lead shielding; the use of such a shielding and anti-coincidence technique (to reject cosmic ray backgound independently measured by a plastic scintillator) allows one to obtain an essentially noiseless detector.

The behaviour of MCP detectors at high count rates has been investigated in recent years [69,70,71,72] to make a comprehensive model [73,74,75] which predicts the count rate characteristics of the MCP detector with known resistance, capacitance, bias voltages, low count rate, peak gain and illuminated area. MCP count rate can be improved reducing the MCP resistance [69,76] and MCPs having gold electrodes rather than nichrome have a much improved (up to 20 fold) count rate performance.

The principle of using the MCP developed at the Leicester University has been proved in 1995 [57] by imaging the "fingerprint" of slime mould DNA, labelled with ^{35}S. Comparison of the original film contact print and the electronic image obtained with the electrophoresis gel in close proximity to the MCP detector in vacuum showed that the DNA could be sequenced equally well from either representation. More recently, MCPs have been used by the same scientists in autoradiographic applications involving ^{14}C [77] and ^{3}H [78,79,80] markers. In particular, the MCP detector is highly sensitive to the biologically important but previously difficult to detect low energy β emission from tritium [81].

The response to ^{14}C has been investigated [77] using two two-stage "chevron" MCP detectors, one with a large active area (93 mm x 93 mm) and the other with a smaller active area (30 mm diameter). The spatial resolution of both detectors was measured using commercially available masks. For the large area detector has been found a FWHM of ~60 μm using a USAF 1951 Test Target placed in close proximity (~100 μm) to the MCP input surface and irradiated with 0.28 keV X-rays; for the small format MCP has been used a photo-etched mesh in direct contact with the front MCP illuminated with a mercury vapour lamp (~2540 angstrom), obtaining a FWHM of ~26 μm. Sensitivity and linearity have been measured using an autoradiography Microscale™ positioned within 20 μm of the input MCP surface. The resulting limiting sensitivity is 0.031 Bq and the detector is linear over the entire range of the Microscale, showing a full dinamic range of six orders of magnitude. Autoradiographic applications obtained with this device and ^{14}C marker include a ^{14}C labelled rat whole body slice showing selective radiopharmaceutical uptake in clearly identifiable organs and ^{14}C-Tamoxifen doped semi-thin tissue sections of rat womb and liver.

The first measurements involving ^{3}H have been [81] made to obtain spatial resolution, sensitivity and linearity range of the device already described. An image of a tritium standard source [82] mounted 0.5 mm from the large area MCP input surface has been accumulated under vacuum (operating pressure $<10^{-6}$ mbar) for 20 hours. From the edge response function of the brightest cell [81] a spatial resolution of 400 μm FWHM was estimated; a minimum detectable activity of 0.001 Bq in 20 hours [79], over a wide (5-6 orders of magnitude) dynamic range have been also obtained. To improve spatial resolution in imaging ^{3}H labelled biological samples, the samples have been placed in direct contact with the input surface of the smaller MCP detector by an automatic sample loading and vacuum load-lock mechanism designed [79] to accommodate a wide variety of sample types. With this technique have been acquired the images of 1 μm thick slices of rat lung [79,81] labelled with tritiated Putrescine (Fig. 6.14 [83]), immunological assays [79] (Fig. 6.15 [83]) and 1-D and 2-D electrophoretic gels [78,79] for proteine identification; comparison with other systems sensitive to tritium (gas proportional counters, imaging plates, avalanche photodiodes) showed very good results. The same authors have also studied [80] three approaches to increase the beta detection efficiency of MCP detectors for autoradiography: reversing the MCP bias polarity, changing the conventional high negative voltage (which repels all betas below about 4 keV, 30% of the total) on the input MCP to a grounded input; a reduction in the MCP pore size from 12.5 μm to 6 μm; using a CsI coating as an efficient secondary electron emitter. Beta efficiency may be enhanced by careful selection of pore size and channel bias angle and further enhancement can be achieved using a CsI photocathode, which also confers dual-tracer capability. The properties of the CsI coated MCP to simultaneously map high and low energy beta particle distributions have been exploited with ^{3}H and ^{14}C (Fig. 6.16 [84]).

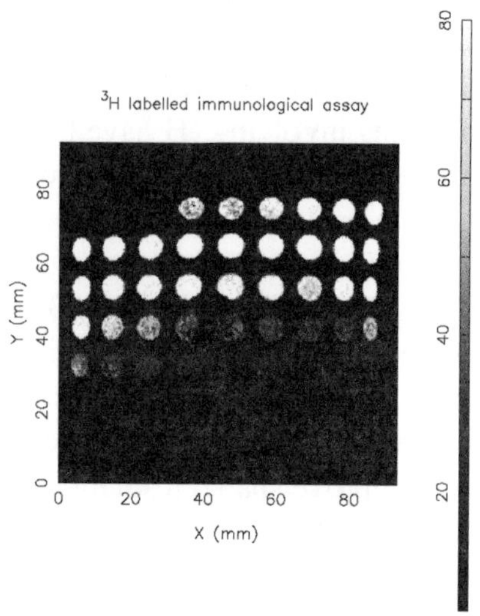

Fig. 6.14. Image of a ^{3}H labelled lymphocyte proliferation assay. From Ref. [83].

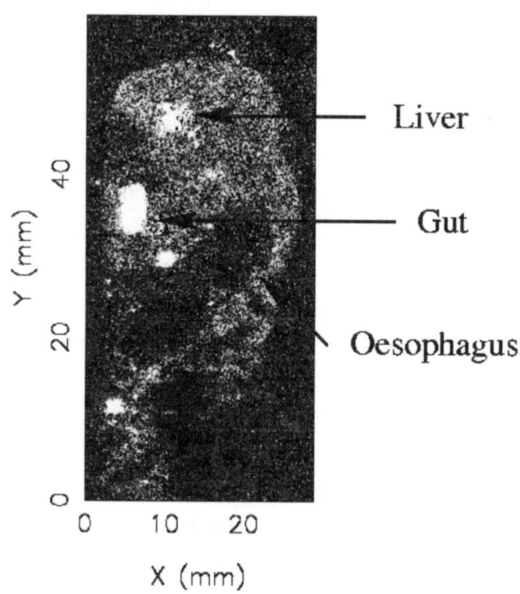

Fig. 6.15. Image of a 3H labelled whole body thin tissue of a rat showing uptake in various organs. From Ref. [83].

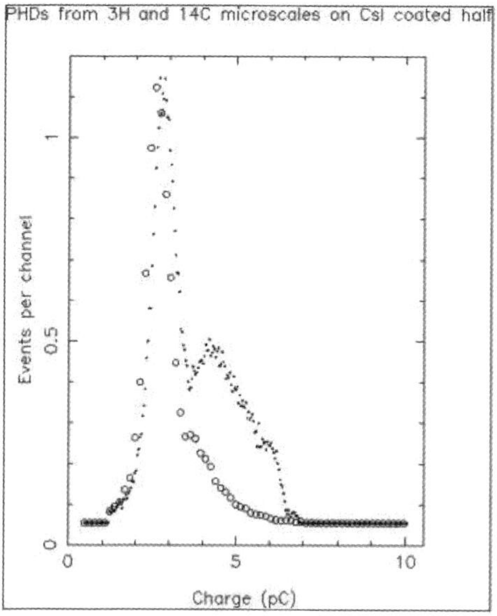

Fig. 6.16. Normalised PHDs from ^{3}H (circles) and ^{14}C (dots) microscales. From Ref. [84].

References

1. www.ambion.com
2. www.med.unc.edu
3. http://www.med.unc.edu/pmbb/MBT2000/Northern%20Blotting.htm
4. http://www.med.unc.edu/pmbb/MBT2000/Western%20Blotting.htm
5. http://www.ilford/com
6. http://www.nirs.go.jp
7. Takahashi et al., *Nucl. Instr. and Meth.* ***A*422**, 751-755 (1999).
8. http:/www.kodak/com
9. http://www.eabiotech.com
10. http://www.mdyn.com
11. http://home.fujifilm.com
12. http://www.bio-rad.com
13. Y. Amemiya, et al., *Imaging plate illuminates many fields*, Nature, 336, November (1988)
14. W. R. Leo, Techniques for Nuclear and Particle Physics Experiments (Berlin Heidelberg: Springer-Verlag), pp.119-148 (1967)
15. R. Cesareo, X-ray physics: Interaction with matter, production, detection (Bologna: Editrice Compositori), pp.89-102 (2000)
16. G. Charpak et al., *Nucl. Instr. and Meth.* **62**, 262 (1968)

17. G. Charpak and F. Sauli., *Nucl. Instr. and Meth.* **113**, 381 (1973)
18. C. W. E. Van Eijk, et al., *Nucl. Instr. and Meth.* ***A*273**, 764-766 (1988)
19. F. Angelini, et al., *Nucl. Instr. and Meth.* ***A*269**, 430-435 (1988)
20. W. Dominik, et al., *Nucl. Instr. and Meth.* ***A*278**, 779-787 (1989)
21. R. Bellazzini, et al., *Nucl. Instr. and Meth.* **204**, 517 (1983)
22. J. E. Bateman, et al., *Nucl. Instr. and Meth.* ***A*264**, 430-435 (1988)
23. J. E. Bateman, et al., *Nucl. Instr. and Meth.* ***A*241**, 275 (1985)
24. http://www.biospace.fr/en/bi.php
25. http://ww2.packardinst.com
26. http://x.biochem.nwu.edu
27. W. R. Leo, Techniques for Nuclear and Particle Physics Experiments (Berlin Heidelberg: Springer-Verlag), pp.207-239 (1967)
28. E. M. Schooneveld, et al., *Nucl. Instr. and Meth.* ***A*288**, 250-253 (1990)
29. C. Ronnqvist, et al., *Nucl. Instr. and Meth.* ***A*348**, 440-443 (1994)
30. R. Bauer, et al., *Nucl. Instr. and Meth.* ***A*376**, 443-450 (1996)
31. E. Bertolucci, et al., *Nucl. Instr. and Meth.* ***A*381**, 527-530 (1996)
32. M. Conti, et al., *Nucl. Instr. and Meth.* ***A*345**, 120 (1994)
33. http://opal.physik.ubi-bonn.de/html/bio_main.htm
34. M. Overdick, et al., *Nucl. Instr. and Meth.* ***A*392**, 173 (1997)
35. S. R. Amendolia, et al., *IEEE Trans. On Nucl. Sci.* **44** (3), 929-933, June (1997)
36. D. Puertolas, et al., *IEEE Trans. on Nucl. Sci* **43** (5), 2477-2487, October (1996)
37. E. H. M. Heijne, et al., *Nucl. Instr. and Meth.* ***A*349**, 138 (1994)
38. P. Russo, private communication (2002)
39. E. Bertolucci, et al., *Nucl. Instr. and Meth.* ***A*422**, 242-246 (1999)
40. E. Bertolucci, et al., Real Time β-imaging with silicon hybrid pixel detectors: kinetic measurements with C-14 amino acids and P-32 nucleotides. Presented at Nuclear Science Symposium and Medical Imaging Conference, San Diego, CA, USA, 4-10 Novembre 2001, and in press on *IEEE Trans. Nucl. Sci.* Vol. **49** (2002)
41. L. Abate, et al., *Nucl. Instr. and Meth.* ***A*460**, 97-106 (2001)
42. M. G. Bisogni, et al., *Proc. SPIE* **3445**, 298-304 (1998)
43. L. E. Antonuk, et al., *Med. Phys.* **19** (6), 1455-1466, Nov/Dec (1992)
44. http://www.sunnybrook.utoronto.ca:8080/~hunter/selenium.html
45. L. E. Antonuk, et al., *IEEE Trans. on Nucl. Sci.* **37** (2), 165-170 (1990)
46. L. E. Antonuk, et al., *IEEE Trans. on Nucl. Sci.* **38** (2), 636-640 (1991)
47. http://www.dpix.com/sensors/sensors.html
48. http://ndt.agfa.com
49. http://www.kodak.com/go/health
50. http://www.ndt.net/abstract/dg_bb67/data/22.html
51. W. R. Leo, Techniques for Nuclear and Particle Physics Experiments (Berlin Heidelberg: Springer-Verlag) pp.231 (1967)
52. Y. Charon, et al., *Nucl. Instr. and Meth.* ***A*273**, 748-753 (1988)
53. A. Karellas, et al., *IEEE Trans. on Nucl. Sci.* **40** (4), 979-982, August (1993)
54. J. H. MacDonald, et al., *Nucl. Instr. and Meth.* ***A*392**, 220-226 (1997)
55. J. S. Gordon, et al., *IEEE Trans. on Nucl. Sci.* **41** (4), 1494-1499, August (1994)

56. http://www.src.le.ac.uk/mcp/optics/mcp-optics.html
57. http://www.src.le.ac.uk/mcp/optics/mcplates.html
58. G. W. Fraser, and J. E. Lees, Conference on Academic and Industrial Cooperation in Space Research 432, 87-90, (1999)
59. http://www.src.le.ac.uk/mcp/betaauto/
60. J. E. Lees, et al., *Nucl. Instr. and Meth.* ***A*384**, 410-424 (1997)
61. Philips photonique, Avenue Roger Roncier, B.P. 520,F-10106 Brive la Gaillarde Cedex, France
62. G. W. Fraser, et al., *Nucl. Instr. and Meth.* **179**, 591 (1981)
63. B. V. Hymec, Dr. Nolenshan 107 B, Postbus 336, 6130 AH Sittard, The Netherlands
64. G. W. Fraser, et al., *Nucl. Instr. and Meth.* ***A*254**, 447 (1987)
65. O. H. W. Siegmund, et al., *IEEE Trans. on Nucl. Sci.* **35** (1), 524-528, February (1988)
66. Poly-Chloro-Tetra-Fluoro-Ethylene, Fluorocarbon Company Ltd, Caxton Hill, Hertford, England, SG 13 7NH
67. G. W. Fraser, et al., *Proc. SPIE* **982**, 98 (1988)
68. J. E. Lees, et al., *Nucl. Instr. and Meth.* ***A*384**, 410 (1997)
69. O. H. W. Siegmund, et al., *Proc. SPIE* **1549**, 81 (1991)
70. D. C. Slater, et al., *Proc. SPIE* **1549**, 68 (1991)
71. G. W. Zombeck, et al., *Proc. SPIE* **1549**, 90 (1991)
72. M. L. Edger, et al., *Rev. Sci. Instr.* 63, 816, (1992)
73. G. W. Fraser, et al., *Nucl. Instr. and Meth.* ***A*306**, 247 (1991)
74. G. W. Fraser, et al., *Nucl. Instr. and Meth.* ***A*324**, 404 (1993)
75. G. W. Fraser, et al., *Nucl. Instr. and Meth.* ***A*327**, 328 (1993)
76. W. B. Feller, et al., *Nucl. Instr. and Meth.* **A310**, 249 (1991)
77. http://www.src.le.ac.uk/projects/bioimaging/publications/c14.html
78. J. E. Lees, et al., *Electrophoresis* **20**, 2139-2143 (1999)
79. O. H. W. Siegmund, et al., *IEEE Trans. on Nucl. Sci.* **46**, 636-638 (1999)
80. J. E. Lees, et G. W. Fraser, 5th International conference of Position Sensitive Detectors, University College, London, 14–15 September (1999)
81. J. E. Lees, et al., *Nucl. Instr. and Meth.* ***A*392**, 349-353 (1997)
82. American radiolabeled chemicals Inc., 11624 Bowling Green, St. Louis, MO 63146
83. http://www.scr.le.ac.uk/mcp/betaauto/paper.htm
84. http://www.scr.le.ac.uk/mcp/betaauto/paper_v3.htm

CHAPTER 7

SPECT and Planar Imaging in Nuclear Medicine

Alessandro Passeri, Andreas R. Formiconi
Department of Clinical Physiopathology - Nuclear Medicine Unit
University of Florence
Italy

7.1. Introduction

Nuclear medicine cameras used for single photon emitters follow a design developed by Anger[1] about four decades ago.

Overall ability to perform accurate data acquisition and reconstruction in single photon imaging tasks depends on a number of clinical and physical factors. The development potential of nuclear medicine techniques relies mainly on the thorough analysis of such factors and, whenever useful, on the possibility of transferring this knowledge into the mathematical algorithms devoted to highlight information contained in the "raw" data set.

The quantity of γ-rays usefully detected depends on *i)* factors related to the morphology of the object (patient or phantom) under investigation, *ii)* procedural factors dealing with the specific acquisition performed, *iii)* technical factors related to the instrumentation involved and *iv)* factors related to physical effects like Compton diffusion and attenuation of the electromagnetic radiation. Once the raw data have been obtained, either in a planar or in a tomographic acquisition, the final image depends on the reconstruction or post processing methods employed.

It is worthwhile noting that imaging hardware limits the information available for any given imaging procedure. For instance, projection data acquired with an inappropriate collimation system may lead to a degraded image whose original characteristics, despite post processing recovery efforts, can never be restored. It is known that the collimator choice determines the trade-off between image spatial resolution and detection efficiency. The spatial resolution may limit the detectability of a lesion present in the processed image, thereby influencing the accuracy in size quantitation. It is also known that the spatial resolution worsens as the distance of the radioactive source from the collimator surface increases, so that the correct post-acquisition management of the collimator blur deals with a spatially non-stationary effect.

On the other hand, the acquisition efficiency is strictly related to the *signal to noise ratio* parameter, whose value determines the numerical reliability of the acquired data. The latter parameter is not uniquely related to the collimator mechanical characteristic, but it also concerns the acquisition efficiency of the active detection system. It is known that the detection efficiency, that is the ability of the detection system to stop incident radiation minimizing the energy escape, improves as the 4th power of the effective atomic number of the detection material. At present, as it was in the original design, most camera detectors consist in a slab of NaI:Tl scintillator, 3/8 or 3/4 inch thick, whose light output is read by an array of photomultiplier tubes (PMT). The summed output of all PMT signals is used to reject events that have undergone Compton diffusion in the body of the object under investigation, and would otherwise contribute to blur the raw data set. With this respect, it is obvious that several groups oriented their research in identifying and investigating new materials whose properties fit as much as possible the "ideal" detector.

As a final point, it must be assessed that accurate imaging potential of emission computed tomography (ECT), relies mainly on the introduction into the reconstruction algorithm of the physical *"a priori"* information concerning the experimental acquisition process. As long as the radiation emission and detection processes in emission tomography can be considered as linear operators acting as blurring functionals over the solution space, three main elements can be highlighted in the solving process of an ECT problem:

(1) The mathematical accuracy of the transformation kernel describing the physical behaviour of the variables involved in the formulation of the problem.
(2) The ill-posedness (see section [7.4.2]) degree of the peculiar problem under examination.
(3) The magnitude of noise effects into the experimental data set.

Once the formation process of projection images could be considered accurately modelled, the choice of the best suited image reconstruction method is oriented toward those classes of algorithms whose structure is able to manage the inversion of the Radon transform taking full account of the system modellization efforts.

The general objective of this chapter is to provide the readers with a basic, but nonetheless accurate understanding of the arguments previously sketched. A paragraph devoted to a few but meaningful examples of imaging from single photon emitters will complete this chapter.

7.2. Collimators

In modern nuclear medicine the scene is dominated by advances in electronics and digital computing technologies. However, image quality is still mostly determined by the collimator. The use of the collimator is unavoidable in single photon emission nuclear medicine since, in order to realize the concept of projection, the only choice is to select photons with the desired direction. This task is realized by rejecting all others photons by means of absorption.

The collimation process involves a great waste of information: approximately 1 photon in 10000 is able to pass through a hole and to contribute to image formation. It is possible to construct a collimator so that more photons are accepted, by making larger holes for instance. However, the negative aspect is that excess photons have directions which differ from the ideal projection direction. We have therefore the following tradeoff: if more photons are acquired, then the spatial resolution of images is degraded; if the only photons whose trajectories are close to the ideal one are selected, then the count statistic is poor.

It is significant that literature continues to be full of contributions related to various aspects of collimation, despite the apparent simplicity of collimators with respect to other components of the cameras.

Collimators are basically very simple pieces of instrumentation but there are many ways to construct them. There are several aspects which in principle have to be determined when designing a collimator: hole shape (circular, triangular, rectangular, hexagonal), hole array pattern (square, rhombic, triangular), hole convergence (parallel, convergent, divergent, fan beam, cone-beam), hole tapering, hole diameter, collimator thickness, hole separation (septa thickness), collimator-detector distance and material.

Not all these features have the same influence on the performances of the collimator; by far the most important parameter is the ratio between the size of the hole and its length. The influence of this parameter is evident if one looks at the basic formulas which give estimates of spatial resolution and efficiency of a generic parallel hole collimator.

The spatial resolution is usually estimated from the Full Width at Half Maximum (FWHM) of the Point Spread Function (PSF) as

$$\text{FWHM} = \frac{D}{L}(z_0 + L + B) \tag{7.1}$$

where D is the diameter of the hole, z_0 is the source distance, L is the thickness of the collimator and B is the distance between the collimator back face and the image plane located in the crystal. The image plane is not a well defined geometric plane; a short discussion of this concept appears later in this chapter. For a multi-hole configuration, efficiency is given by [2]

$$g = k\left(\frac{a_{\text{hole}}}{a_{\text{cell}}}\right)\left(\frac{a_{\text{hole}}}{4\pi L^2}\right) \tag{7.2}$$

where k is the efficiency of transformation of photons in detected events in the digital image, a_{hole} is the area of a hole at the patient side, a_{cell} is the area of a single cell of the periodic hole array and L is the collimator thickness. The first factor in Eq.(7.2) is the so-called "air-to-total" ratio and the second factor estimates the fraction of solid angle subtended by a single hole. In these formulae the shape of the holes is not specified, therefore the exact meaning of D as well as the relationship between D and a_{hole} are undetermined here. However in any case the proportionality $a_{\text{hole}} \propto D^2$ will hold true and, consequently, the same will hold true for the well known relationship between resolution R and efficiency

$$g \propto R^2. \tag{7.3}$$

Other parameters, such as the hole shape or the hole array pattern have a smaller influence and they are chosen by manufacturers for fine tuning of the collimators. Arrangement of holes in convergent geometry patterns is often used for increasing the counting efficiency, the latter being a critical quantity particularly in SPECT studies. The most common convergence configuration is the fan beam, where focusing is provided only along one axis so that the axis of each hole passes through a point belonging to a focal line. In a cone-beam collimator all holes converge to a unique focal point, instead to a focal line. Cone-beam collimators are more efficient. For instance, for a source located at 15 cm from a fan-beam with the focal line at 50 cm, the gain in efficiency is about 1.8 with respect to the same collimator but with parallel holes. With a cone-beam, in the same situation, the gain in efficiency is 2.3. On the other hand, the sampling pattern of cone-beam collimators complicates the reconstruction of SPECT data: thus only fan-beam collimators are commercially available so far.

Hole tapering must not be confused with convergence. Convergence concerns the direction of the axis of holes and the same pattern of convergence may be associated with different kind of tapering. The effect of hole tapering is not large but it may be noticed when accurate descriptions of the collimator response are sought [2; 3]

Table 7.1. Collimator material.

Material	Z	A	$\rho(gr/cm^3)$
Lead	82	207	11.35
Tungsten	74	184	19.30
Tantalum	73	181	16.60
Gold	79	197	19.32
Uranium	92	238	18.95

Hole separation influences the efficiency since, for a given hole aperture a_{hole}, it determines the measure of a_{cell}. Larger hole separation corresponds to thicker septa, a larger value of a_{cell} and, therefore, lower efficiency (Eq.(7.2)). However, septa cannot be made too thin, otherwise the penetration will be prominent. On the other hand septa cannot be too large because hole pattern artifacts may appear. So hole separation is an important parameter for the design of a collimator.

The distance between the collimator and the crystal, called B in Eq.(7.1), is not a parameter strictly dependent on the collimator but it is a fixed parameter for a certain camera-collimator setup and it influences the collimator performances too. The existence of this distance is due to two main reasons. First, the collimator is always kept at a distance of about 3 or 4 mm in order to protect the fragile camera crystal. Second, photons travel through the crystal before they interact; the travelled distance is a statistical quantity spatially distributed according to the attenuation law for electro-magnetic radiation. This means that the concept of image plane cannot be a well defined one. It is however commonly accepted assuming that the equivalent image plane is located inside the crystal at a depth equal to the mean free path of photons. For instance, when the energy of 140 keV of the emission of ^{99m}Tc is concerned, the mean free path is about 2 mm. Thus, typical values of B ranges around 6 mm. Recently, the influence of the B parameter on artifacts caused by the hole array pattern has been studied [4].

The specific density of the employed material is a crucial characteristic of the collimator since it determines the ability of absorbing photons if they have undesired directions. Unfortunately, the choice is very limited since it must be restricted to high density materials with high atomic numbers and the few available alternatives to lead could pose manufacturing or fabrication problems. Table 7.2 reports the materials that have been used so far. In the following, a general theory regarding the multi-hole collimator behaviour will be presented. Single-hole collimator theory will be sketched as a particular case of the multi-hole theory as well.

7.2.1. *Multi-hole theory*

The fluence of photons detected at the point (x, y) of the camera detection plane for a point source of unitary activity placed in the point of coordinates (x_0, y_0, z_0) defines the concept of point spread function (PSF) of the collimator conceived as imaging instrument:

$$\phi(\mathbf{r}, \mathbf{r}_0, z_0) = k \frac{\mathcal{A}_s(\mathbf{r}, \mathbf{r}_0, z_0)}{4\pi \left[(z_0 + L + B)^2 + |\mathbf{r} - \mathbf{r}_0|^2\right]} cos\theta \tag{7.4}$$

where $\mathbf{r} = (x, y)$, $\mathbf{r}_0 = (x_0, y_0)$, $\mathcal{A}_s(\mathbf{r}, \mathbf{r}_0, z_0)$ is the aperture function of the camera collimator system at the point (x, y), k is the efficiency factor of the camera detector and θ is the angle between the direction through the source point and point (x, y) on the detector plane and the direction through the source point perpendicular to the source plane (see Fig.(7.1)). The $1/\left[(z_0 + L + B)^2 + |\mathbf{r} - \mathbf{r}_0|^2\right]$ factor describes the decrease with the square of the distance and the $\cos\theta$ factor the effect of the non perpendicular incidence of the radiation on the detector plane in its peripheral regions. The 4π factor at the denominator tells us that $\phi(\mathbf{r}, \mathbf{r}_0, z_0)$ is the fluence for a unitary activity of the point source, provided that the source emits radiation isotropically in space. In other words, in this formulation, $\phi(\mathbf{r}, \mathbf{r}_0, z_0)$ is the efficiency of detection in the point (x, y) of the detection plane for a point source placed in (x_0, y_0, z_0).

It is worthwhile to observe that the PSF as defined by Eq.(7.4) is spatially non-stationary with respect to both (*i*) the source position, and (*ii*) the coordinate of the detection plane, as a consequence of the $\mathcal{A}_s(\mathbf{r}, \mathbf{r}_0, z_0)$ functional dependence from the $(\mathbf{r}, \mathbf{r}_0)$ coordinates.

The aperture function depends on the geometrical properties of the collimator: shape, size, length, orientation and hole array pattern are the principal parameters to be taken into account. Symbols used for describing

Table 7.2. Symbols describing the geometric characteristics of a camera-collimator system.

Symbol	Parameter
L	hole length
W	hole side
σ_s	holes distance (centre-to-centre)
F_x	focal distance along x-axis
F_y	focal distance along y-axis
B	collimator-image plane distance

the geometric features of a camera-collimation system are resumed in Table 7.2

According to [5] and [6] the aperture function of a hole can be defined by means of a function $a(\mathbf{r})$ with $a(\mathbf{r}) = 1$ for positions $\mathbf{r}$ inside the aperture and $a(\mathbf{r}) = 0$ for $\mathbf{r}$ outside the aperture. Let $a_f(\mathbf{r})$ and $a_b(\mathbf{r})$ represent

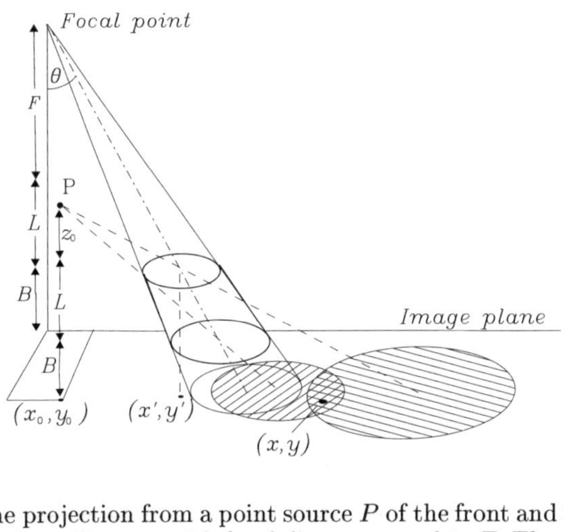

Figure 7.1. The projection from a point source P of the front and back aperture of a tapered hole of a collimator with focal distance equal to F. The source is located at the distance z_0 from the collimator front face and its orthogonal projection onto the image plane is given by $\mathbf{r}_0 = (x_0, y_0)$. $\mathbf{r}' = (x', y')$ is the orthogonal projection of the centre of the hole front aperture and $\mathbf{r} = (x, y)$ is the point of detection on the image plane. The larger shaded area is the projection onto the image plane of the front aperture function $a_f^p(\mathbf{r}, \mathbf{r}'; \mathbf{r}_0, z_0)$ defined by Eq.(7.6), while the smaller shaded area is the projection of the back aperture function $a_b^p(\mathbf{r}, \mathbf{r}'; \mathbf{r}_0, z_0)$ defined by eq.(7.7). The intersection between the two shaded areas is given by the product $a_f^p(\mathbf{r}, \mathbf{r}'; \mathbf{r}_0, z_0)\, a_b^p(\mathbf{r}, \mathbf{r}'; \mathbf{r}_0, z_0)$ and it represents the aperture function of a hole. The aperture function of the collimator $\mathcal{A}(\mathbf{r}, \mathbf{r}_0, z_0)$ is obtained by summing the aperture functions of all the holes (see text).

the aperture functions at the front and back planes of the collimator, respectively. In order to evaluate the aperture function of a collimator's bore for a source located at the point (x_0, y_0, z_0) it is necessary to find out the fraction of detector surface that γ-rays can reach by passing through both the front and back aperture of the bore. It is now possible to define the projection of an aperture: it deals with the fraction of detector's surface that the aperture makes available to the point source. Let $a_{\mathrm{f}}^{\mathrm{p}}(\mathbf{r}, \mathbf{r}'; \mathbf{r}_0, z_0)$ represent the projection of the front aperture and $a_{\mathrm{b}}^{\mathrm{p}}(\mathbf{r}, \mathbf{r}'; \mathbf{r}_0, z_0)$ the projection of the back aperture, respectively.
Thus the collimator aperture can be estimated as

$$\mathcal{A}_{\mathrm{a}}(\mathbf{r}, \mathbf{r}_0, z_0) = \int_{-\infty}^{\infty} \int_{-\infty}^{\infty} a_{\mathrm{f}}^{\mathrm{p}}(\mathbf{r}, \mathbf{r}'; \mathbf{r}_0, z_0)\, a_{\mathrm{b}}^{\mathrm{p}}(\mathbf{r}, \mathbf{r}'; \mathbf{r}_0, z_0)\, d\mathbf{r}' \tag{7.5}$$

where the subscript a in $\mathcal{A}_{\mathrm{a}}$ reminds that the expression (7.5) approximates the true aperture function.

In the previously cited formulations, the projections of the front and back aperture functions onto the image plane were found as

$$\begin{aligned} a_{\mathrm{f}}^{\mathrm{p}}(\mathbf{r}, \mathbf{r}'; \mathbf{r}_0, z_0) = a_{\mathrm{f}} \Bigg(& \frac{z_0}{z_0 + L + B} x + \frac{L + B}{z_0 + L + B} x_0 - x' \\ & , \frac{z_0}{z_0 + L + B} y + \frac{L + B}{z_0 + L + B} y_0 - y' \Bigg) \end{aligned} \tag{7.6}$$

$$\begin{aligned} a_{\mathrm{b}}^{\mathrm{p}}(\mathbf{r}, \mathbf{r}'; \mathbf{r}_0, z_0) = \; & a_{\mathrm{f}} \Bigg[\left(\frac{F_{\mathrm{x}}}{F_{\mathrm{x}} + L} \right) \left(\frac{z_0 + L}{z_0 + L + B} \right) x \\ & + \left(\frac{F_{\mathrm{x}}}{F_{\mathrm{x}} + L} \right) \left(\frac{B}{z_0 + L + B} \right) x_0 - x' \\ & , \left(\frac{F_{\mathrm{y}}}{F_{\mathrm{y}} + L} \right) \left(\frac{z_0 + L}{z_0 + L + B} \right) y \\ & + \left(\frac{F_{\mathrm{y}}}{F_{\mathrm{y}} + L} \right) \left(\frac{B}{z_0 + L + B} \right) y_0 - y' \Bigg] \end{aligned} \tag{7.7}$$

It may be useful to observe that Eq.(7.7) involves parameters for both the F_{x} and F_{y} focal distances. Of course, specific collimator configurations can be obtained as summarized in Table 7.3.

Table 7.3. Collimator types.

Collimator	Focal lenghts
parallel	$F_{\mathrm{x}} \to \infty, F_{\mathrm{y}} \to \infty$
fan beam	$F_{\mathrm{y}} \to \infty$
cone beam	$F_{\mathrm{x}} = F_{\mathrm{y}}$

Considering expressions (7.6) and (7.7), Eq.(7.5) can be written in the following way

$$\mathcal{A}_{\mathrm{a}}(\mathbf{r}, \mathbf{r}_0, z_0) = \int_{-\infty}^{\infty} \int_{-\infty}^{\infty} a_{\mathrm{f}}\left(-\boldsymbol{\sigma}\right) a_{\mathrm{f}}\left(\mathbf{r}_{\mathrm{T}} - \boldsymbol{\sigma}\right) d\boldsymbol{\sigma} \tag{7.8}$$

where

$$\boldsymbol{\sigma} = (\sigma_{\mathrm{x}}, \sigma_{\mathrm{y}})$$

$$\mathbf{r}_{\mathrm{T}} = (x_{\mathrm{T}}, y_{\mathrm{T}})$$

and

$$\boldsymbol{\sigma} = \mathbf{r}' - a\mathbf{r} - b\mathbf{r}_0$$

$$x_{\mathrm{T}} = \alpha_{\mathrm{x}} x - \beta_{\mathrm{x}} x_0$$

$$y_{\mathrm{T}} = \alpha_{\mathrm{y}} y - \beta_{\mathrm{y}} y_0$$

with

$$\begin{cases} a = \frac{z_0}{z_0+L+B} & b = \frac{L+B}{z_0+L+B} \\ \alpha_{\mathrm{x}} = \frac{L(F_{\mathrm{x}}-z_0)}{(F_{\mathrm{x}}+L)(z_0+L+B)} & \beta_{\mathrm{x}} = \frac{L(F_{\mathrm{x}}+L+B)}{(F_{\mathrm{x}}+L)(z_0+L+B)} \\ \alpha_{\mathrm{y}} = \frac{L(F_{\mathrm{y}}-z_0)}{(F_{\mathrm{y}}+L)(z_0+L+B)} & \beta_{\mathrm{y}} = \frac{L(F_{\mathrm{y}}+L+B)}{(F_{\mathrm{y}}+L)(z_0+L+B)} \end{cases}$$

Unfortunately, an analytical description of the $\mathcal{A}_{\mathrm{a}}(\mathbf{r}, \mathbf{r}_0, z_0)$ function cannot be derived in close form in the coordinate space. With this respect, the Fourier domain transform of Eq.(7.4) can be defined as the geometric transfer function (GTF), where the phase component of the formulation in the frequency space reflects the PSF space dependent variability. Nevertheless, the evaluation of the GTF in close form is still not possible. However, realizing that

- the factor $k\cos\theta/4\pi\left[(z_0+L+B)^2+|\mathbf{r}-\mathbf{r}_0|^2\right]$ for a given source position $\mathbf{r}_0$ is substantially constant for all values of $\mathbf{r}$ for which $\phi(\mathbf{r}, \mathbf{r}_0, z_0)$ is significantly non-zero.
- $\cos\theta = (z_0+L+B)/\left[(z_0+L+B)^2+|\mathbf{r}-\mathbf{r}_0|^2\right]^{1/2}$.

It is possible to calculate the Fourier transform of Eq.(7.4) in close form, giving

$$\Phi(\boldsymbol{\nu}, \mathbf{r}_0, z_0) = k \frac{\left| A_f \left(\frac{\nu_x}{\alpha_x}, \frac{\nu_y}{\alpha_y} \right) \right|^2}{4\pi (z_0 + L + B)^2} \cos^3 \theta \qquad (7.9)$$

where $A_f\left(\nu_x/\alpha_x, \nu_y/\alpha_y\right)$ is the Fourier transform of the front aperture function and $\boldsymbol{\nu} = (\nu_x, \nu_y)$ is the two-dimensional spatial frequency.

Thus, the evaluation of the aperture function in the Fourier domain represents the next logical step of the present formulation.

Equation(7.8) deals with aperture functions describing a single hole continuously displaced (as represented by the $\boldsymbol{\sigma}$ variable) throughout the projection space. Actually, the collimator is composed of an array of holes: introducing the information relative to the array pattern of holes, Eq.(7.8) can be replaced by

$$\mathcal{A}(\mathbf{r}, \mathbf{r}_0, z_0) = \sum_{l_1=-\infty}^{\infty} \sum_{l_2=-\infty}^{\infty} a_f\left(-\boldsymbol{\sigma}_{[l_1,l_2]}\right) a_f\left(\mathbf{r}_t - \boldsymbol{\sigma}_{[l_1,l_2]}\right) \qquad (7.10)$$

where we have lost the continuous variation of the variable $\boldsymbol{\sigma}$, addressing a vector $\boldsymbol{\sigma}_{[l_1,l_2]}$ for each hole. The impulse $\delta(x)$ function can be usefully introduced in (7.10) to clarify the matter:

$$\begin{aligned} \mathcal{A}(\mathbf{r}, \mathbf{r}_0, z_0) &= \sum_{l_1=-\infty}^{\infty} \sum_{l_2=-\infty}^{\infty} \int_{-\infty}^{\infty} a_f(-\boldsymbol{\sigma})\, a_f(\mathbf{r}_t - \boldsymbol{\sigma})\, \delta\left(\mathbf{r}' - \mathbf{r}'_{[l_1,l_2]}\right) d\boldsymbol{\sigma} \\ &= \int_{-\infty}^{\infty} a_f(-\boldsymbol{\sigma})\, a_f(\mathbf{r}_t - \boldsymbol{\sigma}) \sum_{l_1=-\infty}^{\infty} \sum_{l_2=-\infty}^{\infty} \delta\left(\mathbf{r}' - \mathbf{r}'_{[l_1,l_2]}\right) d\boldsymbol{\sigma} \end{aligned} \qquad (7.11)$$

where

$$\mathbf{r}'_{[l_1,l_2]} = l_1 \mathbf{r}'_{s1} + l_2 \mathbf{r}'_{s2},\ l_1, l_2 = 0, \pm 1, \pm 2, \ldots. \qquad (7.12)$$

The vectors $\mathbf{r}'_{s1}$ and $\mathbf{r}'_{s2}$ are the basis vectors in the reference system associated with the lattice of holes and they may be non-orthogonal. For instance, in the case of a 120° rhombic lattice oriented as in Fig.(7.2) the components of $\mathbf{r}'$ in the cartesian reference system, (x', y'), and the components in the lattice reference system, (r'_1, r'_2), are related by the following equations

$$\begin{cases} x' = r'_1 \frac{\sqrt{3}}{2} & r'_1 = x' \frac{2}{\sqrt{3}} \\ y' = r'_2 + \frac{r'_1}{2} & r'_2 = y' - \frac{x'}{\sqrt{3}} \end{cases} \qquad (7.13)$$

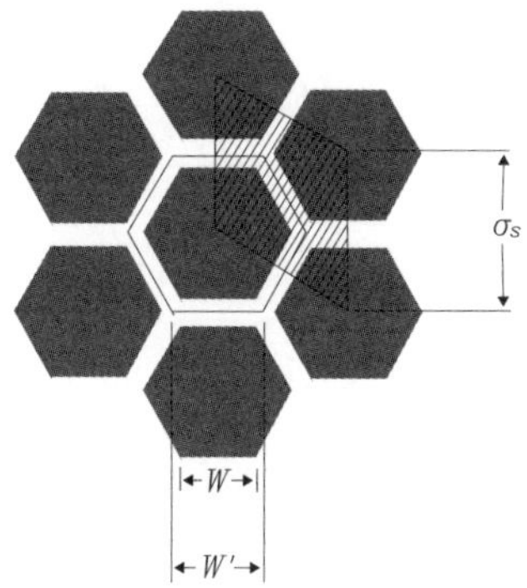

Figure 7.2. This figure shows how a hexagonal array of holes is actually generated by a 120° rhombic lattice of points which are the centres of the holes. It is also useful to note that the area of the 120° rhombus of side σ_s is equal to the area of the hexagon of side W'.

In the general case of a given lattice, the summation of impulse functions as reported in Eq.(7.11) can be expanded in Fourier series

$$\sum_{\substack{l_1=-\infty \\ l_2=-\infty}}^{\infty} \delta\left(\mathbf{r}' - \mathbf{r}'_{[l_1,l_2]}\right) = \frac{1}{Q} \sum_{\substack{m_1=-\infty \\ m_2=-\infty}}^{\infty} e^{-i2\pi \frac{r'_1}{\sigma_s} m_1} e^{-i2\pi \frac{r'_2}{\sigma_s} m_2} \tag{7.14}$$

where we assumed that the sampling distance along the two lattice's axes is the same: $\sigma_s = |\mathbf{r}'_{\mathbf{s1}}| = |\mathbf{r}'_{\mathbf{s2}}|$ and Q is the lattice's elementary cell area.

Actually, modern collimator's lattice configurations usually deal with array of holes generated by a 120° rhombic lattice of points representing the centers of the holes. In this case, substituting Eq.(7.14) (modified according to the rhombic lattice reference system defined by Eq.(7.13)) into Eq.(7.11) and taking its Fourier transform, we obtain:

$$\begin{aligned}
&\int_{-\infty}^{\infty} \mathcal{A}(\mathbf{r}, \mathbf{r}_0, z_0) e^{-i2\pi \mathbf{r}\cdot\boldsymbol{\nu}} d\mathbf{r} = \int_{-\infty}^{\infty} dx dy e^{-i2\pi(x\nu_x + y\nu_y)} \\
&\times \int_{-\infty}^{\infty} d\sigma_x d\sigma_y a_f(-\sigma_x, -\sigma_y) a_f(x\alpha_x - x_0\beta_x - \sigma_x, y\alpha_y - y_0\beta_y - \sigma_y) \\
&\times \frac{1}{Q} \sum_{m_1=-\infty}^{\infty} \sum_{m_2=-\infty}^{\infty} e^{-i2\pi \frac{x'}{\sqrt{3}\sigma_s} m_1} e^{-i2\pi \frac{y'}{\sigma_s}(m_1+2m_2)}
\end{aligned} \tag{7.15}$$

Equation(7.15) can actually be evaluated in close form. By first integrating in $dxdy$ with the change of variables $t_x = \alpha_x x - \beta_x x_0 - \sigma_x$, $t_y = \alpha_y y - \beta_y y_0 - \sigma_y$, and, as a second step, integrating in $d\sigma_x d\sigma_y$ we can write

$$\int_{-\infty}^{\infty} \mathcal{A}(\mathbf{r}, \mathbf{r}_0, z_0) e^{-i2\pi \mathbf{r}\cdot\boldsymbol{\nu}} d\mathbf{r} = \frac{1}{Q\alpha_x\alpha_y} e^{-i2\pi\left(\frac{\beta_x}{\alpha_x} x_0\nu_x + \frac{\beta_y}{\alpha_y} y_0\nu_y\right)}$$
$$\times \sum_{m_1=-\infty}^{+\infty} \sum_{m_2=-\infty}^{+\infty} e^{-i2\pi\left[\frac{x_0}{\sigma_s}\left(b+\frac{\beta_x}{\alpha_x}a\right)\frac{(m_1+2m_2)}{\sqrt{3}} + \frac{y_0}{\sigma_s}\left(b+\frac{\beta_y}{\alpha_y}a\right)m_1\right]}$$
$$\times A_f\left[\frac{\nu_x}{\alpha_x} + \frac{1}{\sigma_s}\frac{a}{\alpha_x}\frac{(m_1+2m_2)}{\sqrt{3}}, \frac{\nu_y}{\alpha_y} + \frac{1}{\sigma_s}\frac{a}{\alpha_y}m_1\right]$$
$$\times A_f\left[-\frac{\nu_x}{\alpha_x} - \frac{1}{\sigma_s}\left(1+\frac{a}{\alpha_x}\right)\frac{(m_1+2m_2)}{\sqrt{3}}\right.$$
$$\left., -\frac{\nu_y}{\alpha_y} - \frac{1}{\sigma_s}\left(1+\frac{a}{\alpha_y}\right)m_1\right] \tag{7.16}$$

where A_f is the two dimensional Fourier transform of the front plane aperture function of a single hole of generic shape.

Then, eq.(7.9) can be re-written in the following general form:

$$\Phi(\boldsymbol{\nu}, \mathbf{r}_0, z_0) = k\frac{\int_{-\infty}^{\infty} \mathcal{A}(\mathbf{r}, \mathbf{r}_0, z_0) e^{-i2\pi\mathbf{r}\cdot\boldsymbol{\nu}} d\mathbf{r}}{4\pi(z_0 + L + B)^2} \cos^3\theta \tag{7.17}$$

and making the opportune substitutions, one finally obtain the geometric transfer function in frequency space of a collimator device with a 120° rhombic lattice of generically shaped holes:

$$\Phi(\boldsymbol{\nu}, \mathbf{r}_0, z_0) = g\left(x_0, y_0, z_0\right) e^{-i2\pi\left(\frac{\beta_x}{\alpha_x} x_0\nu_x + \frac{\beta_y}{\alpha_y} y_0\nu_y\right)}$$
$$\times \sum_{m_1=-\infty}^{+\infty} \sum_{m_2=-\infty}^{+\infty} e^{-i2\pi\left[\frac{x_0}{\sigma_s}\left(b+\frac{\beta_x}{\alpha_x}a\right)\frac{(m_1+2m_2)}{\sqrt{3}} + \frac{y_0}{\sigma_s}\left(b+\frac{\beta_y}{\alpha_y}a\right)m_1\right]}$$
$$\times A_f\left[\frac{\nu_x}{\alpha_x} + \frac{1}{\sigma_s}\frac{a}{\alpha_x}\frac{(m_1+2m_2)}{\sqrt{3}}, \frac{\nu_y}{\alpha_y} + \frac{1}{\sigma_s}\frac{a}{\alpha_y}m_1\right]$$
$$\times A_f\left[-\frac{\nu_x}{\alpha_x} - \frac{1}{\sigma_s}\left(1+\frac{a}{\alpha_x}\right)\frac{(m_1+2m_2)}{\sqrt{3}}\right.$$
$$\left., -\frac{\nu_y}{\alpha_y} - \frac{1}{\sigma_s}\left(1+\frac{a}{\alpha_y}\right)m_1\right] \tag{7.18}$$

where $g(x_0, y_0, z_0)$ represents the collimator's efficiency, and the remaining part of Eq.(7.18) describes shape and position effects related to the lattice. Of course, other kinds of lattice array patterns could be considered (such as the square one for instance) by simply substituting into Eq.(7.12) the

appropriate set of equations for the lattice reference system under investigation.
Moreover, the hole shape should be taken into account, as it determines the actual mathematical formulation for the A_f variable in Eq.(7.18). As an example, let us consider the case where hexagonal holes are involved. Taking W to be the hexagon's side, it can be demonstrated that

$$\begin{aligned} A_{\text{es}}(\nu_{\text{x}}, \nu_{\text{y}}) = & \frac{2}{3W^2} \frac{1}{\nu_{\text{x}}\left(\nu_{\text{x}}^2 + 3\nu_{\text{y}}^2\right)} \\ \times & \left[\nu_{\text{x}} \cos(W\nu_{\text{x}}) \cos\left(W\sqrt{3}\nu_{\text{y}}\right) - \nu_{\text{x}} \cos(2W\nu_{\text{x}}) \right. \\ & \left. - \sqrt{3}\nu_{\text{y}} \sin(W\nu_{\text{x}}) \sin\left(W\sqrt{3}\nu_{\text{y}}\right)\right] \end{aligned} \tag{7.19}$$

so that the GTF in the Fourier space results completely defined. It may be worthwhile to highlight two main items as direct consequences of the present formulation for multi-hole collimators:

- At frequency $\nu = 0$ the value of $\Phi(\nu, \mathbf{r}_0, z_0)$ gives the efficiency of the collimator since it is equal to the integral in $\mathbf{r}$ of its Fourier transform, $\phi(\mathbf{r}, \mathbf{r}_0, z_0)$; moreover, the $g\,(x_0, y_0, z_0)$ variable contains factors related to the efficiency of the collimator. These factors describe the contributions to efficiency which are not related to the hole array pattern; they are the relevant factors in most imaging situations, but, if the source is located in the proximity of the collimator surface, effects due to this pattern appear. Such effects are described by the frequency dependent part in Eq.(7.18).
- The phase factor outside the summation in Eq.(7.18) is related to the projection point of the source onto the image plane. In case of fan beam or convergent collimators this factor takes also into account the displacement with respect to the orthogonal projection caused by the converging geometry. Moreover, the summation in Eq.(7.18) provides information about the shape of the system response and therefore also about the spatial resolution. The terms of the summation are in turn composed of three factors. The first one is a phase factor which depends on the relative position between the point source and the hole array pattern. The last two factors in each term of the summation are given by the the Fourier transform of the front and back apertures of the collimator's holes.The spread of the PSF, and therefore the spatial resolution, is directly related to the constants $1/\alpha_{\text{x}}$ and $1/\alpha_{\text{y}}$.

7.2.2. *Single-hole theory*

Starting from Eq.(7.4), the geometric transfer response of a multi-hole collimator has been formulated pursuing the basic idea of a collimation system conceived as a single hole aperture function distributed on the collimator plane according to a geometrically defined lattice of holes. Actually, as already noticed, Eq.(7.8) defines the aperture function $\mathcal{A}_{\rm a}$ as a continuous displacement of the $a_{\rm f}$ function, the continuity being assessed by the variable σ. As a consequence, the single-hole theory can be derived from the previous formulation by fixing the hole position, that is by defining the **r'** coordinate as a constant parameter. In this case, centering the hole front aperture at the origin of the spatial reference system **r'**$= 0$, and Eq.(7.9) can be restated as

$$\Phi(\boldsymbol{\nu}, \mathbf{r}_0, z_0) = k \frac{\left|A_f\left(\frac{\nu_{\rm x}}{\alpha}, \frac{\nu_{\rm y}}{\alpha}\right)\right|^2}{4\pi(z_0 + L + B)^2} \cos^3\theta \tag{7.20}$$

where

$$\alpha = \alpha_{\rm x} = \alpha_{\rm y} = \frac{L\,(F - z_0)}{(F + L)\,(z_0 + L + B)}$$

and F is the single hole focal distance.
As an example, it can be easily demonstrated that in the case of circular shaped hole, the front aperture function can be expressed as

$$A_f(\boldsymbol{\nu}) = \frac{\mathcal{J}_1(\pi w \boldsymbol{\nu})}{\pi w \boldsymbol{\nu}} \tag{7.21}$$

where w is the hole diameter and $\mathcal{J}_1(\cdot)$ is the first order Bessel function of the first kind.

7.2.3. *Penetration effects*

The outlined theoretical description provides a great deal of information as far as the geometrical response of collimation devices is concerned. Unfortunately, the theory does not tell us anything about the penetration of collimator septa and the scattering of photons in the collimator material. Both these effects may be relevant in clinical practice when energies > 200 keV are involved, and both represent a severe problem.

Photons travelling through the septa are expected to be absorbed by the collimator material. However, due to the statistical nature of the attenuation law, a fraction of them survive and may interact within crystal material giving rise to detected events which are included in the images.

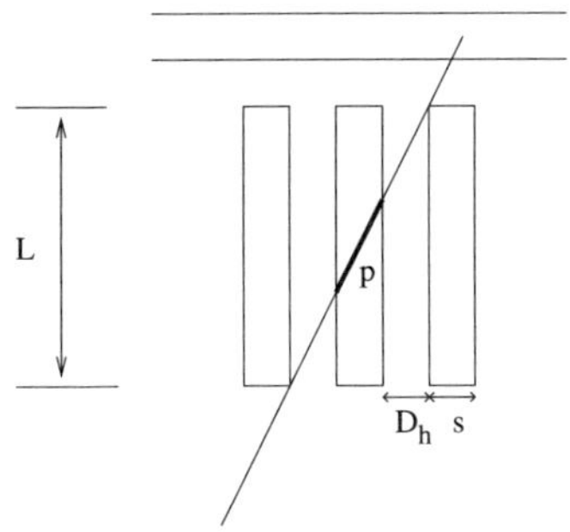

Figure 7.3. Calculation of the minimum path length through a septum. This figure may represent a section taken along one of the principal lattice directions of a rectangular array of rectangular holes or of a 120° rhombic array of hexagonal holes. The segment p is the minimum path length, s is the septal width, L the collimator thickness and $D_{\rm h}$ is the face-to-face hole width.

The unique way to control this penetration effect is to design collimators so that no septa penetration occurs. However, in most circumstances this leads to designs which cause the appearance of the shadow of the collimator holes in the final image, the so called hole pattern artifacts. For a long time, the most used criterion for designing collimators with acceptable penetration effects consisted in allowing a fixed fraction of photons to penetrate along the minimum path length (Fig.7.3) through a single septum [7]. It is easy to see that the minimum path length p can be expressed as

$$p = \frac{sL}{2D_{\rm h} + s} \tag{7.22}$$

provided that the dimension of a single lattice cell is small with respect to the collimator thickness, that is $L >> 2D_{\rm h} + s$. The acceptable penetration effect is set by requiring that along the path p it amounts to a certain small (5 % for instance) fraction $e^{\mu p}$, where μ is the linear attenuation coefficient for the energy of the incident photons. Defining $P_{\rm f} = \mu p$, we can write the following condition

$$p \geq \frac{P_{\rm f}}{\mu}. \tag{7.23}$$

Then

$$\mu L \frac{s}{s + 2D_{\rm h}} \geq P_{\rm f} \tag{7.24}$$

or, assuming that $\sigma = s + D_{\rm h}$ σ represents the center-to-center hole separation,

$$\mu L \frac{\sigma - D_{\mathrm{h}}}{\sigma + D_{\mathrm{h}}} \geq P_{\mathrm{f}}. \tag{7.25}$$

As an alternative approach to the previous formulation, it should be noted that, in more recent times, powerful computer techniques have been made available to study such complex problems. The study of the penetration constraint has been done by means of extensive ray tracing calculations ([8; 9]). A new criterion was deduced from these studies. The criterion, that is called the University of Chicago (UC) penetration criterion, states that the penetration effect will be supportable provided that the following relationship holds true

$$\mu L \frac{V_{\mathrm{a}}}{V} \geq P \tag{7.26}$$

where P is a constant that depends slightly on the hole shape and hole array pattern ($11 < P < 13$), V_{a} is the volume of the absorbing material and V is the total volume of the collimator. For parallel collimators, this relationship can also be written as

$$\mu L \left(1 - \frac{a_{\mathrm{hole}}}{a_{\mathrm{cell}}}\right) \geq P, \tag{7.27}$$

leading, in the case of 120° rhombic array of hexagonal holes, to

$$\mu L \frac{\sigma^2 - D_{\mathrm{h}}^2}{\sigma^2} \geq P. \tag{7.28}$$

Even if Eq.s(7.25) and (7.28) are different, they provides similar conditions in practical cases but the way the UC criterion was validated gives confidence in the fact that the criterion is suitable for current collimator designs. However, as stated in [9], no theoretical explanation for the criterion currently exists.

At high energies, besides the penetration effect, the scattering of radiation in the collimator's material may be also relevant. Unfortunately, this effect is even more difficult to tackle then the penetration one. Only complex and time consuming Monte Carlo methods ([10; 11; 12; 13])are available to make predictions about the scattering effects for a certain collimator. For low-energy collimators, the collimator scatter contribution should be considered a minor problem: it has been shown [11] that for a low-energy all-purpose collimator it amounts to about 2 % in a 20 % energy window centered at 140 keV energy of ^{99m}Tc.

On the contrary, the case of isotopes with higher-energy contaminant photons may be of some concern. For instance, it has been estimated [11] that if ^{123}I is contaminated with 5 % of ^{124}I, the collimator scatter contribution may amount to 6%. In clinical practice, the situation may become even worse since the patient (spatially extended) absorbtion would attenuate more significantly the ^{123}I photons than the higher-energy ^{124}I contaminants [14].

Recently there has been a renewed interest in ^{131}I isotope for tumor quantification with SPECT after radioimmunotherapy. However, acquisition and processing of the data are complicated by the presence of several high energy γ-ray emissions typical of the ^{131}I decay scheme. The most significant emissions are at 364 keV (82 %), 637 keV (7.2%) and 723 keV (1.8%), while the photopeak window is usually centered at 364 keV. Even if 637 keV and 723 keV photons are of low intensity, their scatter contribution may be significant since while suffering low attenuation in the patient's body because of their reduced cross section in water-equivalent density material, they have higher probability of scattering in the collimator. Dewaraja et al [12] reported that for a point source in air acquired with a high-energy collimator and a 20 % window at 364 keV, 43 % of the events in the photopeak are given by penetration and 29 % by scattering in the collimator material. Somewhat comparable figures [13] found that for 300 keV photons, 30 % of the events in the photopeak are due to penetration and 18 % are due to scatter.

Even if modern computer methods allow collimator penetration and scatter contributions assessments in specific cases, efforts to reduce substantially these effects have eluded researchers. When ^{131}I tumor quantification with SPECT is concerned, in order to avoid the spill-over given off from penetration originated from a noticeable background activity, the use of an Ultra High Energy (UHE) collimator (designed for 511 keV imaging) instead of the regular High Energy (HE) one has been advocated [15]. Of course, in this case the presence of clear hole pattern artifacts in the PSF should be considered as an unavoidable counterpart effect.

The use of UHE collimators have also been reported for SPECT imaging with ^{18}F [16; 17]. In this case, it is a common practice to smooth the projection data before the reconstruction in order to eliminate strong hole pattern artifacts. This procedure is usually supported because the cutoff frequency of the low-pass filter needed to eliminate the hole pattern artifacts is higher then the higher frequencies transmitted by the collimator.

7.3. Detectors

Detectors used in nuclear medicine studies have been historically addressed to the category of scintillators.

Despite ongoing attempts to switch to direct converting detectors as, for instance, those represented by semiconductors, scintillators still show competitive characteristics in the field of nuclear medicine imaging.

This is particularly true when inorganic crystals like cerium doped lutetium orthosilicate (LSO), cerium doped gadolinium orthosilicate (GSO) or yttrium aluminium perovskite (YAP) are considered. In fact, because of their good energy resolution and fast fluorescence decay time constant and together with to the possibility of being produced in large quantities, new possibilities of camera design are opened up.

As an additional feature, position sensitive photo multiplier tubes (PSPMT)have been coupled to several scintillator crystals [18] [19], thus allowing a radical improvement of camera performance on a small scale (radiopharmaceutical investigations on little animals as well as scinti-mammography dedicated gamma-camera). At present, the possibility of extending to whole body imaging the PSPMT-scintillator technology is under investigation (see Chapter xxx).

On the other hand, progresses in the development of semiconductor detectors widen the possibilities for improving the intrinsic spatial resolution and the overall performance of SPECT based gamma cameras by using room temperature semiconductor detectors as, for instance, CdTe and CdZnTe [20] [21].

Generally speaking, the direct readout of such a category of detectors make feasible the realization of compact and light-weight camera heads. Moreover, their unreachable energy resolution improves the identification of the isotope full energy peak, with consequent improvement in Compton events suppression task. It is also possible to partition slabs of semiconductor into pixel array with fine pitch, thus realizing precise position sensitive detector [22].

However their high cost and the possibility to be only produced in relatively small sizes should be considered as hindrance factors to their widespread employment in nuclear medicine imaging systems. Other delaying factors for the development of semiconductor based gamma cameras could be addressed as both physical effects like severe charge carrier trapping and technological aspects like the severe electronic complexity necessary to properly manage semiconductor array configurations.

Several attempts have been performed to overcome these difficulties. For example, studies have been done on the segmentation of large semiconductor electrodes, that is the typical position sensitive detector configuration sketched above. In this case, both the hole-trapping effect and the consequent depth dependent signal production seem to have reduced their significativity.

New techniques devoted to the production of integrated electronic circuits have also been developed. Actually they should reduce the complexity of read-out electronics, thus making possible the realization of cost-effective and reliable circuitry.

The aim of the following paragraphs is to describe the most advanced detectors' configurations presently used in Nuclear Medicine applications.

7.3.1. *Scintillators*

The last ten years have seen a growing interest in inorganic scintillators development. The elective [23] properties of the scintillator crystal, that is

(1) fast decay time with negligible delayed phosphorescence tail,
(2) emission wavelength compatible with both efficient read-out, coupling and negligible intensity leakage during crystal crossing,
(3) mass density and atomic number suitable for stopping γ rays of proper energy,
(4) high scintillation efficiency,
(5) energy versus light yield conversion as linear as possible over the widest energy range,

are not fulfilled by a single known scintillator, so that a number of materials have been developed for fitting different application requirements.

The mechanism of scintillation is well described by

$$N_{ph} = \frac{E}{\beta E_{gap}} \Upsilon_{e-h} \Upsilon_{LC} \tag{7.29}$$

where N_{ph} is the number of scintillation photons emitted under γ absorbtion of energy E, E_{gap} is the forbidden band energy amplitude, β is a parameter ($\beta \simeq 2-3$) related to the the average energy E_{e-h} required to produce an electron–hole pair, such that

$$E_{e-h} = \beta E_{\text{gap}}. \tag{7.30}$$

In this case, the fraction on the right-hand side of Eq.(7.29) represents the number N_{e-h} of electron-hole pairs created during the γ interaction process. The term Υ_{e-h} in Eq.(7.29) represents the transfer efficiency of the e-h pair energy on the luminescence centre, while Υ_{LC} represents the efficiency for photon emission of the luminescence centre itself. However, it should be noted that the value assumed by the Υ_{e-h} parameter is hardly predictable, being subjected, for example, to crystal defects that could act such as non-luminescent traps for e–h pairs, with consequent reduction of the scintillation efficiency.

As it is known, Tl-activated scintillators represent the historical background of the modern scintillation spectroscopy when γ-ray are involved, and they have been extensively applied to nuclear medicine devices. However, during the last decade, a number of lanthanide-doped scintillators, particularly in the field of 5d–4f transitions, whose intrinsic properties allow efficient γ-ray spectroscopy at high count rates, have been identified, and their applicability as nuclear medicine detectors has been investigated by several research groups.

Among the number of newly proposed detectors, Ce^{3+} materials seems to show the most promising characteristics, featuring *i)* a relatively small $E_{\rm gap}$ value, *ii)* an Υ_{e-h} value close to unity and *iii)* high specific density or high effective atomic number $Z_{\rm eff}$.

In this context, GSO:Ce, YAP:Ce and LSO:Ce are some of the most interesting scintillators, with potential wide application in nuclear physics, high energy physics and imaging techniques based on nuclear applications as well. All of them are characterized by a high light output and fast decay times of few tenths of nanoseconds. Due to high specific density, GSO:Ce and LSO:Ce are the potential deputies of the BGO scintillator in various applications, while the fast light decay time constant typical of YAP:Ce crystal suggests unexplored possibilities in the pulse shape discrimination field.

7.3.1.1. *$YAlO_3$:Ce*

The properties of $YAlO_3$:Ce scintillator have been studied since 20 years ago. $YAlO_3$:Ce crystal (abbreviated YAP:Ce or simply YAP) is a monocrystal with the reticular structure of perovskite. It shows rugged mechanical characteristics (hardness) and very good chemical stability, non-hygroscopicity and non solubility in inorganic acids [24].

Luminescence

With respect to luminescence properties, YAP shows a main UV emission band with a maximum at $\simeq$ 370 nm as well as minor emission bands centered around the blue and green wavelengths [25; 26]. While under UV excitation YAP shows light emission lifetimes ranging around 15 ns, the typical light decay time constant under γ-ray excitation was measured [24] as 27 ns: the difference between these two values should probably be referred to a delay in energy transfer between the energy band structure of the crystal. Moreover it shows very low afterglow emission (<0.005% at 6 ms) and light output (yield) equal to 55% of the light yield featured by NaI:Tl under the same experimental conditions. It should be highlighted that the featured time decay constant is about 9 times faster than the one registered for the NaI:Tl crystal.

γ-Detection and Energy Resolution

As long as detection properties are involved, YAP shows mass density equal to 5.37 g/cm^3, leading to stopping powers of 1.54 cm^{-1} and 0.46 cm^{-1} at 140 keV and 511 keV, respectively. It should also be noted that the mean energy of fluorescence X-ray of YAP:Ce is as low as 15 keV, leading to a energy deposition due to photoelectric effect very close to the primary interaction point. In particular, Blažek and co-workers [27] developed a Monte Carlo method for investigating the limit of the theoretical extension of the energy transport range related to radiation transport in scintillators. In their work, the maximum value of energy deposition spot diameter obtained for YAP was 300 μm, while, for example, 500 μm, 1 mm and 2 mm were the values obtained for CsI, NaI and BGO respectively.

Several authors performed light yields measurements in order of determining the energy resolution FWHM of the YAP crystal. At present, it appears a difficult matter to determine a range of FHWM values in general agreement with the papers present in literature. Nevertheless, a majority of experiments seem to converge toward energy resolution values ranging from 35% at 32 keV and 11% at 662 keV [24]. It should be noted that in YAP crystal, as well as in other scintillators, the energy resolution behaves in disagreement with the inverse of the square root of the energy dependence, the disagreement becoming more evident for incident energies greater than 100 keV, that is when Compton scattering probability starts increasing toward sensible levels [28].

7.3.1.2. Gd_2SiO_5:Ce

The characteristics of Cerium doped Gadolinium Orthosilicate scintillator (abbreviated GSO:Ce or simply GSO) have been studied since 1983 [29].

Luminescence

With respect to luminescence properties, GSO shows a typical light decay time under γ-ray excitation ranging between 30 to 65 ns, about 4 to 8 time faster than NaI:Tl crystal. As it was for YAP:Ce scintillator, the GSO afterglow emission is lower than 0.005% (at 6 ms) and the typical UV emission band seeks its maximum at $\simeq$ 440 nm.
On the other hand GSO shows light output as lower as about 30% with respect to NaI:Tl crystal. However, it should be highlighted that with NaI:Tl the output light pulse is typically clipped and integrated at 200 ns, thereby using less than 60% of the total pulse.

γ-Detection and Energy Resolution

Featuring a high effective atomic number ($Z_{eff} \approx 64$) and, consequently, high density (6.71 g/cm^3), the γ-ray detection properties GSO reflects in stopping powers of 6.71 cm^{-1} and 0.71 cm^{-1} at 140 keV and 511 keV respectively. Those values collocate the GSO scintillator in an attractive position for Nuclear Medicine applications involving β^+ particles annihilation as well as high-energy research in nuclear physics applications.

Typical values for the photon energy resolution have been reported between 7% and 8% at 662 keV [29].

7.3.1.3. Lu_2SiO_5:Ce

The past decade has seen noticeable efforts directed at developing scintillation materials for both SPECT and PET. The characteristics of Cerium doped Lutetium Orthosilicate scintillator (abbreviated LSO:Ce or simply LSO) seem to fit the required combination of properties better than other scintillating materials. On the other hand, the observed background coming from the natural radioactivity of Lutetium [30] [31] [32] may limit the possible application spectrum.

Luminescence

Under γ-ray excitation, LSO shows a main UV emission band with a maximum at $\simeq$ 420 nm. The related light pulse shows a fast decay time constant of about 40 ns at room temperature. The pulse shape is essentially mono-exponential, without evidence for other time decay constant components.

With respect to light output, LSO:Ce shows values of about 25×10^3 photons/MeV, 0.6 times the typical value referred for NaI:Tl crystal (4.1×10^4 photons/MeV) under comparable experimental conditions.

However, it should be high-lighted that the light output value of the LSO:Ce itself is characterized by a broad variation from one specimen to another, despite that detectable differences in its chemical structure cannot be found. This effect ought be referred to a number of intertwined causes relied to the stability of the transfer efficiency of the scintillation, that is the fraction of the produced electron-hole pairs whose energy actually excites the luminescence centers. Among the possible causes, *i)* the unavailability of one or both carriers (electron-hole pair) to the scintillation mechanism because of out-site anomalous trapping, and *ii)* the presence of shallow Ce^{3+} traps in the forbidden energy band with consequent temperature variation of the transfer efficiency parameter should be highlighted [33].

γ-Detection and Energy Resolution

Featuring an effective atomic number Z_{eff}=66, and a density equal to 7.4 g/cm^3, LSO shows stopping powers of 9.768 cm^{-1} at 140 keV and 0.866 cm^{-1} at 511 keV. The reported values suggest high efficiency in γ-radiation detection and this characteristic of the LSO scintillator could reflect in both SPECT and PET tomographs design. In fact, the hardware improvement of PET spatial resolution is closely related to the possibility of reducing the size of the single detector element without dramatic losing in detection efficiency. Moreover, the possibility of reducing the thickness of a SPECT detector reflects on both the light collection efficiency and on the intrinsic camera spatial resolution.

As previously noticed, the LSO scintillator features a natural radioactive background due to the presence of the radioactive ^{176}Lu in the natural lutetium (β^- decay to ^{176}Hf, relative abundance of 2.6%) whose energy spectrum is characterized by γ transitions whose main energies are 88, 202 and 307 keV. As reported in literature [34], the typical background count rate is about 300 counts/s/cm^3. Energy resolution data as reported in literature, refer to test executed using the ^{137}Cs source: in this case the 662 keV photopeak energy resolution is about 11%.

7.3.2. *Semiconductors*

Semiconductor based γ-ray detectors, both in single and array configurations, are well known and irreplaceable devices in various nuclear spectroscopy based applications. During the past decade, their interest in medical imaging applications has grown appreciably in spite of *i)* their good energy resolution, and *ii)* the possibility of arranging compact position sensible arrays with very small elements. In this respect, it has been demonstrated [35] that the reduction in single detector size allows noticeable improvements in spatial resolution of the experimental image SPECT applications. If the SPECT system was equipped with peculiar pin-hole collimator configuration, a noticeable improvement in sensitivity was also achieved.

Silicon and germanium semiconductor materials represent the most widen used detectors for γ-ray detection. Despite their high spatial resolution and good charge transport properties, their detection efficiency should be considered not well suited for imaging purposes where, because of count statistic (i.e. detection efficiency) reliability, higher atomic number is required. Moreover, because of its intrinsically small forbidden energy band gap, germanium must operate at low (cryogenic) temperature, giving rise to complex detector arrangements, particularly when rotating detector heads are concerned.

Recent developments in semiconductor detectors research address materials with high atomic number and wide forbidden energy band gap. Among the proposed detectors, cadmium telluride (CdTe) and cadmium-zinc telluride ($Cd_{1-x}Zn_xTe$, $x \simeq 0.04 - 0.2$, or simply CZT) have attracted most of the attention. Though these materials feature high atomic number as well as operativity at room temperature, their characteristic carrier mobility coefficients (μ_e and μ_h for electrons and holes respectively) are lower than those featured by germanium or silicon detectors. As a consequence, carriers (most seriously holes with respect to electrons) are trapped at material defects or impurities sites. So, the total charge induced into the external readout circuit behaves as a function of the distance as measured from the site within the detector's sensible volume where the γ-ray interaction takes place, to the collecting anode. As a result, the energy pulse-height spectra of these materials shows a typical tail toward the low energy site. This reflects in two main consequences, that is, *i)* loss of accurate scatter component discrimination possibility, and *ii)* reduction of count efficiency.

The collected fraction of induced charge can be expressed by means of the Hetch equation [36] as follows:

$$\eta(\ell) = \frac{\lambda_e}{T}\left(1 - e^{-\frac{T-x}{\lambda_e}}\right) + \frac{\lambda_h}{T}\left(1 - e^{-\frac{x}{\lambda_h}}\right) \qquad (7.31)$$

where $\eta(\ell)$ is the collected fraction of the induced charge as a function of the distance ℓ from the cathode, T is the detector thickness and λ_e and λ_h are the electron and hole drift lengths (or mean free paths), respectively. It should be noted that for a charge carrier travelling toward the collection electrode, pulled by an electric field $\mathcal{E}$, the drift speed is $v = \mu\mathcal{E}$. The ratio between λ and v

$$\tau = \frac{\lambda}{v} \qquad (7.32)$$

define the trapping-limited carrier lifetime. Equation(7.31) can then be interpreted as a parametric functional, where the λ parameters work as "normalized" drift lengths with respect to the applied electric field. Such "normalized" drift lengths can be directly expressed by means of the product $\mu \cdot \tau$, thus highlighting the relationship between the collected signal and the carrier mobility parameter.

Methods developed to reduce the effects of hole trapping can be grouped in two main items:

- Electronic techniques
- Detector design techniques

electronic techniques

The first item concerns pulse processing methods ([37] [38] [39]). From the point of view of the characteristic charge collection time, CdTe detectors show a $(\mu\tau)_h/(\mu\tau)_e$ ratio smaller than 0.1. An even worst situation originates when CZT is considered, where a 0.01 value for this ratio is standard. Thus, pulses show both a fast (electron) and a slow (hole) components and, as a consequence of Eq.(7.31), they also show different shapes. Available techniques for electronic pulse correction deal with the addition of a peculiar amplitude gap to detect pulses, this gap being calculated on the basis of empirical relationships between charge losses and output pulse rise-time of a 'template' class of detectors. Alternatively, other methods conceive the possibility of subtracting a peculiar magnitude from the pulse amplitude, with this magnitude being defined by the discrepancy of the actual pulse shape from an ideal form.

detector design techniques

Methods falling in the second item deal with the design of detectors where the collected charge depends primarily on the sole electrons moving in the close proximity of the anode, thus reducing the induced charge dependence on the distance ℓ, as defined in Eq.(7.31).

This configuration has been derived from gas ionization chambers technology, where μ_e is much higher than μ_h. In ionization chambers this effect was corrected using a peculiar unipolar charge sensing scheme, known as the Frisch grid technique [40]. In short, a Frisch grid consists of a gridded electrode placed inside the detector at a short distance from the anode. As an appropriate electrostatic bias is applied to the grid, as well as to the other active electrodes, electrons that are being collected may efficiently travel through the grid. In this sense, the grid behaves as an electrostatic shield making the signal collected at the anode site quite insensitive to carrier (electrons) movements crossing the region between the cathode and the grid. As a consequence, the whole signal amplitude is developed into the detector space between the grid and the anode: carriers that are created within the region between the cathode and the grid will produce full-amplitude signals as long as all the electron carriers are collected at the anode site, regardless of whether the hole carriers are collected or not at the cathode site.

The Frisch technique can be applied in semiconductor detectors too, replacing the grid with a coplanar electrodes arrangement (coplanar grid)[41]. The basic structure of a coplanar grid consists of a series of narrow strip electrodes placed on a detector surface, electrically connected in an alternate way, just to form two separate sets of "comb" arranged electrodes as sketched in Fig.(7.4). If $\mathcal{A}$ and $\mathcal{B}$ are the two sets of electrodes and $\mathcal{C}$ is

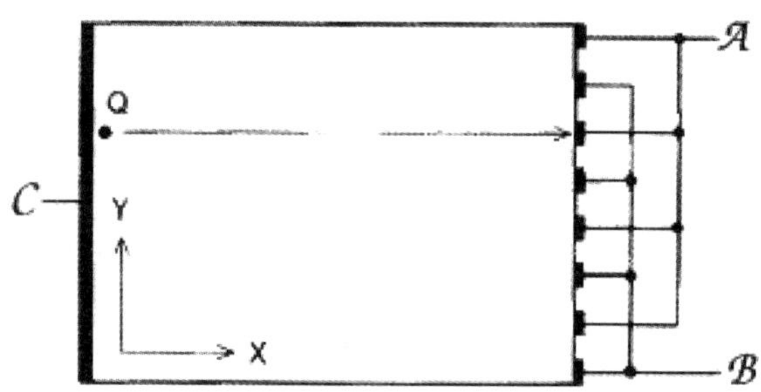

Figure 7.4. Sketch of the basic design for a xcoplanar grid configuration in semiconductor detectors. The relative positions of $\mathcal{A}$, $\mathcal{B}$ and $\mathcal{C}$ sets of electrodes are highlighted.

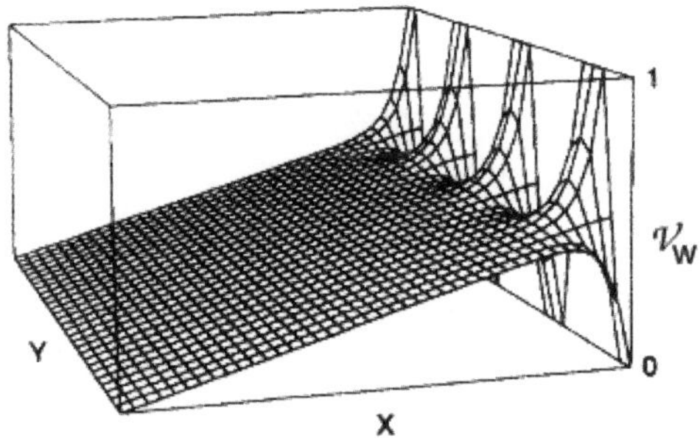

Figure 7.5. Spatial distribution of $\mathcal{V}_w$ in the coplanar grid configuration.

the opponent full-area electrode, it is possible to sustain a uniform electric field inside the detector body setting, for example, $\mathcal{A}$ and $\mathcal{B}$ at the same potential and applying a different potential to $\mathcal{C}$. Ramo's formulation for multiple electrode arrangements defines a "weighing" potential, that is the potential that would exist inside the detection volume in absence of space charge distribution, with a given electrode at "unit" potential and all other electrodes at zero potential.

If $\mathcal{V}_w$ is the weighing potential for a given electrode, the charge variation δq observed because of the presence of a carrier of charge Q is defined as

$$\delta q = Q \delta \mathcal{V}_w$$

where $\delta \mathcal{V}_w$ is the observed variation in the weighing potential.

Figure 7.5 shows the calculated weighing potential spatial distribution for one set of grid electrodes (say $\mathcal{A}$). The weighing potential for the $\mathcal{B}$ set will have the same spatial distribution, except that, in the neighborhood of the electrodes, the spatial distribution is inverted. In other words, within the y axis segments where the weighing potential for the $\mathcal{A}$ set is zero, the weighing potential for the $\mathcal{B}$ set will reach its maximum (unit) value, and *vice-versa*. So, the flat part of the weighing potential distribution remains unchanged: a charge carrier created in the neighborhood of the $\mathcal{C}$ electrode will travel toward the collecting electrode (say $\mathcal{A}$), inducing balanced signals in both $\mathcal{A}$ and $\mathcal{B}$ until it reach the proximity of $\mathcal{A}$. At that point the induced charge is expected to rise abruptly to the Q value in $\mathcal{A}$, while (see Fig.(7.6)) the induced charge in $\mathcal{B}$ steps down toward zero. The difference between the $\mathcal{A}$ and $\mathcal{B}$ signals results in a new signal, that is the target signal of this method, whose shape does not depend significantly from the dynamic behavior of the carrier over the detection volume. This

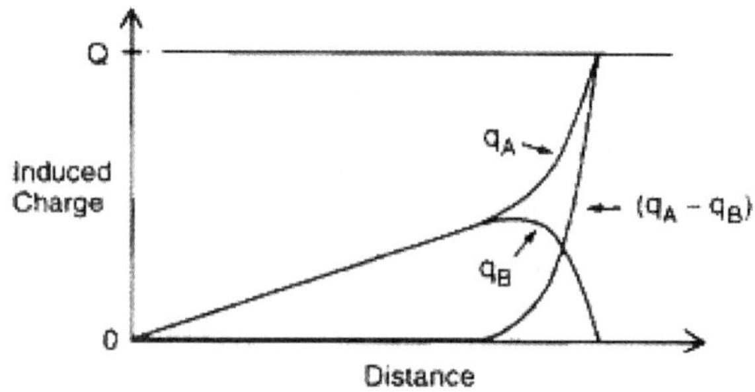

Figure 7.6. Plot of the induced charge spatial distribution into the detector's volume. The difference signal $(q_{\mathcal{A}} - q_{\mathcal{B}})$ is evidenced.

effect should be considered very close to the one previously described for the Frisch grid configuration. This method can also be improved by studying the correlation of the signal from the coplanar grid and the $\mathcal{C}$ electrode.

7.3.2.1. *Materials*

CdTe detectors manufacturing usually starts from a Te rich phase, that is a melt (about 1100°C) composition near stoichiometry with the aim of minimizing inclusions and defects. During the growth process, the material is doped by using chlorine. Chlorine doped detectors behave as p-type materials whose typical resistivity is $10^9\ \Omega \cdot cm$ order of magnitude.

$Cd_{1-x}Zn_xTe$ detectors are usually produced using a growth method employing high pressure (up to several hundred atmospheres) containment, allowing growth temperature as high as 1600°C [42]. This method prevent the contamination of the melt by oxygen, silicon and other impurities usually present. Oxygen contaminants are of particular concern, because they tend to form isoelectronic traps as well as to slow down mobilities and reduce lifetimes of charge carries. Using the high pressure method, CZT features an intrinsic resistivity of $5 \times 10^{10} - 10^{11}\ \Omega \cdot cm$ (order of magnitude). This high resistivity implies lower leakage currents for CZT material as compared to CdTe, allowing the employment of larger volume detectors.

As previously highlighted, CdTe and CZT semiconductors behave as high impedance devices. Their typical equivalent capacity is small of the order $1pF$. In this respect, leakage current can affect the energy resolution value more than the electronic noise superimposed on the signals from other components of the electronic chain can do. Moreover, one should remember that the intensity of the leakage current increases as the temperature increases. When medical imaging instruments are considered, as long as the operational temperature increases, this effect reflects in a sensible variation of the energy resolution and of the photopeak efficiency. Particularly [43], it has been shown that the the energy FWHM for a 3 mm tick planar CZT detector operating at 450 V raise from 4.5% at 25°C to 8% at 40°C.

Considering a constant electric field spanning the detector active volume, as it is the case of detectors designed using the coplanar grid configuration, the charge carrier mobility μ is given by

$$\mu = \frac{T^2}{t_r V} \tag{7.33}$$

where V is the applied potential between active electrodes and t_r is the pulse rise time. For electron (unipolar charge scheme) pulses, mobility coefficients for both CdTe and CZT can be directly extracted from experimental data using Eq.(7.33). On the other hand, in the case of different pulses, it is possible to extract the same information from the experimental data by fitting Eq.(7.31) for electrons and holes independently, stressing its dependence from (μ_e, τ_e) or (μ_h, τ_h) parameters. Using this method [43], the average mobility for CdTe detectors were found to be (1000±50) and (70±10) $\frac{cm^2}{(V \cdot s)}$ for electrons and holes respectively. In parallel, the average mobilities for CZT detectors were found to be (1100±50) and (45±10) $\frac{cm^2}{(V \cdot s)}$.

7.3.2.2. *Nuclear Medicine applications*

Digirad Corporation has developed a prototype design for a portable semiconductor detector, the Digirad 2020tc Imager^{TM}. This camera has a 21.6 cm × 21.6 cm field of view, composed of 64 modules of 2.5 cm × 2.5 cm × 0.5 cm CZT detectors each, arranged in an 8×8 array. Each module is electrically partitioned in an 8×8 array of pixel, with an electrode configuration similar to the coplanar grig arrangement previously described for minimizing trapping problems. The Digirad camera has a mean energy resolution of 8% at 140 keV, to be compared with ∼ 11% for NaI:Tl gamma camera configurations.

Eisen and co-workers [44] developed a semiconductor camera containing CdTe detectors arranged in a 40×32 matrix. The energy resolution achieved is 5% FWHM at 140 keV and 8% FWHM at 60 keV. Actually, their results regarding cardiac perfusion studies show competitive features with respect to traditional cameras. Moreover it should be noted that the camera head weights about 4 kg. This datum seems widen the possibilities for nuclear medicine imaging diagnosis to those patients that cannot be transported to a stationary camera. Again, test results have been reported by Chambron [45] and co-workers, regarding CdTe detectors.

As general remarks, the following items can be fixed:

- The energy resolution of a semiconductor based γ-camera allows better scatter rejection and, in turns, better image contrast with respect to conventional tomographs.
- The image spatial resolution is not affected by the detector type, being strongly dependent on the collimator used.
- Large area semiconductor detectors are not easily realized: leakage currents, electronic read-out requirements, high cost and growth hindrances seem slow down their clinical development.
- Since the molteplicity of independent detection elements forming the camera head, higher count rates than conventional cameras can be achieved .

Whenever possible, images regarding test-oriented or clinically assessed results obtained with these cameras will be reported in Section 7.5.

7.4. Reconstruction Algorithms

During the last fifteen years, image reconstruction and processing techniques involved in diagnostic problems have seen a noticeable development. This should be mainly addressed to the recent availability of powerful digital computers as well as to the development of faster technologies for data acquisition.

Image reconstruction from tomographic acquisition can be concerned as a mathematical structure that lies into the category of the so-called linear *inverse problems*, whose field of application involve a broad spectrum of image acquisition techniques. Space-variant imaging systems, inverse diffraction, inverse source problems are other examples of linear inverse problems. Even if the Fourier transform is very useful to understand important aspects of tomography, the basic mathematical tool to have an insight on inversion issues is the Singular Value Decomposition (SVD). This holds true for all linear inverse problems. The SVD allows one to describe in detail the regularization properties of the inversion methods in a manner similar to the Fourier transform in the case of image deconvolution (space-invariant systems).

In the next following paragraphs, the main items concerned with inverse problems and their application in tomographic techniques will be discussed.

7.4.1. *Inverse problems*

Generally speaking, a typical direct stated mathematical problem requires to determine the values of a set of unknowns, starting from a set of known data. Of course, it is possible to conceive the possibility of an inverse problem by a simple inversion of the known and unknown data sets. To solve the intrinsic ambiguity of this formulation (where what is direct and what is inverse is hard to determine), it is necessary to take into account the physical meaning of the problem under examination.

Usually, when the mathematical formulation of a physical problem follows the cause-effect path, a direct stated problem is conceived. As an example, the '*a priori*' determination of the 2D activity distribution as resulting from the projection of a 3D-radioactive source of known spatially-variant activity distribution, is a direct stated problem. On the other hand, the determination of the 3D activity distribution of a radioactive source, starting from a suitable set of experimentally acquired 2D projections of the source itself, falls into the cathegory of the inverse problems.

In this sense, when direct form stated problems are considered, we always find that an information shrinkage appears during the path from the formulation toward the solution of the problem. In other words, when direct stated problems are considered, the solution is characterized by a minor amount of information with respect to the amplitude of the known data set. So when inverse stated problems are involved, we face with an unavoidable lack of information: the physical characteristics that lead from the space of unknowns toward the space of experimental data (known data set) are missed. Then the correct inverse path can no more be backward followed. The previous described characteristic of the inverse problems in known as "*ill-posedness*", and it could be addressed as a general problem for biomedical imaging, particularly when emission tomography is concerned.

7.4.2. *Ill-posed problems*

Let us consider the Fredholm integral equation of first kind

$$g(y) = \int_a^b K(x,y) f(x) dx, \quad c \le y \le d \tag{7.34}$$

where $K(x,y)$ is the kernel of the integral equation. Here we assume that Eq.(7.34) characterize a physical problem where $f(x)$ represents a physical quantity defined over a spatial domain, the integration kernel $K(x,y)$ represents the detector behaviour and $g(y)$ represents the measured data. Let us suppose $f(x)$ as a possible solution for (7.34) and let us apply a perturbative (oscillating) term:

$$g(y) = \int_a^b K(x,y)[f(x) + \sin(mx)]dx, \quad c \leq y \leq d. \tag{7.35}$$

As long as the oscillation frequency increase its value, we can write

$$\lim_{m\to\infty} \int_a^b K(x,y)\sin(mx)dx = 0 \tag{7.36}$$

so that for $m \to \infty$, $f(x) + \sin(mx)$ too is a solution for Eq.(7.34): finite numerical variations in $f(x)$ reflect in infinitesimal variations in $g(y)$. When we deal with the inverse problem (i.e. trying to derive $f(x)$ from an experimental knowledge of $g(y)$), small (infinitesimal) numerical variations on $g(y)$ can induce wide variations in $f(x)$. In this case the problem (7.34) is said *ill-posed*. Generally speaking we can state that a problem is said to be *well-posed* if

- the solution is unique,
- the solution exists for all data,
- the dependence of the solution on the data is continuous.

If only one of the previous requirements is false then the problem falls into the cathegory of *ill-posed* problems.

7.4.3. *Ill-Conditioning and regularization*

Usually, the mathematical formulation of an inverse problem is performed by means of an first kind integral operator defined on an Hilbert space. Particularly, if $K : f \to g$, with $f \in \mathcal{X}$ and $g \in \mathcal{Y}$, is an integral operator with square-integrable kernel, then the problem can be described as

$$g(y) = h(y)f(y) - \lambda \int_a^b K(x,y)f(x)dx, \; c \leq y \leq d. \tag{7.37}$$

However, when real problems are concerned, the experimental data are collected in a discrete form, so that the problem itself must be translated in a discrete space. As it is known, the discretization of an inverse linear problem reflects in a system of linear equations. For the existence of a unique solution, a linear system of equations should have the same number of equations and unknowns. In practice this point becames quite irrelevant with respect to the intrinsic loss of stability of the solution because of the *ill-posed* nature of the original problem. In the real case, the loss of stability is due to small variations on experimental data (i.e. statistical noise fluctuations or experimental errors) that reflect in possible wide numerical variations of the solution. In case of discrete formulation, this characteristic, typical of *ill-posed* problems, is called *ill-conditioning.*

However, it is not generally true that an ill-posed problem is necessarily ill-conditioned in its discrete form. Sometimes the discrete version of an ill-posed problem turns out to be well-conditioned. The clearest way to study the ill-posedness of problem is to study the behaviour of the singular values of its SVD. The same is true with discrete problems since the transform of the direct problem is described by a matrix and the SVD of any rectangular matrix can also be calculated.

As a short reminder, the SVD technique may be defined for integral operators, matrices and even semi-discrete mappings [46]. For instance, if $K : f \to g$, with $f \in \mathcal{X}$ and $g \in \mathcal{Y}$, is an integral operator with square-integrable kernel, then its SVD allows to write

$$Kf = \sum_{k=1}^{\infty} \sigma_k (f, v_k)_{\mathcal{X}} u_k \tag{7.38}$$

where v_k and u_k are the singular function in $\mathcal{X}$ and $\mathcal{Y}$, respectively; σ_k are the singular values; $(\cdot,\cdot)_{\mathcal{X}}$ and $(\cdot,\cdot)_{\mathcal{Y}}$ are the scalar products in $\mathcal{X}$ and $\mathcal{Y}$ respectively. It is easy to express the solution of the problem $Kf = g$ in terms of the SVD of K:

$$f = \sum_{k=1}^{\infty} \frac{1}{\sigma_k} (g, u_k)_{\mathcal{Y}} v_k . \tag{7.39}$$

The behaviour of the σ_k with $k \to \infty$ is very important to determine in which measure a problem is *ill-posed.* If the $\sigma_k \to 0$ as $k \to \infty$ then the problem is *ill-posed.* The way $\sigma_k \to 0$ determines the severity of the illness. Of course , in the case of discrete sampling, the previous problem can be stated in the form

$$\mathbf{K}f = \sum_{k=1}^{p} \sigma_k (f, v_k)_N u_k \tag{7.40}$$

where $\mathbf{K}$ is a matrix, $M \times N$, transforming a vector of the object space $\mathcal{X}_N$, of dimension N, into a vector belonging to the data space $\mathcal{Y}_M$, of dimension M; p is the rank of $\mathbf{K}$, v_k and u_k are the singular vectors of $\mathcal{X}_N$ and $\mathcal{Y}_M$, respectively. Again, the singular values σ_k determine the basic properties of the problem, in particular its ill-conditioning. The severity of the conditioning is given by the *condition number* which is given by

$$\alpha = \frac{\sigma_1}{\sigma_p} \tag{7.41}$$

A large value of the condition number means that the problem is ill-conditioned whereas a small value means that the problem is well-conditioned. The ill-conditioning tends to be less severe as the dimension of the problem decreases; so it happens that the 'finesse' of the sampling transforming continuous quantities in discrete ones may influence the ill-conditioning of a problem, leading to a discrete problem that features a condition number lower than the one we could expect observing its continuous formulation.

As previously stated, difficulties related to the solution of an inverse problem originate from a lack of information in the direct stated problem. An attempt to reduce the *ill-conditioning* deals with the introduction of external information (called *a priori* information) in order of shrinking the class of possible solutions as close as possible to a class of solutions whose physical meaning matches the problem under consideration.

In other words we can seek a *constrained* solution. Possible constrains are bounds on the total energy (integral of square) of the solution, on the derivative of a certain order, limitations of the regions where the objects are non-zero, positivity. The practice of constraining the possible set of solutions to control the ill-conditioning of the system is called *regularization.* The term was proposed in 1963 by the Russian mathematician Tikhonov as a general approach to control ill-posed problems. The basic idea consists of considering a family of approximate solutions depending on a positive parameter called the *regularization parameter.*

7.4.4. *The Radon transform*

The mathematical description of the direct problem of tomography is given by the Radon transform (Radon, 1917 [47]):

$$g(s,\theta) = \iint\limits_{\mathcal{D}\in\mathbb{R}^2} f(\boldsymbol{x})\delta(s - \boldsymbol{x}\cdot\boldsymbol{\theta})d\boldsymbol{x} \tag{7.42}$$

where $\boldsymbol{x} = \{x, y\}$ and $\boldsymbol{\theta} = \{\cos\theta, \sin\theta\}$ with $\theta \in [0, 2\pi]$; $\delta(\cdot)$ is the Dirac impulse function.

Equation (7.42) establishes the mathematical relationship between a function in $\mathbb{R}^n$ and all its projections in $\mathbb{R}^{n-1}$. Radon found the inverse transform which in the two-dimension version can be written as

$$f(r,\phi) = \frac{1}{2\pi^2}\int_0^{\pi}\int_{-\infty}^{\infty}\frac{\partial g(s,\theta)}{\partial s}\frac{1}{r\sin(\phi-\theta) - s}dsd\theta \tag{7.43}$$

where f is expressed in function of the polar coordinates $r = \sqrt{(x^2+y^2)}$, $\phi = \arctan(y/x)$. This form of the inverse transform is not well suited for computer implementations. Nevertheless, it is possible to derive an analytical solution for Eq.(7.43) whose general mathematical form is widely used and known as *filtered back-projection.*

7.4.5. *Analytical methods: filtered back-projection*

To derive a computer better-suited form of Eq.(7.43), let us observe that the integration along ds reflects the convolution among $\frac{\partial g(s,\theta)}{\partial s}$ and $1/s$ in $r\sin(\theta-\phi)$. Naming $g'(s,\theta)$ such a convolution, i.e.

$$g'(s,\theta) = \frac{\partial g(s,\theta)}{\partial s} * \frac{1}{s} \tag{7.44}$$

(where $*$ indicates the convolution operator), we can write Eq.(7.43) in the following form:

$$f(r,\phi) = \frac{1}{2\pi^2}\int_0^{\pi} g'(r\sin(\phi-\theta),\theta)d\theta \tag{7.45}$$

The integral in eq.(7.45) represents the back-projection operator. It is known that, if $h(x)$ is a continuos funcion over a spatial domain, the Fourier transform of its derivative can be written as

$$\mathcal{F}\left[\frac{dh(x)}{dx}\right] = 2\pi\imath X\tilde{h}(X) \tag{7.46}$$

where X is the spatial frequency and $\tilde{h}(X)$ represents the Fourier transform of $h(x)$. Moreover, the Fourier transform $1/x$ can be written

$$\mathcal{F}\left(\frac{1}{x}\right) = -\pi\imath\mathrm{sgn}(X) \tag{7.47}$$

where $\mathrm{sgn}(X)$ is the sign function. So, we can write the Fourier transform of $g'(s,\theta)$ as

$$\mathcal{F}\left[g'(s,\theta)\right] = 2\pi^2|S|\tilde{g}(S,\theta) \tag{7.48}$$

where $\tilde{g}$ is the Fourier transform of g and S is the spatial frequency. Then, Eq.(7.45) can be restated as

$$f(r,\phi) = \int_0^{\pi}\int_{-\infty}^{\infty} |S|\tilde{g}(S,\theta)\mathrm{e}^{-2\pi\imath Sr\sin(\phi-\theta)}dSd\theta \tag{7.49}$$

Unfortunately, the previous equation is no longer integrable in close form. To let Eq.(7.49) became integrable over the R domain, it is a necessary requirement to replace $|S|$ with a function defined over a limited domain: it is the case of $\tilde{c}(S) = |S|w(S)$, where $w(S)$ is a window shaped function (the so-called *apodisation window*) whose meaning is to control noise amplification. The latter requirement is related to the physics of the experimental acquisition of projection data, whose Fourier transform is a frequency limited function. Using the Shannon theorem, the most obvious choice for the high frequency limit is defined by the Nyquist frequency $S_t = 1/2a$, where a is the spatial sampling period. In this case, the inversion formula can be restated as

$$f(r,\phi) = \int_0^{\pi} \int_{-\infty}^{\infty} \tilde{c}(S)\tilde{g}(S,\theta)\mathrm{e}^{-2\pi\imath Sr\sin(\phi-\theta)}dSd\theta \qquad (7.50)$$

This expression reflects, among the possible computer-suited algorithms for image reconstruction, the filtered backprojection. In this case the filtering operation is expressed in the Fourier domain. Equation(7.50) can be shortly and usefully sketched by the following relatioship

$$f = \mathcal{B}\left\{\mathcal{F}^{-1}\left[\tilde{c}\mathcal{F}(g)\right]\right\} \qquad (7.51)$$

where four main steps can be recognized:

(1) take the Fourier transform $\mathcal{F}(g)$ of the projection set g,
(2) filter the projection set using the filter function $\tilde{c}$,
(3) take the Fourier inverse transform,
(4) back-project the projection set.

In the case that the $w(S)$ is represented by a simple step function, i.e. the apodisation window does not perform any more control on noise amplification, we have

$$w(S) = \begin{cases} 0 & S > |S_t| \\ 1/2 & S = |S_t|. \\ 1 & S < |S_t| \end{cases} \qquad (7.52)$$

The filter function $\tilde{c}(S) = |S|w(S)$ is the well known *ramp*-filter, whose spatial domain representation is easily obtained as

$$c(x) = \int_{-S_t}^{S_t} |S|\mathrm{e}^{-2\pi\imath Sx}dS, \qquad (7.53)$$

that is

$$c(s) = S_t^2\left[2\mathrm{sinc}(S_t s) - \mathrm{sinc}^2(S_t\frac{s}{2})\right]. \qquad (7.54)$$

Naming a the spatial sampling period on the projection space, the discrete form of Eq.(7.54) reflects the well-known weigth set by Ramachandran and

Lakshminarayanan:

$$c_n = \begin{cases} \frac{1}{4} & n\ = 0 \\ -\frac{1}{\pi^2 n^2} & n \text{ odd} \\ 0 & n \text{ even} \end{cases} \tag{7.55}$$

The inverse Radon transform is *ill-posed.* This can be easily seen from its singular value decomposition whose singular values are given by ([46])

$$\sigma_m = \left(\frac{4\pi}{m+1}\right)^{1/2}. \tag{7.56}$$

Since $\sigma_k \to 0$ as $\frac{1}{\sqrt{k}}$ with $k \to \infty$ the problem is *ill-posed.* This feature is crucial when describing the issues related to the reconstruction of noisy data in SPECT, PET and CT. In fact, Eq.(7.39) shows that a solution for f is sought by dividing $(g, u_k)_{\mathcal{Y}}$ by σ_k, whose values tend to 0: small numerical variations in experimental data (g, u_k) will reflect in numerical instability conditions for the solution values.

To reduce the ill-positioning of the problem, we could implement an apodisation window $w(S)$ whose behaviour in the high frequency range follows a smoother trend toward 0 with respect the one featured by the rectangular shaped window. Among the number of proposed window-functions it could be useful to report two examples:

(1) Hann filter function

$$w(S) = 0.5 + 0.5\cos\left(\pi \frac{S}{S_t}\right) \tag{7.57}$$

whose shape is determined by the cut-off frequency S_t.

(2) Butterworth filter function

$$w(S) = \frac{S}{\sqrt{1 + \left(\frac{S}{S_c}\right)^{2n}}}. \tag{7.58}$$

whose shape is determined by the cut-off frequency S_t as well as by the filter order n.

Generally speaking, the definition of a window function $w(S)$ and its application on experimental data can be seen as a mathematical constrain whose aim is to control the amplitude of the statistical noise on the reconstructed image. In other words, it deals with a regularization technique defined by numerical parameters whose variation reduce or widen the range of possible solutions. In the case of Hann and Butterworth filter functions, the regularization parameter is represented by *i*) the cut-off frequency S_t, and *ii*) both the cut-off frequency S_t and the filter order n. It should be noted that the right value for the regularization parameter is hard to determine in the clinical practice: the object to recontruct is unknown and the severe possibility to lose useful information from (as well as, on the other hand, to introduce high noise level into) the solution-image should be carefully taken into account.

7.4.6. *Iterative algorithms*

From a physical point of view, the Radon transform kernel represented in Eq.(7.42) should be considered a rough reduction of the actual projection process. The SPECT projection differs from the simple geometrical projection in spite of the following factors:

- *Geometrical variable system response*
  As previously discussed (collimators), the geometrical system response of collimated SPECT cameras produces spatially non-stationary but object independent effects. So, the mathematical description of the collimator effect on the projection data is related to the physical modelling of the spatial dependent shape smearing process of the camera point spread function (PSF).
- *Attenuation of γ-radiation*
  In this case too, it should be necessary to introduce some modifications in the Radon transform kernel. Tretiak and Metz ([48]) found the exact Radon inversion formula under the following assumptions:

  - The attenuation coefficient has a constant value over the whole solution domain.
  - The outer profile of the transaxial image under reconstruction has a convex shape.

  Unfortunately, the previous assumptions result simplificative with respect to the experimental conditions. So, the usefulness of this mathematical approach to the *attenuated* Radon inversion formulation, should be considered heavily compromised.
- *Compton effect.*
  The Compton effect is responsible for the overlapping of *i*) spatially diffused distribution of γ-radiation, and *ii*) collimator PSF. Mathematical approaches to the Radon transform kernel that could take into account this physical effect, have not been formulated yet.

An alternative approach to the Radon inversion problem when physical effects are involved, could be found solving, using a numerical approach, a discretized version of the direct Radon transform. This means that the integration operator is substituted with a summation, the functions f and g are substituted with vectors and the transform kernel is replaced with a matrix which we will call A. The elements of matrix A determine which image elements contribute to a certain projection element. In this case, Eq.(7.42) can be written as

$$g_{klm} + w_{klm} = \sum_{i=1}^{N}\sum_{j=1}^{N}\sum_{l=1}^{L} A_{ijl'}^{klm} f_{ijl'} \tag{7.59}$$

where g_{klm} is the number of counts registered into the $(k,l)^{th}$ pixel of the mth two-dimensional projection, w_{klm} is the related statistical noise, $f_{ijl'}$ in the unknown content of the $(i,j,l')^{th}$ voxel of the solution space and $A_{ijl'}^{klm}$ represents the fraction of photons emitted from the $(i,j,l')^{th}$ voxel detected by the $(k,l)^{th}$ pixel of the mth projection. In this formulation, the physical information (*a priori*) relative to the acquisition process are involved in the $A_{ijl'}^{klm}$ matrix elements.

As previously stated (see Section 7.4.3) the discretization of an inverse linear problem reflects in a large system of linear equations. In this case, the choice for reconstruction algorithms falls in the class of iterative methods. It should be noted, however, that the large dimension of the tomographic problem as well as the computational burden typical of iterative techniques reduce the number of possible computational choices. A brief description of the reconstruction techniques most widely studied and implemented will follow.

least squares algorithms The problem solution is sought for by means of a least squares method where a variational problem of the following type is posed

$$\Phi(f) = ||\mathbf{A}f - g|| = \text{minimum} \tag{7.60}$$

where the statistical noise component w_{klm} in Eq.(7.59) is included into the g term and $||\cdot||$ is the Euclidean norm. It is easy to see that the f satisfying this variational problem is solution of the so-called Euler equations:

$$\mathbf{A}^T\mathbf{A}f = \mathbf{A}^T g. \tag{7.61}$$

The least squares solution

$$f^+ = (\mathbf{A}^T\mathbf{A})^{-1}\mathbf{A}^T g. \tag{7.62}$$

of this problem is also called *generalized inverse* or *Moore-Penrose inverse.* It is not possible to apply Eq.(7.62) to such large problems but there are many iterative methods which allow one to calculate the least squares solution: Landweber, steepest-descent and conjugate gradients are some of these methods:

- ***Landweber*** $f^{n+1} = f^n + \gamma \nabla \Phi(f^n)$,
- ***Steepest-descent*** $f^{n+1} = f^n + \alpha_n \nabla \Phi(f^n)$,
- ***Conjugate gradients*** $f^{n+1} = f^n + \alpha_n p^{n+1}$,

where it may be useful to remember that

$$\nabla \Phi(f) = A^T A f - A^T g, \tag{7.63}$$

i.e. the gradient of the least-squares functional is equal to the negative of the residual associated with the least-squares equation $\bar{A}f = \bar{g}$, with $\bar{A} = A^T A$ and $\bar{g} = A^T g$.

The regularization properties of the methods used to solve large least squares problems are very well known since for all these methods the solution can be expressed in form of a *filtered* version of the SVD solution, the filter shape depending on the specific method. So, if the SVD solution in the discrete case is written as

$$f = \sum_{k=1}^{p} \frac{1}{\sigma_k} (g, u_k)_M v_k \tag{7.64}$$

the solution of all the least squares regularized methods can be expressed as

$$f = \sum_{k=1}^{p} F_\gamma(\sigma_k) \frac{1}{\sigma_k} (g, u_k)_M v_k \tag{7.65}$$

where $F_\gamma(\sigma_k)$ is a filter function depending on a positive real number γ. Of course, the mathematical formulation of $F_\gamma(\sigma_k)$ is related to the minimization method implemented.

statistical algorithms Statistical methods are based on a completely different approach. In the previously discussed methods no mention has been done about the statistical properties of noise as well as about its influence on the measured data. In the frame of the statistical approach, the random nature of the noise in considered; moreover measured data are conceived as a particular realization of a random process. It is possible to identify two main approaches. The first assumes that the object is deterministic so that it plays the role of a set of parameters characterizing the data probability distribution. Methods devoted to the assessment of parameter distribution such as the *method of maximum likelihood* are used for image reconstruction. In the second approach, the object is assumed to be a realization of a random process with a given probability distribution. The methods for reconstructing the images in such a context are the *Bayesian methods.*

Maximum likelihood methods It is assumed that g and w are a realization of two random vectors η and ν, respectively, so that Eq.(7.59) writes

$$\eta = \mathbf{A}f + \nu. \tag{7.66}$$

It is assumed that the statistical properties of η are known via its probability density function $p_\eta(g|f)$ which gives the probability of obtaining the realization g of the data, given a certain set of f values. As it is very appropriate for emission tomography, it is assumed that the data follow the Poisson statistics, so that we can write

$$p_\eta(g|f) = \prod_{m=1}^{M} e^{-(\mathbf{A}f)_m} \frac{(\mathbf{A}f)_m^{g_m}}{(g_m)!} \tag{7.67}$$

where M is the number of the data. A logarithmic likelihood function is formed $l(f) = \ln(p_\eta(g|f))$ and an algorithm is sought in order to minimize it with respect to f. A very popular algorithm to minimize the log-likelihood is the *Expectation-Maximization* (EM) iterative algorithm which was derived for emission tomography by Shepp and Vardi in 1982.

Even if the basic assumptions of this method are very pleasant in the context of emission tomography, its properties are more difficult to study and less known. In particular the consequences of the ill-posedness of the tomographic problem are not described so clearly as with the previous methods. From the practical and numerical point of view the same behaviour of other iterative algorithms is observed and in the particular case where the Poisson statistics can be assimilated to the Gaussian one, the ML method is equivalent to the LS method. Thus in this case the same description of ill-conditioning is valid.

For the sake of completeness, a few among the most used mathematical formulations of statistical methods will follow:

ML-EM: For iterations $n = 1, 2, \ldots$

$$f_j^{n+1} = \frac{f_j^n}{\sum_i A_{ij}} \sum_i \frac{A_{ij} g_i}{\sum_k A_{ik} f_k^n} \tag{7.68}$$

MAP-EM OSL: For iterations $n = 1, 2, \ldots$

$$f_j^{n+1} = f_j^n + \frac{f_j^n}{\sum_i A_{ij} + \frac{\partial \log p_\phi(f)}{\partial f_j}} \times \left[\sum_i A_{ij}\left(\frac{g_i}{\sum_k A_{ik} f_k^n} - 1\right) - \frac{\partial \log p_\phi(f)}{\partial f_j}\right] \tag{7.69}$$

OS-EM: For iterations $n = 1, 2, \ldots$; for subsets $m = 1, 2, \ldots, M$

$$f_j^{n,m+1} = \frac{f_j^{n,m}}{\sum_{i \in S_m} A_{ij}} \sum_{i \in S_m} \frac{A_{ij} g_i}{\sum_k A_{ik} f_k^{n,m}} \tag{7.70}$$

with

$$f_j^{n+1,0} = f_j^{n,M} \tag{7.71}$$

RBI-EM: For iterations $n = 1, 2, \ldots$; for subsets $m = 1, 2, \ldots, M$

$$f_j^{n+1,m} = f_j^{n,m} + \left[\frac{1}{t_m}\right]\left[\frac{f_j^{n,m}}{\sum_i A_{ij}}\right] \times \left[\sum_{i \in S_m} A_{ij}\left(\frac{g_i}{\sum_k A_{ik} f_k^{n,m}} - 1\right)\right] \tag{7.72}$$

with

$$t_m = \max_j \left[\frac{\sum_{i \in S_m} A_{ij}}{\sum_i A_{ij}}\right]$$

and

$$f_j^{n+1,0} = f_j^{n,M}$$

RBI-MAP: For iterations $n = 1, 2, \ldots$; for subsets $m = 1, 2, \ldots, M$

$$f_j^{n+1,m} = f_j^{n,m} + \left[\frac{1}{t_m}\right]\left[\frac{f_j^{n,m}}{\sum_i A_{ij} + \frac{\partial \log p_\phi(f)}{\partial f_j}}\right] \times \left[\sum_{i \in S_m} A_{ij}\left(\frac{g_i}{\sum_k A_{ik} f_k^{n,m}} - 1\right) - \frac{\partial \log p_\phi(f)}{\partial f_j}\right] \tag{7.73}$$

with

$$t_m = \max_j \left[\frac{\sum_{i \in S_m} A_{ij} + \frac{\partial \log p_\phi(f)}{\partial f_j}}{\sum_i A_{ij} + \frac{\partial \log p_\phi(f)}{\partial f_j}}\right]$$

and

$$f_j^{n+1,0} = f_j^{n,M}$$

Bayesian methods Bayesian methods can be considered a kind of regularized version of ML methods where the *a priori* information is introduced in form of the knowledge of the probability density function of the object. Here the basic fact is that the (unknown) object f is also a realization of a random vector ϕ whose probability density function $p_\phi(f)$ is given. Therefore the tomographic problem is stated as follows

$$\eta = \mathbf{A}\phi + \nu. \tag{7.74}$$

Now the problem is to seek information about the random vector ϕ given a realization g of the random vector η. The problem is faced by evaluating the *conditional probability density function*, or *a posteriori probability density function* of the random vector ϕ:

$$p_\phi(f|g) = \frac{p_\eta(g|f)p_\phi(f)}{\int p_\eta(g|f)p_\phi(f)} \quad . \tag{7.75}$$

where $p_\phi(f)$ is the probability density function of ϕ, i.e. the *a priori* information, and $p_\eta(g|f)$ is the conditional probability density function of η, i.e. the probability density function of η for a given realization f of ϕ. Reconstruction algorithms require the determination of a particular value f from the density function $p_\phi(f|g)$; they are not easy to derive and, usually, they are very expensive from the computational point of view. Among the most used in emission tomography there are the GEM MAP (Generalized EM Maximum *a posteriori*, [49]) and OSL EM (One Step Late EM, [50]) algorithms. It is interesting to note that, when applied to image deconvolution in the hypothesis that the noise ν and the object ϕ are independent gaussian random vectors with zero expectation value ([46]), the Bayesian approach leads to the well-known Wiener filter.

The great flexibility of all the iterative approaches consists in the fact that, potentially, in the elements of the system (or transition) matrix all kind of effects can be embedded, thus improving the consistency of the mathematical formulation with the data. In principle, there is no limit in the information one can include in the elements of matrix $\mathbf{A}$. However, there are practical limits related to computing requirements and to the complexity of object-dependent models.

Researchers began to pay attention to the spatially variable PSF since the end of the eighties [51; 52]: the inclusion of the PSF blurring into the system matrix $\mathbf{A}$ results a simpler approach with respect to the management of attenuation and scatter effect. In fact, the PSF is only instrument dependent whereas the other effects are also object-dependent. In [52] an effort was made in order to establish a feasible link between the experimental setup of the SPECT system and the reconstruction algorithm and therefore an appropriate acquisition of a line source was made to calculate a lookup table of weighting factors needed by the reconstruction program. Since the need of an additional experimental procedure turned out to be rather annoying, more recently a theoretical methods for the determination of the weighting factors was proposed [53]. This work was based on a revision of the theoretical representation of geometrical response function of multihole collimators [2].

The inclusion of attenuation compensation into the system matrix elements is rather easy with iterative algorithms. The main limitation is in the lack of information about the spatial distribution of attenuation coefficients. A reasonable solution to the attenuation problem is now available thanks to the new transmission option offered with many SPECT systems. The reconstruction of transmission data allows one to take into account the non uniform distribution of the attenuation coefficient, a feature which is particularly relevant in heart studies.

As previously sketched, the compensation for the scatter effect is probably the most difficult problem to tackle in SPECT. The object-dependent nature and the complexity of the scatter effect is such that no analytic approximate description exists till now nor there will probably exist. Iterative methods are best suited for attempting corrections but there are two kind of problems for practical applications: the difficulty in defining a sufficiently accurate but manageable model, and the computational requirements caused by the extension and variability of the scatter spread function.

7.5. Clinical Imaging

In the following, a few examples of nuclear medicine imaging will be presented. Aim of this section is to provide the reader with some results whose peculiarity consist in deriving from the software and hardware imaging techniques discussed so far.

Moreover, although the problem of attenuation correction was already addressed in Section 7.4, a general introduction to the main basic approaches oriented toward the study and management of the attenuation of γ-radiation effect in SPECT will also be reported.

7.5.1. *High-resolution SPECT imaging*

The transformation kernel of the direct problem can be interpreted as the mathematical expression of the knowledge of the physics of the acquisition process. On the other hand, its physical complexity directly reflects into the mathematical formulation of the transformation kernel: the reconstruction algorithm appointed for the solution of the inverse problem should be able to manage kernels as complete as possible in order of reducing incongruities between acquisition model and experimental data.

When the previous requirements can be fulfilled, the image quality results improved in the fields of both lesion detectability and tracer uptake quantification. Figure 7.7 refers to a SPECT *in vivo* study: starting from the same set of experimental data, images obtained using a reconstruction algorithm where the variable geometrical system response of the collimator system is accounted for can be compared to a commercially available one (see figure caption for details).

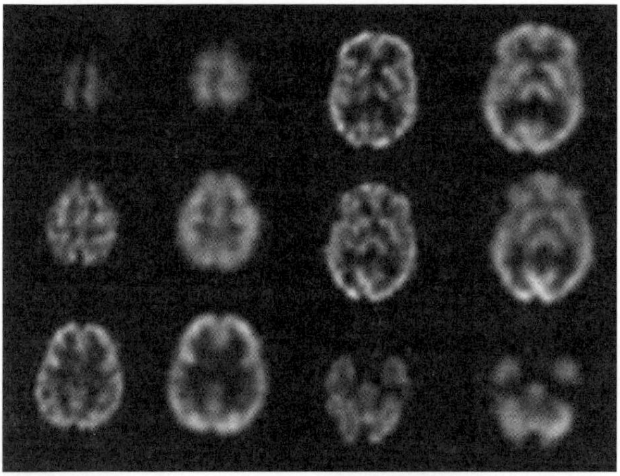

Figure 7.7. Reconstruction of a SPECT HM-PAO brain perfusion study. Six axial slices are shown. First and third column from the left: conjugate gradient iterative technique with 3D model of the geometrical system response. Second and fourth column from left: backprojection of Wiener filtered projections.

7.5.2. *Planar imaging from Semiconductor detectors*

Figure 7.8 refers to breast planar scan acquired with the Digirad 2020tc Imager (see Section 7.3.2.2). As shown by the reported images, the signal to noise ratio value seems actually improved by the superior energy resolution of the semiconductor detector with respect to a standard scintillator device. However, it should be noted that the reported example falls into a category of imaging tasks that, better than others, can take advantage from the high energy resolution features of the semiconductor detectors: high specific radiotracer uptake, low tissutal thickness, low distance of the target organ from the collimator surface should be considered the experimental conditions that highlight the imaging benefits obtainable from semiconductor based γ-cameras.

SPECT imaging could also take advantage of semiconductor characteristics, particularly when brain imaging is involved. Actually, simulation studies [55] demonstrated that a fixed gantry, multiple-head camera equipped with pin-hole collimators (Fig.7.9) ought to feature both interesting spatial resolution and efficiency in experimental studies. However, to the author's knowledge, a camera prototype fulfilling this design was not still produced.

7.5.3. *Attenuation corrected imaging*

As previously discussed, soft tissue attenuation as well as Compton scattering degrade SPECT image quality, thereby decreasing the overall sensitivity of the tomographic method. Several post processing as well as indirect estimation algorithms have been proposed to minimize the impact of the attenuation on the diagnostic image. During the last decade, however, technology efforts have been oriented toward the design of hardware solutions that, as it happens in positron emission tomography (PET), could make an actual transmission map of the 3D activity source (patient or phantom)

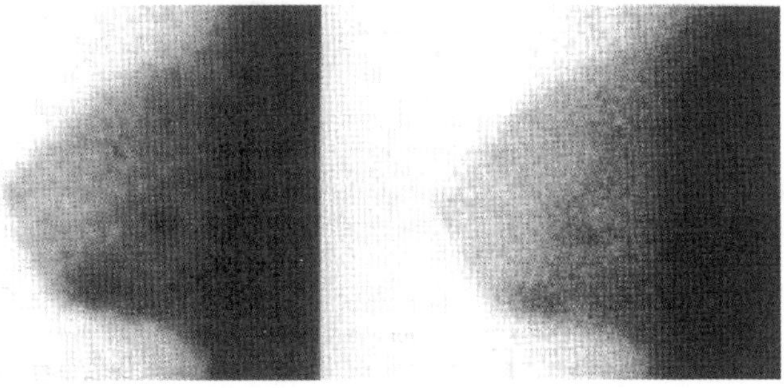

Figure 7.8. Comparison of images obtained with (left) Digirad2020tc Imager and (right) with conventional γ-camera.

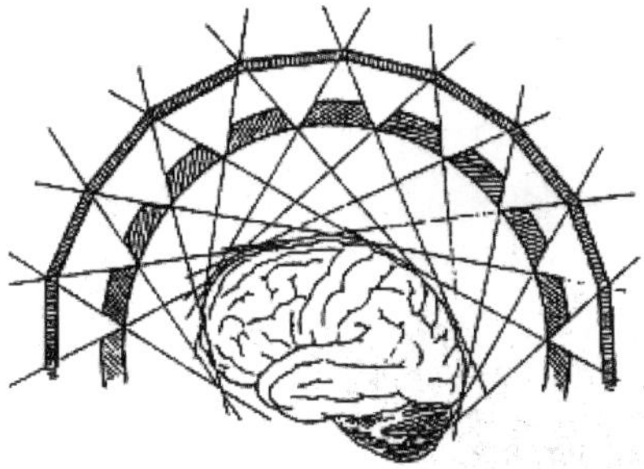

Figure 7.9. Multiple pinhole design for high resolution SPECT.

under investigation available.

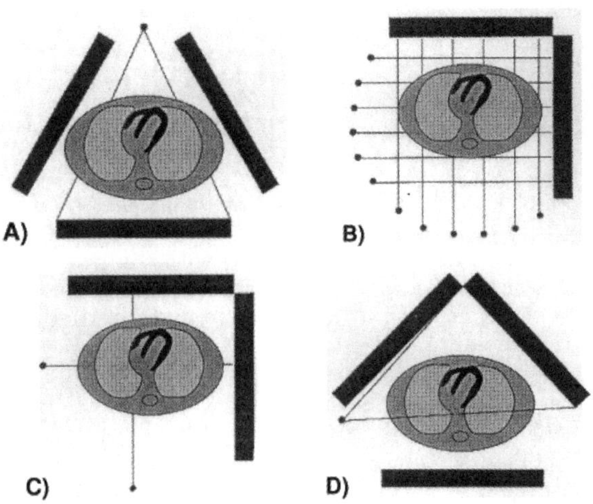

Figure 7.10. SPECT transmission imaging configurations: a) fixed source, fan-beam collimation; b) multiple fixed sources, parallel collimation; c) sweeping source, parallel collimation; d) sweeping source, anyhow collimation.

Several types of transmission systems equipped with external radioactive sources have emerged: ^{153}Gd, ^{57}Co, ^{133}Ba, ^{241}Am and ^{99m}Tc are the most widely used. The main transmission configurations are sketched in Fig.7.10 and can be resumed as:

– Fixed line source with convergent collimation on triple-120°-head camera configuration (Fig.7.10 A):

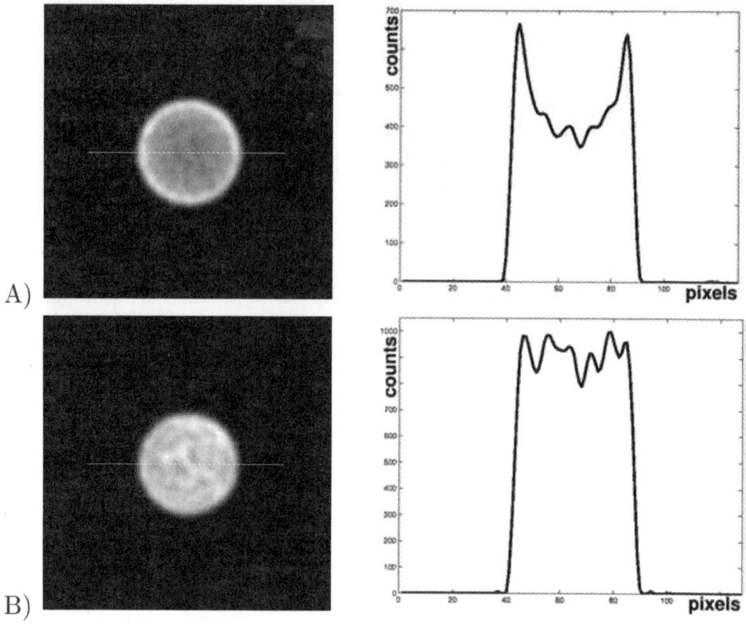

Figure 7.11. Slices of a cylindrical homogeneous phantom reconstructed along a transaxial plane using a FBP reconstruction method with Wiener apodization window: A) without attenuation correction; B) with compensation of the attenuation effect as sketched in fig.7.10-D). The continuous horizontal marker on the transaxial images highlights where the activity profiles related to these images, reported on the right column, are drawn.

- *positive impact*
  High count efficiency allowing comparatively low source strength; high mechanical stability.
- *negative impact*
  Quite high probability of truncation artifacts because of the limited field of view of convergent collimation.

– Scanning line sources with parallel collimation on dual-90°-head camera configuration (Fig.7.10 C):

- *positive impact*
  The camera field of view is completely exploited so that the likelihood of patient truncation is minimized.
- *negative impact*
  Prone to mechanical instability and available with parallel collimation

geometry only.

– Multiple line source array with parallel collimation on dual-90°-head camera configuration (Fig.7.10 B):

- *positive impact*
  High transmission count efficiency without supplementary mechanical motion; moreover the opposite detector field of view is completely exploited.
- *negative impact*
  The continuous incidence of transmission source γ-rays over the camera detectors may reflects in significant energy cross-talk between scatter and full energy peak: in this case sequential transmission-emission imaging may be required; moreover the parallel collimation geometry is required.

– Scanning point sources regardless the collimation type on dual- or triple-head camera configuration (Fig.7.10 D):

- *positive impact*
  Point sources of ^{133}Ba are used, so that its high energy emission peak can penetrate low-energy collimator septa, producing a transmission map regardless the collimation geometry involved.
- *negative impact*
  The point source asymmetry with respect to the target detector require accurate patient positioning just to minimize the likelihood of truncation artifacts.

Some examples of phantom studies performed with and without attenuation correction are reported in Fig.7.11.

References

1. H.O.Anger, J.Nucl.Med. **5** 515-531 (1964).
2. A.R.Formiconi, Phys.Med.Biol. **43** 3359-3379 (1998).
3. D.Pareto, Pav Jía, C.Falcón *et al.*, Eur.J.Nucl.Med., **28** 144-149 (2000).
4. A.R.Formiconi, F.Di Martino, D.Volterrani, A.Passeri, IEEE Trans.Nucl.Sci., **46** 1075-1080 (1999).
5. C.E.Metz, F.B.Atkins, R.N.Beck, Phys.Med.Biol. **25** 1059-1070 (1980).
6. B.M.W.Tsui, G.T.Gullberg, Phys.Med.Biol. **35** 81-93 (1990).
7. B.M.W.Tsui, D.L.Gunter, R.Beck, and J.Patton, in: *Diagnostic nuclear medicine*, M.P.Sandler, J.A.Patton, R.E.Coleman, A.Gottschalk, F.J.Th.Wackers, and P.B.Hoffer, Eds., pp. 67–79. Williams and Wilkins, Baltimore, (1988).
8. R.N.Beck and L.D.Redtung, IEEE Trans.Nucl.Sci., **32** 865-869 (1985).

9. D.L.Gunter, in: *Nuclear medicine*, R.E.Henkin, M.A.Boles, G.L.Dillehay, J.R.Halama, S.M.Karesh, R.H.Wagner, and A.M.Zimmer, Eds., pp. 96-124. Mosby, St Louis, (1996).
10. C.E.Floyd, R.J.Jaszczak, C.C.Harris, and R.E.Coleman, Phys.Med.Biol., **29** 1217–1230 (1984).
11. D.J.deVries, S.C.Moore, R.E.Zimmerman *et al.*, IEEE Trans.Med.Imag., **9** 430-438 (1990).
12. Y.K.Dewaraja, M.Ljungberg, and K.F.Koral, J.Nucl.Med., **41** 123-130 (2000).
13. D.P.Lewis, B.M.W.Tsui, C.Tocharoenchai, and E.C.Frey, *Characterization of medium and high energy collimators using ray-tracing and monte carlo methods.* in: *Conference record of 2000 IEEE Nuclear Science Sympsium and Medical Imaging Conference*, (2000).
14. S.C.Moore, K.Kouris, I.Cullum, Eur.J.Nucl.Med., **19** 138-150 (1992).
15. Y.K.Dewaraja, M.Ljungberg, and K.F.Koral, J.Nucl.Med., **41** 1760-1767 (2000).
16. P.K.Leichner, H.T.Morgan, K.P.Holdeman *et al.*, J.Nucl.Med., **36** 1472-1475 (1995).
17. G.L.Zeng, G.T.Gullberg, *et al.*, J.Nucl.Med., **39** 124-130 (1998).
18. F.de Notaristefani, R.Pani *et al.*, Proc IEEE NSS Synposium and Medical Imaging Conference; San Francisco 1801 (1995).
19. R.Pani, R.Pellegrini *et al.*, Nucl.Inst. and Meth. A **409** 524-528 (1998).
20. G.Entine *et al.*, IEEE Trans.Nucl.Sci. **NS- 26(1)** 552 (1979).
21. J.F.Butler *et al.*, SPIE **1896** 30 (1993).
22. F.P.Doty *et al.*, Nucl.Inst. and Met. A **353** 361 (1994).
23. G.F.Knoll *Radiation detection and measurement*, John Wiley and Sons (1979).
24. S.Baccaro, K.Blažek, F.de Notariatefani *et al.*, Nucl.Inst. and Meth. A **361** (1995).
25. J.A.Mares *et al.*, *Heavy Scintillators and Industrial Applications*, Proc. CRYSTALL 2000 Int.Work-shop, Chamonix, September 1992, France, Editions Frontiers, p. 207 (1992).
26. J.A.Mares *et al.*, Mater.Chem.Phys. **32** 342 (1992).
27. K.Blažek, F.de Notariatefani *et al.*, Eur.J.Nucl.Med. **20(10)** 854 (1993)
28. E.Sakai, IEEE Trans.Nucl.Sci. NS **34** 418 (1987).
29. K.Takagi and T.Fukazawa, App.Phys.Lett., **42** (1983).
30. M.Kobayashi, M.Ishii, C.L.Melcher, Nucl.Inst. and Met. A **335** 509 (1993).
31. H.Suzuki, T.A.Tombrello *et al.*, IEEE Trans.Nucl.Sci. **40** 380 (1993).
32. F.Daghighian *et al.*, IEEE Trans.Nucl.Sci. **40** 1045 (1993).
33. A.Lempicki, J.Glodo, Nucl.Inst. and Met. A **416** 333 (1998).
34. T.Ludziejewski, M. Moszyňska *et al.*, IEEE Trans.Nucl.Sci. **42** 328 (1995).
35. M.M.Rogulski, H.B.Barber *et al.*, IEEE Trans.Nucl.Sci. **26** 552(1979).
36. K.Hetch, Zeits.Phys. **77** 235 (1932).
37. Y.Eisen, Y.Horovitz, Nucl.Inst. and Met. A **353** 60 (1994).
38. M.Richter, P.Siffert, Nucl.Inst. and Met. A **322** 529 (1992).
39. V.I.Ivanov *et al.*, IEEE Trans.Nucl.Sci. **42** 258 (1995).
40. O.Frisch, British Atomic Energy Report, **49** (1944).
41. P.N. Luke, IEEE Trans.Nucl.Sci. **42** 207 (1995).

42. E.Raiskin, J.F.Butler, IEEE Trans.Nucl.Sci. **35** 81 (1988).
43. Y.Eisen, A.Shor, I.Mardor, Nucl.Inst. and Met. A **428** 158 (1999).
44. Y.Eisen, A.Shor *et al.*, Nucl.Inst. and Met. A **380** 474 (1996).
45. J.Chambron *et al.*, Nucl.Inst. and Met. A **428** 138 (1999).
46. M.Bertero & P.Boccacci, *Introduction to Inverse Problems in Imaging*, Institute of Physics Publishing (IOP), Bristol and Philadelphia, (1998).
47. J.Radon, *'Über die bestimmung von funktionen durch ihre integralwerte längs gewisser nmannigfaltigkeiten'*, Berichte Sächsische Akademie der Wissenshaften, Math.Phys.Kl. **69** 262 (1917).
48. O.Tretiak, C.E. Metz, SIAM J.Appl.Math. **39** 341-354 (1980).
49. T. J.Hebert, & R. M.Leahy, IEEE Trans.Med.Imag. **8** 194–202 (1989).
50. P.Green, IEEE Trans.Med.Imag. **9** 84–93 (1990).
51. B.M.W.Tsui, H.Hu, D.Gilland, & G.T.Gullberg, IEEE Trans.Nucl.Sci. **35** 778–783 (1988).
52. A.R.Formiconi, A.Passeri & A.Pupi, Phys.Med.Biol. **34** 69–84 (1989).
53. A.R.Formiconi, A.Passeri & P.Calvini, IEEE Trans.Nucl.Sci. **46** 1075-1080 (1999).
54. H.B.Barber, Nucl.Inst. and Met. A **436** 102 (1999).
55. M.M.Rogulski *et al.*,IEEE Trans.Nucl.Sci. **40** 1123 (1993).
56. R.C.Hendel, J.R.Corbett *et al.*, J.Nucl.Card. **9** 35 (2002).

CHAPTER 8

POSITRON EMISSION TOMOGRAPHY

Alberto Del Guerra, Alfonso Motta

Department of Physics, University of Pisa, and INFN, Sezione di Pisa, Pisa, Italy

8.1 Introduction to Emission Imaging

Positron Emission Tomography (PET) is a powerful and sensitive technique for functional imaging in the Field of Nuclear Medicine. The principle of operation is different from most other transmission technologies. These tend to be based on the detection of a beam of radiation transmitted through the patient: the image represents the three-dimensional (3D) attenuation properties of biological tissues and gives density and morphology information.

Conversely in PET, a positron emitting radiopharmaceutical is injected into the patient and spreads physiologically within the body: the activity distribution will be proportional to the drug concentration. The emitted positrons annihilate with electrons in the tissue producing two back-to-back 511 keV photons. These are measured in electronic coincidence by using opposing pairs of detectors, which are mounted on a rotating gantry or in a stationary configuration, as either a circular or a polygonal ring (Fig. 8.1). This gives thus an intrinsically tomographic imaging (Fig. 8.2). The positron emission is isotropic: there is no bias in the emission directions. The electronic coincidence detection intrinsically implies alignment, so anti-scattering and passive collimators are not needed and a higher detection efficiency is achieved.

PET system development has been characterized by the need to enhance the axial dimension of the detection area, accomplished either by using large area detectors, or by increasing the number of high-efficiency detector rings, as in commercial tomographs with tens of thousands of scintillating crystals.

An "historical" approach restricted the collection of the emission radiation from each slice reducing as much as possible scattering and random coincidences between different rings by using septa, with a consequent reduction in the sensitivity. Today, tomographs recover the isotropic properties of radiation emission: the acquisition is no longer limited to a planar slice, but is performed on the whole solid angle between the two detectors, therefore achieving a fully 3D image. The removal of the inter-ring shield results in a remarkable increase in the sensitivity, obviously with a corresponding increase in the scatter contribution [1].

There are three modalities of PET imaging that each require different normalizations and corrections for scattered and random events. A purely two-dimensional (2D) PET collects the radiation in planar slices only:

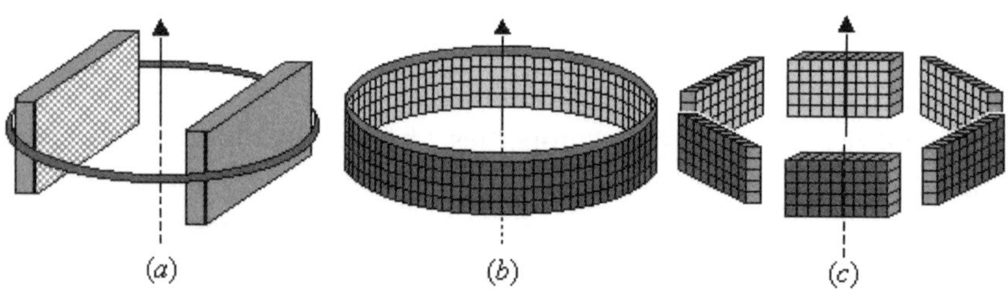

Fig. 8.1. PET systems: detectors mounted on a rotating gantry (*a*), or circular or polygonal detector rings (*b* and *c*). Commercial cameras generally offer a multi-ring geometry, with tens of thousands of scintillation elements.

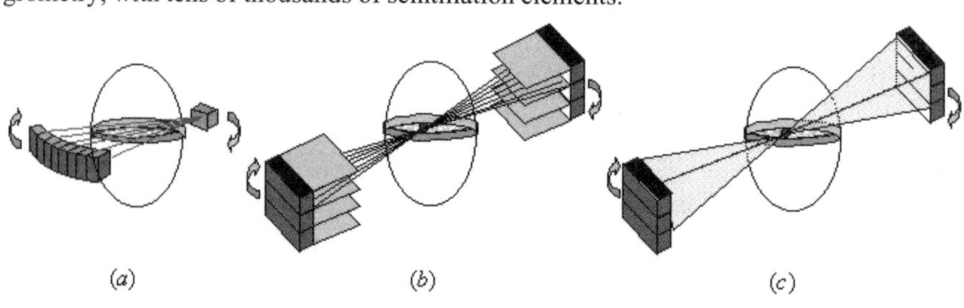

Fig. 8.2. Tomography: transmissive (*a*) and emissive (*b*, *c*) schemes. Inter-ring septa allow separation of slices (*b*); without septa, acquisition is intrinsically 3D, with an increase in the sensitivity (*c*).

the first commercial, multi-ring scanners have been equipped with non-removable inter-ring septa, and work in this 2D modality only. In 3D multi-slice (3D-MS) PET, tomographs acquire *lines-of-flight's* that may cross several planar slices: it is a 3D acquisition but with a 2D limitation in the reconstruction, where a rebinning process redistributes the non-planar acquisitions on planar slices. Since 1990 several multi-ring cameras have been commercialized with retractable septa, thus allowing 2D and 3D capabilities. Data rebinning avoids high computation time and allows the adoption of correction modes and reconstruction routines from other methodologies. The positron volume imaging (PVI) is intrinsically a 3D modality, which considers all the emission directions in the whole volume of interest in both the detection and reconstruction. Large area rotating NaI(Tl) detectors, multi-wire PET chambers, and the last generation of multi-ring commercial cameras without septa (since 1994) can all use this 3D modality. Volume imaging is possible today because of advances in scatter and random corrections and especially in 3D reconstruction techniques, that were not available until the 1990's.

8.1.1 *Tomography procedures and terminologies*

PET is based on the coincident detection of the opposite photons resulting from the annihilation of an electron-positron pair [2-8].

A positron emitting radiopharmaceutical is administered to a patient, and physiologic processes cause the distribution of the drug throughout the body. Emitted positrons slow down and annihilate with the electrons of the biological tissue. PET measures the activity distribution by detecting the annihilation photons in electronic coincidence. Elements such as carbon, nitrogen and oxygen have positron-emitting isotopes with short half-life ($T_{1/2}$, Table 8.1).

Table 8.1. The most important radioisotopes used in PET and their half-life.

Isotope	^{11}C	^{13}N	^{15}O	^{18}F
half-life (min)	20.4	9.96	2.07	109.7

Their short half-lives imply an "optimal" utilization of the radiation dose administered to the patient, but there is the requirement for an on-site cyclotron for production. These chemical elements are characteristic in human metabolism, and their radioisotopes can be used to label several substrate molecules in trace amounts, without perturbation of the phenomenon to study. A fluorine isotope, ^{18}F, can be included by radiochemical synthesis in several natural substances, such as carbohydrates, amino acids, and lipids, in addition to a number of pharmaceuticals in order to study a wide variety of metabolic and physiologic functions. The ^{18}F isotope is commonly used to tag a glucose analogue, deoxyglucose (DG), producing ^{18}F-fluorodeoxyglucose (^{18}F-FDG). This is one of the most widely used tracers for studying glucose metabolism. The popularity of ^{18}F is also due to its half-life that allows its use at PET centres that are remote from the isotope production facility.

Such isotopes are used in PET for binding site studies (such as ligand-receptor interaction), kinetic analysis of radiopharmaceuticals, and biological research. Many clinical studies of human disease are possible. Brain PET can study schizophrenia, addiction, anxiety and depression in the field of psychiatry, blood flow activation and neurochemical processes in the area of cerebral functionality, and dementia, movement disorders and stroke recovery in the field of neurological disease. PET is also used in studies of cardiac disease, such as ischemia and infarct study, myocardial viability, and so on. PET is particularly important in oncology where there is the need for both detection and location of the tumour, in addition to follow-up studies after the performed therapy. There is also the branch of research on animals (mice, rats, primates and pigs), for which dedicated tomographs have been built (see Chapter 10 in this book).

When a pair of detectors sees simultaneously two 511 keV photons in electronic coincidence, it is assumed that the annihilation point lies inside the volume between the two detectors. Typical PET terminology therefore includes concepts such as the *Line of Flight* (LOF) or *Line of Response* (LOR), *Field of View* (FOV) and *Time of Flight* (TOF). The LOF indicates the line along which the annihilation event might lie: with small area detectors, it is generally simplified as the line connecting the

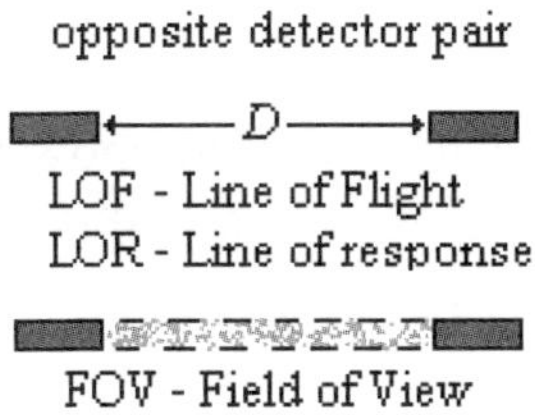

Fig. 8.3. Common PET terminology: the LOF is the ideal line of flight of the photon pair; the LOR is the detected line of flight, shown as the line between the centres of two opposite detection elements; a *coincidence tube* is the volume between two opposite elements; the FOV is the union of all the tubes.

centres of the two detectors in coincidence. The LOR is the detected LOF. Ideally LOF and LOR should be the same, but several errors can occur in the actual detection: therefore, the LOF is the unknown ideal flight direction, while the LOR is the known detected line, the one that has a real physical meaning.[a] In detectors with separate components, or that allow pixel identification, the volume between a pair of elements in electronic coincidence constitutes a *tube*.[b] The LOR is the tube axis. The "union" of all tubes gives the FOV, which is therefore the volume between the two opposing detectors. The TOF is the *time of flight* from the pair annihilation to the detection: were it possible to discriminate the time difference in detection between the two annihilation photons, it could be possible to determine the annihilation point along the LOR. In this respect, there are some difficulties in realizing a TOF system. Firstly, there is the finite duration of the scintillation emission of the detection crystal; secondly, there is the technological issue of measuring a very short time difference that is of the order of a few hundred picoseconds. However, TOF information acquired on a dedicated camera could enhance the spatial resolution: a 100 ps difference in the detection time for a photon pair implies that a gamma ray has covered a distance of 3 cm longer than the other one along the LOR. This information could help to reduce the background during the subsequent image reconstruction [9].

---

[a] Hereafter, we will refer to LOF as being synonymous with LOR.

[b] *Tube*, *coincidence tube* or *detection tube* are also synonymous.

8.2 Physics of Positron Emission Tomography

The acquisition of the information required to reconstruct the activity distribution in the patient can be divided into several stages: the positron emission, its annihilation, the interaction of the two annihilation photons in the biological tissue, and their detection.

8.2.1 *Positron emission and radionuclides*

The *positron* is the antiparticle of the electron: it is identical to an electron in all of its properties, except that the sign of its electric charge and its magnetic moment is opposite to that of an electron. Positrons are denoted by the symbol e^+ or more frequently β^+: they can be generated via the decay of several nuclei whose instability is due to an excessive number of protons with respect to the number of neutrons. These kinds of nuclei are inclined to adopt a more stable energetic configuration, by reducing the number of protons through the β^+ decay (Fig. 8.4)

$$_{Z}N \rightarrow _{Z-1}N^* + \beta^+ + \nu_e \,.$$

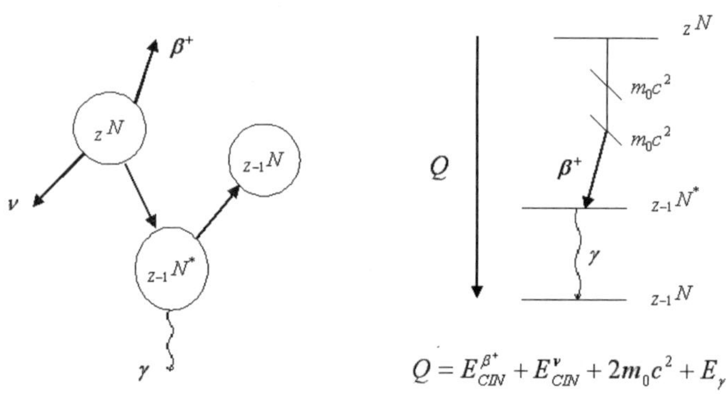

Fig. 8.4. Diagrams of β^+ decay: the nucleus of the β^+ source $_ZN$ is unstable, by having an excessive number of protons with respect to neutrons. The nucleus reaches a more stable and lower energy configuration by reducing the proton number via positron emission; a neutrino is also emitted. The mass balance Q, which represents the released energy, is given by the sum of the kinetic energies of the positron and of the neutrino plus twice the electron rest mass. If the daughter nucleus is created in an isomeric state $_{Z-1}N^*$, the γ decay energy has also to be considered.

Often the daughter nucleus shows up in an isomeric state ${}_{Z-1}N^*$ and a successive γ decay can follow. In a β^+ decay the positron is emitted with a *neutrino*, an uncharged particle with spin 1/2 and zero rest mass. This is a three-particle decay, where the kinetic energy is mainly shared between the lighter particles: the positron and the neutrino. The emission spectrum, which indicates the probability that a β^+ particle is emitted with kinetic energy in the range $[E, E + dE]$ with respect to its kinetic energy *E*, is continuous (Fig. 8.5 and Fig. 8.6) [10].

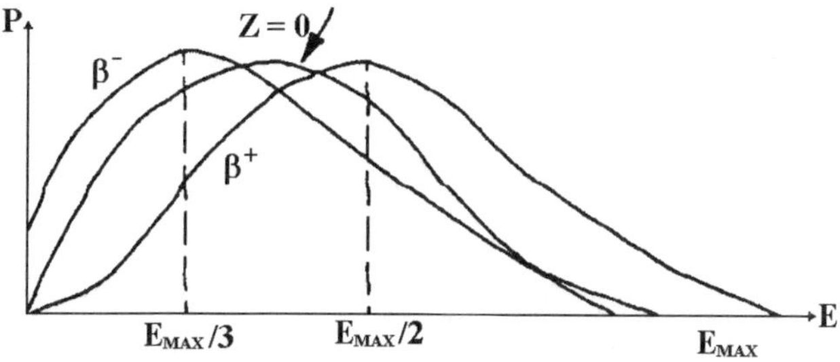

Fig. 8.5. Energy spectra of β decays: positron-neutrino (β^+) or electron-antineutrino (β^-) pairs are ejected. These three-particle decays present continuous energy spectra. identical in the mathematic case of atomic number $Z = 0$: the positron (or the electron) can be emitted with any energy from zero to E_{max}, a value characterizing the radioisotope. In real emission, ($Z \neq 0$), Coulomb law perturbations by residual nucleus charge have to be considered: electron is pulled to a slightly lower energy, while the positron is pushed to a higher energy; this differentiates the two spectra.

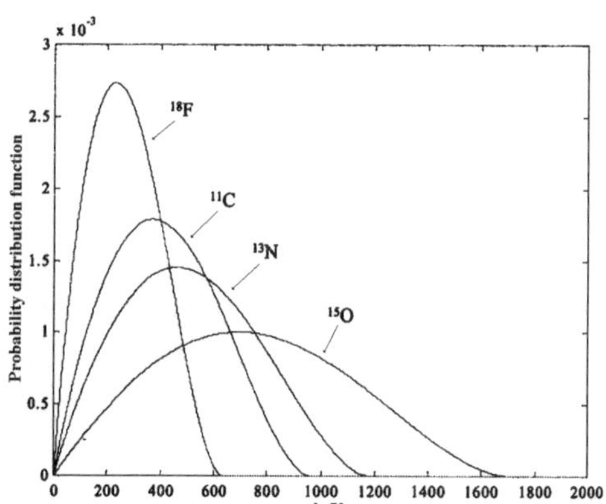

Fig. 8.6. Energy spectra for the β^+ radioisotopes most commonly used in PET. The most probable emission is always close to the centre of the spectrum. The maximum energy value characterizes the radioisotope.

The exponential decay law governs the emission of a positron source,

$$N = N_0 e^{-\lambda t},$$

where N_0 and N represent the number of nuclei at the reference time ($t = 0$) and at the instant t ($t > 0$) respectively, while λ is the decay constant of the radioisotope. The number of disintegrations per second is the *source activity* A and is given by

$$A = -\lambda N = \frac{dN}{dt}.$$

The time which is necessary to halve the number of initial radioactive nuclei is called *half-life* $T_{1/2}$ and is given by

$$T_{1/2} = \frac{1}{\lambda}\ln(2).$$

After a number n of *half-lives* from $t = 0$, the number of remaining radioactive nuclei is given by

$$N = N_0\left(\frac{1}{2}\right)^n.$$

The tracer mass administered to the patient in PET is an amount of the order of nanograms: such a quantity causes neither pharmacologic al nor toxicological effects in the patient [11].

Positron sources are not stable in nature, so have to be artificially produced by bombarding stable inert matter with positively charged particles. Some reactions to obtain PET radioisotopes are reported in Table 8.2 [11]. The incident particles must have enough kinetic energy to overcome the Coulomb repulsion of the target nuclei. Two basic types of accelerators are used: the linear accelerator and the cyclotron. For the production of medical radionuclides, incident particle energies in the range 1 – 100 MeV per nucleon are typical. However, radionuclides used in PET can be produced with about 10 MeV. The produced nucleus has a different Z value from the target, so the two species are chemically separable. The specific activity is very high and the final product contains a minimum amount of impurity.

Table 8.2. Reactions used for β^+ source production: the column of energies shows the energy of the incident nucleon, which strikes the target; the last two columns refer to the final isotope and not to the reaction; the small quantity of product makes the separation of it from the target harder because the yield of a chemical reaction depends on the mass amount; N_A indicates the Avogadro number.

produced radionuclide	**nuclear reaction**	**energy** (MeV)	**atoms**/mCi	N_A/mCi
^{11}C	$^{14}N(p,\alpha)^{11}C$	10 – 20	$6.5\cdot10^{10}$	$1.0\cdot10^{-13}$
^{13}N	$^{13}C(p,n)^{13}N$ $^{16}O(p,\alpha)^{13}N$	10 – 11 10 – 16	$3.2\cdot10^{10}$	$5.0\cdot10^{-14}$
^{15}O	$^{14}N(d,n)^{15}O$ $^{15}N(p,n)^{15}O$ $^{16}O(p,pn)^{15}O$ $^{16}O(^3He,\alpha)^{15}O$	6 – 10 10 > 17 8	$6.6\cdot10^{9}$	$1.0\cdot10^{-14}$
^{18}F	$^{20}Ne(d,\alpha)^{18}F$ $^{16}O(^3He,p)^{18}F$ $^{16}O(^4He,pn)^{18}F$ $^{18}O(p,n)^{18}F$	> 6 > 8 > 25 > 10	$3.5\cdot10^{11}$	$6.0\cdot10^{-13}$

The isotopes produced can then be used to tag the radiopharmaceuticals. The marker needs an optimal site on the molecule to be labelled, the availability and stability of both marker and inert matter, and a sufficient time to synthesize, to purify and to sterilize. The labelled site can be in a natural substrate (labelled amino acids), in a substrate analogue or in a substrate antagonist. Although water is the fundamental component of biological tissues and carbon is involved in metabolic and physiologic functions, there are several difficulties with using the ^{15}O and ^{11}C isotopes. A half-life of ~2 min implies the use of ^{15}O only in centres with an *on site* cyclotron. The ^{11}C isotope has ~20 min half-life and it is more practical for a radiochemical synthesis, but significantly, it gives the possibility of performing a second test on a patient in the same day. It is mainly used in research on neuronal receptors and new radiopharmaceuticals.

The importance of the ^{18}F radioisotope is due to ^{18}F-FDG, whose uptake can be immediately interpreted as the glucose metabolic rate by using the Sokoloff model: this only requires the measurement of the time course of radioactivity in arterial blood [12]. This procedure allows imaging and quantification of metabolic disturbances in cardiology, oncology, and in neurological disease or disorder. The ^{13}N isotope is especially used as $^{13}N\text{-}NH_3$ in cardiac flow studies.

8.2.2 *Annihilation of positron*

Positrons are emitted with a kinetic energy greater than zero, and are slowed down through multiple Coulomb interactions in the biological tissue. Energy loss continues in this way until the positron reaches thermal equilibrium with the surrounding medium, at which point, annihilation with an electron can occur.

Usually, the positron-electron *in-flight-annihilation* is not considered (~2% of occurrences). This event happens before the positron is thermalized, and has two competing processes: Coulomb interaction and bremsstrahlung, at low and high energy values, respectively; by means of these the positron loses its energy quickly (~1 ps) [13]. The in-flight annihilation has the greatest probability of occurring when the following condition is satisfied:

$$E_{kin}^{\beta^+} \approx m_0 c^2 .$$

The range of the positron depends on the electron density of the medium. In water, which is the major component of biological tissue (70%), the range of the positrons emitted from PET radionuclides is about 1–2 mm, as reported in Table 8.3 [3, 4, 14]. An ideal tomograph should accurately measure the activity distribution in the body, i.e., the positron emission points. Actually, a tomograph can only detect the annihilation point: the range of the positron separates these two points. This *range effect* introduces a blurring in the image, resulting in a degradation of the spatial resolution (Fig. 8.7). The contribution of this effect to the total spatial resolution is reported in Table 8.4 for several isotopes [5].

The positron annihilation can be easily described in a Lorentz system where the centre of mass of the electron-positron pair is at rest: here, the two particles have momenta of equal magnitude and opposite direction. The momentum and energy conservation laws imply that the two emitted gamma-rays have the same energy, equal to the energy of one of the initial particles, and opposite propagation directions, "exactly" at 180°.

Table 8.3. Numerical data for the radioisotopes most commonly used in PET. Electrons and positrons do not move along a straight line in matter because of the high number of interactions with small energy exchange; therefore their path length is always longer than their range [3, 4, 14].

isotope	**half-life** (min)	**average kinetic energy** (Mev)	**maximum kinetic energy** (Mev)	**average range in water** (mm)	**path length** (mm)
^{11}C	20.4	0.385	0.960	1.7	4.1
^{13}N	10.0	0.491	1.198	2.0	5.1
^{15}O	2.0	0.735	1.732	2.7	7.3
^{18}F	109.8	0.242	0.633	1.4	2.4

Table 8.4. Estimation of the *range effect* contribution to the total spatial resolution [5].

isotope	**FWHM** (mm)	**FWTM**[c] (mm)
^{18}F	0.22	1.09
^{11}C	0.28	1.86
^{68}Ga	1.35	5.92
^{82m}Rb	2.60	13.20

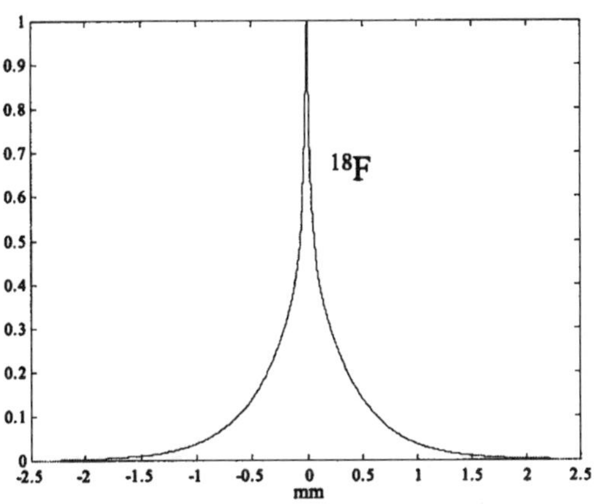

Fig. 8.7. Monte Carlo simulation of the range distribution in water for the ^{18}F isotope.

[c] With the Full Width Half Maximum (FWHM), the Full Width Tenth Maximum (FWTM) is also given to estimate the deviation of range distributions from a Gaussian of standard deviation σ, where $FWTM = 1.82 \times FWHM$. ($FWHM = 2\sigma\sqrt{2 \ln 2}$, while $FWTM = 2\sigma\sqrt{2 \ln 10}$)

In the laboratory frame (tomograph), the centre of mass has a nonzero momentum; this can be taken into consideration by applying a Lorentz transformation: in the new coordinate system the positron is moving and the electron is at rest. The Lorentz transformation from the centre of mass to the laboratory frame implies two consequences: the annihilation photons have different frequencies, and their flight directions are deviated from 180°. In addition, this model of a positron slowing down and catching a practically at rest electron, directly implies that the two gamma-rays are orthogonally polarized. Experimental evidence of polarization was provided in 1950 [15]. In the laboratory frame, for high energies of the positron, the two quanta are emitted one forward and one backward: the forward emitted photon has almost all the energy of the positron, while the other has energy of about m_0c^2; no marked polarization effects are present. At positron energy lower than m_0c^2, the two photons have both energy of about m_0c^2, are orthogonally polarized, and are emitted co-linearly.

The deviation from co-linearity is shown in Fig. 8.8: in (*a*) the positron strikes an electron at rest, in (*b*) the emerging photon momenta do not lie on a straight line. The energy and momentum conservation laws give the photon deviation θ

$$\theta \approx \frac{p \sin \phi}{m_0 c},$$

where p is the incident particle momentum, and ϕ is the angle between the incident direction and that of an emerging photon [16].

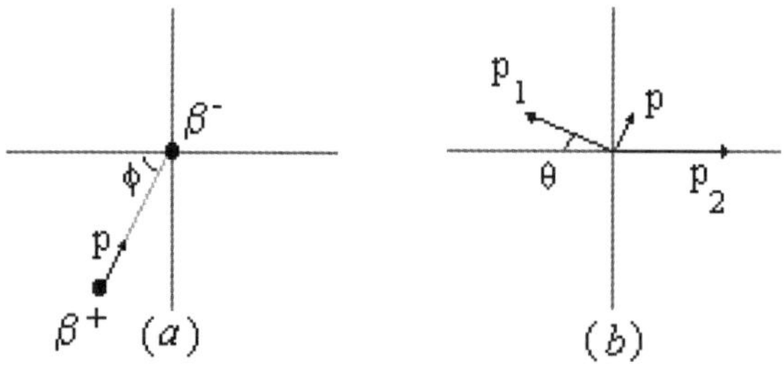

Fig. 8.8. A slight deviation from 180° of the angle between the two annihilation photons in the laboratory frame: the positron, which has been slowed down, strikes an electron at rest; p is the incident particle momentum, p_1 and p_2 are the annihilation photon momenta. (Figure not in scale, from Ref. [16], Chapter 13, p. 465)

The greatest deviation occurs when one photon is emitted orthogonally to the incident positron direction; there is no deviation if a photon travels on the same direction as the positron (a photon forward and the other backward).

The cross-section for the annihilation of an *at rest* positron via 2γ emission was given by Dirac, as [13, 17-20]

$$\sigma_{2\gamma} = \pi\ r_0^2 \frac{1}{\gamma+1}\left[\frac{\gamma^2+4\gamma+1}{\gamma^2-1}\log\left(\gamma+\sqrt{\gamma^2-1}\right)-\frac{\gamma+3}{\sqrt{\gamma^2-1}}\right]$$

where r_0 is the classic radius of the electron, and $\gamma = E/m_0c^2$, with E the positron energy given by the kinetic energy plus the rest mass.

An annihilation via 3γ emission is also possible, but has a lower probability, with a cross section [17, 18]

$$\sigma_{3\gamma} = \frac{\sigma_{2\gamma}}{372} << \sigma_{2\gamma},$$

and it is usually not considered.

Generally, the annihilation occurs in 1.8 ns, when the positron does not perfectly reach the thermal equilibrium and the electron is not at rest. Electrons constitute a statistical system of particles (Fermi gas) with an isotropic distribution of momentum in space. A calculation of the deviation from the co-linearity of the emitted photons can be made by just considering the conservation of the Fermi momentum and the thermal motion of the particles: this gives 180° ± 0.25° (*De Benedetti et al*, 1950), a numerically lower result than the experimental. More correctly, the Fermi momenta generate the spreading of the energy levels of the two light quanta, and not the co-linearity deviation: the larger measured deviations can be only explained with the existence of a particle with higher momentum. The research of this particle led to the discovery that in most organic substances, amorphous and perhaps even ionic solids (but never in metals), the positron can be in a bound system with an electron: this hydrogenoid atom is called *positronium* [13, 15, 18-20]. Therefore, the measured deviation are due to the higher momentum of the bound system. The *positronium* system has two minimum energy configurations: *para-positronium*, that is a singlet state with opposite spin (S = 0, ↑↓), and half-life $T_{1/2}$ ~ 0.1 ns; *ortho-positronium*, a triplet state with parallel spin (S = 1, ↑↑), half-life $T_{1/2}$ ~ 1 ns in organic tissue, and $T_{1/2}$ ~ 140 ns in inorganic substance, which constitutes about 75% of the number of bound states.

The *para-positronium* almost exclusively decays by self-annihilation in 0.1 ns (the bound system collapses and the two particles annihilate), but the annihilation with free electrons is not prohibited (*pick-off*, 1.8 ns).

The *ortho-positronium* naturally decays by self-annihilation in 140 ns with a three photons generation, but a so long decay time makes more probable a 1.8 ns pick-off process with free electrons.

In the emission spectrum (Fig. 8.9), where the emission refers to the angular deviation, two different behaviours are present:

- a narrow component, $|\Delta\theta| < 4$ mrad at ~300 K, which is dependent on the temperature,
- a large component, $|\Delta\theta| > 4$ mrad, which is temperature-independent.

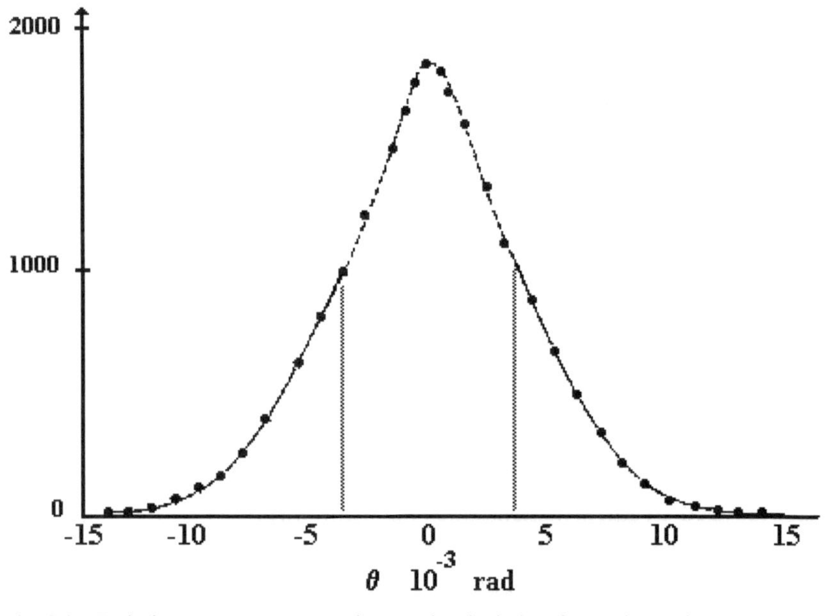

Fig. 8.9. Emission spectrum versus the angular deviation for positrons in water at 4°C. Two different behaviours are identifiable: a narrow ($|\Delta\theta| < 4$ mrad) due to lower momenta, and a large one ($|\Delta\theta| > 4$ mrad) due to the triplet of positronium [18].

The narrow component is given by particles with low momenta: annihilation via 2γ production of free positrons that have been slowed bound, *para-positronium* self-annihilation, and *para-positronium* pick-off. The large component is predominantly composed of *ortho-positronium* pick-off events. A summary scheme of positron annihilation is given in Table 8.5.

Table 8.5. Summary of the positron annihilation processes.

state	**annihilation process**	**comments**	**angular deviation**
no bound (62% in H_2O)	in-flight	just 2%, coulomb interactions and bremsstrahlung preferred (~1 ps)	
	via 1γ emission	in heavy elements with inner shell electrons, 20% of occurrence, with $E \sim 10\ m_0c^2$; non PET situation	
	via 2γ emission	standard PET situation (~1.8 ns)	narrow
	via 3γ emission	improbable	
	via more than 3γ emission	more and more improbable	
Positronium (38% in H_2O)	para-positronium self-annihilation	25% of bound states, preferred annihilation for para-p in 0.1 ns	narrow
	para-positronium pick-off	It annihilates in about 1.8 ns	narrow
	ortho-positronium self-annihilation	via 3γ, in 140 ns, it is anticipated by pick-off	
	ortho-positronium pick-off	75% of bound states, it annihilates in about 1.8 ns	large

The distribution of the angular deviation in water (e.g. biological tissue) is a Gaussian with FWHM $\approx 0.5°$. The reconstruction process considers a LOR as a straight line, and therefore the distribution of the flight angle generates a distortion in the image. By setting $\Delta\theta$ to a value of $0.5°$ (8.7 mrad), the contribution to the spatial distribution in the centre of a detector ring of diameter D is

$$FWHM \approx \Delta\theta \times \frac{D}{4} = 0.0022\ \ D,$$

which means a FWHM of 2.2 mm per meter of detector separation.

The *range effect* and the *angular distribution*, as described above, are fundamental physical limits in PET (Fig. 8.10): the effect is a blurring that degrades the spatial resolution in the reconstructed image. This is especially evident when using isotopes with a high E_{max} that are characterized by higher range, or whole body PET systems where the detector separation is greater.

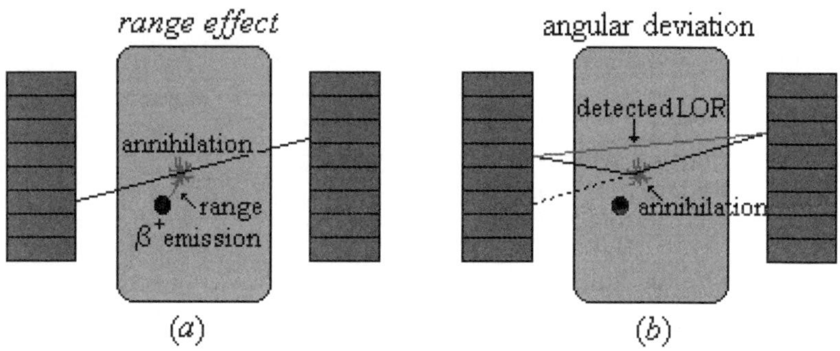

Fig. 8.10. Fundamental limits in PET. (*a*) *Range effect*: the detected LOR contains the annihilation point, not the emission point. (*b*) Angular deviation: the tomograph records a LOR on which the annihilation point does not lie.

8.2.3 *Interaction of gamma rays in biological tissue*

The two 511 keV annihilation photons have a "*mean free path*" of about 10.4 cm in water. The human head and chest have dimensions of the order of 20 cm, and this means that in whole body PET approximately 15% of photon pairs are detected without any interaction within biological tissue. In the remaining pairs, one or both photons will be either Compton-scattered or absorbed. The Compton scattering of the photons in tissue results in inaccurate LOR counting (Fig. 8.11), generating noise in the reconstructed image. On the other hand, absorption of gamma rays produces an *attenuation*, which diminishes the number of detected photons. Hence, an adequate correction is necessary to obtain quantitative information. While the range effect and the angular deviation are intrinsic physical limits, attenuation and scattering can be considered in the design phase of the tomograph and during the image reconstruction process, with adequate containment and correction.

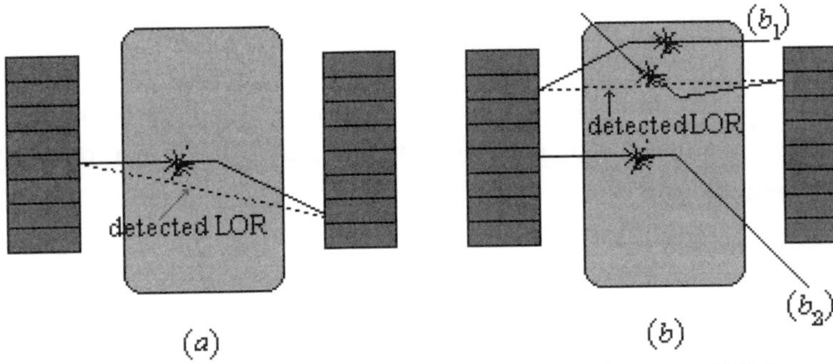

Fig. 8.11. Effects of Compton scatter in the FOV: (*a*) an incorrect LOR is recorded; (b_1) the detected LOR is given by two photons emerging from different annihilation points, or (b_2) the coincidence is lost. Symbol * indicates the annihilation point.

8.3 Detection of Annihilation Photon

Annihilation photons interact within the detector through Rayleigh, Compton and photoelectric effects: only the latter two release energy to the detector, making the photons "detectable". Rayleigh scattering is not usually considered at typical PET energies.

In most detection systems in the Field of Nuclear Medicine, the response is not constant over the whole detection area: the efficiency varies from a maximum value at the centre of the FOV down to a minimum near the edges. There is also non-uniformity in the sensitivity: the signal shape of the response to a reference source varies with its position. This behaviour can be quantified by putting a linear source of

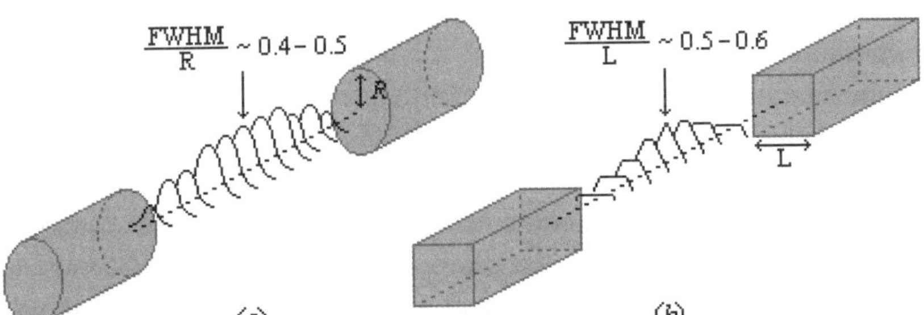

Fig. 8.12. Response uniformity and LSF shape for PET detection elements. (*a*) Circular detectors show a Gaussian LSF, which is practically uniform in the FOV. (*b*) Square detectors have a trapezoidal LSF, from triangular in the centre to rectangular near the edges; the response is uniform in the central zone only (from Ref. [5], Chapter 6, pg. 259).

activity in various positions in the FOV: the resulting response constitutes the Line Spread Function (LSF) of the system, and indicates how much the source image spreads (Fig. 8.12). In a PET system, the linear source is moved along the central line between the two detectors (radial sampling), and along the two orthogonal directions (transaxial and axial sampling). Differently, PET systems show notable response uniformity: the FWHM value of the LSF is constant along most part of the detection volume.

8.3.1 *Photon detection with inorganic scintillator crystals*

Inorganic scintillator crystals are the most commonly used detectors in PET. The annihilation photon is detected through the deposition of a fraction of or all its energy within the crystal. The deposition can occur in one location by photoelectric effect, or at several points within the same crystal by successive Compton interactions, or in different (most of time adjacent) crystals.

A crystal that absorbs a photon makes a transition to a higher energy state, from which it may undergo a decay after a characteristic time by emitting other lower energy photons, the "*scintillation photons*". Re-emitted light is detected by the photocathode of a phototube, which produces an electronic signal of an amplitude proportional to the number of scintillation photons produced, and thus to the energy deposited within the crystal. Energy thresholds can then allow the rejection of events scattered within the patient, which have lover energy.

Scintillation photons are due to *fluorescence*, a luminescence phenomenon, in which light is emitted in presence of electromagnetic excitation. Incident radiation can excite either the atoms or the molecules of the luminescent material. An excited system returns to its ground state, directly or through intermediate energy levels: the photon emitted between intermediate levels has lower energy, i.e. lower frequency and higher wavelength than the incident radiation. This converts photons from a non-visible region of the electromagnetic spectrum to another that is either in the near infrared, the visible spectrum, or in the near ultraviolet. Fluorescence has very short decay time, characterizing the material, in the order of nanoseconds, and stops with the end of external irradiation. Below ~100° C, it is independent from temperature.[d]

In the 30 years of PET development, several inorganic scintillator crystals have been employed, most of them only in experimental studies and tomograph prototypes (Table 8.6). Only a few are employed in commercial cameras, such as NaI(Tl), BGO, GSO, and LSO; also BaF_2 which had a very limited commercial use. Physical properties of scintillator crystals are shown in Table 8.7; their performance parameters as PET detectors in Table 8.8.

Table 8.6. Nomenclature and chemical formula of some scintillators that are commonly used in PET crystals. In the last column, the dopant impurity substitutes the element in the bracket. The percentage amount of impurity is related to the peak wavelength that is reported in Table 8.8; variations of impurity concentration sensibly change the scintillation properties [21-26].

crystal	nomenclature	chemical formula	main impurity		
NaI(Tl)	Sodium Iodide	NaI:Tl	0.1%	Thallium	(Na^+)
BGO	Bismuth Germanate	$Bi_4Ge_3O_{12}$		-	
CsF	Cesium Fluoride	CsF		-	
LSO	Lutetium Oxyorthosilicate	$Lu_2SiO_5:Ce^{3+}$	0.2%	Cerium	(Lu^{3+})
GSO	Gadolinium Silicate	$GdSiO_5:Ce^{3+}$	0.5%	Cerium	(Gd^{3+})
BaF_2	Barium Fluoride	BaF_2		-	
YAP	Yttrium Aluminate	$YAlO_3:Ce^{3+}$	0.7%	Cerium	(Y^{3+})
CsI(Tl)	Cesium Iodide	CsI:Tl	0.06%	Thallium	(Cs^+)
CeF_3	Cerium Fluoride	CeF_3		-	
LuAP	Lutetium Aluminate	$LuAlO_3:Ce^{3+}$	0.5%	Cerium	(Lu^{3+})

[d] Scintillation crystals can also show *phosphorescence*, another luminescent phenomenon, which goes on after the irradiation depending on the presence of metastable levels of the atoms or of the molecules of the substance. From excited to final state the system goes through a succession of intermediate levels whose decay time ranges from the order of seconds to days, therefore enhancing the emission time. This is a drawback for some PET scintillators. Phosphorescence results inversely proportional to the temperature: a higher temperature implies a greater thermal agitation, which enhances the decay probability of metastable levels. This phenomenon is more intense if incident radiation lies in the range of X rays or in the ultraviolet spectrum.

Table 8.7. Physical properties of some crystals commonly used in PET. The linear attenuation coefficients are given for 511 keV [21-26].

crystal	**density** (g/cm3)	**Zeff**	**hygroscopic**	**linear attenuation coefficient** (cm-1)	**index of refraction**
NaI(Tl)	3.67	51	yes	0.34	1.85
BGO	7.13	75	no	0.94	2.15
CsF	4.64	53	highly	0.88	1.48
LSO	7.40	66	no	0.88	1.82
GSO	6.71	56	no	0.72	1.85
BaF2	4.89	53	no	0.47	1.54
YAP	5.35	34	no	0.45	1.95
CsI(Tl)	4.51	54	slightly	0.54	1.78
CeF3	6.16	53	no	0.52	1.68
LuAP	8.34	65	no	0.91	1.94

The effective atomic number (Z_{eff}) for a crystal is given by $Z_{eff} = (\Sigma_i a_i Z_i^m)^{1/m}$, where $a_i = (f_i Z_i/A_i)/(\Sigma_k f_k Z_k/A_k)$ is the fraction of the electrons present in the crystal that belong to atoms of atomic number and mass Z_i and A_i, respectively, and f_i is the weight fraction of the ith element: if h_i is the multiplicity of the ith atom in the formula of the crystal, then $f_i = (h_i A_i)/(\Sigma_k h_k A_k)$. The used exponent m is 4.5 when considering both photoelectric and Compton interactions (from Ref. [7], formula (12.13), with a change of the m value).

Table 8.8. Performances of scintillators used in PET. The time resolution is dependent on the electronic readout. The energy resolution is strongly dependent on the coupling method, the signal integration time, and so on; values are given for 511 keV [21-27].

crystal	**decay time** (ns)	**time resolution FWHM** (ns)	**energy resolution FWHM** (%)	**relative yield**	**peak wave-length** (nm)
NaI(Tl)	230	1.5	3	100	410
BGO	300-370 (84%) 94 (16%)	7	14.7 32	15	480
CsF	3-5 2.5	0.4	30	3-6	390
LSO	12 (30%) 48 (70%)	1.4	12	75	420
GSO	30-60 (84%) 600 (14%)		10	20 14-20	440
BaF2	0.8 700	0.24	7.5	5 16	195-220 300
YAP	27 (98%) 246 (2%)	1.1	6.7	65	370
CsI(Tl)	1000		8.5	45	560
CeF3	2 31	27 (UV glass)	0.54	3	310 340
LuAP	17	0.41	23	30	365

As shown in the Table 8.6, some crystals are doped: this enhances their scintillation properties. In a pure crystal, electrons can be bound to crystal lattice sites (valence band) or can have a sufficient amount of energy to migrate within the crystal (conduction band). Both the photon capture and the emission decay produce transitions between energy levels of the scintillator lattice: the photon capture from the valence to the conduction band, the decay the opposite transition. Between the valence and conduction band there is a forbidden gap, which electrons cannot occupy in pure crystals. A doping atom, also called an *impurity* or *activator*, introduces energy states within the forbidden gap through which electrons can de-excite back to the valence band (Fig. 8.13). The introduced impurity has levels whose energy gaps are shorter than those of the pure crystal. The new levels often offer faster transitions, with shorter decay times, and forbid the re-absorption of the scintillation photons. The re-absorption limitation in the crystal lattice increases the light yield. Sometimes, the re-emitted spectrum is more useful: Thallium impurity in Sodium Iodide shifts the emission from ultraviolet to visible, where the sensitivity of most PMT is maximal.

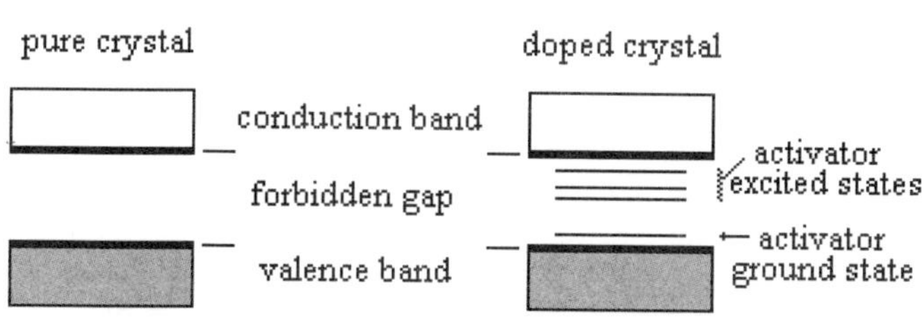

Fig. 8.13. Comparison between the energy levels for pure and doped scintillation crystals; the impurity introduces smaller energy gaps with faster transitions; therefore, the scintillator is transparent to the photons from the impurity gap, and the light yield is increased.

A commonly used doping impurity is Cerium, whose electronic configuration is (Xe)$4f^2 5d^0$. The photon absorption leads to the state (Xe)$4f^1 5d^1$, which rapidly decays in about 30 ns, with a wavelength ranging from 400 to 500 nm: this is an electric dipole transition whose probability is much higher than those of pure elements.[e] In GSO, Gadolinium configuration is (Xe)$4f^7 5d^1$, with six electrons more than Cerium in the same shell. The consequence is a lower transition probability because a smaller number of holes are available and local magnetic interactions forbid several transitions. There is an identical situation for the Lutetium in LSO, whose configuration is (Xe)$4f^{14} 5d^1$.[f] It is even worse in the case of pure Yttrium, (Kr)$4d^1 5s^2$, because the shell $5s$ is complete and this hinders electric dipole transitions. But not all crystals require impurities for efficient scintillation: all of the ionic compounds with Fluorine have high probabilities of fast decays thanks to its electronegativity, which perturbs the electronic shells of the other element, by favouring the level transitions.

The energy absorption of the photon in the crystal occurs through both photoelectric and Compton effects, whose cross-sections are function of the density (ρ) and of the effective atomic number (Z_{eff}) of the crystal. The atomic cross-section for photoelectric effect is proportional to $\rho\, Z_{\text{eff}}^{\,n}/E_\gamma^{\,m}$, where n is about 4 at 100 keV and gradually rises to 4.6 at 3 MeV, and m decreases slightly from 3 at 100 keV to 1 at 5 MeV. The cross-section for Compton scattering results proportional to $\rho\, Z_{\text{eff}}/A$, where A is the mass number; so $\rho\, Z_{\text{eff}}/A$ is about $\rho\,(0.45 \pm 0.05)$, except for hydrogen.

A high density favours the interaction of the photon in the crystal, whilst a higher Z_{eff} value increases the number of photoelectric

---

[e] The symbol (Xe) indicates the electronic configuration of inert gas Xenon (symbol Xe), with 54 electrons in the shells $1s^2 2s^2 2p^6 3s^2 3p^6 3d^{10} 4s^2 4p^6 4d^{10} 5s^2 5p^6$; it is a stable electronic configuration, without empty levels. The Cerium (symbol Ce) has 4 electrons more than Xenon. The symbols (Xe)$4f^2 5d^0$ and (Xe)$4f^1 5d^1$ of the ground and excited levels of Cerium, respectively, mean that this element has two free-to-transition electrons more than Xenon; at ground level, both of the two electrons stay in the shell $4f$, which can contain 14 electrons; at higher energy level one of the two electrons remains at $4f$, whilst the other occupies the higher shell $5d$, with 15 holes at $4f$. Such a high number of holes favors the electric dipole transition of de-excitation in ~30 ns. The other two electrons of Cerium occupy the shell $6s$, so the ground level is exactly (Xe)$4f^2 5d^0 6s^2$; the shell $6s$ is complete, with two electrons with opposite spin: these electrons are magnetically bound and are not free for electrical transitions, therefore the shell $6s$ is neglected in transition symbolism.

[f] Both the elements Gadolinium and Lutetium have a filled shell $6s$, whose electrons are not free to electrical transitions.

occurrences with respect to Compton scattering. Therefore, high Z_{eff} crystals are preferred: if a photon, scattered in the patient, is absorbed via photoelectric effect, the event can be rejected, by using an energy discriminator; Compton events in the crystal reduce the effectiveness of this discrimination.

The probability of a photon interaction within the crystal also depends on the scintillator dimension. Table 8.9 reports the intrinsic detection efficiencies of several crystals as a function of their depth, at typical PET and SPECT energies of 511 and 140 keV respectively.

Long crystals are more efficient, but excessively long ones can present several drawbacks (Fig. 8.14). Firstly, a photon can perform a higher number of Compton interactions, depositing energy in several locations: generally this degrades the spatial resolution, because the probability of multiple penetration (where the photon deposits energy

Table 8.9. Comparison of intrinsic detection efficiencies among the most commonly used scintillator crystals with respect to the crystal length, at typical PET and SPECT energies (from Ref. [28], Table 9III, page 271).

photon energy (keV)	**thickness** (mm)	**intrinsic detection efficiency** (%)		
		NaI(Tl)	**BGO**	**LSO**
511	10	29	62	58
	25	58	91	89
140	10	93	100	100
	25	100	100	100

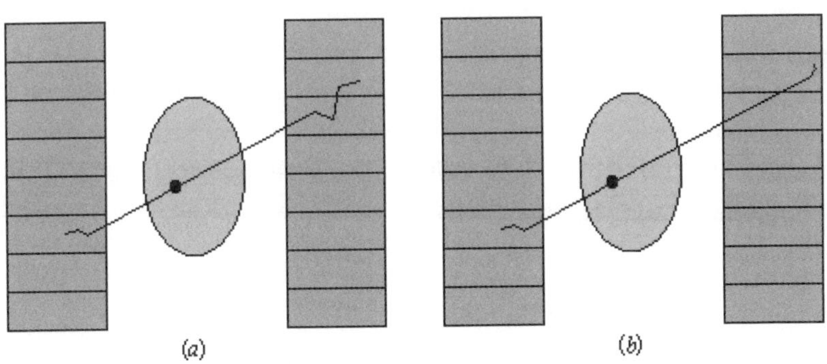

Fig. 8.14. Multiple penetrations (*a*) of an annihilation photon, and parallax error (*b*): both effects degrade the spatial resolution of the image; an excessive length of the crystal enhances their probability of occurrence.

within the correct and in adjacent crystals, Fig. 8.14*a*), and the parallax error (where the photon deposits energy only in crystals adjacent to the correct one, Fig. 8.14*b*) increase. Finally, the attenuation of light within an excessively long scintillator reduces the amount of light to the phototube, with a degradation in the energy and timing resolution.

The crystal width (*w*) determines the thickness of the tomographic slice, and, thus directly affects the spatial resolution, through an estimated contribution of $w/2$: by increasing the width, the sensitivity (*cps/slice*) also increases, while the axial resolution is degraded. However, very small crystals can be hard to package and will result in multiple penetrations.

Another crystal property that should be considered is the behaviour with environmental humidity: hygroscopic materials deteriorate and require airtight-containments, with additional packaging constraints.

The decay-time and the light-yield are other important physical properties of the crystal. The first is the most important parameter to choose the temporal coincidence window: slow decay needs a longer window, with increase of random coincidences. The light-yield has to be appropriate to the chosen phototube: a greater amount of light in the photoemission zone implies a more linear response, a better energy resolution and a more accurate spatial resolution. This is particularly relevant for the new position-sensitive phototubes, which work by weighting the energy depositions within the crystal. Unfortunately, fast emission is often coupled to a low light yield, while crystals with high light emission could be up to 100 times slower.

The index of refraction is important when coupling the crystal to the phototube with the best possible optical continuity: the scintillation light has to cross the phototube window without suffering appreciable reflection and refraction. The third law of geometrical optics (Snell's Law) regulates this process

$$\frac{\sin(\varphi_1)}{\sin(\varphi_2)} = \frac{n_2}{n_1},$$

where n_1 and n_2 are the refraction indices of the two media. The smallest deviation ($\varphi_1 \sim \varphi_2$) is obtained when $n_2 \sim n_1$. Typical refractive indices of the windows of available phototubes are 1.5 – 1.6.

By laterally wrapping the crystal with a substance with index n_2 smaller than the crystal index n_1, there is a limit angle of incidence of the light, φ_0, given by

$$\sin(\varphi_0) = \frac{n_1}{n_2},$$

corresponding to the disappearance of the refraction ray (*total reflection*). Scintillation light that is incident on the crystal face with an angle larger than φ_0, will be trapped by total reflection. This can allow efficient transportation of the light to the photodetector, hence increasing the light yield.

Among the available scintillator crystals, NaI(Tl) offers the highest light yield, whilst BGO is the most efficient. All of the fluorine compounds have fast decays. The fastest is the BaF_2, used in TOF systems: it also has a slow emission (300 nm, 700 ns), which can be substantially reduced with La^{3+} impurities; if the crystal is doped with Ce^{3+}, the fast emission vanishes and the compound emits only the typical Cerium emission component. LSO is particularly promising since it offers a high light yield (almost 75% relative to NaI), high density and fast decay. It is slightly radioactive (~12.5 cps/cm^3 at 140 keV, in a 20% energy window), which would be a drawback in combined multi-modality PET-SPECT cameras.

8.3.2 *Inorganic scintillator readout*

In order to be useful, the scintillation light must be converted into an electronic signal. Although several devices have been studied, only *photomultiplier tubes* (PMT's) have actually been adopted in commercial PET cameras. Their use has been optimized, by studying different crystal-PMT couplings.

Historically, large NaI(Tl) detectors presented the advantages of being both economic and simple. Of all the crystals used for PET (Table 8.7), NaI(Tl) has the lowest density and a rather low effective atomic number; therefore, it results in a poorer spatial resolution. In spite of this, NaI(Tl) technology is still widely used for large area tomographs with planar geometry (Fig. 8.15). The interaction position of the annihilation photon within a large crystal is computed by weighting the light deposited with respect to the energy, by using

$$X = \frac{\sum_i x_i A_i}{\sum_i A_i}$$

where A_i is the signal of the ith photomultiplier tube and x_i the corresponding position. With this configuration, the error ΔX of the position X is the spatial resolution of the system, and depends on Compton scattering within the crystal. This is a complex function of the energy of the photons, the width of the crystal and of its stopping power (not very high in NaI:Tl). However, overall this crystal assures the high light yield that this coupling requires to limit the statistical variance: all statistical fluctuations have to be quadratically summed to compute ΔX; therefore, high fluctuations degrade the spatial resolution. High count-rates can generate several problems; mainly, they increase the dead time until the system is paralyzed: for 100 – 200 kcps half of the arriving photons are lost. The count leakage increases the statistical variance, degrades the spatial and energy resolution, and generates arte facts in the image and loss of position linearity. These problems depend on the pulse pile-up, the gain instability of the phototube and on the amplifier baseline shift. Pulse cut techniques can allow counting rates of about 1 Mcps [29-30]. NaI(Tl) also shows a 0.15 s phosphorescent component, whose light yield is 9% with respect to the fluorescence, with a further increase in the dead time at high rates.

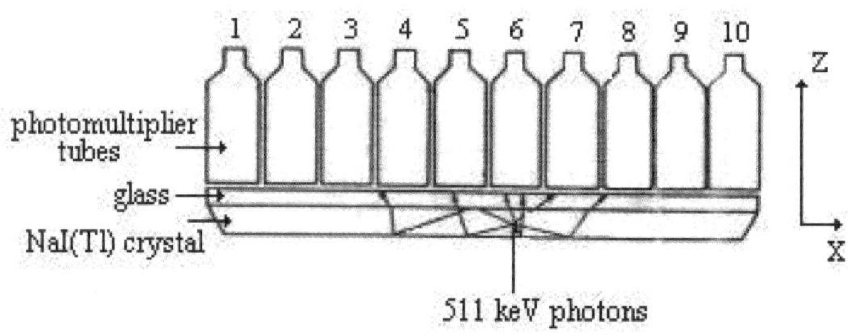

Fig. 8.15. Large NaI(Tl) crystal readout [31]. The scintillation light is converted into electrical signals by several phototubes, whose response is proportional to the number of incident scintillation photons. The scintillation position is determined by energy weighting of the photomultiplier responses whose signals exceed a threshold value.

Pure NaI is now being investigated as a scintillation detector: at the temperature of -90°C, NaI loses its high light yield, but it becomes a fast crystal, with a decay time of 35 ns, a light yield of 20% with respect to NaI(Tl) at Standard Temperature and Pressure (STP), and an energy resolution of about 15% [32].

Small dimension individual crystals, or small blocks of crystals coupled to one or more PMT's, achieved greater success (Fig. 8.16). PET development has shown the progressive reduction of the sensitive area of each detection element, with the exponential growth of the number of elements composing a multi-ring configuration. More efficient crystals have been found, initially BGO and today LSO. Finally, the "block detector" has been introduced, which drastically reduces commercial camera costs.

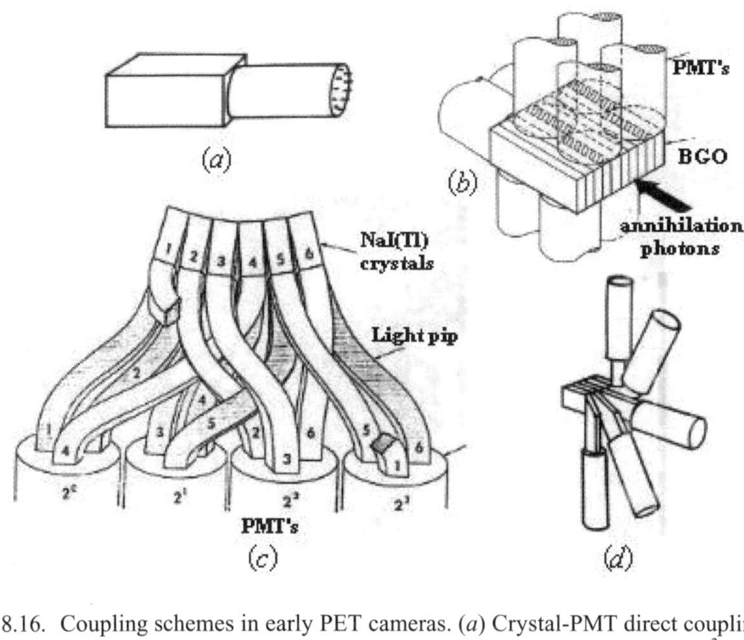

Fig. 8.16. Coupling schemes in early PET cameras. (*a*) Crystal-PMT direct coupling. (*b*) Coupling on old tomographs with only one ring: 10 BGO crystals (3×3×10 mm^3) coupled to 10 circular PMT's of 14 mm diameter, placed 4 on top, 4 on bottom and 2 on front. (*c*) An example of code coupling scheme: scintillation crystal identification is obtained by analyzing the readout combinations. (*d*) Scheme for single ring of the Donner tomograph (1979), where 280 BGO crystals 10 mm depth are read by 40 PMT's; each PMT is coupled to 7 crystals (from ref [33], with permission).

The first human PET, realized by Phelps and Hoffman in 1974, was a hexagonal ring with 48 NaI(Tl) crystals, each one read out by a PMT; the crystals were circular with a diameter of 5 cm. In the seventies, NaI(Tl) was the only available crystal, until the arrival of BGO: the first commercial camera with BGO was NeuroECAT in 1978, again with 1:1 crystal-PMT coupling [34]. In 1984, Scanditronix introduced the first block detector, with two crystals, one of BGO and the other of GSO, coupled to the same PMT: the crystals having different decay constants, the measurement of which allowed the identification of the activated crystal in the block.

Burnham introduced the "block detector" idea, by adapting Anger type logic: a great number of small crystals are coupled to a circular lightguide. The light is read by PMT's placed on the other guide side. The measurement of the ratios between PMT outputs (Fig. 8.17) allows the identification of the scintillating crystal. This interesting technique was still complicated and costly.

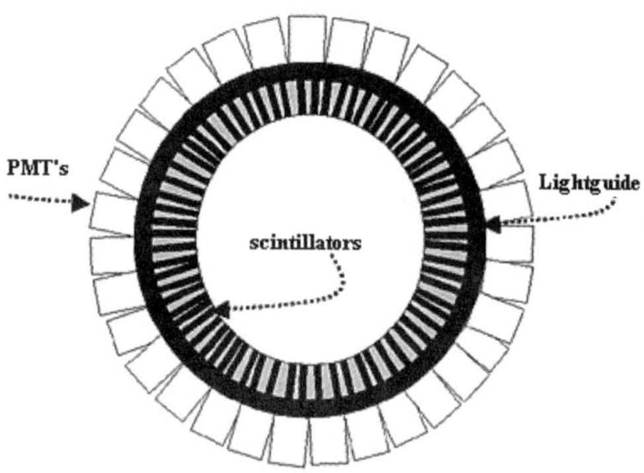

Fig. 8.17. Burnham coupling scheme (1984): a great number of small crystals are coupled to a circular lightguide, whose emissions are readout by PMT's placed on the other guide side. The scintillating crystal identification is obtained by measuring the ratios between PMT outputs. This complicated and expensive technique anticipated the "block detector".

A more efficient improvement was introduced by Casey and Nutt's "block detector", composed of a package of 32 crystals readout by 4 PMT's, i.e. an 8:1 coupling: this solution combines high efficiency, high resolution and relatively low cost (Fig. 8.18). The block detector configuration, with an increasing number of crystals coupled to each PMT is, since 1985, the most widely used in commercial PET cameras.

Fig. 8.18. Casey and Nutt "*block detector*" (1985). (*a*) Picture of the *block detector*. (*b*) The matrix of 32 BGO crystals and the 4 PMT package. (*c*) Coupling scheme: 8 crystals are coupled to each PMT.

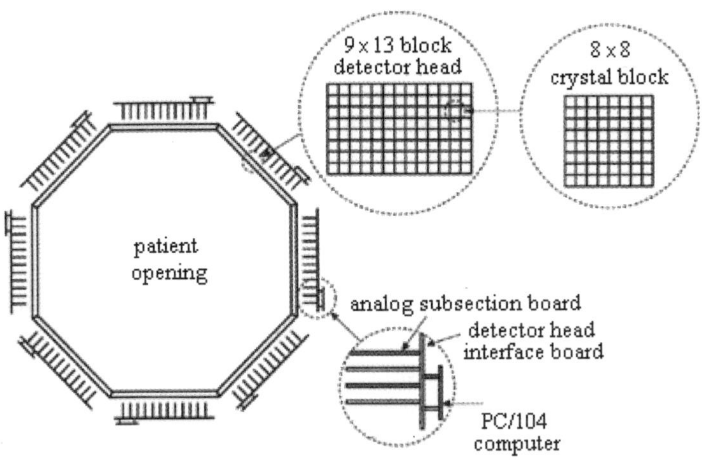

Fig. 8.19. Scheme of the ECAT High Resolution Research Tomograph (HRRT), for brain studies, installed at Max Planck Institute of Köln, Germany. The camera is an octagonal ring with 8 detection panels, of 25.2 cm length and 46.9 cm apart. Each panel has 117 block detectors disposed as a 9×13 matrix; in each block there are 2 overlapped 8×8 LSO crystal matrices; each crystal is $2.1 \times 2.1 \times 7.5$ mm^3. In total, the tomograph has 119808 crystals. The two LSO layers have different light decay times, so *depth-of-interaction* capability is given by pulse shape discrimination. The system has a spatial resolution of 2.5 mm FWHM, uniform over the FOV.

The introduction of position-sensitive photomultipliers (PS-PMT), which allow the identification of the scintillation position by energy weighting of outputs, implied another increase in the number of crystals for each tube. The LSO crystal, with higher light yield and faster decay time than BGO, is the last notable improvement. Initially it suffered from high cost and low availability. However, today's techniques can make LSO the same price as BGO. In a block, BGO allows a maximum of 16 crystals per PMT; above this number crystal identification is not feasible. The superior light yield of LSO allows 144 crystals to be coupled to each photomultiplier. The first LSO-based human PET camera was the HRRT brain tomograph at the Max Planck Institute of Köln, in Germany, with 120000 LSO crystals (Fig. 8.19): it has a spatial resolution of 2.5 mm FWHM, uniform over the FOV. In March 2000 the first LSO-NaI(Tl) combined camera for PET-SPECT was produced. In April 2001 the first LSO-based whole body human PET, ACCEL was installed at Northern California PET Center, with can acquire a torso image with 4 mm FWHM in a 10 minute scan, or with 2 mm FWHM in a half hour [34].

8.3.3 *Parallax error, radial distortion and depth of interaction*

Parallax error occurs when photons hit the scintillation crystals at an oblique angle, and penetrate into adjacent elements before interaction: incorrect LORs are recorded, and the spatial resolution worsens. This error also affects the detector efficiency, with a change in the apparent dimension of the crystal with incidence angle: the more skewed the photon's arrival, more probable the error becomes. An angular dependence shows up both in the spatial resolution and in the efficiency, with loss of space-invariance, especially in ring tomographs. Clearly, the distribution of incidence angles depends on the dimensions of the object and on its scattering properties.

Skew photons can also cause *radial distortion*: the scintillation light could be generally non-uniformly distributed around the centre of the crystal. This generates artefacts in the image: a point source will be reconstructed as an ellipse.

The measurement of the *depth of interaction* (DOI) allows the reduction of both errors. Without this information, a first order correction could consider a tomograph effective radius for mean photon penetration in the crystal.

Initial DOI measurements have been obtained with the Pulse Height Discrimination method: energy depositions near the PMT generate sharp and narrow output signals, while far depositions correspond to larger impulses.

One approach is to arrange several silicon photodiodes (PD's) along the scintillation crystal (Fig. 8.20*a*): PD's perform DOI measure by detecting scintillation light. A temperature below minus 100°C is required, which consequently causes a 10-fold increase in the decay time [35].

The relocation of PD's in front of the scintillation crystal improved the technique (Fig. 8.20*b*). Photodiodes identify the first interaction crystal and can work at the temperature of 25°C (optimum at -40°C). A lateral reflecting wrapping contains light dispersion. The PD matrix detect the scintillation light, and the signal ratio between PD's and PMT signals provides continuous DOI information:

$$DOI = \frac{PD_{output}}{PD_{output} + PMT_{output}}.$$

This is a first example of the *light sharing approach*: the scintillation crystal is coupled to two photo-detectors in different positions; the amount of light observed by each detector depends on the interaction point; the ratio of the two signals estimates the DOI.

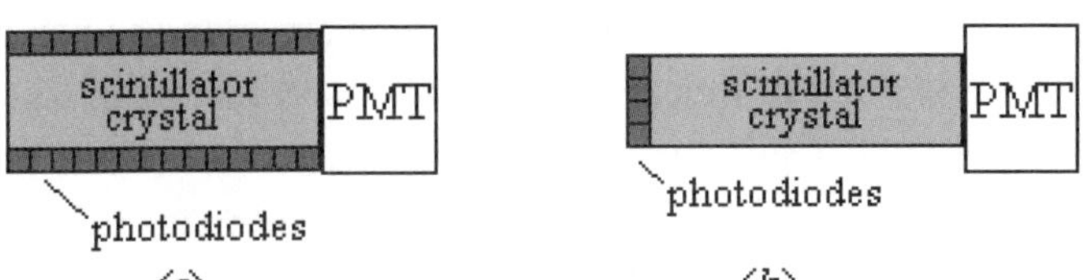

Fig. 8.20. (*a*) Photodiodes along the crystal detect the scintillation light and perform DOI measurement. This scheme requires heavy circuitry. (*b*) Photodiodes in front of the scintillation crystal: this scheme is less sensitive to object scatter, and provides continuous DOI information by the use of a signal ratio technique (or light sharing approach), but requires additional readout channels.

A new method is the *Phoswich approach*, which requires the coupling of decay-time-graded scintillators, so as to introduce a DOI sampling gradient. The technique is based on pulse shape discrimination by differentiating the energy signal summed with a shaping amplifier, sensing the zero cross time with a constant fraction discriminator, and measuring the time with an additional time-to-amplitude converter. The implementation is made by coupling two crystals with different decay-times: the identification of the crystal in which the energy is deposited is obtained by measuring the impulse shape. The phoswich was initially proposed by using BGO and GSO crystals. Since BGO allows count rates higher than 16 times with respect to GSO, there is an efficiency lack in GSO; this generates an excessive asymmetry in data collection [36]. Another way proposed was to apply a temperature gradient along the crystal, so that the decay time is depending on depth [37]. An LSO/NaI prototype is under development, and should also allow simultaneous PET/SPECT capabilities. The first full implementation of DOI in PET has been obtained in the HRRT, which uses two LSO layers with two different decay times.

8.4 Image Reconstruction

The aim of PET image reconstruction is to achieve an image of the activity distribution within the patient from the recorded sets of acquired data. In a tomographic study, data are acquired by performing several sets of linear scannings of the examined slice at different angles: each set gives an activity profile. The acquired profiles are then used to generate an appropriate number of independent equations to mathematically solve, where the unknowns are the activity elements of a grid (the image matrix).

In 2D, the image matrix elements are called *pixels*, and each slice is a plane, without thickness. The number of detected events in a LOR is the sum of activity along the line of scanning in the plane (*line integral*), and the activity is uniform in each pixel (the value associated to the pixel is the mean activity). In addition, the reconstruction program considers the resolution and sensitivity as independent from the depth of the annihilation event. The reconstruction should include a method for attenuation correction. In 3D-MS (*multi-slice*), the image is obtained by simply overlapping 2D images of adjacent slices, therefore the assumptions used are the same. In 3D-PVI (*volume imaging*), the slice thickness is considered and voxels (*volume elements*) substitute pixels: the uniformity assumption is then true for the voxel.

A profile or a line integral does not contain the information on the depth of the annihilation events. An attempt to recover this information is the *linear superposition of the backprojections* (LSBP), where profiles are projected and overlapped on the image grid (Fig. 8.21, *top row*). This method exalts the larger structures, corresponding to the lower spatial frequencies. To correct this, the backprojection is filtered (FBP). The filtering is usually applied in the frequency domain, after the application of a Fourier transform (Fig. 8.21, *bottom row*).

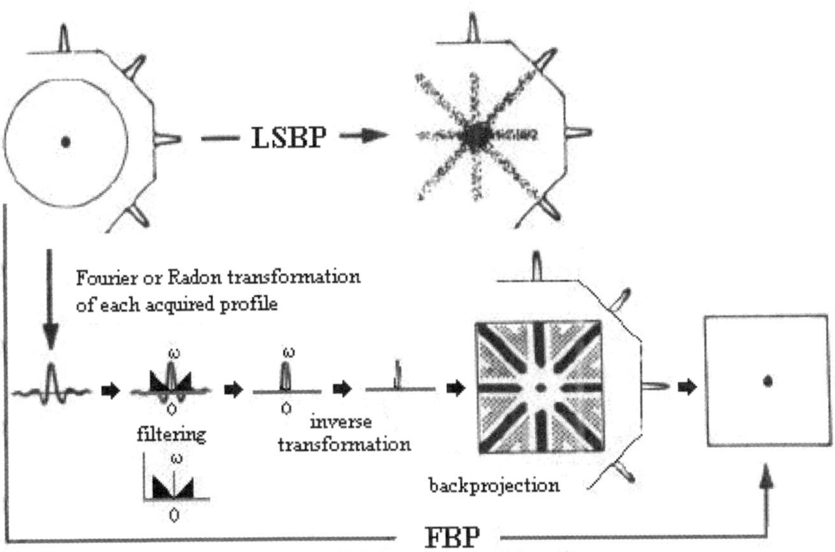

Fig. 8.21. LSBP (*top*) and FBP (*bottom*): a ramp filter recovers spatial resolution in the image (from Ref. [5], Chapter 6, page 242, with permission).

Filtering can remove the low or the high frequencies, can average the pixel values, and so on. Low frequencies are usually magnified so as to reduce the image noise; on the contrary, enhancement of the high frequencies contributes to increasing the spatial resolution in the image or to emphasize a structure. FBP uses the concept of the *line integral*, which contains everything; true signal, scattered and untrue counts, noise, background and so on, and nothing can be *a priori* isolated. The filtering is *a posteriori* rescue procedure.

Alternative reconstruction methods are represented by probabilistic iterative techniques, which *a priori* bind emission and detection with a model of the tomograph response. This allows the recovery of sensitivity and spatial resolution.

8.4.1 *The Filtered Backprojection*

The reconstruction of the image firstly requires the acquisition of several sets of projections p of the object f. By using the Central-Section Theorem, the projections p give the function $F = \Im(f)$, the Fourier transform $\Im$ of f. Therefore, the image f is achieved by using the inverse Fourier transform $\Im^{-1}$:

$$f = \Im^{-1}(F(p(f))).$$

This procedure generates the *Filtered Backprojection* (FBP) algorithm, the technique most used for image reconstruction.

The acquisition procedure consists of the collection of several profiles (or projections) of the object at several angles. By first considering a 2D case, the situation is shown in Fig. 8.22, where a projection

$$p(x_r, \phi) = \int_{-\infty}^{+\infty} dy_r\, f(x, y) \tag{8.1}$$

of the object $f(x,y)$ is obtained at angle ϕ. The spatial sampling is given by the number of x_r points, while the angular by the number of angle ϕ. The rotated coordinate system (x_r, y_r) of the detector and the (x,y) frame of the object are related:

$$\begin{pmatrix} x_r \\ y_r \end{pmatrix} = \begin{pmatrix} \cos\phi & \sin\phi \\ -\sin\phi & \cos\phi \end{pmatrix} \begin{pmatrix} x \\ y \end{pmatrix} \tag{8.2}$$

To better derive the FBP algorithm, let us consider Fig. 8.23, which visualizes the geometry of the Fourier transforms, both of the projection p and of the object f. Let $F(u_x, u_y)$ be the Fourier transform of the function

$f(x,y)$ in the frequency space, defined as

$$F(u_x,u_y) \equiv \Im_2\{f(x,y)\} = \int_{-\infty}^{+\infty}\int_{-\infty}^{+\infty} dxdy f(x,y) e^{-2\pi i(x\cdot u_x + y\cdot u_y)} \quad (8.3)$$

where $\Im_2$ indicates the 2D transform with respect to both the two variables.

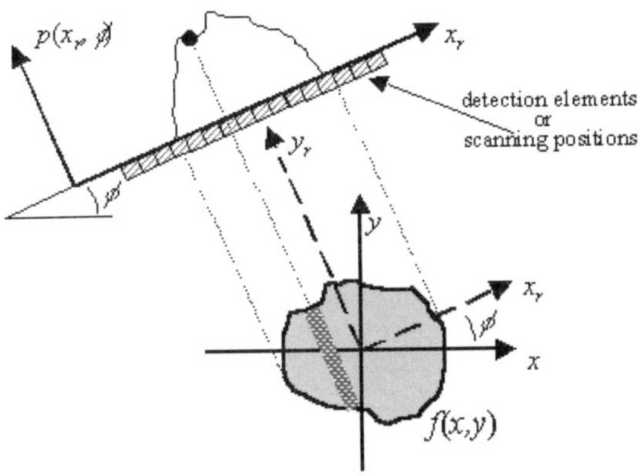

Fig. 8.22. Projection geometry in 2D tomography: the profile or projection $p(x_r, \phi)$ can be acquired by parallel scanning with a detector, or by unique scanning of a composed system. The spatial resolution is a function of the distance between the centres of two adjacent detection elements. If the profile is acquired with a large area detector, the spatial resolution is given by the error in energy weighting of positions. The number of angular position ϕ constitutes the angular sampling.

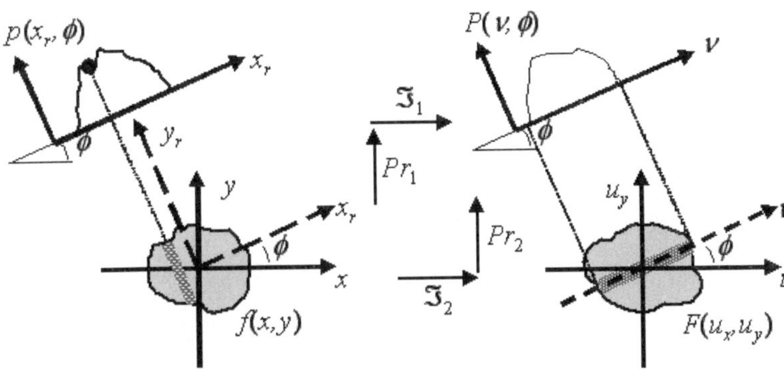

Fig. 8.23. Geometry of the Central Section Theorem: the operation sequences $(Pr_1+\Im_1)$ and $(\Im_2+Pr_2)$ are equivalent. Pr_1 is the acquisition data procedure, when a projection $p(x_r, \phi)$ of the object $f(x,y)$ is taken at angle ϕ, $\Im_1$ is the 1D Fourier transform of $p(x_r, \phi)$ with respect to the spatial coordinate. $\Im_2$ is the 2D Fourier transform of the object $f(x,y)$; Pr_2 is the extraction of a profile $P(v, \phi)$ of the function $F(u_x,u_y)$ along a line passing through the origin and inclined at the angle ϕ. The theorem states that $P(v,\phi)=\Im_1\{p(x_r, \phi)\}$.

By changing from Cartesian (u_x, u_y) to polar (v,ϕ) coordinates in frequency space, by using

$$\begin{cases} u_x = v\cos\phi \\ u_y = v\sin\phi \end{cases} \qquad \begin{cases} v = \sqrt{u_x^2 + u_y^2} \\ \phi = arctg\left(\dfrac{u_y}{u_x}\right) \end{cases}, \tag{8.4}$$

expression (8.3) becomes

$$F(u_x, u_y) = \int_{-\infty}^{+\infty}\int_{-\infty}^{+\infty} dxdy\ f(x,y)\ e^{-2\pi i(x\cdot v\cos\phi + y\cdot v\sin\phi)}$$

$$= F(v\cos\phi, v\sin\phi). \tag{8.5}$$

In transformation (4), it is useful to assume the variable ranges:

$$\begin{cases} -\infty < u_x < +\infty \\ -\infty < u_y < +\infty \end{cases} \qquad \begin{cases} -\infty < v < +\infty \\ 0 \le \phi < \pi \end{cases}.$$

Now, at each angle ϕ, let $P(v,\phi)$ be the Fourier transform of the projection $p(x_r,\phi)$ with respect to the first variable only:

$$P(v,\phi) \equiv \Im_1\{p(x_r,\phi)\} = \int_{-\infty}^{+\infty} dx_r\ p(x_r,\phi)\ e^{-2\pi i(x_r\cdot v)}$$

which, by developing $p(x_r,\phi)$ as defined in (8.1), is

$$P(v,\phi) = \int_{-\infty}^{+\infty} dx_r \int_{-\infty}^{+\infty} dy_r\ f(x,y)\ e^{-2\pi i(x_r\cdot v)}\ ?$$

Using rotation (8.2) and remembering that $dx_r dy_r = dxdy$, it becomes

$$P(v,\phi) = \int_{-\infty}^{+\infty}\int_{-\infty}^{+\infty} dxdy\ f(x,y)\ e^{-2\pi i(x\cdot v\cos\phi + y\cdot v\sin\phi)}$$

which implies, by comparing with result (8.5), that

$$P(v,\phi) = F(v\cos\phi, v\sin\phi). \tag{8.6}$$

Equation (8.6) constitutes the demonstration of the Central-Section Theorem, which states that $P(v,\phi)$ is the profile of the $F(u_x, u_y)$ passing through the origin along the direction fixed by the angle ϕ.

The significance of the theorem is shown in Fig. 8.24: the knowledge of several sets of $p(x_r,\phi)$ projections allows the construction of the $F(u_x,u_y)$ as a set of lines through the origin. It is evident that there is an oversampling at the origin in the frequency domain. Therefore, the reconstruction problem is resolvable if the projections $p(x_r,\phi)$ are known at each x_r and at each angle ϕ in the range [0, π].

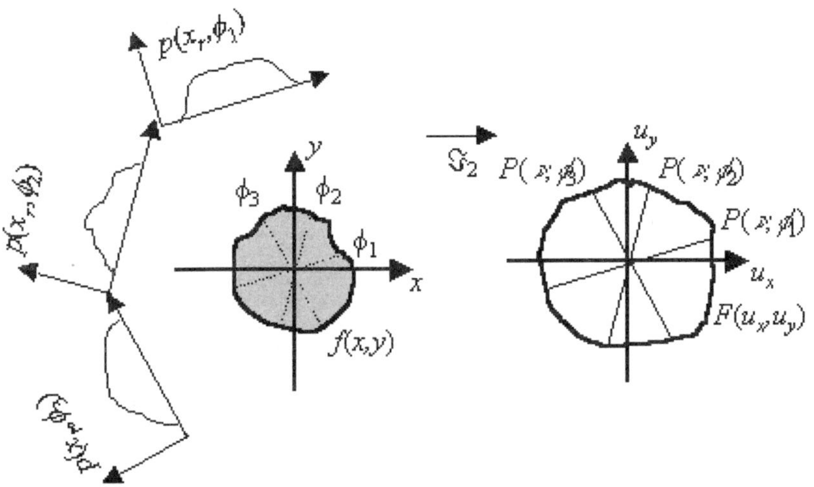

Fig. 8.24. Effect of the Central-Section Theorem: the collection of the projection $p(x_r,\phi)$ allows the knowledge of the $P(v,\phi)$, which are profiles through the origin of the $F(u_x,u_y)$. The function $\Im_2\{f(x,y)\}$ can therefore be obtained as a set of lines through the origin. This procedure generates an over-sampling of the frequencies around the origin.

The FBP solution is given by

$$f(x,y) \equiv \Im_2^{-1}\{F(u_x,u_y)\} = \int_{-\infty}^{+\infty}\int_{-\infty}^{+\infty} du_x du_y\ F(u_x,u_y)\ e^{+2\pi i(x\cdot u_x + y\cdot u_y)}.$$

Passing from rectangular to polar coordinates (with the substitution $du_x du_y = v dv d\phi$), it becomes

$$f(x,y) = \int_{-\infty}^{+\infty} dv \int_0^{\pi} d\phi\ |v|\ F(v\cos\phi, v\sin\phi)\ e^{+2\pi i(x\cdot v\cos\phi + y\cdot v\sin\phi)}$$

where the absolute value $|v|$ allows the integration over all of the values. By applying the Central-Section theorem, result (8.6), one obtains

$$f(x,y) = \int_{-\infty}^{+\infty} dv \int_0^{\pi} d\phi\ |v|\ P(v,\phi)\ e^{+2\pi i(x\cdot v\cos\phi + y\cdot v\sin\phi)}$$

and by using an additional rotation (8.2)

$$f(x,y)=\int_{-\infty}^{+\infty}d\nu\int_0^{\pi}d\phi\ |\nu|\ P(\nu,\phi)\ e^{+2\pi i(x_r\cdot\nu)}\ ,$$

which is known as the FBP algorithm.

This formulation underlines that the absolute value $|\nu|$ exalts the higher frequencies of the $P(\nu,\phi)$ values: the operation is called *filtering* (hence FBP) and the $|\nu|$ is the well-known *ramp filter*. The filtering also adjusts the over sampling of the origin.

There are two principal limitations of the FBP algorithm. Firstly, the sampling of all of the spatial frequencies is not possible: Shannon's sampling theorem states that the maximum recoverable frequency is the Nyquist frequency, given by

$$\nu_N=\frac{1}{2\cdot\Delta x_r}\ ,$$

where Δx_r is the interval of spatial sampling (when using small or pixilated crystals this is the detector width). Reconstruction of frequencies higher than ν_N generates *aliasing* artefacts. Furthermore, the ramp filter also amplifies the statistical noise, characterized by a high frequency spectrum.

This latter problem can be solved by additional filtering using a low-frequency pass window, which dampens the frequency amplitude above a cut-off ν_c, whose value depends on both the spatial sampling Δx_r and the collected statistic. The low pass filter and the cut-off frequency affect the spatial resolution and the statistical noise of the reconstructed image. The new filtering does not generally substitute the ramp filter but multiplies it. As an example, a filter commonly used in PET is the Hamming window,

$$H(\nu)=\begin{cases}k+(1-k)\cos\left(\dfrac{\pi\nu}{\nu_c}\right) & |\nu|<\nu_c\\ 0 & |\nu|\geq\nu_c\end{cases}$$

where k ($k\in[0.5,1]$) determines the filter curvature, and $\nu_c\leq\nu_N$ (Fig. 8.25). The FBP algorithm expression then becomes

$$f(x,y)\approx\int_{-\infty}^{+\infty}d\nu\int_0^{\pi}d\phi\ |\nu|\ H(\nu)\ P(\nu,\phi)\ e^{+2\pi i(x_r\cdot\nu)}\ .$$

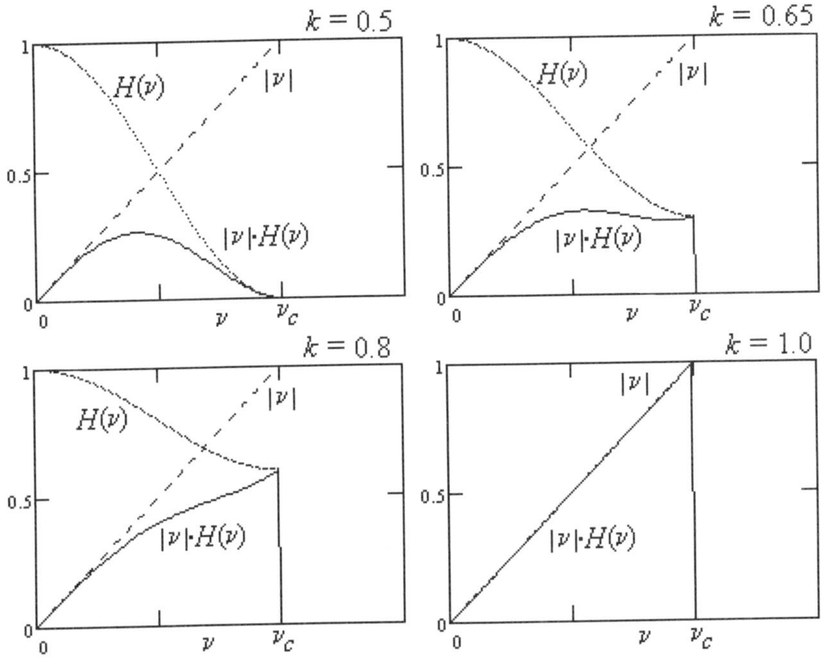

Fig. 8.25. Modification of the ramp filter by the Hamming window, at several k values. At the highest value of $k = 1$, the ramp filter is unaltered.

There are reconstruction procedures other than the FBP algorithm. *Direct Fourier Methods* use computerized interpolation from $P(v,\phi)$ to $F(u_x,u_y)$, by changing polar to rectangular coordinates in frequency space, and then the inverse Fourier transform to find $f(x,y)$. The *Convolution-Backprojection* method (CBP) filters the projections in the (x,y) space by convolution with the filter product, and so on.

In 3D, reconstruction suffers from space variance and data redundancy. This latter is used to reduce the statistical noise but the data storage is more difficult, and the high number of LORs to backproject requires a lot of computing time. The most important problem is the lack in shift-invariance: several projections are not complete and standard Fourier techniques cannot be directly used. In 2D acquisitions, septa insure the shift invariance, which means that each detector sees the FOV with constant solid angle; at most, there are small parallax errors, which are not taken into account or can be corrected for. In 3D mode, the detectors see a solid angle that decreases from the centre of the FOV to the edges. The consequence is that there are incomplete or truncated projections: there are angular positions that collect a lesser number of LORs. In this situation, FBP cannot be used because its application requires that all projections are complete.

In theory, with all complete projections, the 3D FBP is an extension of the 2D method described previously. The 3D projection p of the object $f(x,y,z)$ is the plane

$$p(x_r, y_r, \phi, \theta) = \int_{-\infty}^{+\infty} dz_r \; f(x,y,z).$$

The Fourier transform of the object is

$$F(u_x, u_y, u_z) = \int_{-\infty}^{+\infty} \int_{-\infty}^{+\infty} \int_{-\infty}^{+\infty} dxdydz \; f(x,y,z) \; e^{-2\pi \, i(x \cdot u_x + y \cdot u_y + z \cdot u_z)},$$

while that of the projection is the plane

$$P(v_x, v_y, \phi, \theta) = \int_{-\infty}^{+\infty} \int_{-\infty}^{+\infty} dx_r dy_r \; p(x_r, y_r, \phi, \theta) \; e^{-2\pi \, i(x_r \cdot v_x + y_r \cdot v_y)}.$$

This is the passage that requires non-truncated projection. With these expressions, the Central-Section Theorem (Fig. 8.26) gives:

$$P(v_x, v_y, \phi, \theta) = F(u_x, u_y, u_z)\Big|_{v_z=0}.$$

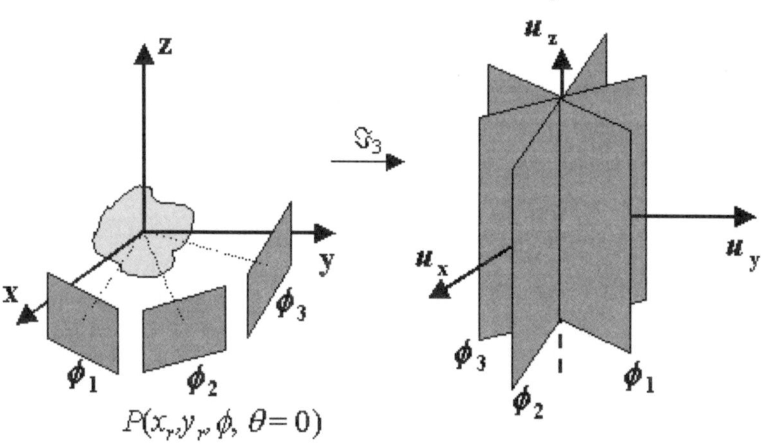

Fig. 8.26. Central Section Theorem in 3D: by acquiring the projection planes $p(x_r, y_r, \phi, \theta)$ it is obtainable the $F(u_x, u_y, u_z) = \Im_3\{f(x,y,z)\}$ as a set of planes through the origin in the frequency domain; the drawing is simplified by only considering some projections p at the angle $\theta = 0$; the 3D acquisition leads to data redundancy.

Beside the integrability and derivability of the functions, one must define the reconstruction conditions, more complex than in 2D geometry, which can be derived using the notation introduced in 3D reconstruction by Orlov for electron microscopy in 1976.

For each acquired projection plane, a versor (an unit vector from the plane centre) individuates a point on a unit sphere (a sphere with radius 1), as shown in Fig. 8.27.

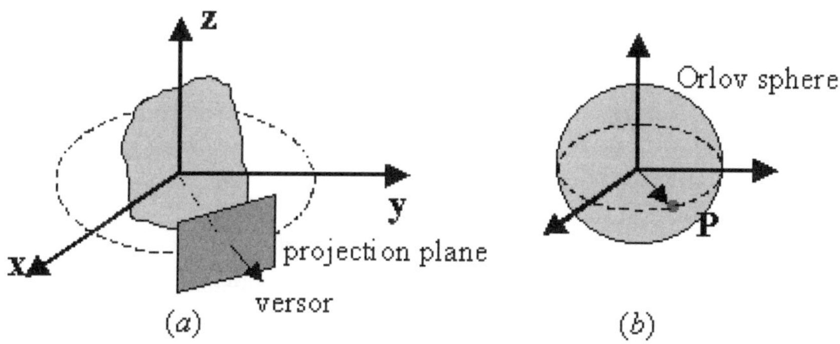

Fig. 8.27. The versor of a projection plane (*a*) specifies a point **P** on the Orlov sphere (*b*).

The envelope of all the points of the acquired planes constitutes a region Ω on the unit sphere, called the Orlov surface. A circular set of projection planes about the object which describe a great circle, forms an Orlov surface (Fig. 8.28). A multi-ring scanner, without gaps and septa, generates an equatorial band (Fig. 8.29).

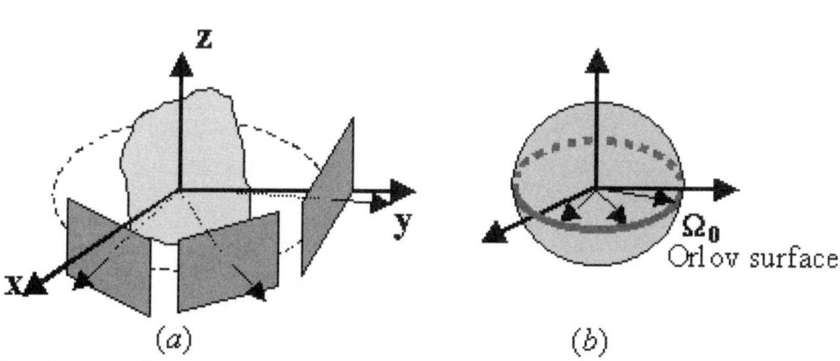

Fig. 8.28. A circular set of projection planes around the central slice of an object (*a*) describes a great circle as Orlov surface (*b*).

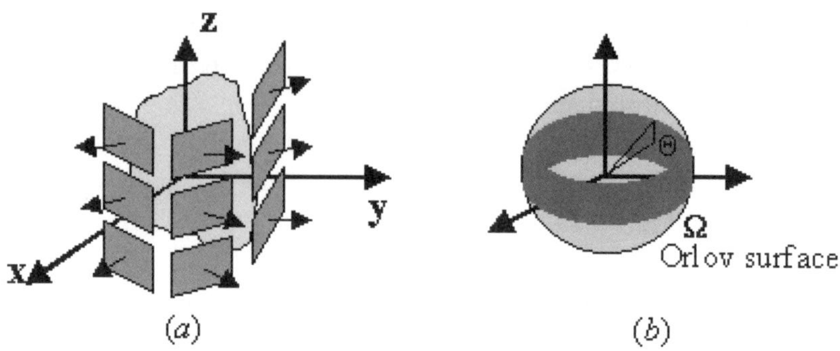

Fig. 8.29. The Orlov surface (*b*) of a multi-ring scanner (*a*).

The Orlov theorem states that a 3D data set is complete, and therefore the 3D image obtainable, if the generated Orlov surface intersects all possible great circles on the unit sphere (or more exactly: "there are no great circles which do not intersect the Orlov surface").

Generally, the sampling does not satisfy this condition.

Once the condition for the reconstruction has been defined, the FBP algorithm simply requires the filter definitions in 3D. The *Colsher filter* substitutes the ramp filter

$$H_C(v_x, v_y, \vartheta) = \sqrt{v_x^2 + v_y^2}\;\; H(\Theta, \Psi) \;,$$

where

$$\frac{1}{H(\Theta,\Psi)} = \begin{cases} 2\pi & \cos\Psi \le \sin\Theta \\ 4\arcsin\left(\dfrac{\sin\Theta}{\cos\Psi}\right) & \cos\Psi > \sin\Theta \end{cases}$$

and

$$\tan\Psi = \sqrt{\frac{v_y^2\cos^2\vartheta}{v_x^2 + v_y^2\sin^2\vartheta}} \;.$$

This expression also performs a correction in (ϕ,θ) to increase the uniformity of the solid angle sampling, and (like the ramp filter) amplifies the statistical noise. The 2D Hamming window is obtained by just multiplying two 1D ones

$$H(v_x, v_y) = H(v_x)\ H(v_y)\,.$$

The practical implementation can generate several algorithms, which differ for the application sequence of the integral operators, for the choice of the adopted filters or for their application order, and so on. Some techniques can also require interpolation methods, since theory considers spherical coordinates, while computers use cubic matrices. However, these methods differ mainly in the way that they complete the projections.

The resulting 3D FBP algorithm is:

$$f(x,y,z) \approx \int_{-\infty}^{+\infty} dv_x \int_{-\infty}^{+\infty} dv_y \int_0^{\pi} d\phi \int_{-\Theta}^{\Theta} d\theta \sin\theta\ H\ H_C\ P\, e^{+2\pi i(x_r \cdot v_x + y_r \cdot v_y)}$$

where the integration limit Θ (shown in Fig. 8.29) is determined by the Orlov surface. Truncated projections generate artefacts, which reflect the lack of uniformity of the sensitivity in the 3D cameras. Unfortunately, a correction based on FBP normalization with respect to local sensitivity is not possible: both 2D and 3D tomographic reconstructions are not local, because the value in a point is not obtained by only using the LORs, which cross the point or its neighbourhood.

The 3D-reprojection algorithm (3DRP) extracts from the entire set of collected LORs, a 2D data set, as if the tomograph is equipped with inter-planar septa. Extracted data are then used to reconstruct an image by 2D FBP, from which the unmeasured data are evaluated. The 3D data set is then complete and the 3D FBP can be performed. All 3D techniques require a lot of computation time to backproject the number of LORs, much greater with respect to the 2D case. There are several rebinning methods, which transform 3D data in 2D sets: the Single-slice rebinning algorithm (SSRB), the Multi-slice rebinning (MSRB) and the Fourier rebinning algorithm (FORE). After the rebinning, each slice is rapidly reconstructed by 2D FBP, without the characteristic 3D problems, such as the truncated projections.

8.4.2 *The Expectation Maximization algorithm*

Traditional methods of image reconstruction, such as the widely used FBP, are mathematical solutions, where is not possible to isolate *a priori* the true signal in the line integral. To recover the true information, they need some kind of integral transform operator, with additional filtering. However, the main problem is that every operation performed on the acquired data, operates both on true and incorrect data in the same way. Moreover, every operation can introduce new errors and artefacts in the image.

Alternatively, iterative methods model the response of the tomograph, by drawing on the relationship between emission and detection. The reconstruction can be *a priori* optimized by modelling the detector, the scattering, several noise sources, and so on.

There are several iterative algorithm; the most important is the *Expectation Maximization* (EM) [38-42].

The core of the iterative reconstruction is a probability matrix, $\mathbf{P}=\{p_{ij}\}$, which correlates positron emission and photon detection. Generally, it is a sparse matrix, and requires a lot of computation time. Theoretically, the **P** matrix should include all the parameters to exactly describe the tomograph: system geometries, detector properties, scattering evaluation, attenuation correction, and so on. This means that **P** should be a product of matrices:

$$\mathbf{P}_{\text{tomograph}} = \mathbf{P}_{\text{geometry}} \cdot \mathbf{P}_{\text{detector}} \cdot \mathbf{P}_{\text{scattering}} \cdot \mathbf{P}_{\text{attenuation}}$$

If one of the **P** terms is not considered or not computed, it is substituted with the unit matrix. In the simplest models, the p_{ij} values include only geometric considerations, such as the length of the j-LOR that lies within the i-voxel; this is equivalent to only considering the $\mathbf{P}_{\text{geometry}}$ term.

It is possible to include in the model a differentiation of the voxels in a LOR according to their depth within the tube, for instance by including a parameterization of the solid angle subtended from the centre of each voxel to the ends of the coincidence tube. Then, the model is deconvolved and enhances the spatial resolution in the reconstructed image, as if the tomograph had an ideal response.

Let λ(x, y, z) be the spatial distribution of activity in the region of interest. The goal of the reconstruction is to obtain the discrete distribution $\{\lambda\}$ of $\beta+$ sources in the FOV, where the i-index covers the FOV voxels:

$$\lambda_i = \int_{voxel\ i} \lambda(x,y,z)dxdydz \qquad i = 1,\dots,N.$$

Let $\{n^*\}$ be the detected distributions of counts, where n^*_j is the number of pairs of coincidence photons recorded in the tube j; the j-index covers the detection tubes, $j = 1,\dots, N_t$. A tube is the volume between two opposite detection elements, and a LOR is its axis. Let now p_{ij} be the probability that photon pairs emitted from the ith voxel are detected in the jth tube; this implies $p_{ij} \geq 0$. The probability that the emission from the i-voxel is detected in a generic tube is

$$p_i = \sum_j^{N_t} p_{ij} \leq 1. \tag{8.7}$$

If n_{ij} indicates the number of photon pairs emitted from the ith voxel and detected in the j-tube, its mean value is λ_{ij}

$$\langle n_{ij} \rangle = \lambda_{ij} = p_{ij}\lambda_i. \tag{8.8}$$

The $\{n_{ij}\}$ is a very useful set of linearly independent Poisson variables, correlated both to the voxel emissions distribution $\{n_i\}$ and to the acquired distribution $\{n^*_j\}$: the number of photons emitted from the voxel i and detected in all tubes j is given by

$$n_i = \sum_j^{N_t} n_{ij},$$

while the number of photons detected in the tube j (emitted from all the voxels i) is

$$n^*_j = \sum_i^{N} n_{ij}. \tag{8.9}$$

With this symbolism, the fraction of counts recorded for the tube j and due to a fixed voxel i_0 is given by

$$\frac{\lambda_{i_0 j}}{\sum_i^N \lambda_{ij}}.$$

Thus, one arrives at the expression

$$n_{i_0 j} = n^*_j \frac{\lambda_{i_0 j}}{\sum_i^N \lambda_{ij}} = n^*_j \frac{\lambda_{i_0} p_{i_0 j}}{\sum_i^N \lambda_i p_{ij}} \tag{8.10}$$

which represents the number of counts assigned to the tube j and due to the voxel i_0; the second expression is obtained using definition (8.8). Result (8.10) is an important step in deriving the final result.

The probability $\boldsymbol{P} = \boldsymbol{P}(n^*|\lambda)$ of acquiring the recorded $\{n^*\}$ distribution from the unknown activity $\{\lambda\}$ can be obtained as the product of several independent probabilities. At fixed i and j indices, as a consequence of the Poisson nature of the radiation, the distribution of the n_{ij} values with a mean value λ_{ij}, is

$$P_{ij} = e^{-\lambda_{ij}} \frac{\lambda_{ij}^{\ n_{ij}}}{n_{ij}!}.$$

The P_{ij} probabilities are all linearly independent; therefore the probability P_i that the photons emitted from a voxel i are detected in any tube is given by the product of all the P_{ij} terms

$$P_i = \prod_j^{N_t} P_{ij} = \prod_j^{N_t} e^{-\lambda_{ij}} \frac{\lambda_{ij}^{\ n_{ij}}}{n_{ij}!}.$$

The P_i are linearly independent from each other, so the probability $P(n_{ij})$ to obtain the distribution $\{n_{ij}\}$ is again the product

$$P(n_{ij}) = \prod_i^N P_i = \prod_i^N \prod_j^N P_{ij} = \prod_i^N \prod_j^N e^{-\lambda_{ij}} \frac{\lambda_{ij}^{\ n_{ij}}}{n_{ij}!}.$$

This could appear as the final result, but one needs to take into consideration that expressions (8.9) are not unique: the n_{ij}, within a tube j, can assume different values which bring to the same n^*_j result. For example, if a tube j is composed by six voxels and its associated count is $n^*_j = 8$, this can be obtained from several voxel value sequences, like 1+1+1+1+4+0, or 0+4+0+4+0+0, and so on. Only the distribution $\{n^*\}$ has a physical validity; therefore it is necessary to consider all the sets of the n_{ij} values which can produce the same results: for each n_{ij} set a $P(n_{ij})$ probability has to be computed, and then all these probabilities have to be summed up. The necessary expression is a sum and not a product, because the $P(n_{ij})$ are not all linearly independent: the detected $\{n^*\}$ values constitute the boundary condition (the n_{ij} are linearly independent in each set, but the n_{ij} sets are bound by the $\{n^*\}$ distribution). By considering all of the possibilities of how to decompose a n^*_j value in the independent set of variables n_{ij}, the final result is a function $L(\lambda)$ of the activity distribution $\{\lambda\}$

$$L(\lambda) = P(n^* \mid \lambda) = \sum_{n_{ij}} \prod_{i}^{N} \prod_{j}^{N} e^{-\lambda_{ij}} \frac{\lambda_{ij}^{\,n_{ij}}}{n_{ij}!}.$$

The aim is to find the distribution $\{\lambda^{M}{}_{i}\}$ which maximizes the probability $\boldsymbol{P}(n^*|\lambda)$ with the known $\{n^*\}$ distribution: practically, this means that we try to estimate the distribution $\{\lambda^{M}{}_{i}\}$ which has the maximum probability of generating the detected counts $\{n^*\}$. Therefore, the unknown distribution is given by studying the zeros of $\boldsymbol{P}(n^*|\lambda)$ derivates.

In practice, by having a function composed by several multipliers, it is easier to study the zeros of its logarithm

$$l(\lambda) = \ln\big(L(\lambda)\big).$$

The derivative of the $l(\lambda)$ results

$$\frac{\partial l(\lambda)}{\partial \lambda_{i_0}} = -\sum_{j}^{N_t} p_{i_0 j} + \sum_{j}^{N_t} \frac{n^*{}_j \, p_{i_0 j}}{\sum_{i}^{N} \lambda_i p_{ij}} \tag{8.11}$$

and the second derivative is

$$\frac{\partial^2 l(\lambda)}{\partial \lambda_{i_0} \partial \lambda_{i_1}} = -\sum_{j}^{N_t} \frac{n^*{}_j \cdot p_{i_0 j} p_{i_1 j}}{\left(\sum_{i}^{N} \lambda_i p_{ij}\right)^2}. \tag{8.12}$$

The two distributions $\{n^*\}$ and $\{\lambda\}$ and the probability matrix $\{p_{ij}\}$ are semi-defined positive, i.e. their elements do not have negative values. This implies that (8.12) is semi-defined negative: the function $l(\lambda)$ is concave and all of the existing zeros of (8.11) are maximum points. By changing the indices ($i \rightarrow k$ and $i_0 \rightarrow i$) and using (8.7), a zero of (8.11) is given by

$$\sum_j^{N_t} \frac{n^*_j \, p_{ij}}{\sum_k^N \lambda_k \, p_{kj}} \rightarrow \sum_j^{N_t} p_{ij} = p_i \, . \tag{8.13}$$

The result (8.13) is always a zero of (8.11) if both the terms are multiplied by the same λ_i value

$$\lambda_i \cdot \sum_j^{N_t} \frac{n^*_j \, p_{ij}}{\sum_k^N \lambda_k \, p_{kj}} \rightarrow \lambda_i \, p_i \, .$$

This last result suggests that it is possible to make a succession of distributions $\{\lambda^n{}_i\}$ that tend towards the result $\{\lambda^{\mathrm{M}}{}_i\}$ by following the scheme

$$\lambda_i^{new} = \frac{\lambda_i^{old}}{p_i} \sum_j^{N_t} \frac{n^*_j \, p_{ij}}{\sum_k^N \lambda_k^{old} \, p_{kj}} \, . \tag{8.14}$$

Expression (8.14) constitutes the *Expectation Maximization* (EM) algorithm, introduced by L. A. Shepp and Y. Vardi in 1982 [38]. This method is known as Maximum Likelihood (ML family). In the algorithm (8.14), the denominator is the proje ction step, while the summation is the backprojection.

Generally, the initial distribution $\{\lambda^0\}$, known as the *test distribution*, is chosen uniform and equal to unity. However, the chosen test distribution has no effects on the final result $\{\lambda^{\mathrm{M}}\}$, but only on the convergence speed. The algorithm presented in (8.14) is not the only one obtainable, but rather it belongs to a wide family of solutions, all showing little differences. Other iterative algorithms can be found by adding boundary conditions to the problem.

The solution (8.14), besides concavity and convergence, has the following properties:

- monotony: $P(n^*|\lambda^{\mathrm{k}}) \geq P(n^*|\lambda^{\mathrm{k-1}})$ $\forall$k>1; theoretically, the equal sign is only valid in the case $P(n^*|\lambda^{\mathrm{k}}) = \max_\lambda P(n^*|\lambda)$;
- constancy of the count number: $\sum_i \lambda_i p_i = \sum_j n^*_j$ after each iteration; this property is immediately obtainable by applying a summation over the index i to both members of algorithm (8.14);
- semi-defined positiveness: $\lambda_i \geq 0$ $\forall$ i after each iteration;
- speed: it is rapidly converging.

The algorithm works in a simple way. During each iteration, it computes the expected data collection from the last obtained image by following the modelled probabilities; at the first iteration it uses the test distribution. Then, the expected and acquired data are compared LOR by LOR: if the counts in the expected LOR exceed the acquired one, the count in each voxel of the LOR are diminished on the basis of the probability model; otherwise, if the counts in the expected LOR are less than in the corresponding acquired LOR, the value in each voxel of the LOR is increased. The iteration process theoretically stops when the difference between the expected and acquired LORs is smaller than a fixed value. In practice, because the solution is not local, a similar termination criterion is not useful, and the terminal image has to be determined in a more complex way.

The EM algorithm needs to compute the p_{ij} elements, whose number is given by the product between the sinogram space (number of coincidence tubes) and the number of voxels that are used to divide the FOV. The **P** matrix has typically a size from several hundreds of gigabytes to tens of terabytes that normally does not allow the storage of all of the p_{ij} values on a disk: probabilities have to be computed on line for each tube during the iteration. A first computation is necessary for the normalization step, and successively is repeated twice, both in the projection and in the backprojection steps. This makes the reconstruction very time-consuming.

This problem is typical of all the statistical models, which require very high computation time. Fourier rebinning to reduce the computation space dimensions or other Fourier applications to statistical model save computation time but tend to introduce the imperfections of the non-statistical methods without achieving a full resolution recovery [43]. Each reconstruction method requires a backprojection step to the image matrix, and this step generally takes from 45% to 60% of the computation time [44]. By using symmetry properties and sparse properties of the $\{p_{ij}\}$ matrix, this time can be reduced by a factor that depends on the adopted procedures. In response models, one has to consider also these topics regardless whether they are based on calculation of the length of the intersections between voxels and detection tube or on solid angle calculation [45-50].

Another problem is the termination, that is, when the iterative process has to be stopped. For its own nature, the EM algorithm works in two steps. The first is rapid: the algorithm distributes the counts of each tube to all its inner voxels. By starting from a uniform test distribution, iterations first localize positions and dimensions of large structures: the reconstruction is practically completed for the low spatial frequency structures. In the second phase, the algorithm searches all the local solutions in the image. Practically, it adapts the result to higher and higher spatial frequencies by accumulating the counts of each tube around the inner voxels that show the higher values. The problems arise since the low frequency solution is then not conserved. Obviously, it is necessary to stop the process someway "in the middle"; the difficulty is to be able to specify when to stop. Iterative methods are "able" to present false emission source points in a uniformly empty field, if the number of iterations is excessive.

8.4.3 *The OSEM algorithm*

In 1994, H. M. Hudson and R. S. Larkin developed the *Ordered Subset-Expectation Maximization* (OSEM) algorithm [51].

The OSEM algorithm is a relatively straightforward adaptation of the conventional EM. With OSEM, the projection data are grouped in ordered subsets; the OSEM level is defined as the number of these subsets. The subsets normally consist of projection views separated by some fixed angle within the FOV. The standard EM algorithm (8.14), i.e. projection followed by backprojection, is then applied to each of the subsets in turn: the resulting reconstruction of a subset then becomes the starting image to be used with the next subset. An iteration of OSEM is then defined as a single pass through all the specified subsets. Further iterations may be performed by passing through the same ordered subsets, using the reconstruction provided by the previous iteration as a starting image. By adopting mutually exclusive and exhaustive subsets, an OSEM iteration will have a similar computation time to a standard EM iteration. However, the conditions on the chosen subsets are not binding: the subsets can be non-overlapping or cumulative, that means that each subset is entirely contained in the successive one, and that the last one is the whole sinogram space. Obviously, the non-overlapping choice implies a faster OSEM iteration. The EM is equivalent to a first level OSEM. It is advantageous to select each subset in a balanced way, so that either pixel activity contributes equally to any subset, i.e. the emission probability is equally divided between all of the subsets. Since several PET cameras can collect all tube counts simultaneously, OSEM can best be applied after a full collection of counts. The order in which subsets are processed is arbitrary; though it may be useful to the quality of the reconstructed image to choose a special order: one might encourage substantial new information to be introduced as quickly as possible by choosing first the projection corresponding to the direction of greatest variability in the image, a second one perpendicular to the first, and third and fourth projections midway between these, and so on.

For the case of noiseless projections, each OSEM estimate on a single subset of projections converges towards a maximum likelihood solution as fast as a full iteration EM that uses all projections. In other words, if the projection data are divided into m subsets, then once all projections have been used in a single iteration of OSEM, a result has been produced which is similar to m iterations of EM. This is the property that gives OSEM a considerable "acceleration" as compared to EM. We note that the OSEM result is at most similar to the EM one, not identical. The final images deteriorate as noise increases, or if data are divided into too many subsets. In an actual reconstruction, the background prevents the convergence, and the results could become increasingly worse with the number of iterations.

8.5 Correction and Normalization Procedures

8.5.1 *Attenuation*

Ideal PET imaging requires that neither of the annihilation photons undergo any interaction in the biological tissue. By neglecting the photon absorption within the FOV, quantitative measurements are not possible: the number of counts does not correspond to the number of emitted positrons. In PET, an *attenuation correction factor* (ACF) should be used for each LOR: it is theoretically equal to the inverse of the attenuation probability along the LOR.

The ACF is easily measured for a phantom that presents a uniform linear attenuation coefficient μ, by using the exponential attenuation law. The first photon covers the distance d_1 and its probability to hit the detector without any interaction in the object is proportional to $e^{-\mu d_1}$; the second one covers the distance d_2 with an associated probability proportional to $e^{-\mu d_2}$. Thus, the probability that both the photons are detected without any interaction is proportional to $e^{-\mu(d_2+d_2)} = e^{-\mu d}$. The ACF is $e^{\mu d}$, which is independent from the position of the annihilation point, and is a function only of the thickness d of the phantom along the LOR. Thus, the correction needs only a sampling of the object thickness in the FOV (Fig. 8.30). This fast correction is used in examinations where the μ value is only approximately uniform, as in human head and abdomen studies. In thorax studies, the μ values sharply change, for instance from tissue to lung, and the collection of the information on the d values is not enough. In this case, a transmission image of the body is required in order to sample the $e^{-\mu d}$ values for all of the collected LORs.

In 2D cameras with septa, this sampling is usually performed using a ^{68}Ge linear source (Fig. 8.31). The positron source orbits between detection ring and patient body: by performing two acquisitions, a blank one without the patient (I_0) and a second one with the patient (I), the attenuation factors are given by inverting the resulting

$$e^{-\mu d} = \frac{I}{I_0}.$$

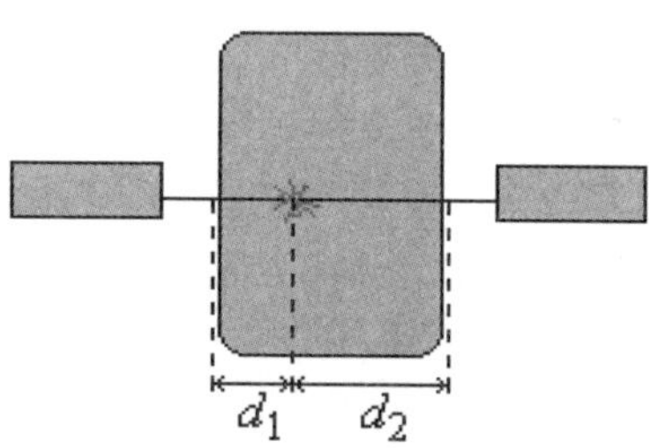

Fig. 8.30. Attenuation correction factor geometry for a uniform phantom: the correction factor is a function of the sum of the two distances d_1 and d_2, and does not depend on the annihilation position along the LOR.

This procedure requires a long examination time, especially in whole body cameras: on a tomograph with 16 rings, a complete sampling needs about 20 minutes per ring, i.e. more than 5 hours, and it is not practically possible. The reduction of the acquisition time increases the statistical noise of the ACF, and this noise also affects the reconstructed image after the correction. Conventionally, the time has to be chosen so as to minimize the ratio between the variances of transmission and emissive data; practically, it is presumed that emissive noise covers that of ACF.

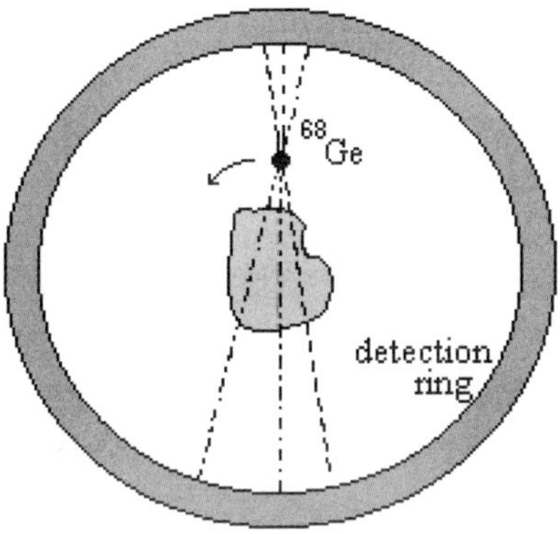

Fig. 8.31. The orbiting linear sources of ^{68}Ge: a software acceptance condition of co-linearity among three points (two on crystals and the source position) allows one to perform the ACF measurement during the patient examination, and also to reject scatter in transmission data.

The ^{68}Ge source, which has a half-life of 275 days and generates the positron emitter ^{68}Ga (half-life of 68.3 minutes), is expensive and requires periodical replacement. The two acquisitions can be made simultaneously: a condition of co-linearity among the three points (two interactions within crystals and the ^{68}Ge source position) discriminates transmission from emissive data, and allows one to reject scatter. However, interleaved procedures are also used, where a fast transmission acquisition follows the emissive.

New segmentation techniques are a very fast alternative. A very short scan gives a transmission image that is affected by high statistical noise. This image is segmented to localize the main anatomical component, by identifying just two different media, such as lung and tissue (where tissue includes bones), or three (lung, soft tissues and muscles, air) [52]. Attenuation coefficient (*a priori*) known is so assigned to the identified areas, by obtaining a noiseless attenuation map.

Transmission scanning can be also performed with an orbiting focused quasi-point-like source of ^{137}Cs, which emits 662 keV photons (Fig. 8.32). The source position is known and the detection point specifies the LORs. This source cannot be used during the emission data acquisition to avoid the *down scatter*, where the 662 keV photons interfere with those from the annihilation. The ^{137}Cs source is less expensive than the ^{68}Ge and does not require periodical replacement, by having half life of ~30 years. However, it has the drawback of giving the μ value at 662 keV instead of at 511 keV as required, and so must be scaled down [53].

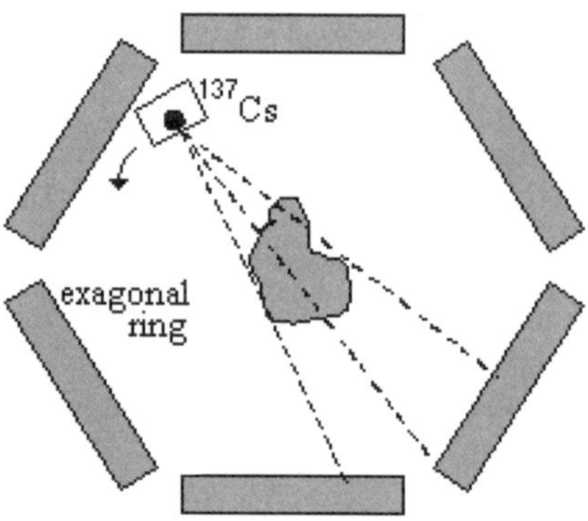

Fig. 8.32. The almost point-like source of ^{137}Cs: the transmission acquisition has to be performed before or after the emissive one to avoid the *down scatter* due to interference of the 662 keV photons of the source with those of the annihilation.

In 3D tomographs without septa, scatter photons can be a problem, especially with the ^{137}Cs source, which is a single photon emitter. Segmentation techniques, which use known attenuation coefficients, solve the scatter problem, give noiseless attenuation map, and do not need to scale the identified μ values from 662 keV to 511 keV [54-57].

In modern PET systems, the attenuation map can be given by a CT device. Transmission acquisition does not give the same qualitative and quantitative information of a dedicated CT system, but a high quality attenuation map. Moreover, these hybrid PET-CT cameras allow a morphologically and anatomically correct positioning of functional data.

8.5.2 *Scattering*

The Compton effect is the dominant interaction for 511 keV photons within the biological tissue, where the mean free path is about 10 cm. Most of the scattered photons escape and are not detected: only a small fraction interact within the detector. However, in 3D this small amount could be numerically compared to the unscattered events, i.e. the amount of scatter could be as much as 50% of the total counts. In 2D PET, septa reduce the amount to about the 15%. This implies that most of the 3D scatter is inter-planar. Scatter introduces a diffused noise into the image, in the lower and central regions of the spatial frequency spectrum (*broad noise*). When the energy of a scattered photon exceeds the discriminator threshold in detection, it generates an incorrect counting of the LOR. At the energy of 511 keV, scatter diffusion is mainly forward, with a small angular deviation; therefore incorrect counts group around the correct, with a blurring effect in the image (*blooming*). An increase of the energy discrimination threshold reduces scatter, but practically, it is not possible to achieve an energy resolution better than 10% with scintillation crystals. Thus, scattered photons whose deviation is less than 30° cannot be rejected because the energy difference for these photons is less than about 60 keV.

By labelling with S the number of scattered events, with U the amount of unscattered, and with T their sum ($T = U + S$), the ratio S/T is known as "*scatter fraction*", while S/U as "*scatter ratio*". The scatter fraction concerns the *amount* of scatter in the collected data, but does not give any information about the spatial distribution with respect to unscattered data: this information is given by the "*scatter function*". The scatter fraction depends on the dimensions and density of the object in the FOV, on the scintillation crystal, and on the geometrical dimension of the tomograph.

The Society of Nuclear Medicine (SNM) and the US National Electrical Manufacturers' Association (NEMA) have defined a standard phantom for measurement and comparison of scatter in different tomographs. The SNM/NEMA phantom is a 20 cm diameter water-filled cylinder, with a central hole to insert a needle with the positron emitter. Typical profiles of this phantom are shown in Fig. 8.33, obtained with 2D (*a*) and 3D tomographs (*b* and *c*) [58]. These profiles show that the majority of scattered counts are reconstructed outside of the object.

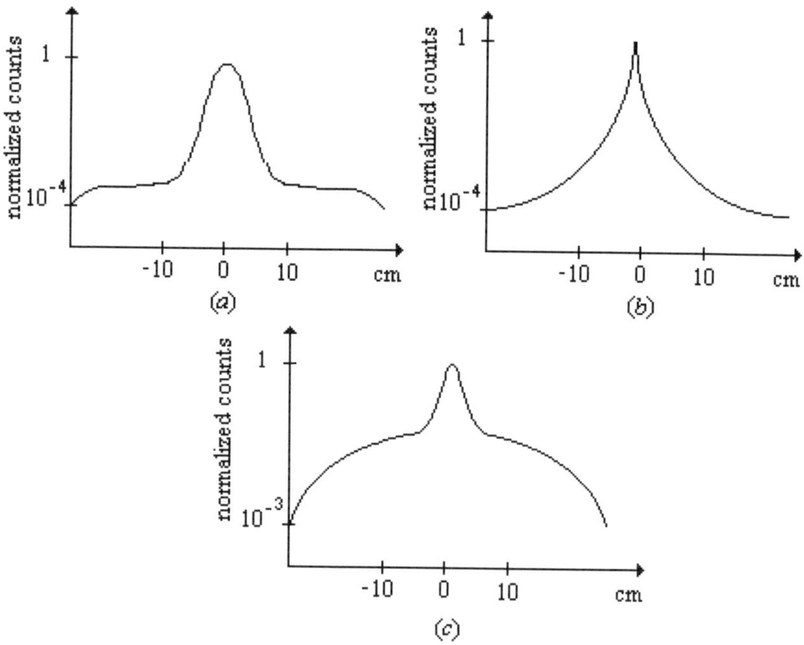

Fig. 8.33. Profiles of a linear source: in the SNM/NEMA phantom for a 2D tomograph (*a*), in air for a 3D tomograph (*b*), and in the SNM/NEMA phantom for a 3D tomograph (*c*). The profiles show that the most of counts due to scatter are reconstructed outside of the phantom.

The scatter effects in the images can be corrected with subtraction methods, or with convolution-subtraction method, which have different theoretical bases. Correction methods generally require the knowledge of both the scatter fraction and of the scatter function. Several SNM/NEMA protocols indicate how to measure the scatter fraction, by putting the phantom in several positions, and how to determinate the scatter function, by fitting and interpolating the phantom profiles with Gaussian functions.

An alternative method is to make a tomographic image of a needle in air and of the same needle in the phantom: these two measures only differ in scatter, thus giving the scatter fraction. This approach is the basis of convolution-subtraction methods [59].

A tomograph with retractable septa can give 2D and 3D images of the phantom, by allowing one to evaluate the inter-plane scatter [60]. This method cannot be extrapolated, especially to situations in which the activity distribution rapidly changes. But these measurements allow one to identify scatter components due to camera structure or to extra FOV deviation.

If the geometry and the activity distribution are known, a Monte Carlo simulation gives exactly both S and U. Obviously, this is only possible for phantom studies and never in a clinical situation.

The corrections of scatter have to satisfy several requirements: they have to work both for hot and cold areas, both for uniform and distributed activity sources, both with symmetric and asymmetric objects. The broadest possible count rate range also has to be covered: in particular, extra FOV scatter sources and multiple window methods can suffer at high-count rates.

The simplest method for scatter correction is the "dual energy windows" (DEW) method (Fig. 8.34). Two adjacent windows are used: an upper around the photopeak, at 380 – 850 keV (index 1), and a lower at 150 – 380 keV (index 2). The counts in both windows, C_1 and C_2 respectively, contain in-FOV unscattered (U_i) and in-FOV scattered (S_i) events:

$$\begin{cases} C_1 = U_1 + S_1 \\ C_2 = U_2 + S_2 \end{cases}.$$

The known quantities are the C_i, so the system cannot be resolved without additional information. The quantity that this method seeks to determine is S_1. The ratio R_S between the scattered components in the two windows is defined by

$$R_S = S_2 / S_1 .$$

In the same way one can define the ratio R_U between the unscattered counts

$$R_U = U_2 / U_1.$$

The initial system can be solved in R terms with respect to S_1, by obtaining

$$S_1 = \frac{R_U}{R_U - R_S} C_1 - \frac{1}{R_U - R_S} C_2.$$

The additional information is given by experimental measurements of R. The R_U is measured by using a line source in air; it is spatially invariant. The R_S is obtained by putting the line source inside the 20 cm diameter phantom; R_S slowly changes with respect to the position, so an average value is calculated and used. The obtained S_1 values are filtered to reduce the statistical noise and the result is subtracted to the data so as to obtain scatter-free data. The correction efficiency is within 2% for a cerebral phantom, and within 7% for a torso phantom.

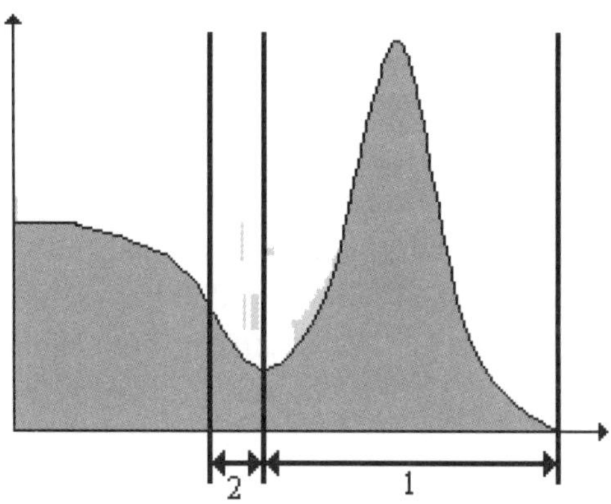

Fig. 8.34. The DEW method windows: upper (1) and lower (2) windows are separate but adjacent; the windows count scattered and unscattered data.

Another simple procedure is the "estimation of true method" (ETM), which uses two energy windows (Fig. 8.35): a large (1) around the photopeak, and a tight (2) above the photopeak; the two windows have the same upper energy level of the discriminator. Beyond a critical threshold, characteristic of the tomograph detector system, there is no more scatter data in the window (2) [61]. Besides, at a fixed temperature and without pile-up effects, the ratio f between the unscattered counts in the two windows depends on the lower and the upper levels of the discriminators, on the radial position of the LORs, and on the aperture angle for each radial position:

$$\frac{C_{1,unscattered}}{C_{2,unscattered}} = f .$$

Therefore, the function f does not depends neither on the object nor on the source distribution, and can be experimentally measured by using a line source in air. By using a specific threshold value, the sensitivity can also be reduced by a factor of 10^3; therefore, a compromise between sensitivity and residual scatter in window 2 has to be made. The ETM method uses the function f and the value $C_{2,unscattered}$ of the window 2 to find the value $C_{1,unscattered}$ of the window 1. This is then subtracted by the acquired data to obtain the sinogram of the scatter, which is firstly smoothed to reduce the statistical noise and finally subtracted from the acquired data. Without the smoothing, the final set of data would be affected by the same statistical noise as the counts acquired in the window 2, which is highly noisy because of the low sensitivity. This method is sensitive to temperature variations in the detectors, and suffers from pile-up effects.

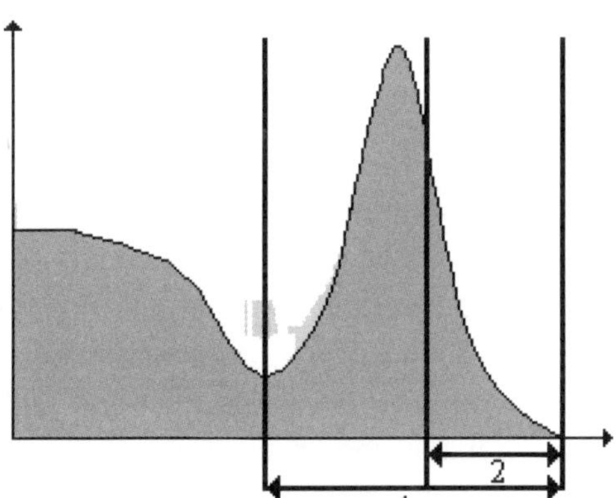

Fig. 8.35. Scheme of the ETM method windows: a small window (2) is positioned above the photopeak, within a larger window covering the whole photopeak (1); beyond a threshold, characteristic of the detection system, there are no scattered data in the small window.

Both the DEW and ETM methods require smoothing: this additional filtering does not degrade the spatial resolution in the final image, since scatter distribution is characterized by low spatial frequencies.

The "triple energy window method" (TEWM), uses the ratio C_2/C_3 between the counts in the two windows under the photopeak (Fig. 8.36). The ratio is obtained for the object to study and for a uniform cylindrical phantom of reference: the comparison of these two measurements gives a calibration function, which describes the scatter distribution at low energies. The function obtained is applied to the counts of window 3 in order to obtain the correction for window 1, with the same procedure as the DEW method.

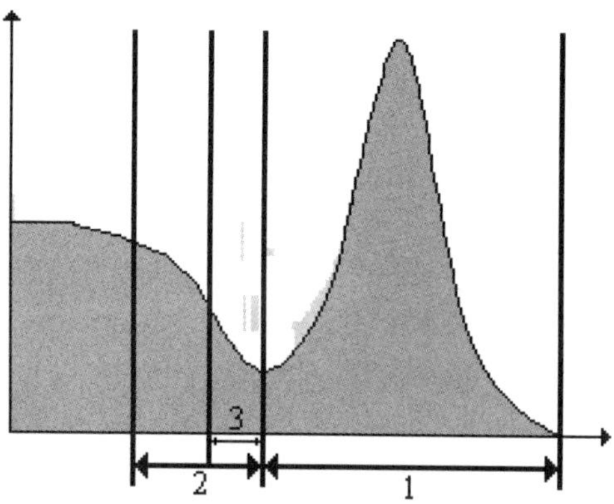

Fig. 8.36. The TEWM method's windows: windows 2 and 3 have the same upper energy level; the ratios C_2/C_3 between the counts in these two windows for the object and for a reference phantom give the correction for window 1.

There are also multi-spectral methods. The acquired spectrum is split into a high number of energy windows, up to 256, all of the same width. In each window, by using simple exponential functions, the spatial distributions of three different scatter components are identified: from the object, from the structure of the tomograph, and within the detector itself. This method assigns the correct position to the scattered photons: therefore, the sensitivity of the system increases. Practically, events are moved from the set S to the set U. Some difficulties can arise due to the high noise of each window, which requires smoothing. Besides, the tomograph must have the hardware to use so many spectral windows.

A different method is the convolution-subtraction technique, which requires an *a priori* knowledge of both the scatter fraction sc and the scatter function $SC(x)$, experimentally obtainable as described for the DEW method. The data distribution in the photopeak can be interpreted in a very simplified model, like the composition of three terms

$$C_0(x) = C_U(x) + C_S(x) + \mathrm{N}(x)$$

where C_U is the distribution of the true (or unscattered) data, C_S of the scattered, and N is the statistical noise (ignored in the following); C_0 is the acquired count distribution. The aim is to derive C_S from C_0 and then to use the result to obtain C_U. The hypothesis is that C_S is given by the convolution of the true data C_U with the scatter function SC

$$C_S(\vec{x}) = \int C_U(\vec{x}') \cdot SC(\vec{x} - \vec{x}')d\vec{x}'$$

or, in dense notation

$$C_S = C_U \otimes SC .$$

True data, by considering the scatter fraction as the correct amount of the scatter, is given by

$$C_U = C_0 - sc \cdot (C_U \otimes SC).$$

It is possible to iteratively use the result, by assuming C_0 as the first approximation of C_U

$$C_U^{(1)} = C_0 - sc \cdot (C_0 \otimes SC),$$

$$C_U^{(n)} = C_0 - sc \cdot \left(C_U^{(n-1)} \otimes SC\right) \text{ for } n > 1,$$

with n as the number of iterations. The method can be applied to the

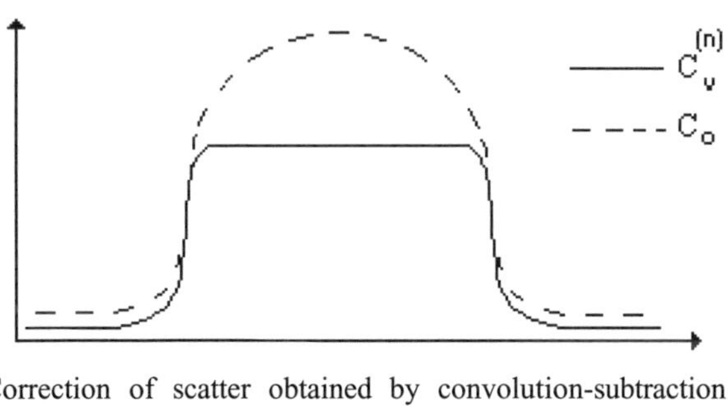

Fig. 8.37. Correction of scatter obtained by convolution-subtraction method for a uniform phantom.

acquired data, to reconstruct C_0-C_S. Alternatively, it is possible to obtain a scatter image by reconstructing C_S, which is then subtracted from the reconstructed C_0 image. An example of the method application is shown in Fig. 8.37.

There are also fitting methods for scatter correction, which consider that the activity distribution reconstructed outside the SNM/NEMA phantom are only scattered ones: their spatial spectrum is characterized by low frequencies and the intensities depend on the activity distribution. The scatter distribution from outside of the phantom is extrapolated by fitting it with a simple Gaussian function, in order to obtain the scatter for inside the phantom. The obtained scatter distribution is smoothed and then subtracted from the acquired data. The fitted scatter distribution depends on the smoothing, the interpolation range and the number of interpolated points. The limit of the method is the use of Gaussian functions to interpolate, because the scatter profile is not independent of the activity distribution. Particularly, Gaussian fitting is not adaptable to situations with isolated spots of activity. The advantages of this method are the simplicity, the absence of auxiliary measurements, and the fact that it does not need an independent knowledge both of the scatter fraction and scatter function.

8.5.3 *Random coincidences*

A random occurrence happens when two photons are in time coincidence but they originate from two different annihilation events. These counts cause a decrease in the contrast of the image and inaccuracy in the quantitative measures of the activity distribution. They give a uniform and elevate background in the entire image and along the whole spectrum of the spatial frequencies. Random events cannot be differentiated during the acquisition; they have to be evaluated and subtracted in reconstruction. They are due to the finite extension of the temporal window: when a photon strikes a detector, a temporal window with width τ is opened; if a second photon hits in an opposite detector in this window, a coincidence is recorded. The time τ depends on the scintillation decay time, the light output and the electronic capability. The first is the dominant component: longer decay times imply larger time windows, and so the number of random counts increases.

In a detector pair, by indicating with c_1 and c_2 the count rates of the two detection elements, the number of random counts within the time τ is given by $\tau c_1 c_2$. The expression becomes τc^2 if there is no need to differentiate the two detectors. Thus, the random counts are given by $C_r = f\tau c^2$ in a tomograph with f number of opposite detector pairs [62]. This result can be simplified in PET, by considering the count rate of a whole ring of detector (C_s)

$$C_r = f\tau C_s^2. \tag{8.15}$$

Formally the expression is unchanged, but now f is the fraction of solid angle that each detector of the ring covers with its coincidences. The result (8.15) implies that all the detectors have the same efficiency; this is not true, so an accurate ring calibration needs. A second hypothesis assumes a uniform density of detectors along the ring: this is approximately true just for circular rings; in other geometries with detector gaps, the factor f is not simple to derive. However, the expression in (8.15) is very useful because it connects directly and simply, the random count rate (C_r) to the true (C_t) count rate: experimental evidence indicates that C_s is directly proportional to C_t, therefore

$$C_r \,/\, C_t \propto f\tau C_s. \tag{8.16}$$

This expression emphasizes that the number of random events increases with the count rate and the width τ of the time window. At higher count rates the direct proportionality between C_s and C_t is lost, and saturation phenomena become apparent. In 2D cameras, septa also reduce the random counts: a smaller factor f results from the acceptance angle reduction, i.e. the random count-rate decreases as the septa length increases.

The *delayed coincidence* method is a standard procedure to evaluate the number of random events: the coincidence window is opened not when the first photon hits the detector, but with a time delay that prevents any true coincidence. This method can be applied with the delayed window that follows the coincidence one (Fig. 8.38*a*), or with the coincidence inside the delayed (Fig. 8.38*b*). The first case directly gives an evaluation of random counts; if the two windows have different widths, one should consider that the statistics are also different. With the second method, the ratio between the count-rates in the two windows gives the fraction of random counts with respect to the true ones; but this needs the capacity to simultaneously use both the windows.

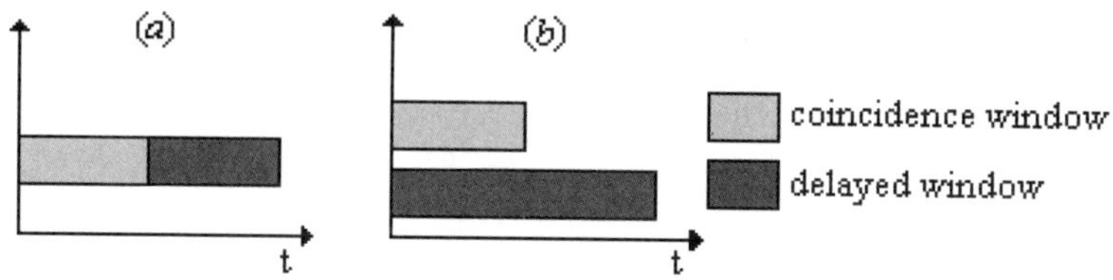

Fig. 8.38. Schemes of the delayed coincidences method; (*a*) sequential and (*b*) simultaneous applications. The width of the two windows can be different.

It is also possible to evaluate the count rate of a single detector (c) or of a single ring (C_s) and then to derive C_r using the formula (8.15) and (8.16). The difficulty is the correct evaluation of the factor f, typical of each tomograph.

8.5.4 *Partial volume effect*

Quantitative measurements require direct proportionality between the number of counts assigned to the pixel (or to the voxel) and the activity concentration in the object. This direct proportionality is generally verified if the object dimension is greater than two times the FWHM of the spatial resolution. For smaller structures the proportionality is lost: the relationship between the image and the radioisotope concentration is no longer linear but depends on the dimensions of the same object. This lack of linearity is knows as the "*partial volume effect*" (PVE), and causes a reduction in the image contrast. This effect is especially present in the axial direction. The object contrast decreases independently from the camera efficiency, because the object only partially fills the voxel, but counts are assigned to the whole voxel, with consequent under-estimation. The PVE arises when the object size is between one times the FWHM and two times the FWHM; structures smaller than one FWHM are not detectable [63, 64].

Quantitative examinations in the nonlinearity region need to know the *recovery curve* (RC), which is given by the ratio between the activity measured in the image and the true radioisotope activity as a function of the thickness of the object (Fig. 8.39). The curve should be experimentally measured for each tomograph.

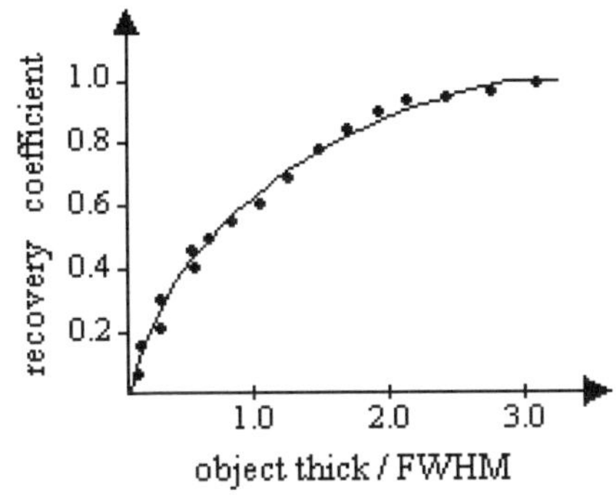

Fig. 8.39. Graph of a typical recovery curve for the PVE: the curve has to be experimentally determined for each tomograph.

8.5.5 *Normalization*

The normalization correction is due to the lack of uniformity in the detection response. This has several causes. Primarily, the efficiency of a single crystal is not uniform: different LORs see variations of the crystal depth and the detection area changes. There are also differences in efficiency from one crystal to another. In tomographs with detector blocks, there are differences among the blocks and among each block component. The geometry of the detection could introduce other non-uniformities, especially the presence of gaps between detectors. Problems can also arise due to the inaccuracy of the time coincidence windows.

The direct method of performing normalization is to make a *blank* acquisition of a source in the FOV: the inversion of the image constitutes the correction. To limit the scatter and to avoid the deviation from linearity that high count rates can generate, the source used must be small (a point or a line source) and with rather low activity; obviously this increases the acquisition time.

There are also indirect methods, which consider the response of each single detector, but it is very hard to make a complete evaluation of all the angular dependencies, both azimuthal and polar. The problem is mainly in the procedure used to test the sensitivity of each single detector: by using transmission and emissive techniques, the obtained results are often different. Generally, the same source is used for attenuation correction, or sometimes a 20 cm diameter uniform cylinder, usually filled with ^{68}Ge is used.

Scatter and normalization interfere and affect one another: true and scattered data require different normalizations. If more energy windows are used, normalization for each window has to be performed.

The normalization of each possible LOR requires a lot of time to reach a meaningful statistic, especially in 3D cameras. The measurement of groups of LORs can reduce the time but introduces some inaccuracies. However, normalization has to be performed just once, unless there is a significant drift in the tomograph functioning. An incorrect normalization can introduce geometrical and quantitative artefacts in the image. True coincidences, scattered and random events require different normalizations (Fig. 8.40): true coincidence can only arise from the volume between the two opposite detectors; the scattered events from the area of the intersection between the acceptance volumes of the detectors; random events from each point in the union of the two acceptance volumes.

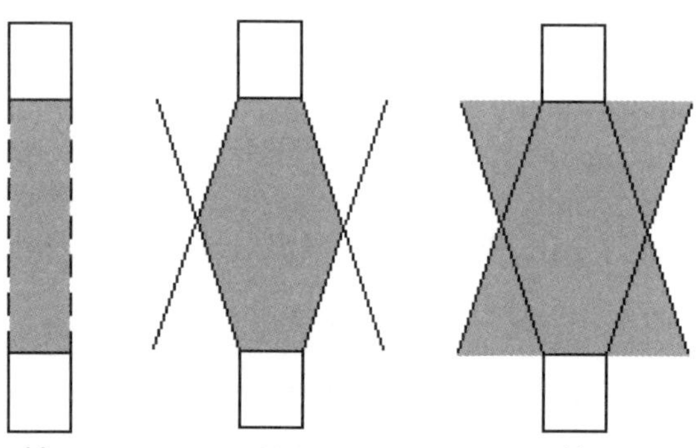

Fig. 8.40 Geometric considerations about normalization: true data arise from the volume between the two opposite detectors (*a*), the scattered from the intersection area of the two acceptance volumes (*b*), the random from the union of the two acceptance volumes (*c*).

8.6 Commercial Camera Overview

The ring tomograph market is led by CTI PET systems, who are the worlds largest manufacturer of PET equipment. They produce four different scanners of differing cost and performance ranging from the ECAT ART, which utilizes two incomplete rings of BGO that require rotation in order to obtain complete imaging, up to the ECAT EXACT HR+ (*High Resolution plus*) which has both high-resolution and high count-rate capabilities for the high-end clinical research market. However, an important point to note is that all of these systems are based on the same block detector design. In fact, none of the main commercial cameras differ substantially from each other. This can be seen from Table 8.10. The few differences seen are due to the scintillator material where an alternative to BGO has been used. For instance, the improved energy resolution of GSO and the consequently tighter energy window. Both LSO and GSO allow a much shorter coincidence window due to their faster decay times and higher light yields. The use of a faster scintillator can also reduce the scanning time required, down to one half of the time for LSO as compared with BGO. Also, both LSO and GSO are said to make the resulting image crisper due to the higher light yield, thus allowing a more accurate centroiding. It should be noted that the EXACT scanner is now also available in LSO, called the ACCEL.

All models of commercial scanner come complete with computer, screens, patient platform, software and standard control devices. The operation of these systems is designed to be fairly automated. Such functions as centring, efficiency control, timing and other calibrations are all automated. Corrections for random and scattered events, and attenuation are also automatic. The operator can select the desired reconstruction and the acquired data is presented as a corrected image on the screen.

One feature that is starting to differentiate the scanners on the market, is the addition of a second modality scanner, such as CT, MR and SPECT. The Allegro, GE Advance and the ECAT EXACT are all now available with a 'built-in' CT scanner. However, even if the second scanner is not present, there is the possibility to merge images from separate scans via software. Such techniques can be useful for increased diagnostic certainty, tumour localization for surgery or treatment, or with appropriate software such data can be used for attenuation correction.

Table 8.10 Characteristics of some commercial PET camera.

Parameter	CTI ECAT EXACT HR+	GE Advance	Hitachi Sceptre	Philips Allegro
Number of rings	32	18	24 (equivalent)	N / A
Pixel size (mm^3)	4.05×4.39×30	3.9×8.2×30	6.75×6.75×20	4×6×20
Number of pixels	18432	12096	4224	17864
Number of phototubes	1152	672 dual	264	420
Crystal material	BGO	BGO	LSO	GSO
Transaxial FOV (cm)	58.5	50	58.5	57.6
Axial FOV (cm)	15.5	15.2	16.2	18
Axial resolution (mm)	4.2	4	4.9	4.2
Transaxial resolution (mm)	4.6	4.8	6.2	4.8
Sensitivity 3D (Mcps/μCi/cc)*	1.4	1.9	1.0	1.0
Energy threshold (keV)	350 – 650	300 – 650	350 – 650	450 – 580
Coincidence window (ns)	12	12	8	8
Minimum slice thickness (mm)	2.46	4.25	3.375	2
Energy resolution (%)	25	20	25	14
Scatter fraction, 2D (%)	18	9	N / A	N / A
Time for whole body scan (min)	45 – 90	32 – 45	45 – 70	25

* true coincidence, after scatter correction with NEMA 94 phantom

References

1. D. W. Townsend, T. J. Spinks, T. Jones, A. Geissbuhler, M. Defrise, M. C. Gilardi and J. Heather, *Three-dimensional reconstruction of PET data from a multi-ring camera*, IEEE Trans. Nucl. Sci., **36**, 1056–1065 (1989)
2. M. E. Phelps, E. J. Hoffman, N. A. Mullani and M. M. Ter-Pogossian, *Application of annihilation coincidence detection to transaxial reconstruction tomography*, J. Nucl. Med., **16**, 210–224 (1975)
3. Y. Yamamoto, C. J. Thompson, M. Diksic, M. Meyer and W. H. Feindel, *Positron Emission Tomography*, Journal of Radiation Physics and Chemistry, **24,** 385–403 (1984)
4. A. Del Guerra, *Tomografia ad emissione di positroni*, Enciclopedia del Novecento, VIII (1989)
5. M. E. Phelps, J. C. Mazziotta and H. R. Schelbert, *Positron emission tomography and autoradiography*. (Raven Press, New York, 1986)
6. T. F. Budinger, S. E. Derenzo, R. H. Huesman and J. L. Cahoon, *PET : instrumentation perspectives*. (1982)
7. F. H. Attix, Introduction to radiological physics and radiation dosimetry, Ed. (John Wiley & Sons , NY, 1986)
8. S. Webb, *The physics of medical imaging*. (Institute of Physics Publishing, Bristol and Philadelphia, 1993)
9. B. Mazoyer, R. Trebossen, C. Schoukroun, B. Verrey, A. Syrota, J. Vacher, P. Lemasson, O. Monnet, A. Bouvier and J. L. Lecomte, *Physical characteristic of TTV03, a new high spatial resolution time-of-flight positron tomography*, IEEE Trans. Nucl. Sci., **37**, 778–782 (1990)
10. E. Fermi, *Nuclear Physics*. (The University of Chicago Press, 1949)
11. S. Webb, *The physics of medical imaging*. (Institute of Physics Publishing, Bristol and Philadelphia, 1993)
12. M. E. Phelps, S. C. Huang, E. J. Hoffman, C. Selin, L. Sokoloff and D. E. Kuhl, *Tomographic measurement of local cerebral glucose metabolic rate in humans, with (F-18)2-fluoro-2-deoxy-D-glucose:Validation of method*, Ann. Neurol. **6**:371–388 (1979)
13. W. Heitler, *The quantum theory of radiation*. (Dover Publications, Inc., NY, 1953)
14. Z. H. Cho, J. K. Chan, L. Ericksson, M. Singh, S. Graham, N. S. MacDonald and Y. Yano, *Positron ranges obtained from biomedically important positron-emitting radionuclides*, J. Nucl. Med., **16,** 1174–1176 (1975)
15. C. S. Wu and I. Shaknov, Phys. Rev., 77, 136 (1950)
16. R. Eisberg and R. Resnick, *Quantum Physics of atoms, molecules, solids, nuclei, and particles*, Ed. (John Wiley & sons, NY, 1985)
17. G. L. Brownell and W. H. Sweet, *Localitation of brain tumours with positron emitters*. Nucleonics, **11,** 40–45 (1953)
18. P. Colombino, B. Fiscella and L. Trossi, *Study of positronium in water and ice from 22 to -144 °C by annihilation quanta measurements*, Il Nuovo Cimento **XXXVIII** (2), 707–723 (1965)
19. S. Berko and F. L. Hereford, *Experimental studies of positron interactions in solids and liquids*. Reviews of Modern Physics, **28** (3), 299–307 (1956)

20. M. .D. Harpen, *Positronium: Review of symmetry, conserved quantities and decay for the radiological physicist*, Med. Phys. **31** (1), 57–61 (2004)
21. F. Daghighian, P. Shenderov, K. S. Pentlow, M. C. Graham, B. Eshaghian, C. L. Melcher and J. S. Schweitzer, *Evaluation of LSO :Ce scintillation crystal for PET*, IEEE Trans. Nucl. Sci., **40**, 1045–1047 (1993)
22. Y. Kurata, K. Kurashige, H. Ishibashi and K. Susa, *Scintillation characteristics of GSO single crystal grown under O2 containing atmosphere*, IEEE Trans. Nucl. Sci., **42**, 1038–1040 (1995)
23. S. I. Ziegler, J. G. Rogers, V. Selivanov and I. Sinitzin, *Characteristics of the new YAlO3 :Ce compared with BGO and GSO*, IEEE Trans. Nucl. Sci., **40**, 194–197 (1993)
24. R. S. Miyaoka and T. K. Lewellen, *Evaluation of CeF3 as a scintillator for high speed dynamic PET imaging*, IEEE Trans. Nucl. Sci., **41**, 2743–2747 (1994)
25. A. G. Petrosyan, G. O. Shirinyan, C. Pedrini, C. Durjardin, K. L. Ovanesyan, R. G. Manucharyan, T. I. Butaeva and M. V. Derzyan, *Bridgman growth and characterization of LuAlO3-Ce3+ scintillator crystals*, Cryst. Res. Technol., **33** (2), 241–248 (1998)
26. W. W. Moses, S. E. Derenzo, A. Fyodorov, M. Korzhik, A. Gektin, B. Minkov and Aslanov, *LuAlO3 :Ce - a high density , high speed scintillator for gamma detection*, IEEE Trans. Nucl. Sci., **42**, 275–279 (1995)
27. E. J. Hoffman, M. Dahlbom, A. R. Ricci and I. N. Weinberg, *Examination of the role of detection systems in quantitation and image quality in PET*, IEEE Trans. Nucl. Sci., **33**, 420–424 (1986)
28. M. Marengo, *La Fisica in Medicina Nucleare.* (Patron Editore Bologna, 2001)
29. G. Muehllelner, *Positron camera with extended counting range capability*, J. Nucl. Med., **16**, 653–657 (1975)
30. E. Tanaka, N. Nohara and H. Murayama, *Variable sampling-time technique for improving count rate performances of scintillation detectors*, Nucl. Instr. Meth., **158**, 459–466 (1979)
31. J. S. Karp and G. Muehllehner, *Performance of a position-sensitive scintillation detector*, Phys. Med. Biol., **30**, 643–655 (1985)
32. J. A. Wear, J. S. Karp, A. T. Haigh and R. Freifelder, *Evaluation of moderately cooled pure NaI as a scintillator for position sensitive PET detectors*, IEEE Trans. Nucl. Sci., **43**, 1945–1951 (1996)
33. J. L. Cahoon, R. H. Huesman, S. Derenzo, A. B. Geyer, D. C. Uber, B. T. Turko and T. Budinger, *The electronics for the Donner 600 crystal positron tomograph.* Trans. Nucl. Sci., **33**, 570–574 (1986)
34. R. Nutt, *The History of Positron Emission Tomography*, Molec. Imag. Bio., **4** (1), 11–26 (2002)
35. S. E. Derenzo, *Gamma-ray spectroscopy using small cooled bismuth germanate scintillators and silicon photodiodes*, Nucl. Instr. Meth., **219**, 117–122 (1984)
36. C. Carrier, C. Martel, D. Schmitt et al. *Design of a high resolution positron emission tomograph using solid state scintillation detectors*, IEEE Trans. Nucl. Sci., **35**, 685–690 (1988)

37. J. S. Karp and M. E. Daube Witherspoon, *Determination of depth-of-interaction in scintillation crystals using a temperature gradient*, Nucl. Instr. Meth., **A260**, 509–517 (1987)
38. L. A. Shepp and Y. Vardi, *Maximum Likelihood Reconstruction for Emission Tomography*, IEEE Trans. Med. Imag., **MI-1** (2), 113–122 (1982)
39. L. Parra and H. H. Barrett, *List-Mode Likelihood: EM Algorithm and Image Quality Estimation Demonstrated on 2-D PET*, IEEE Trans. Med. Imag., **17** (2), 228–235 (1998)
40. K. Rajan, L. M. Patnaik and J. Ramakrishna, *Linear array implementation of the EM algorithm for PET Image Reconstruction*. IEEE Trans. Nucl. Sci., **42** (4), 1439–1444 (1995)
41. K. Rajan, L. M. Patnaik and J. Ramakrishna, *High-speed computation of the EM algorithm for PET Image Reconstruction*, IEEE Trans. Nucl. Sci., **41** (5), 1721–1728 (1994)
42. A. J. Reader, K. Erlandsson, M. A. Flower and R. J. Ott, *Fast accurate iterative reconstruction for low statistics positron volume imaging*, Phys. Med. Biol., **43**, 835–846 (1998)
43. P. Kinahan, C. Michel and M. Defrise, *Fast iterative image reconstruction of 3D PET data*, Conference Records of the IEEE Nuclear Science Symposium and Medical Imaging Conference (Anaheim, CA, 1996) Ed A Del Guerra *(Piscataway, NJ:IEEE)* 1918–22 (1996)
44. M. L. Egger, C. Joseph and C. Morel, *Incremental beamwise backprojection using geometrical symmetries for 3D PET reconstruction in a cylindrical scanner geometry*, Phys. Med. Biol., **43**, 3009–3024 (1998)
45. C. Chen, S. Lee and Z. Cho, *Parallelization of the EM algorithm for 3D PET image reconstruction*, IEEE Trans. Med. Imaging **110**, 513–522 (1991)
46. C. A. Johnson, J. Seidel, R. E. Carson, W. R. Gandler, A. Sofer, M. Green and M. E. Daube-Witherspoon, *Evaluation of 3D reconstruction algorithms for a small animal PET camera*, IEEE Trans. Nucl. Sci., **44**, 1303–8 (1997)
47. C. A. Johnson, Y. Yan, R. E. Carson, R. Martino and M. E. Daube-Witherspoon, *A system for the 3D reconstruction of retracted-septa PET data using the EM algorithm*, IEEE Trans. Nucl. Sci., **42**, 1223–7 (1995)
48. J. M. Ollinger and A. S. Goggin, *Maximum Likelihood Reconstruction in fully 3D PET via the SAGE algorithm*, Conference Records of the IEEE Nuclear Science Symposium and Medical Imaging Conference (Anaheim, CA, 1996) Ed. A. Del Guerra (Piscataway, NJ:IEEE) 1594–8 (1996)
49. A. Terstegge, S. Weber, H. Herzog, H. W. Muller-Gartner and H. Hailling, *High resolution and better quantification by tube of response modelling in 3D PET reconstruction*, Conference Record of the IEEE Nuclear Science Symposium and Medical Imaging Conference (Anaheim, CA, 1996) Ed. A. Del Guerra (Piscataway, NJ:IEEE) 1603–7 (1996)
50. J. Qi, R. M. Leahy, S. R. Cherry, A. Chatziioannou and T. H. Farquhar, *High-resolution 3D Bayesian image reconstruction using the microPET small-animal scanner*, Phys. Med. Biol., **43**, 1001–1013 (1998)

51. H. M. Hudson and R. S. Larkin, *Accelerated image reconstruction using ordered subset of projection data*, IEEE Trans. Med. Imag., **13**, 601–609 (1994)
52. S. R. Meikle, M. Dahlbom and S. R. Cherry, *Attenuation correction using count-limited transmission data in positron emission tomography*, J. Nucl. Med., **34** (1), 143–150 (1993)
53. W. F. Jones, K. Vaigneur, J. Young, C. Moyers and C. Nahmias, *The Architectural Impact of Single Photon Transmission Measurements on Full Ring 3D Positron Tomography*, In Monier PA, ed Proceedings of the IEEE Nuclear Science Symposium and Medical Imaging Conference, San Francisco, **2**, 1026–1030 (1995)
54. S. C. Huang, R. Carson, M. E. Phelps, E. J. Hoffman, H. Schelbert and D. Kuhl, *A boundary method for attenuation correction in positron emission tomography*, IEEE Trans. Nucl. Sci., **22**, 627–637 (1981)
55. E. Z. Xu, N. A. Mullani, K. L. Gould and W. L. Anderson, *A segmented attenuation correction for PET*, J. Nucl. Med., **32**, 161–165 (1991)
56. M. Xu, W. K. Luk, P. D. Cutler and W. M. Digby, *Local threshold for segmented attenuation correction of PET imaging of the thorax*, IEEE Trans. Nucl. Sci., **41**, 1532–1537 (1994)
57. S. K. Yu and C. Nahmias, *Segmented attenuation correction using artificial neural networks in positron emission tomography*, Phys. Med. Biol., **41**, 2189–2206 (1996)
58. B. Bendriem and D. W. Townsend, *The theory and practice of 3D PET*. (Kluwer Academic Publishers, 1998)
59. D. L. Bailey and S. R. Meikle, *A convolution-subtraction scatter correction method for 3D-PET*, Phys. Med. Biol., **39** (3), 411–424 (1994)
60. S. R. Cherry, S. R. Meikle and E. J. Hoffman, *Correction and Characterization of Scattered Events in Three-Dimensional PET Using Scanners with Retractable Septa*, J. Nucl. Med., **34**, 671–678 (1993)
61. C. J. Thompson, *The problem of Scatter Correction in Positron Volume Imaging*, IEEE Trans. Med. Imag., **MI-10**, 234–239 (1993)
62. G. F. Knoll, *Radiation detection and measurement*. Ed (John Wiley & Sons, II ed, 1989)
63. E. J. Hoffman, S. C. Huang and M. E. Phelps, *Quantitation in positron emission computed tomography : 1. Effect of object size*, J. Comput. Assist. Tomogr., **3**, 299–308 (1979)
64. J. C. Mazziotta, M. E. Phelps, D. Plummer and D. E. Kuhl, *Quantitation in positron emission computed tomography : 5. Physical-anatomical effects*, J. Comput. Assist. Tomogr., **5**, 734–743 (1981)

CHAPTER 9

NUCLEAR MEDICINE: SPECIAL APPLICATIONS IN FUNCTIONAL IMAGING

R. Pani

Department of Experimental Medicine and Pathology

University of Rome "La Sapienza", Roma, Italy

9.1 Introduction

In recent years there has been a growing interest in developing compact gamma cameras to improve gamma ray imaging. Conventional full size gamma cameras using NaI(Tl) scintillator block coupled to a bulky array of photomultiplier tubes are precluded from use in all applications where light weight, small size, easy handling and positioning is required. Furthermore the limit of the standard Anger camera is the poor intrinsic spatial and energy resolution. This has led to the design and testing of new gamma cameras which generally fall into three types:

– semiconductor imagers (like Ge, CdTe or CdZnTe) in which the gamma rays interact directly in a position sensitive solid state detector [1, 2];
– scintillation crystals coupled to an array of solid state photodetectors (photodiode) [3-7];
– scintillation crystals coupled to a Position Sensitive Photo-Multiplier Tube.

Other interesting detectors are currently under investigation, such as Avalanche PhotoDiode (APD) [8] and Silicon Drift Detector (SDD) [9, 10] but suitable detection area are not jet available.

The gamma cameras based on position sensitive photomultipliers could be the best chance to obtain a realistic and low cost compact gamma camera.

Over the last 15 years, starting from first generation PSPMT built by traditional PMT manufacturing, the technological enhancement allow to achieve very compact size ($25\times25\times20$ mm^3) by a novel charge multiplication system (see Fig. 9.1).

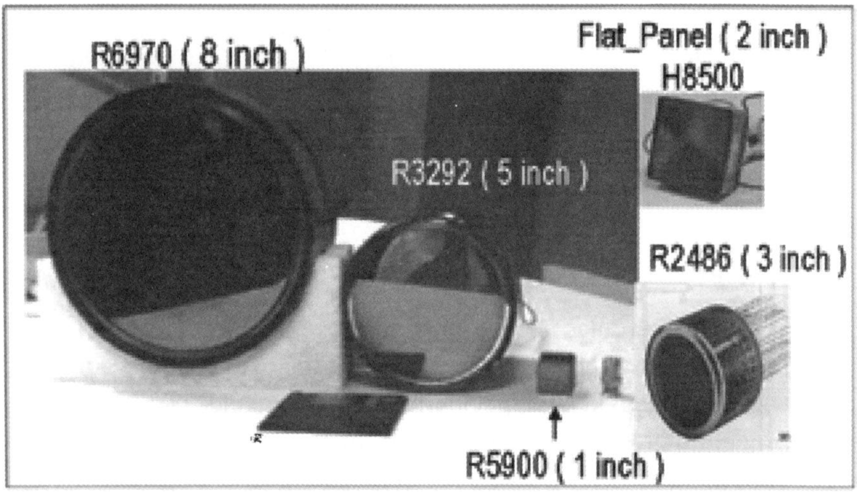

Fig. 9.1 Different Hamamatsu PSPMT generations (from Pani, [11]).

The PMT anode side is usually less than 6 mm instead of 60 mm of the Anger Camera. The PSPMT works under the Anger camera basic principle with the additional feature of using scintillation arrays with pixel dimension less than 1 mm, thus achieving sub-millimeter spatial resolution values. Moreover it is possible to choose low cost geometries with suitable and optimized detection features for specific applications. Furthermore the recent introduction of NaI (Tl) scintillation arrays allows one to carry out an energy resolution of about 10% at 140 keV photon energy. The last technological development is a Hamamatsu PMT based on Flat Panel structure, named H8500. This PSPMT seems to solve some limitations of those of the previous generation. Its dimension are 50×50 mm^2 and the external dead zone is less than 1 mm, thus allowing one to place closely different modules so as to achieve large detection areas. The metal channel dynode technology allowed one to reduce the thickness of the multiplication section down to 5 mm so the Flat Panel final thickness is 12 mm, which is about half the height of second generation PSPMT. Compactness is going to allow a drastic shielding weight reduction at 140 keV photon energy (12 kg for 20×20 cm^2 FOV camera). Some preliminary results of a Flat Panel PMT show an intrinsic spatial resolution limit near 0.5 mm.

9.2 Position Sensitive Photo Multiplier Tube

9.2.1 *Hamamatsu First PSPMT Generation*

The first 3 inch position sensitive PMT, Hamamatsu R2486, developed in 1985 [12, 13], represented a strong technological advance for gamma ray imaging. The first generation was based on proximity mesh dynode by which the charge was multiplied around the original position of the light photon striking on the photocathode. In Fig. 9.2 the dynode structure is shown.

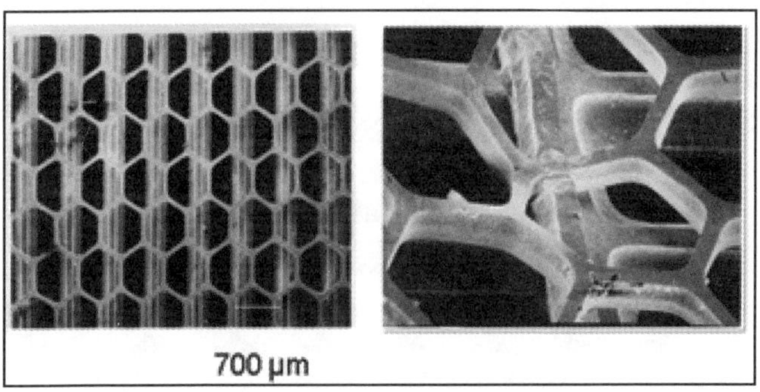

Fig. 9.2 Dynode structure of PSPMT first generation (from Pani, [15]).

The charge shower had a wide intrinsic spread, the whole internal multiplication process is schematically represented in Fig. 9.3. A number of factors affect such spreading like the interstage voltage of dynodes, (in particular between cathode and first one) additional focusing grids between dynodes and finally the intrinsic spreading of the light spot due to the photocathode glass window optical guiding. The first measurement of charge spread resulted 4 mm FWHM at the last dynode for 3 inch PSPMT [12] Values reported by other authors on commercial tubes ranged between 7 and 11 mm FWHM for 3 inch PSPMT and between 11 mm and 17 mm for 5 inch PSPMT respectively [11, 14, 16].

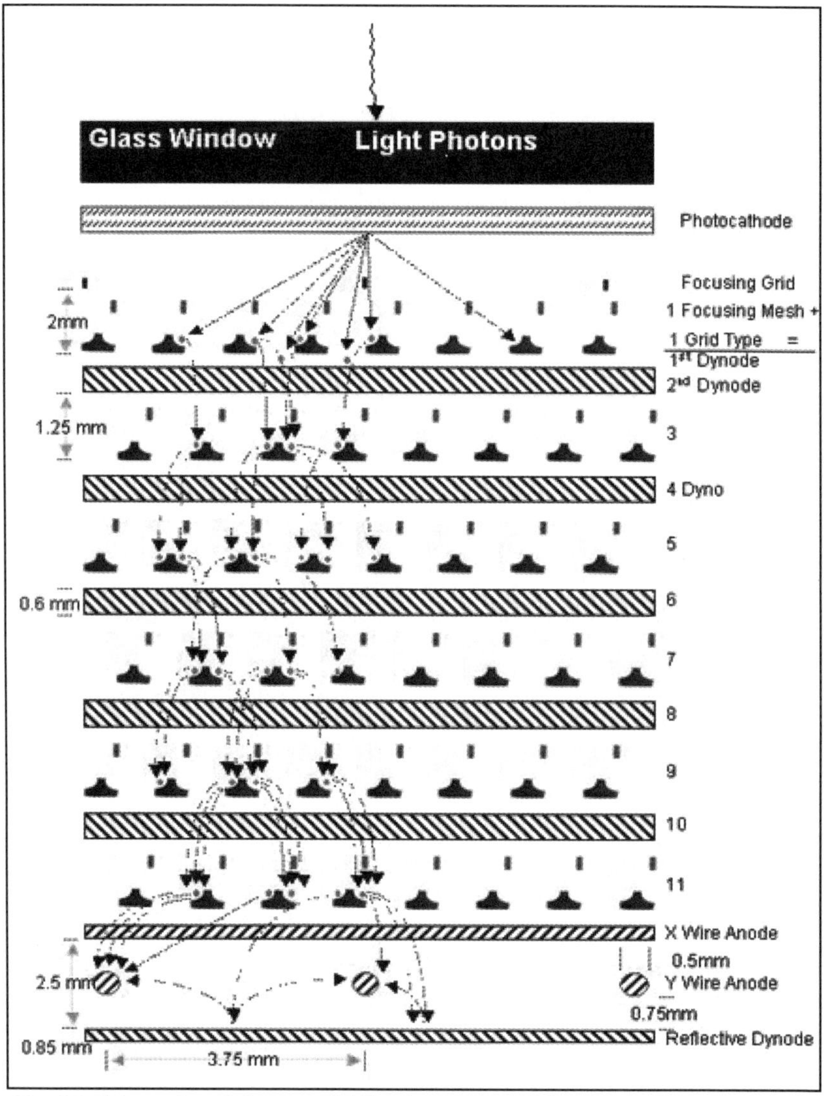

Fig. 9.3 Schematics of charge multiplication process of first generation PSPMT [15].

Such intrinsic wide charge spreading favors the use of crossed wire anodes and the centroide method for position determination rather than application as hodoscope where pixellated anodes with minimum cross-talk is required.

Furthermore a wide intrinsic charge distribution can limit the spatial resolution values of narrow light spots. A detailed analysis of spatial resolution values obtained by pillars and planar crystals with different shapes, size and scintillation materials coupled to a 3 inch PSPMT with crossed wire anode is reported elsewhere [11, 14, 16].

The first generation of PSPMT had a standard glass envelopment and was characterized by large active areas (5 inch) [17] useful for a number of applications in physics, but not large enough for medical imaging where small organs like breast and brain need at least 20 cm FoV.

Unfortunately large PSPMT peripheral dead zones (1 cm or more) did not allow to assemble arrays of tubes as in Anger camera.

9.2.2 *Hamamatsu Second PSPMT Generation*

The 2nd PSPMT generation was based on metal channel dynode for charge multiplication combined to a photocathode glass window of thickness less than 1 mm. The intrinsic light spread was reduced down to 0.5 mm FWHM [18, 19]. The new dynode structure (metal channel dynode shown in Fig. 9.4) of PSPMTs consists of electron multiplier layers that channeling the charge do not change the coordinates relative to the photon interaction on the photocathode. The charge is collected by a multi-anode with array or multi-wire structure. Contrary to the first generation, the narrower charge distribution potentially increases the number of anodes; in fact it would require less than 1 mm anode side for the optimum light sampling.

Actually the intrinsic charge spread poorly affects the scintillation light distribution after charge conversion and 1 mm anode side could be redundant, involving an over sampling of light and useless additional electronic chains for signal read out. Hamamatsu R7600-C12 series (crossed wire anode) [20] and R5900-M64 series (array anode) have anode size well optimized, to 3.6 mm and 2 mm, respectively. Such light sampling size allows one to well identify scintillation crystal pixel size as small as 1 mm (C-12) or less (M64) without strong image position distortions at 140 keV gamma ray irradiation.

A further technological improvement consists of the metal housing that allows very compact size (about 1 cubic inch) and reduced peripheral dead zones down to 2 mm [21, 22].

Fig. 9.4 Metal channel dynode structure.

9.2.3 *Hamamatsu 3rd Generation PSPMT*

Hamamatsu H8500 Flat Panel PMT [23] is the last generation position sensitive PMT. The major advance results from the housing that compacts 2 inch active area with the narrowest dead boundary (less than 1 mm) and a height of only 12 mm. This tube is based on the same principle of metal channel dynode for charge multiplication with 8×8 anode array for charge collection and position calculation. Taking into account a photocathode glass window thickness of 2 mm, such anode side would be able to correctly sample the light spread produced by a 2 mm pixel size of a scintillation array. In order to clearly identify 1 mm crystal pixel size, the optimum anode side would be 3–4 mm. It would compensate the spatial resolution worsening due to photocathode light guide with a better position linearity. The present anode structure choice was primarily conditioned by practical reasons like the lowest anode number for a 50 mm × 50 mm active area to improve the PSPMT gain minimizing its variation between anodes and to simplify the read out electronic arrangement.

Over the last years the technological efforts were focused on the improvement of the following PSPMT characteristics: a narrow intrinsic spread of charge to minimize the cross-talk between anodes; a very compact housing and narrow boundary dead zones to closely pack PSPMTs with a fraction of active area greater than 80%. Gamma camera with 20 cm diameter is the useful size for medical applications. The first attempt in this direction was the development of a 8 inch PSPMT based on the first generation technology [25].

The first limitation arising from a single large area PSPMT was the large thickness of photocathode glass window (7.5 mm) that produces a further broadening of the light spot on the photocathode and as a consequence, on the intrinsic spread of charge. It was the main limitation in obtaining spatial resolution values better than 2 mm with CsI(Tl) scintillating array irradiated at 140 kev photon energy. In 1997, Pani [26] proposed a novel gamma camera assembly based on 1 inch PSPMT array of second generation. The advantage of this novel assembly was in the fixed overall photocathode glass window thickness (less than 1 mm), compactness and the potential unlimited detection areas with regular or irregular shapes. On the contrary the relevant fraction of non-active area (30%) requires the use of an additional light guide (usually 3 mm thickness) to allow the detection in dead zones between two neighboring PSPMTs and to reduce the light loosing. This partially reduces the high imaging performance of individual PSPMTs.

The best performance of this gamma camera assembly can be obtained only by a NaI(Tl) scintillating array with a pixel size of 1.8 mm, to allow the full crystal pixel identification in dead zone and to limit the worst energy resolution value down to 20%. The Flat panel PMT is designed to solve the limitation arisen from the previous PSPMT generations. In fact, it can be assembled in array with an improved effective area up to 97% and fixing the photocathode glass window thickness down to 2 mm. In Fig. 9.5 the upgrading steps in obtaining large detection areas are schematically shown.

In Table 9.1 we summarized and compared the main characteristics of the three PSPMT generations. It is worthy to note the superior performance of the second generation and that Flat panel PMT could be the best trade-off between compactness, large detection areas, anode size (6 mm), effective area (packing density), spatial resolution and position distortion.

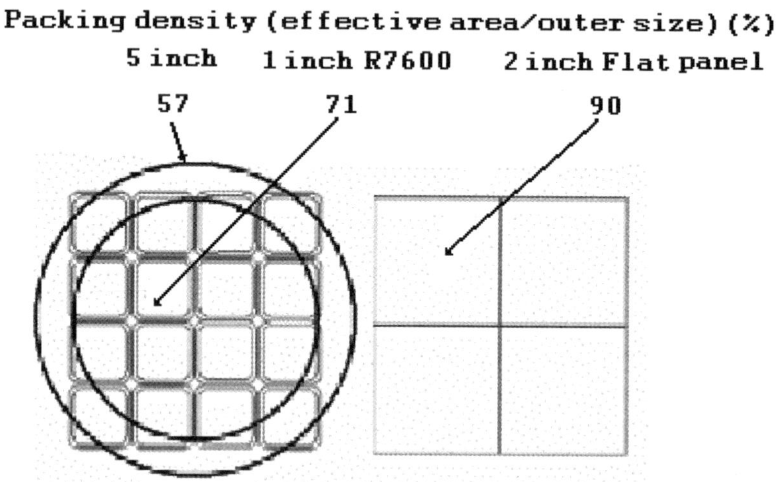

Fig. 9.5 Schematics of the upgrading steps in large detection PSPMT of different generations (from Pani, [24]).

9.3 Signal Read Out Methods and Scintillation Crystals

Two methods are usually employed for reading out the charge collected by PSPMT anodes: independent anode (crossed wire or array) and resistive chain. The first method is based on connecting each anode to an individual independent electronic chain consisting of preamplifier, amplifier and shaper[27, 28, 30]. All anode signals are AD converted and usually stored in an acquisition system that provides data analysis and image processing.

The advantage offered by this method is the best image quality by correcting, distortions due to anode gain variation and light distribution truncation. The main disadvantages are the count rate limitation and high cost arising from the number of electronic channels. In the resistive chain method the anodes are directly connected through a resistive network (see Fig. 9.6) and position signals are read out from the two ends for both the X and Y directions [31].

Table 9.1 : Summary of main characteristics of PSPMT generations (from Pani, [24]).

Parameter	**Flat_Panel**	**3" R2486**	**5" R3292**	**R5900 M64**	**R7600 C8/C12**
Window thickness (mm)	2.8	3.2	6	1.5	0.8
Dynode structure	Metal channel	Proxi mesh	Proxi mesh	Metal channel	Metal channel
Dynode number of stage	12	12	12	12	11
Anode structure	Multi anode	Crossed wire	Crossed wire	Multi anode	Crossed wire
Number of anodes	64 (8×8)	16+16	28+28	64 (8×8)	4+4/6+6
Pixel size/ pitch (mm)	5.6/6.0	3.75	3.75	2/2.25	5/5.5/3.6
Effective area (mm^2)	49×49	50 diam	100 diam	18.1×18.1	22×22
Outer size (mm^2)	51.7×51.7	76 diam	132 diam	26×26	26×26
PMT thickness (mm)	15.5	55	113	20.1	20.1
Intr. charge spread (mm – FWHM)	<1?	7	11	<1	<1
Packing density (%) (effective area/outer size)	90	43	57	48	71.6
Cathode luminous sensitivity (mA/Lm)	80	80	80	70	70
Gain	$3 \cdot 10^6$	10^5	10^5	$3 \cdot 10^5$	$7 \cdot 10^5$
Anode dark current (nA)	96	20	40	12.8	2
Time response Transit time (ns)	5	17	?	5	5
Uniformity among anodes	1:3	1:4	1:4	1:5	1:4

The advantages are a very simple electronic read out (only four channels) and a high count rate. Unfortunately the resistive method introduces noise, usually depending on the number of resistances, producing a worsening of spatial resolution values and of position linearity response in particular for low scintillation signals and for low signal to noise ratio of PSPMT response. An alternative resistive chain method consists of applying the resistive network after having amplified and shaped all anode signals respectively [24, 32]. It strongly improves the image quality.

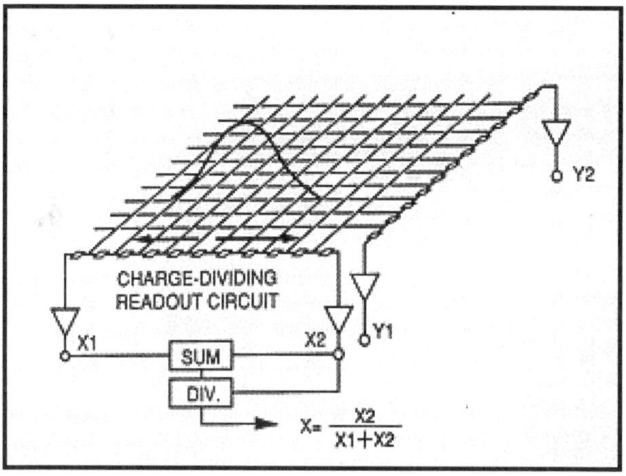

Fig. 9.6 Resistive chain read out (from Hamamatsu, [29]).

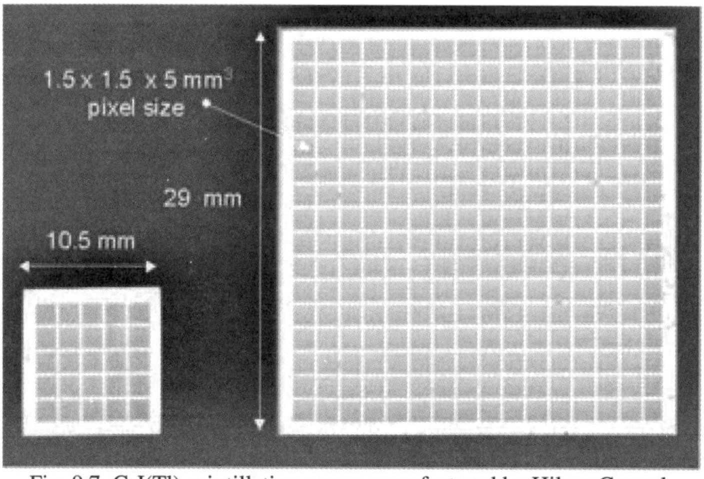

Fig. 9.7 CsI(Tl) scintillation arrays manufactured by Hilger Crystal.

The best image performances are obtained by coupling scintillation arrays to PSPMT. Though they were firstly introduced for PET applications by using standard small PMT, the first application at 140 keV on PSPMT was carried out by Truman et al [33] in 1993, utilizing a multi anode electronic read out. In the same year Pani et al [34] proposed the use of YAP:Ce scintillating arrays with pixel size as small as 0.6 mm and obtaining submillimeter spatial resolution values for SPECT applications. Afterwards a number of manufacturers were involved in manufacturing scintillation arrays with different areas, pixel sizes, thicknesses and scintillation materials. Measurements obtained coupling them to different PSPMT generations are reported elsewhere [35-38, 41, 42]. An example of such arrays is show in Fig. 9.7.

They led to a distribution of scintillation light close to the pixel size. The intrinsic spatial resolution values linearly depend on the pixel size and on the additional thickness of light guide [19, 43]. The spatial resolution is also related to light output as the square root as, in principle, the energy resolution. The array scintillation light output is affected by a number of factors, like light trapping effects produced by the geometrical form factor (pixel side to height ratio), pillar surface treatment and reflective material. Due to the trade off between machining costs and array light output performance, relative energy resolution values are often worse than 20% at 140 keV gamma ray irradiation.

The last advance is represented by the recent development of NaI(Tl) array with pixel size varying between 1 mm and 3 mm. Such production was previously obstructed by machining difficulties related to the hygroscopicity and cleavage plane of the material. Such arrays offer a higher light output allowing one to obtain energy resolution close to the intrinsic ones (10% at 140 keV). Unfortunately they need a housing with a crystal glass window to protect the array from humidity, that further spreads the light distribution on photocathode. Crystal pixel identification (ID) means that the intrinsic spatial resolution of the detector or the error by which the light centroide position is measured, is always better than the crystal pixel side. A good pixel ID allows an accurate correction of the gamma camera response only by a flood field irradiation measurement, for gain uniformity, position distortions as well as for uniformity counting. The spatial resolution of the gamma camera is then related to the crystal pixel size and the FOV usually almost coincident with the PSPMT active area. For a better understanding of the PSPMT technological advances and their consequence on the quality of gamma ray imaging, measurements obtained from Flat Panel PMT are presented. They are also compared with the analogous ones obtained from previous generations of PSPMT. A multi-anode read out electronics are utilized with an acquisition system based on 1.5 MHz National Instruments AT-MIO Analogue to Digital Converter (ADC) mounted on host PC. The irradiation tests were performed by a Co57 radioactive point source.

To compare imaging performances of different generations of PSPMT, NaI(Tl) scintillation array with 48×48 mm^2 area, corresponding to a 1.8×1.8×6 mm^3 pixel size, was coupled to five inch PSPMT (R3292 first generation), to one-inch PSPMT 2x2 assembling (R7600-C12 second generation) and finally to the Flat Panel PSPMT. In Fig. 9.8 are shown the raw images obtained by Co57 flood field irradiation.

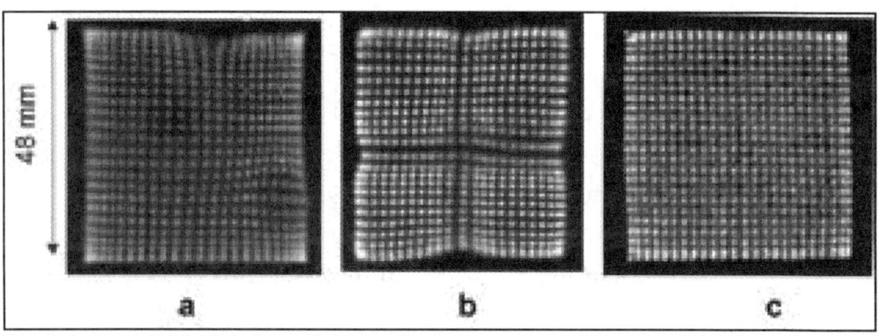

Fig. 9.8 Overall NaI(Tl) flood field irradiation images. (a) Flat Panel PMT, (b) R7600-12 2x2 PMT array, (c) R3292 PMT (from Pani, [39]).

These images clearly show how Flat panel represents a trade off between the two previous generations. In fact five-inch PSPMT image has the worst pixel ID and the best position linearity (due to the single large area photocathode). On the contrary 2x2 assembling (R7600-C12 second generation) shows the best pixel ID but large image distortions arising from the lack of photocathode between PMT neighbors.

The final image reconstruction is based on the crystal pixel identification (ID) on the raw image by which, individuating ROIs related to interaction events in a fixed pixel crystal, it is possible to reconstruct an image corrected for position distortions and anode gain variation. The final reconstructed image, has an intrinsic spatial resolution value corresponding to the scintillator array pixel size.

To understand the process of image formation, the charge distributions generated by a crystal pixel irradiation and collected by Flat panel PMT anodes are shown in Fig. 9.9 for NaI(Tl) and CsI(Tl) array respectively. The NaI(Tl) pixel shows a 8 mm FWHM spread versus 5 mm FWHM spread of the CsI(Tl) pixel. The raw image formation is strongly affected by anode size, charge spread and light output intensity.

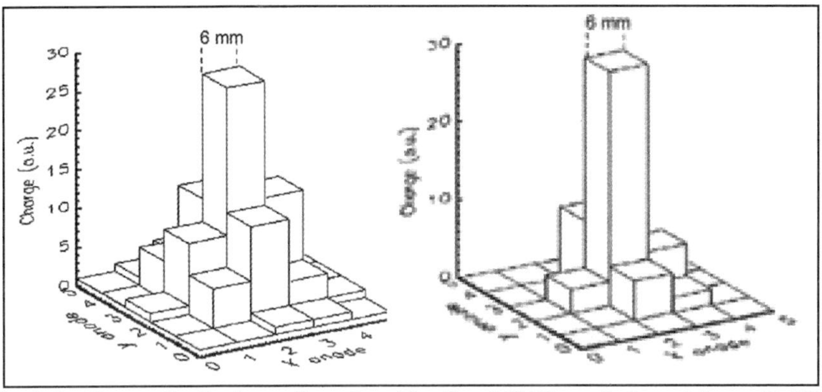

Fig. 9.9 Charge distribution collected on Flat Panel PMT anode from 1.8 mm NaI(Tl) pixel side with 3 mm optical guide (left) and from 3 mm CsI(Tl) pixel side (right) (from Pani, [24]).

Another imaging performance comparison between different PSPMT generation is shown in Fig. 9.10 in which 16×16 CsI(Tl) scintillation array (1.4×1.4 mm^2 pixel size) was coupled to both Flat Panel and R2486 3 inch PSPMT and irradiated by ^{57}Co flood field. Raw images show the superior performances of the last generation PSPMT in term of spatial resolution and crystal pixel identification.

In conclusion Flat Panel shows good image performances, generally better than the previous generation of PSPMT. Considering present technological limits, Flat Panel PSPMT could be the best trade off between gamma camera imaging performances, compactness and large detection areas. Compactness represents the major advance making it attractive for the development of new photodetector applications in the future.

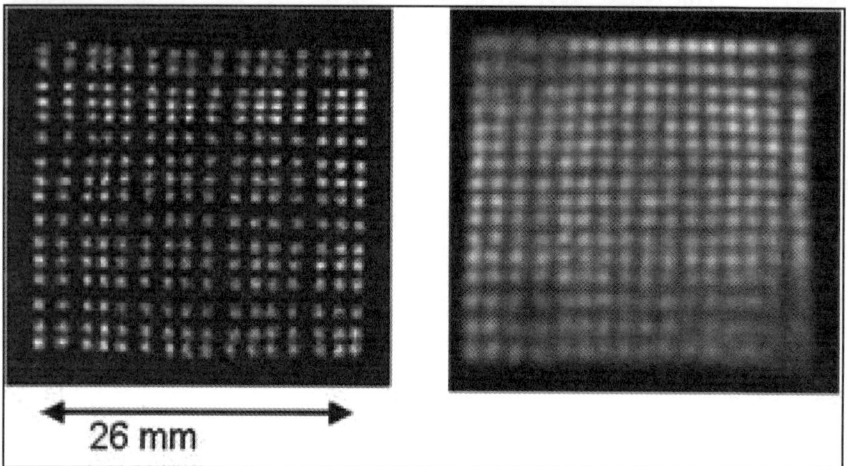

Fig. 9.10 ^{57}Co flood field irradiation raw image of 16×16 CsI(Tl) scintillation array - 1.4×1.4 mm^2 pixel size. Flat Panel PMT (left) and R2486 PSPMT (right) (from Pani, [40]).

9.4 The Role of Compact Imagers in Clinical Application

Diagnosis and treatment of cancer at early stage of development, increases outlook of long term survival. Currently, X-ray mammography represents the principal method of detecting breast cancer. Unfortunately, X-ray mammography is not an ideal examination, because its diagnostic accuracy is below 100% and therefore many patients are undergoing unnecessary biopsies. Some of them can be avoided by exploiting the capabilities of ^{99m}Tc–MIBI Scintimammography (SM) [44, 45]. Breast imaging with radionuclides was being explored as a possible secondary method of screening for cancer. The imaging of breast cancer with radionuclides depends on the increased uptake of a radiopharmaceutical by malignant lesions, when compared with surrounding tissue or benign masses. The present standard scintimammographic technique was introduced for the first time by Khalkhali in 1993 [44]. It is named Prone ScintiMammography (PSM) and consists of positioning the gamma camera in lateral view of the body with the patient in prone position and the breast pendulant. Although very encouraging initial results pointed to the great potential of scintimammography in aiding the diagnosis of breast abnormalities, the detection limit of invasive carcinoma by standard prone technique appears to be ≥ 1 cm diameter. The limited sensitivity for small cancers is mainly due to the inadequate positioning capability of standard scintillation camera, not ideal for breast imaging. In particular the low spatial resolution (3 mm typical resolution limit) and the lack of a close tumor/collimator positioning (see Fig. 9.11) reduce breast lesion contrast in current imaging system. In whole body or brain imaging, the spatial resolution of a typical gamma camera system is limited by the collimator hole size, since major portions of the object are more than 5 cm, and up to 15 cm, from the surface of the collimator. However, as shown in Fig. 9.11, an imager designed specifically for scintimammography can reduce the source-detector distance, increasing the sensitivity of SM to lesions with diameter less than 1 cm (T1a and T1b cancers).

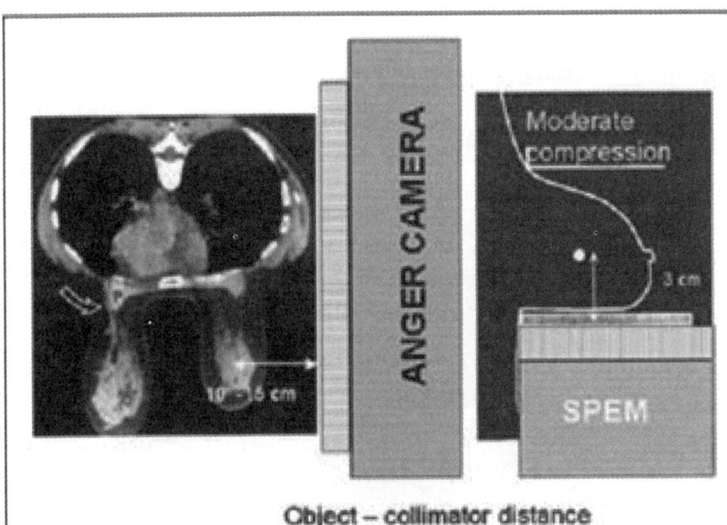

Fig. 9.11 Prone Scintimammography (left) and SPEM (right) techniques (from Garibaldi, [52]).

To this aim Scopinaro and Pani [46-51] demonstrated how a small Field of View (FoV) gamma camera, with very high intrinsic spatial resolution

(1.7 mm), located close to the tumor by breast compression, increased the sensitivity up to 80% in tumors sized ≤ 1 cm [7] without apparently reducing the scintimammography specificity. On the contrary the results obtained in smaller lesions (T1a) remained unsatisfactory. The new small FOV scintigraphic detector was specifically developed by Pani in 1995 [46, 54-58] for a new technique called Single Photon Emission Mammography (SPEM) that allows one to obtain scintigraphies in similar view and technical conditions (as breast compression) of X-Ray mammography.

The small FOV gamma camera consists of an array of CsI(Tl) crystals coupled to 5" Position Sensitive Photomultiplier Tube (PSPMT) Hamamatsu R3292. The crystal was especially studied and designed by Pani et al [48] and developed by Hilger Crystal (Great Britain). It has a circular shape, 11 cm diameter and 3 mm thickness with an epoxy ring of 5 mm. It consists of more than 2300 elements with a pixel size of 2×2 mm^2. Each individual is covered by diffusive white reflector (epoxy) and an overall dead zone of 0.25 mm allows a complete optical insulation between crystal elements. The camera was positioned with the crystals upward. Images were acquired with the patient positioned in front of the camera. The breast, after having been positioned on the collimator, was mildly compressed with a dedicated lead shield, as shown in Fig. 9.12 [54].

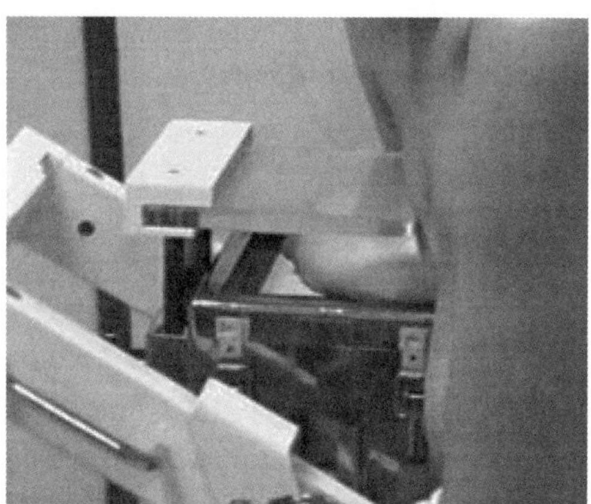

Fig. 9.12 The Single Photon Emission Mammography (SPEM) gamma camera.

In addition to compression, this lead plate also provided the shielding of the camera from radiation arising from the thyroid and upper torso. The spatial resolution of the detector results are better than 2.7 mm at 0 cm Source-Collimator-Distance (SCD) and better than 4 mm between 2 and 3 cm. The intrinsic spatial resolution values ranged between 1.6 and 1.8 mm.

The relative energy resolution of a single spot irradiation was 17%. Final pulse height correction of the flood field spectrum produced a relative energy resolution better than 19% depending on PSPMT selection.

To demonstrate the improvement introduced by SPEM gamma camera, the same patients were imaged by prone SM and SPEM respectively. To this aim 14 patients were enrolled for in vivo measurements, five patients of those with breast tumors sized between 5 and 7 mm, were correctly diagnosed by SPEM camera, in contrast to (2/5) by prone SM. The results have been analyzed, using the histological findings as a gold standard. Breast images and tumor SNR values were then compared. Furthermore some patients were imaged with breast compressed and uncompressed under craniocaudal projection in SPEM obtaining 7.5 ± 0.5 cm and 5.0 ± 0.2 cm mean thickness values respectively. The results of clinical data are shown in Fig. 9.13.

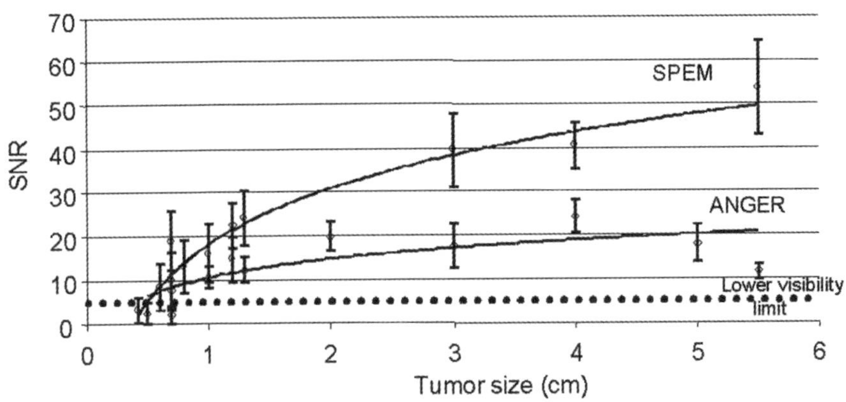

Fig. 9.13 Tumor SNR values obtained from clinical measurements by SPEM camera (craniocaudal projection) and Anger camera (PSM) respectively (from Pani, [53]).

The SNR values obtained by SPEM with the breast compressed in craniocaudal projection are higher than prone SM of about a factor two. In particular tumors less 1 cm sized, imaged by SPEM, show SNR values between 7 and 20, largely distributed over the visibility limit (SNR=5). On the contrary, prone SM tumor SNR values ranged between 3 and 10 confirming the lack of sensitivity for less than one cm in size. Furthermore, compression allows the separation of overlapped objects as it is clearly visible in the images shown in Fig. 9.14 for 7 mm tumor detection. A comparison with PSM image shows the lack of tumor detection of standard gamma camera for the same patient.

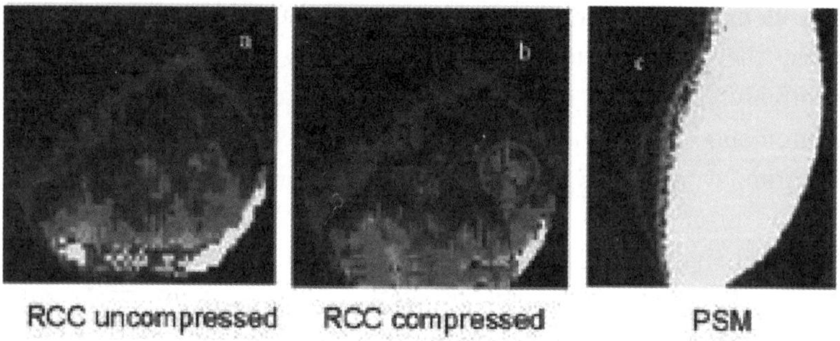

Fig. 9.14 Right breast carcinoma 7 mm sized: SPEM camera ^{99m}Tc Sestamibi scintimammograms with uncompressed breast (a) and mildly compressed (b). Into the circle the lesion is shown as a more enhanced area of increased uptake of tracer. PSM image (c) (from Pani, [53]).

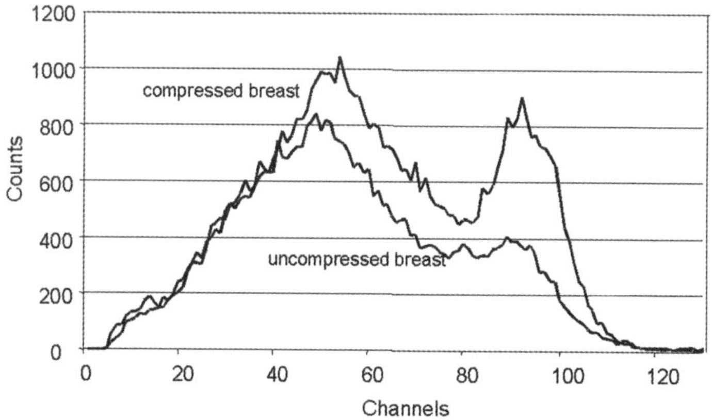

Fig. 9.15 Energy spectrum from the same tumor ROI of compressed and uncompressed breast. Tumor size: 5.5 cm (from Pani, [53]).

A further analysis on the effectiveness of breast compression is shown in Fig. 9.15 where the detected spectra are shown. In this case the large tumor size combined with the large variations in breast thickness (4.5 cm) shows an impressive counts increasing of the signal (full energy peak) with respect to Compton background. The lack of breast full FOV is the main limitation of this gamma camera. To overcome this limitation, Pani recently proposed a novel gamma camera based on that of Anger Camera but substituting conventional PMTs with compact PSPMTs [26]. In this way it is possible to utilize narrow scintillation light distributions and the position detection in the dead space between PMT neighbors. The main advantages offered by the proposed new design are: very low detector thickness (< 3 cm), light weight, small dead boundary zone (3 mm) and no limitation in shape and size of active area. Large compact gamma cameras based on one inch PSPMT up to 7×8 array have been currently developing.

To overcome both the limit of poor X-Ray mammography specificity and the poor scintimammography sensitivity, an Italian group proposed [59, 60] a combined SPECT-CT tomographic imager working in the same geometrical condition. Combined modalities allow both the functional and the morphological image that could be extremely important for increasing the diagnostic information of each modality. The basic idea is to integrate the SPECT technique with a CT scanner by using a high-resolution compact gamma camera and a quasi-monochromatic X-ray system. The tomographic technique used for both CT and SPECT was around the Vertical Axis Of Rotation (VAOR) [61] with the breast in prone position. It seems to be a promising alternative method with respect to traditional SPECT and to planar PSM. SPECT scanner was based on compact gamma ray detector modules located in close proximity to the breast and lodging in the breast inter-space, to work close to the chest wall and to minimize the tumor to collimator front distance. Furthermore the existence of an optimum energy range for X-ray CT has been theoretically demonstrated by determining the energies that give the highest signal-to-noise ratio (SNR) for fixed absorbed dose [62]. The optimum energy depends on object size and composition and on the detector efficiency; hence monochromatic (or quasi-monochromatic) X-ray sources with tuneable energy was used [63,

64]. The schematic diagram of the combined SPECT-CT prototype is shown in Fig. 9.16. The tomographic system consisted of a toroidal ring with a useful diameter of about 13 cm. The VaoR SPECT system consisted of two detector module positioned in opposite position each one consisting of a new compact Position Sensitive Photo Multiplier Tube (PSPMT) Hamamatsu R7600-00-C8 coupled to Hilger CsI(T1) scintillating array. The polychromatic X-ray beam produced by the W-anode X-ray tube was monochromatized via Bragg diffraction using a single highly oriented pyrolytic graphite mosaic crystal. With this assembly, the crystal-to-X-ray tube focus distance was 130 mm. The X-ray detector was based on an array of ultra fast ceramic scintillators seen by photodiodes, manufactured by Siemens for last-generation CT scanners [65].

Fig. 9.16 Photograph of the SPECT-TC scanner. (From Gambaccini, [69])

The optimal energy to detect a tumor of 5 mm was found at 28 keV, and the required statistics was achieved by a tube current of 10 mA, with an exposure of 0.66 s for each angular step, to obtain a scan time of about 1 min per slice. The imaging capability of the full-size prototype were tested with a phantom specially designed and built consisting of two plexi-glass cylinders. The outer cylinder contained a Tecnetiated hydroalcoholic solution whose X-ray attenuation properties are similar to those of breast tissue; the inner cylinder was filled with Tecnetiated water to simulate a tumor. The characteristics of such a liquid phantom enables one to obtain tomograms from both transmission and emission signals. Reconstructed images by the compact modules strongly enhanced the contrast of tumor less 1 cm sized, with respect to planar geometry. The preliminary results shown that single photon compact ring tomography can allow an exact tumor 3D localization if combined with a transmission tomography.

Several techniques of radioguided surgery, including sentinel node biopsy and laparotomic radioguided or radioimmunoguided surgery, start to investigate very compact imaging devices instead of the traditional Anger cameras or single-channel probes that only detect or count radioactivity. When images are needed, nuclear physicians generally use a combined approach: first scintigraphy with a gamma camera, then intraoperative localization and counting with a probe. PSPMT has allowed the reconstruction of a miniaturized high-resolution gamma camera which can produce one -square-inch images; as an intraoperative probe portable and light [66, 67,68]. It actually is small enough to be used as a probe and large enough to give an image. The weight of the entire head is 0.8 kg. It is possible to achieve a scintigraphic image with energy window selection which could be important for scattering rejection in particular working conditions. The sensitivity of the imaging probe is about 1325 cpm/pCi at 140 keV. Its spatial resolution is 3 mm FWHM, with 20% energy window centered on 140 keV. The miniaturized gamma camera called imaging probe (IP) that can be held in only one hand, has been used for a short time on a number of patients to improve the lymphoscintigraphic localization of the sentinel node (SN) in breast cancer. In 5 patients over 53 studied, standard breast lymphoscintigraphy, with ultra-sound or X-ray-guided, peritumoral injection of 0.5 mCi of 99mTC nanocolloids in 0.2-0.5 ml failed to show the SN on gamma camera images that had been acquired in anterior. A common problem using 99mTc-nanocolloids instead of Albures 99mTc, in breast cancer SN localization, is the detection of more than one node rather than the lack of them.. The short diagnostic study suggested a certain degree of specific experience to correctly and usefully work with the intraoperative probe.

Reference

1. R. Amrami, G. Shani, Y. Hefetz, I. Blevis, and A. Pansky: Engineering in Medicine and Biology Society, 2000. Proceedings of the 22nd Annual International Conference of the IEEE , **1**, pp: 352–355 (2000)
2. J. F. Butler, C. L. Lingren, S. J. Friesenhahn, F. P. Doty, W. L. Ashburn, R. L. Conwell, F. L. Augustine, B. Apotovsky, B. Pi, T. Collins, S. Zhao, and C. Isaacson: Nuclear Science, IEEE Transactions on , **45,** (3), Part: 1, June, pp: 359–363 (1998)
3. J. S. Hube, W. W. Moses, S. E. Derenzo, M. H. Ho, M. S. Andreaco, M. J. Paulus, and R. Nutt: IEEE Trans Nucl Sci; **44**, pp: 1197–1201 (1997)
4. C. S. Levin and E. J. Hoffman: IEEE Trans Nucl Sci; **44**, pp: 1208–1213 (1997)
5. C. S. Levin, E. J. Hoffman, M. P. Tornai, and L. R. MacDonald: IEEE Trans Nucl Sci, **44**, pp: 1513–1520 (1997)
6. O. Fries, S. M. Bradbury, J. Gebauer, I. Holl, E. Lorenz, D. Renker, and S. I. Ziegler: Nucl Instr & Meth A, **387**, pp: 220–224 (1997)
7. G. J. Gruber, W. W. Moses, S. E. Derenzo, N. W. Wang, E. Beuville, and M. H. Ho: IEEE Trans Nucl Sci, **45,** pp: 1063–1068 (1998)
8. R. Farrell, K. Shah, K. Vanderpuye, R. Grazioso, R. Myers, and G. Entine: Nucl. Instr. & Meth A, **442**, pp: 171–78 (2000)
9. C. Fiorini, A. Longoni, F. Perotri, C. Labanti, E. Rossi, P. Lechner, and L. Strudel: Nuclear Science, IEEE Transactions on , **48,** (3), Part: 1, June, pp: 346–350 (2001)
10. C. Fiorini, A. Longoni, F. Perotri, C. Labanti, E. Rossi, P. Lechner, H. Soltau, and L. A. Struder: Nuclear Science Symposium Conference Record, 2001 IEEE, **1**, pp: 25–28 (2002)
11. K. Blazek, F. De Notaristefani, P. Maly, R. Pani, R. Pellegrini, A. Pergola, F. Scopinaro, and A. Soluri: IEEE Trans. Nucl. Sci., **425**, pp: 1474–1482 (1995)
12. H. Kume, S. Suzuki, J. Taleuchi et al.: IEEE-Transactions-on-Nuclear-Science, **32**, p.448 (1985)
13. HAMAMATSU Technical Data, Position sensitive photomultiplier tubes with a crossed-wire anode R2486 series, August 1989 Rev., Supersedes October 1987, CR 2000, printed in Japan
14. R. Pani, R. Pellegrini, A. Soluri, G. De Vincentis, R. Scafé and A. Pergola: Nucl. Instr. & Meth. A., **409,** pp: 524–528 (1998)
15. L. H. Barone, K. Blazek, D. Bollini, A. Del Guerra, F. De Notaristefani, G. De Vincentis, et. al: Nucl. Instr. & Meth. A., **360**, pp: 302–306 (1995)
16. F. De Notaristefani, R. Pani, L. M. Barone, K. Blazer, D. Bollini, A. Del Guerra, G. De Vincentis, G. Di Domenico, M. Galli, M. Giganti, T. Malatesta, P. Maly, R. Pellegrini, A. Pergola, A. Piffanelli, F. Scopinaro, A. Soluri, F. Vittori, and G. Zavattini: Proceedings of the International Conference on Inorganic Scintillators and their applications Scint 95. Delft,The Netherlands, pp: 559–566 (1995)

17. HAMAMATSU Technical Data Sheet R3292, tentative data, printed in Japan, October (1988).
18. D. Grigoriev , A. Johnson, W. Worstell, and V. Zavarzin: IEEE Trans Nucl Sci; **44**, pp: 990–993 (1997)
19. R. Pani, R. Scafè, R. Pellegrini, A. Soluri, G. Trotta, L. Indovina, M. N. Cinti, and G. De Vincentis: Nucl Instr & Meth A, **477**, pp: 72–76 (2002)
20. S. Nagai, M. Watanabe, H. Shimoi, H. Liu, Y. Yoshizawa: IEEE NSS/MIC (1998), Conf.Rec., **I**, pp: 199-203 also in IEEE Trans Nucl Sci., **46**, pp: 354–358 (1999)
21. Hamamatsu Technical Data Sheet. Position Sensitive Photomultiplier Tube R5900U-00-C8 Preliminary, printed in Japan, December (1995)
22. R. Pani, A. Soluri, A. Pergola, R. Pellegrini, R. Scafè, G. De Vincentis, and F. Scopinaro: IEEE Trans. on Nuclear Science, **46,** (3), pp: 702–708 (1999)
23. H. Kyushima, H. Shimoi, A. Atsumi, M. Ito, K. Oba, and Y. Yoshizawa: IEEE Nuclear Science Symposium. Conference Record, NJ, USA; **1**, pp: 7/3–7/7 (2000)
24. R. Pani, R. Pellegrini, M. N. Cinti, C. M. Mattioli, C. Trotta, L. Montani, G. Iurlaro, G. Trotta, L. D'Addio, S. Ridolfi, G. De Vincentis, and I. N. Weinberg: Nucl. Instr. & Meth. B., **123**, pp: 197–203 (2004)
25. R. Pani, R. Pellegrini, A. Soluri, G. Trotta, A. Pergola, P. T. Durrant, M. Dallimore, D. Ramsden, R. Scafè, G. Vizzini, and G. De Vincentis: IEEE Nuclear Science Simposium, Conference Record, Albuquerque,New Mexico USA, **2**, November 9-15, pp: 1654–1658 (1997)
26. R. Pani, A. Soluri, R. Scafè, A. Pergola, R. Pellegrini, G. De Vincentis, G. Trotta, and F. Scopinaro: IEEE NSS/MIC; (1997), Conf.Rec., **II**, 1068-1072 also in IEEE Trans Nucl Sci, **46**, pp: 702–708 (1999)
27. S. L. Thomas, P. Seller, and P. H. Sharp: IEEE-Transactions-on-Nuclear-Science, **42**, (4), pp: 830–834 (1995)
28. Rutherford Appleton Laboratories Microelectronics Group HXURAIJSS Technical Data Sheet
29. Hamamatsu photonics Photomultiplier tube Principle to application Hamamatsu photonics KK. March (1994) cat. N° TOTH9001E01. Printed in Japan
30. A.J. Bird and Z. He: NIM A, **348**, pp: 668–672 (1994)
31. S. Siegel, R. W. Silverman, Yiping Shao, S. R. Cherry: IEEE Nuclear Science, **43**, (3), pp: 1634–1641 (1996)
32. R. Pani, A. Pergola, R. Pellegrini, A. Soluri, G. De Vincentis, S. Filippi, G. Di Domenico, A. Del Guerra, and F. Scopinaro: Nucl Instr & Meth A; **392**, pp: 319–323 (1997)
33. A. Truman, A. J. Bird, D. Ramsden, and Z. He: Nuclear Instruments and Methods in Physics Research A, **353**, pp: 375–378 (1994)
34. R. Pani, K. Blazek, F. De Notaristefani, P. Maly, S. Pani, R. Pellegrini, A. Pergola, F. Scopinaro, and A. Soluri: Nucl. Instr. & Meth. A, **348**, pp: 551 (1994)
35. M. Watanabe, T. Omura, H. Kyushima, Y. Hasegawa, and T. Yamashita: IEEE Trans Nucl Sci; **42**, pp: 1090–1094 (1995)

36. S. Avery, C. Keppell, S. Majewski, A. G. Weisenberger, R. Wojcik, and C. Zorn: IEEE NSS/MIC, Conf.Rec., **II**, pp: 1073–1076 (1997)
37. M. Watanabe, H. Okada, K. Shimizu, T. Omura, E. Yoshikawa, T. Kosugi, S. Mori and T. Yamashita: IEEE Trans Nucl Sci., **44**, pp: 1277–1282 (1997)
38. Y. Yoshizawa and J. Takeuchi: Nucl Instr & Meth A; **387**, pp: 33–37 (1997)
39. R. Pani, R. Pellegrini, M. N. Cinti, C. Trotta, G. Trotta, R. Scafè, M. Betti, F. Cusanno, L. Montani, G. Iurlaro, F. Garibaldi, and A. Del Guerra: Nucl. Instr. & Meth. A, **513**, pp: 36–41 (2003)
40. R. Pani, M. N. Cinti, R. Pellegrini, C. Trotta, G. Trotta, L. Montani, S. Ridolfi, F. Garibaldi, R. Scafè, N. Belcari and A. Del Guerra: Nucl. Instr. & Meth. A, **504**, pp: 262–268 (2003)
41. J. J. Vaquero, J. Seidel, S. Siegel, W. R. Gandler, and M. V. Green: IEEE Trans Med Imaging, **17**, pp: 967–978 (1998)
42. R. F. Brem, S. Majewski, J. M. Links, J. M. Schoonjans, A. C. Civelek. RSNA 1999
43. R. Scafè, R. Pellegrini, A. Soluri, N. Burgio, G. Trotta, A. Tatì, M. N. Cinti, G. De Vincentis, and R. Pani: Nucl Instr & Meth A; **497**, pp: 249–254 (2003)
44. I. Khalkali, I. Mena, E. Jouanna, et al: J. Am. Coll. Surg., **78**, pp: 491–497 (1994)
45. F. Scopinaro, O. Schillaci, M. Scarpini, et al: Eur. J. Nucl. Med, **21**,(9), pp: 984–987 (1994)
46. R. Pani, F. De Notaristefani, F. Scopinaro, R. Pellegrini, A. Pergola, A. Soluri, G. De Vincentis, T. Malatesta, F. Vittori. M. Meoni, A. Grammatico and M. Morelli: European Journal of Nuclear Medicine, **22**,(8), pp: 876 (1995)
47. F. Scopinaro, R. Pani, R. Pellegrini, G. De Vincentis, A. Pergola, A. Soluri, F. Iacopi, M. Ierardi, and M. Meoni: Quarterly Journal of Nuclear Medicine, **2**, (1996)
48. F. Scopinaro, G. De Vincentis, R. Pani, R. Pellegrini, A. Soluri, M. Ierardi, R. Danieli, N. S. Tiberio and L. Ballesio: Eur. J. Nucl.Med, (Sept. 1997).
49. G. De Vincentis, F. Scopinaro, R. Pani, R. Pellegrini, A. Soluri, M. Ierardi, L. Ballesio, I. N. Weinberg, and A. Pergola: Anticancer Research, **17** (3B), pp: 1627–1630 (1997)
50. G. De Vincentis, W. Gianni, R. Pani, M. Cacciafesta, R. Pellegrini, A. Soluri, G. Troisi, V. Marigliano, and F. Scopinaro: Breast cancer research and treatment, **48**, pp: 159–163 (1998)
51. F. Scopinaro, G. De Vincentis, R. Pani, W. Gianni, R. Pellegrini, A. Soluri, E. Di Luzio, T. Gigliotti, V. Marigliano, and A.Centi Colella: Annals of Oncology, **9**, (3), pp: 65 (1998)
52. F. Garibaldi, E. Cisbani, F. Cusanno, R. Iommi, G. M. Urciuoli, R. Pani, R. Pellegrini, R. Scafe', et. al.: Nucl. Instr. & Meth. A, **471**, pp: 222–228 (2003)
53. R. Pani, R. Pellegrini, G. De Vincentis, M. N. Cinti, I. N. Weinberg, A. Soluri, A. M. Betti, et. al.: Nucl. Instr. & Meth. A, **497**, pp: 90–97 (2003)

54. F. Scopinaro, R. Pani, G. De Vincentis, A. Soluri, R. Pellegrini and L. M. Porfiri: Europen Journal of Nuclear Medicine, **26**, pp: 1279–1288 (1999)
55. R. Pani, R. Pellegrini, F. Scopinaro, A. Pergola, G. De Vincentis, F. Iacopi, F. De Notaristefani, A. Soluri, A. Grammatico, and A. Del Guerra: IEEE Nuclear Science Simposium, Conference Record, (November 2-9), Anaheim, California USA, **2**, pp: 1170–1174 (1996)
56. R. Pani, F. Scopinaro, R. Pellegrini, A. Soluri, A. Pergola, G. De Vincentis, M. Ierardi, and I. N. Weinberg: Anticancer Research, **17,** (3B), pp: 1651–1654 (May-June 1997)
57. R. Pani, R. Pellegrini, F. Scopinaro, G. De Vincentis, A. Pergola, A. Soluri, A. Corona, S. Filippi, P. L. Ballesio, and A. Grammatico: Nucl. Instr. & Meth. A, **392**, pp. 295–298 (1997)
58. R. Pani, G. De Vincentis, F. Scopinaro, R. Pellegrini, A. Soluri, I. N. Weinberg and A. Pergola: IEEE Trans. on Nuclear Science, **45,** (6), pp: 3127–3133 (1998)
59. M. Gambaccini, A. Fantini, D. Bollini, E. Castelli, A. Del Guerra, G. Di Domenico, R. Pani, R. Pellegrini, N. Sabba, R. Scafe, A. Soluri, A. Taibi, A. Tartari, A. Tuffanelli, and G. Zavattini: IEEE Transactions on Nuclear Science, **48,** (3), Part 2, pp: 703–706 (June 2001)
60. R. Pani, A. Soluri, R. Scafe, R. Pellegrini, G. De Vincentis, M. N. Cinti, M. Betti, R. Inches, G. Garibaldi, F. Cusanno, M. Gambaccini, A. Fantini, A. Taibi, A. Olivo, S. Pani, L. Rigon, D. Bollini, N. Lanconelli, and A. Del Guerra: Nuclear Science Symposium Conference Record- IEEE, **3**, pp: 21/36–21/39 (2000)
61. H. Wang, C. Scarfone, and L. K. Greer: IEEE Trans Nucl.Science, **44,** (3), pp: 1271–1276 (1997)
62. P. Spanne: Phys. Med Biol.,**34** , (6), pp. 679–690 (1989)
63. M. Gambaccini, A. Tuffanelli, A. Taibi, A. Fantini, and A. Del Guerra, in *Proc. SPI*E, **3770**, pp: 174–184 (1999)
64. M. Gambaccini et al.: Phys. Med., **XVI**, (3), pp. 161–164 (July–Sept. 2000)
65. R. Hupke and C. Doubrava: Phys. Med., **XV,** pp: 315–318 (1999)
66. A. Bolognese, M. Ierardi, M. Cardi, I. Muttillo, F. Scopinaro, G. S. Limouris, H. J. Biersack, and H. Bender: MediterraPublishers, Athens, pp: 33–38 (1998)
67. A. Soluri, R. Scafè, F. Scopinaro, V. Casasanta, R. Latini, G. Strangola, R. Pani, G. S. Limouris, H. J. Biersack, and K. C. Kouros, MediterraPublishers, Athens, pp: 67–72 (1999)
68. R. Pani, A. Soluri, R. Scafè, R. Pellegrini, A. Tatì, F. Scopinaro, G. De Vincentis, T. Gigliotti, A. Festinesi, F. Garibaldi, and A. Del Guerra: Nucl. Instr.and Meth. A, **477**, (1-3), pp: 509–513(2002)
69. M. Gambaccini, G. Di Domenico, A. Fantini, N. Sabba, A. Taibi, A. Tartari, A. Tuffanelli, G. Zavattini, C. Baraldi, E. Casnati, R. Pani et al.: Physica Medica, **XVII**, (4), pp: 249–252 (October-December 2001)

CHAPTER 10

SMALL ANIMAL SCANNERS

Guido Zavattini

Department of Physics, University of Ferrara, and INFN, Sezione di Ferrara, Ferrara, Italy

Alberto Del Guerra

Department of Physics, University of Pisa, and INFN, Sezione di Pisa, Pisa, Italy

10.1 Introduction

The development of dedicated tomographic scanners for experimental animal studies is a field in great expansion. The potential of PET and SPECT imaging in small animals [1-11] has led to the design of several specialized small animal imaging systems [12-35]. Some of these will be described in detail below, emphasizing the differences in their design and reasons of the choices made. Both PET and SPECT provide non-invasive methods to perform in-vivo radiopharmaceutical studies, molecular imaging, gene therapy and to evaluate tumor therapy results. Due to the reduced dimensions of small animals, sensitivity and spatial resolution must be very good. Typically the organs under study are small, the animal's metabolic rate is high and the activity used is relatively low.

Although sensitivity and spatial resolution in PET is generally better than in SPECT, PET has the drawback of needing a cyclotron nearby or ideally in the same research centre for producing the necessary short lived radioisotopes. This is particularly true for studies with ^{11}C and ^{15}O whose half lives are respectively 20 and 2 minutes. On the other hand the radioisotopes necessary for SPECT are easier to obtain. Both techniques are therefore interesting and the design of such small animal scanners will be discussed in the following sections.

High resolution position sensitive detectors will first be briefly overviewed to have a better understanding of the choices made by the various research groups.

10.2 Position Sensitive Detectors

10.2.1 *Gamma-Ray Detection*

Firstly let us review the most important aspects of scintillator detectors. Although some existing scanners have adopted other detector techniques, scintillator cameras are the more widely used detectors.

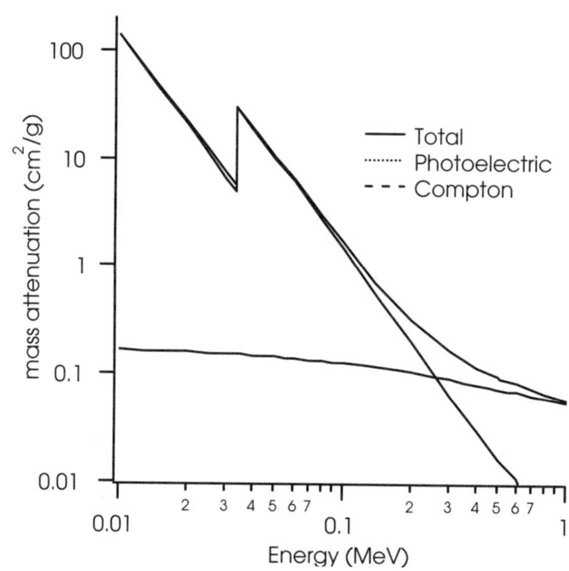

Fig. 10.1. Gamma ray mass attenuation coefficient in NaI(Tl) as a function of gamma ray energy. The density of NaI(Tl) is 3.76 g/cm^2. Data are taken from EGSnrc simulation package.

The dominant interactions of a gamma ray with matter, at the energies of interest (100 – 600 keV), are the photoelectric and Compton interactions. As an example, in Fig. 10.1 we plot the photoelectric, Compton and total mass attenuation coefficients for NaI(Tl) scintillator, in cm^2/g, as a function of energy. The dashed line represents the Compton mass attenuation, the dotted line represents the photoelectric mass attenuation and the continuous line represents their sum.

From Fig. 10.1 one can see how the Compton interaction does not depend strongly on the energy of the gamma ray whereas the photoelectric interaction decreases rapidly with increasing energy of the gamma ray. It must also be noted that there is a discontinuity in the photoelectric mass attenuation coefficient. In this example the discontinuity falls at 33.169 keV, which corresponds to the K-shell energy for iodine present in the scintillator material. The higher the atomic number Z of the elements composing the scintillator the higher will be the energy at which the photoelectric mass attenuation will have a discontinuity.

In the example, the energy dependence shows that at energies below 250 keV the photoelectric effect dominates whereas at energies above about 250 keV Compton scattering dominates. Away from the K-edge discontinuity, the photoelectric probability, as a function of energy, can be approximated by the expression

$$P(E) \propto E^{-5/2} \tag{10.1}$$

For a given gamma ray energy, the photoelectric interaction also depends strongly on the atomic number Z of the material and can be approximated with the expression

$$P(Z) \propto Z^n \tag{10.2}$$

where n is a number between 3 and 4. To enhance the photoelectric probability, high Z materials must therefore be used.

After a photoelectric interaction the atom will have a vacancy in an electron shell normally complete and must reorganize itself. It will do so by either emitting characteristic X-rays, corresponding to atomic level transitions, or by the emission of an electron (called Auger electron) due to the direct transfer of the atomic excitation energy to one of the outer electrons. In high Z materials the X-ray emission is dominant as can be seen in Fig. 10.2. These X-rays will have an energy value just below the K-edge where the linear mass attenuation is at a local minimum and will therefore travel relatively far compared to X-rays just above the K-edge discontinuity. In BGO scintillator (see Table 10.1) these X-rays have energies of 76 keV and 88 keV and their gamma mean free path in BGO is about 1 mm.

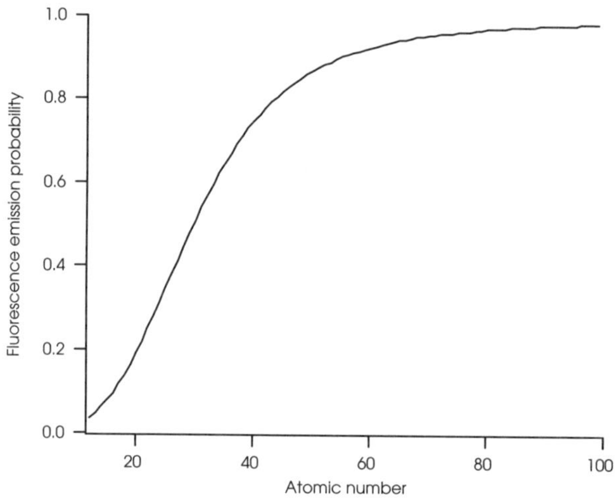

Fig. 10.2. Probability to get fluorescent photon following a K-shell vacancy due to a photoelectric interaction as a function of atomic number, Z. Data are taken from EGSnrc simulation package.

The Compton interaction, on the other hand, depends principally on the electron density n_e in the material, which can be expressed as

$$n_e = \rho\left(\frac{Z}{A}\right)\mathrm{N}_A \tag{10.3}$$

where ρ is the scintillator mass density, N_A is Avogadro's number and Z/A is the ratio of the atomic number and the atomic mass, which is near 0.5 for all materials. The important physical parameter for Compton interaction is therefore the mass density.

In general, the higher the Z of the material, the higher is the energy at which the Compton interaction probability crosses the photoelectric interaction probability.

In Fig. 10.3 are shown the Compton and photoelectric fractions as a function of the atomic number for two energies of interest: 140.5 keV and 511 keV.

As can be seen, above Z ≈ 30, more than 50 % of the 140.5 keV gamma rays will undergo a photoelectric interaction whereas one needs to go as high as Z ≈ 80 to find the cross over between photoelectric and Compton interaction at 511 keV. From these graphs it is clear that detecting the 511 keV gamma rays is somewhat more troublesome than detecting lower energy gamma rays.

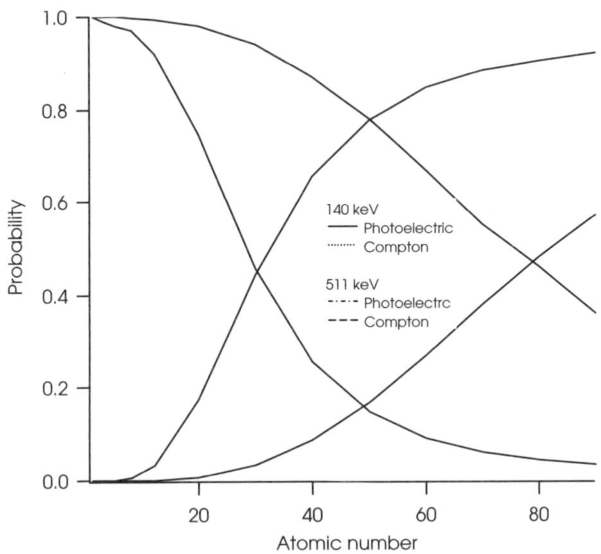

Fig. 10.3. Photoelectric and Compton scattering fractions as a function of atomic number for two energies of interest: 140.5 keV and 511 keV. Data are taken from EGSnrc simulation package.

In high spatial resolution imaging systems the incoming gamma must be absorbed in a very small region. Ideally, there should only be a single interaction within the detector. Two strategies can be adopted for this: use high Z materials (usually also rather dense) to increase the photoelectric probability and confine scattered gamma rays; use high density, low Z materials to take advantage of a single Compton interaction within the detector.

Once gamma rays have participated in any interaction, two fundamental quantities which must be considered for gamma ray imaging are the number of scintillation photons detected per MeV of energy deposited in the scintillator and the scintillation decay time.

In gamma spectroscopy the amount of light generated in the scintillation process is proportional to the energy E deposited by the ionizing electron (this is not usually true for highly ionizing particles such as protons and alpha particles) and, when characterizing a scintillator, is referred to as the photon yield. The amount of light generated is directly related to important measurable quantities such as the energy resolution, spatial resolution and timing resolution of a detector.

Since Poisson statistics governs most scintillator materials' light emission [42], the relative energy resolution can be expressed by

$$\frac{\Delta E}{E} \propto \frac{1}{\sqrt{N_{p.e.}(E)}} \tag{10.4}$$

where $N_{p.e.}(E)$ is the total number of primary photoelectrons generated at the photocathode of the light readout system, for a given energy deposit E.

Usually in the measurement of the position of an interaction, the centroid of the light distribution is at the basis for the determination of the position of any interaction, and since the light distribution will be populated by a finite number of scintillation photons, the spatial resolution will statistically fluctuate. From the central limit theorem [42], the error on the centroid's coordinates depends inversely on the square root of the total number of primary photoelectrons $N_{p.e.}(E)$, generated at the photocathode and depends linearly on the size a of the light distribution reaching the photocathode itself. This can be written as:

$$\Delta x \propto \frac{a}{\sqrt{N_{p.e.}(E)}} \tag{10.5}$$

In Fig. 10.4 is shown the spatial resolution of a position sensitive photomultiplier as a function of the number of incident photons in a given small area, showing clearly the statistical dependency of the spatial resolution which results in a straight line with slope 1/2 on a log-log graph.

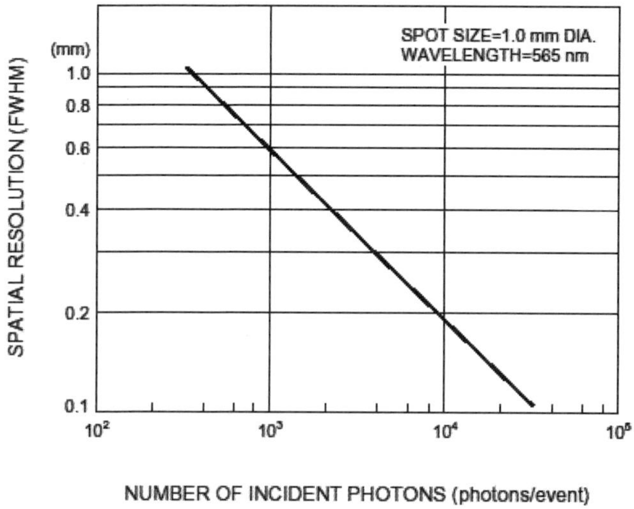

Fig. 10.4. Spatial resolution of Hamamatsu R2486 position sensitive photomultiplier [43] as a function of the number of incident photons on a 1 mm diameter surface. The $1/\sqrt{N}$ dependency is clearly shown by the slope of the curve on a log-log graph. Graph taken from Hamamatsu R2486 position sensitive photomultiplier tube data sheet. (Reprint authorized by Hamamatsu Photonics, Japan.)

Lastly, the ultimate timing resolution which can be obtained with a scintillator depends both on its light yield and on its decay time. By triggering on the first photoelectron the smallest timing uncertainty which can be achieved with a scintillator is given by the expression [46]

$$\Delta t = \frac{\tau}{N_{p.e.}(E)} \tag{10.6}$$

where τ is the decay constant of the scintillator and $N_{p.e.}(E)$ is the total number of photoelectrons generated at the photocathode by the light readout system for a given energy deposit E. This expression derives from the time fluctuation between the interaction instant t_0 and the emission of the first photoelectron.

Summarizing, the characteristics for an imaging scintillator detector should be:

- High detection efficiency.
- High fraction of single interaction.
- High light output for good energy resolution.
- Fast scintillation decay time and high light output for fast timing and event throughput (if needed).

Table 10.1 lists some commonly used inorganic scintillators in PET and SPECT. The photoelectric fractions for 140.5 keV and 511 keV photons are also reported. As can be seen, the scintillator with the highest photoelectric fraction is BGO. This is due to its bismuth content with an atomic number Z of 83 resulting in an effective Z of 75. Even in this case, though, the photoelectric fraction is only 44%. The rest of the interactions are Compton scattering. Unfortunately this scintillator does not produce a large quantity of light, secondly its decay time is rather long and thirdly the high index of refraction makes it difficult to efficiently collect the light onto a readout system. These characteristics limit the energy resolution, the spatial resolution and the timing performances of a BGO system. At the opposite extreme of the atomic number scale is YAP:Ce with an effective Z of 32. At 511 keV only 4.4% of incident gammas will undergo a photoelectric interaction.

Table 10.1. Physical properties of some scintillator commonly used in PET and SPECT

Material	NaI(Tl	BGO	LSO	CsI(Tl)	YAP:Ce
Density g/cm^3	3.76	7.13	7.4	4.51	5.37
Atomic numbers	11,53	83,32,8	71,32,8	55,53	39,13,8
Photoelectric fraction					
@ 140.5 keV	84%	94%	92%	87%	46%
@ 511 keV	18%	44%	34%	22%	4.4%
Light yield %NaI(Tl)	100	15	75	45	55
Decay time (ns)	230	300	40	1000	27
Peak emission wavelength (nm)	410	480	480	565	370
Index of refraction	1.85	2.15	1.82	1.80	1.95
Comments	Hygroscopic	low light and slow	intrinsic background 400 cps/cm^3	high light but slow for PET	medium Z scintillator

*The light yield for NaI(Tl) is 38000 photons/MeV

10.2.2 *Scintillator Based Position Sensitive Detectors*

Ideally, for the correct position determination of an event, a gamma ray must interact only once within the detector. As discussed above, at 140.5 keV the situation is quite favorable for the photoelectric interaction but at 511 keV even BGO scintillator will not satisfy this condition. The scattered photons will then travel a certain distance, interact a second time, and so on. Furthermore K-shell fluorescences will almost always be emitted in high Z materials. These X-rays will also contribute to the deconfinement of the deposited energy. The different interactions, distributed within the detector, will each generate light, which will then be collected by the readout.

The readout system usually will not distinguish between the different interactions and the reconstructed position will be the weighted average of the position of each single interaction where the weight of each interaction is given by the deposited energy E_i.

$$X = \frac{\Sigma E_i x_i}{\Sigma E_i}$$
$$Y = \frac{\Sigma E_i y_i}{\Sigma E_i} \tag{10.7}$$

These coordinates will generally differ from the first interaction coordinate because of scattering and photoelectron statistics.

If the dimensions of the light spot incident on the photocathode has standard deviations σ_x and σ_y, along two perpendicular axes, and the finite number of photoelectrons generated at the photocathode is $N_{p.e.}(E)$, the statistical indetermination in the reconstructed coordinates can be approximated by

$$\Delta X_{FWHM} = 2.35\frac{\sigma_x}{\sqrt{N_{p.e.}(E)}}$$
$$\Delta Y_{FWHM} = 2.35\frac{\sigma_y}{\sqrt{N_{p.e.}(E)}} \tag{10.8}$$

From these expressions it is clear that the more concentrated is the light spot and the higher the number of photoelectrons collected, the better will be the coordinate determination.

To determine the coordinates of a gamma ray interaction using a scintillator detector there are two commonly used crystal configurations: a continuous scintillator and a crystal matrix.

10.2.2.1 *Continuous scintillators*

The first configuration is to use a thin continuous large area crystal. The thickness of the crystal will determine how much the light spot will expand before reaching the window of the light readout system. Since the index of refraction of inorganic scintillators is usually higher than the index of refraction of the light readout entrance window, only the light emitted within a cone whose semi-aperture is given by the total internal reflection angle will be collected. If the backside of the crystal is made reflective, then a second cone with the same angular aperture will also reach the readout window.

A schematic drawing showing such a situation is shown in Fig. 10.5. Considering that the window has the lower refractive index, the total internal reflection angle $\vartheta_{t.i.r}$ will be given by

$$\vartheta_{t.i.r.} = \arcsin\left(\frac{n_{Scin}}{n_{window}}\right) \tag{10.9}$$

where n_{Scin} is the index of refraction of the scintillator and n_{window} is the index of refraction of the window material (we are assuming that the refractive index of the optical coupling grease is greater than n_{window}). The rest of the light produced in the scintillation process will remain trapped inside the scintillator and eventually be absorbed. The average diameter of the light cone entering the photomultiplier window is therefore a function of the thickness of the scintillator and of the depth of interaction. An estimate of this diameter is

$$\Phi = 2d\tan\vartheta_{t.i.r} \tag{10.10}$$

where d is the thickness of the scintillator.

In PET detectors the thickness of the crystal scintillator must be at least a centimeter (in the case of BGO) to maintain a reasonably high efficiency and therefore the light cone reaching the photomultiplier window will also be of that order. In SPECT, though, thinner crystals, of the order of 1–2 mm, can be used allowing good spatial resolution.

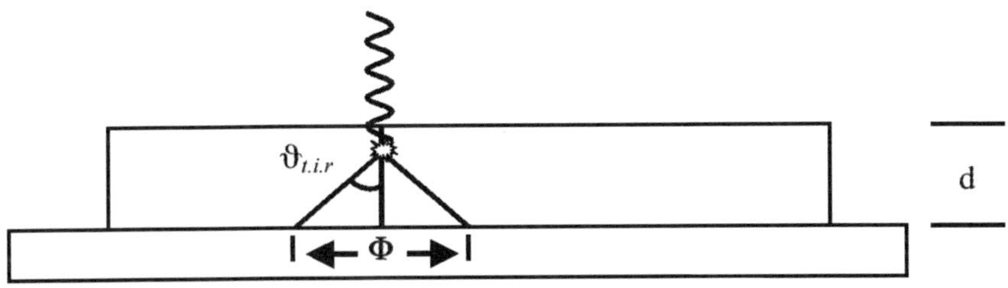

Fig. 10.5. Schematic drawing showing how the thickness of a scintillator will affect the light spot reaching the photocathode window.

A difficulty in using this kind on position sensitive detector is its spatial calibration. All readout systems introduce image spatial distortions and must be corrected for. In addition to this, the response of both the crystal and the photocathode are never uniform and must also be calibrated. An external source must therefore be placed in front of the detector in various accurately known positions and the response determined point by point. This process is long and tedious and must be regularly performed for quality control.

10.2.2.2 *Matrix crystals*

The second configuration is to use a crystal matrix composed of many small section, optically isolated crystals. In this case each crystal element acts as a light guide and keeps the light concentrated within a small surface area. There is a limit to the aspect ratio which can be used due to the large number of reflections which the light must undergo to reach the readout: eventually all the light will be lost. A schematic drawing of the light propagation and a photograph of a 6×6 cm^2 crystal matrix are shown Fig. 10.6.

One of the properties of crystals with a high aspect ratio (small section respect to the length) is that an interaction within a single matrix element will produce light that will reach the photocathode with a uniform distribution over its section. The light spot reaching the readout system will therefore have the dimensions of the single pixel. Pixels with sections as small as 1×1 mm^2 can now be obtained, to be compared with the light spot obtained with a continuous crystal. In SPECT the systems are comparable but in PET it is almost a necessity to use pixellated crystals.

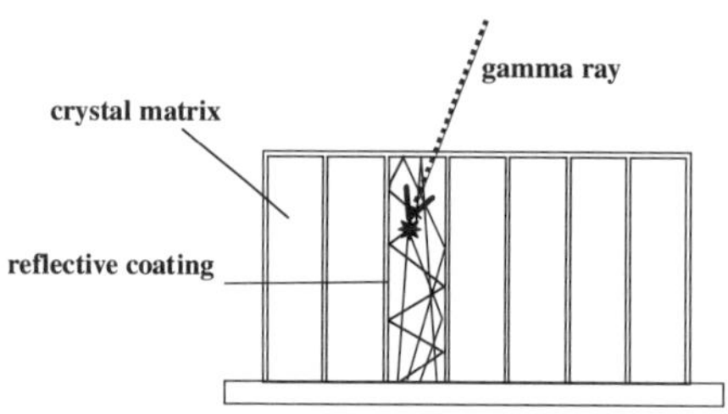

Fig. 10.6. Schematic drawing showing the light confinement in a matrix element and a photograph of a 6×6 cm^2, 3 cm thick YAP:Ce matrix with a 2 mm pixel size.

Since the light distribution reaching the readout window is uniform over the section of an element and is independent of the interaction position within the element, the image as seen by centroid based light readout system is a matrix of discrete spots. The dimensions of the spots in the image are governed by expression (10.8) and determine the spatial resolution of the photomultiplier, *not* the spatial resolution of the detector block. Their dimensions depend on the amount of light reaching the photocathode window and the dimensions of the light spot.

This is a great advantage in calibrating the system both for spatial distortions and for energy response. In fact any source illuminating the entire matrix will allow spatial distortion corrections, energy calibration of each element and efficiency determination. An example of the image generated by a 8×8 crystal matrix of CsI(Tl) with a ^{57}Co source is shown in Fig. 10.7. The pixel pitch is 2.5 mm and the thickness of the matrix is 5 mm.

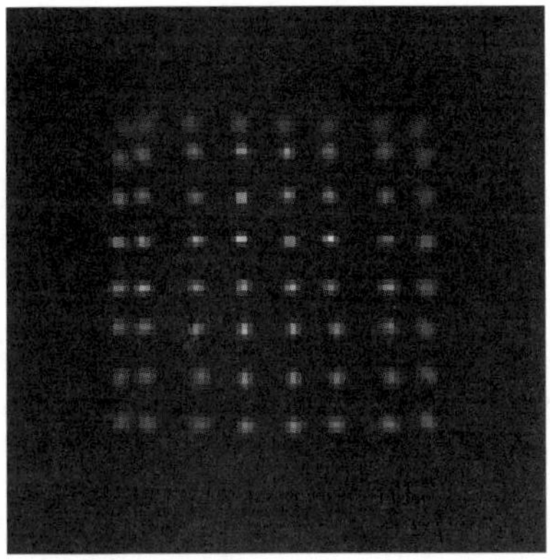

Fig. 10.7. Image of a CsI(Tl) crystal matrix as seen by a position sensitive photomultiplier. The pitch of the matrix is 2.5 mm.

10.3 Single Photon Emission Computerized Tomography (SPECT)

In single photon emission computerized tomography, the radiotracer emits a single gamma ray isotropically. When a gamma ray is detected, one has no information about its direction or line of flight. To determine the line of flight of the detected photons the simplest technique is to place a collimator in front of the detector. Different kinds of collimators can be designed based on the geometry of the holes: pinhole, parallel-holes, converging or diverging holes.

Not many dedicated small animal SPECT scanners have been developed as yet. At the moment there is greater attention towards the development of small animal PET scanners maybe because of its potentially higher sensitivity and generally good spatial resolution. Pinhole SPECT scanners have actually shown very high spatial resolution capabilities, even below 1 mm. Due to their very small useful field of view, though, the use of these scanners is usually limited to imaging mice.

Technically SPECT scanners are somewhat simpler than PET scanners because of several reasons:

- The lower energy of the gamma rays to be detected, with respect to PET;
- The simpler electronics because of the absence of coincidences;

- Low acquisition rate.

In small animal imaging the spatial resolution must be very good to be able to distinguish small details. As we will see below, the spatial resolution depends on the solid angle subtended by the collimator holes. Since the radioisotopes used in SPECT emit radiation isotropically a problem emerges: a high spatial resolution implies a small solid angle and therefore a small fraction of the emitted radiation will be selected through the collimator. For very high spatial resolution collimators the efficiency can be as low as a few parts in 10^{-5}.

The design of a collimator depends on the energy of the radiation to be detected. One of the most commonly used radioisotopes for SPECT imaging is ^{99m}Tc (Technetium 99m), which emits principally a single 140.5 keV gamma ray isotropically. At this energy the absorption length in lead, for example, is about 0.3 mm allowing the production of collimators with thin septa and small holes. A figure of the linear attenuation length of gamma rays in lead as a function of their energy is shown in Fig. 10.8. The curves for photoelectric, Compton and total mass attenuation coefficient, in g/cm^2 are reported. As one can see at 140.5 keV the dominant interaction in lead is the photoelectric effect.

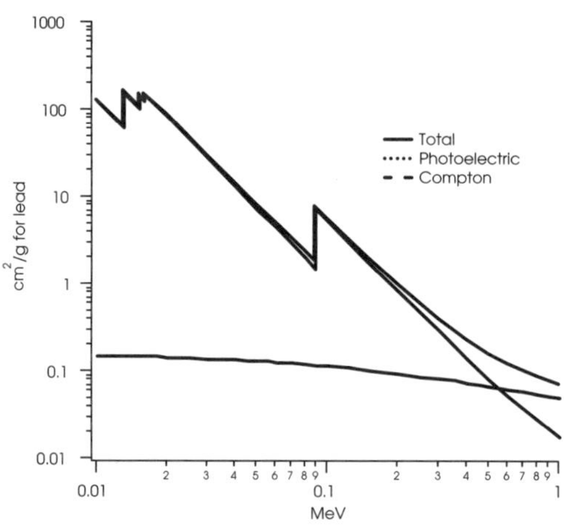

Fig. 10.8. Mass attenuation coefficients for lead, in cm^2/g. The density of lead is 11.34 g/cm^3. Data are taken from EGSnrc simulation package.

As discussed below, a relatively large choice of detector materials can be found for detecting 140.5 keV gamma rays because of the high photoelectric probability for materials with atomic number above about 30. The choice of the detector material can be made depending on which parameter needs to be optimized: in this case high intrinsic spatial resolution.

Other radioisotopes are also used in SPECT such as ^{201}Tl and ^{125}I, which emit low energy gamma rays and ^{131}I which emits higher energy gamma rays. At higher energies the collimator design becomes more difficult due to the increased penetration of the gamma rays.

10.3.1 *The Detector*

In Fig. 10.3, the photoelectric probability for 140.5 keV gamma rays is reported as a function of the atomic number of the absorbing material. The photoelectric effect is dominant from atomic numbers above about 30 giving a large choice for the materials to be used as the detector. Those reported in Table 10.1, in fact, all have high photofractions at this energy. Most events will therefore be single photoelectric interactions allowing a good position and energy determination.

10.3.1.1 *Intrinsic spatial resolution in SPECT*

When defining the spatial resolution of a gamma detector one must be careful. This is especially true for pixellated detectors where a small ideally collimated source will not generate a continuous coordinate distribution but discrete coordinate values corresponding to the centers of the pixels (Fig. 10.7).

It is clear therefore that even if a profile of Fig. 10.7 would give very narrow peaks, two sources, which generate events in the same pixel, will not be distinguishable.

We have already seen that in tomography both the position of an interaction and the direction of the incoming photon must be known. Let us therefore fix a coordinate on a detector and let us take a direction perpendicular to the detector surface. In a pixellated detector the coordinates will correspond to the centers of each pixel and the bin size will be the pitch of the matrix, whereas in a continuous detector any coordinate can be chosen with a bin size equal to the point spread function.

Assuming we have a large number of events and an ideally collimated source, which emits photons only in the chosen direction, we can measure the number of events, occupying the chosen coordinate and direction, as a function of the source position in the detector plane. This will generate a distribution of events with a certain width. We will call the full width at half maximum (FWHM) of this distribution the intrinsic spatial resolution of the detector. In a continuous detector this will correspond to the point spread function. For a pixellated detector this distribution should ideally be a rectangle whose width is the pixel size but in practice there will generally be tails in the distribution extending beyond the pixel size. Therefore in a pixellated detector, given that the pixels are distinguishable, the intrinsic spatial resolution will be determined principally by the pitch of the matrix elements. With a continuous crystal the limiting parameters are the thickness of the crystal and the total number of photoelectrons generated by the light spot.

Spatial resolution is not the only parameter to be taken into account in the choice of a scintillator material for SPECT. Energy resolution and scintillator decay must also be considered.

10.3.1.2 *Energy resolution*

Due to the low acceptance of SPECT collimators, background events generated by the source must be reduced by at least a factor equal to the collimator efficiency. Screening of the detector head is therefore of fundamental importance.

Some photons, though, emitted by the source and scattered by the environment, will enter through the collimator holes. Moreover the number of these photons will depend on the geometry of the whole system, including the source/animal configuration. In the case of ^{99m}Tc the energy of these scattered gamma rays extends to almost 140.5 keV. To distinguish these scattered photons from the original 140.5 keV gammas, energy information can be used both for event selection and for scatter correction. Good energy resolution is therefore desirable. In

Fig. 10.9 one can see the spectra obtained with a YAP:Ce matrix and a capillary filled with ^{99m}Tc.

In Fig. 10.9a the capillary is placed in air whereas in Fig. 10.9b the capillary is at the center of a 4 cm diameter lucite cylinder. The effect of the scattered photons due to the phantom can clearly be seen. The dotted curve in Fig. 10.9b is the difference between the spectra obtained in the configuration with and without the lucite cylinder. An important result is that the number of photons which are scattered within the acceptance angle of the collimator depends on the geometry of the source emitting the radiation.

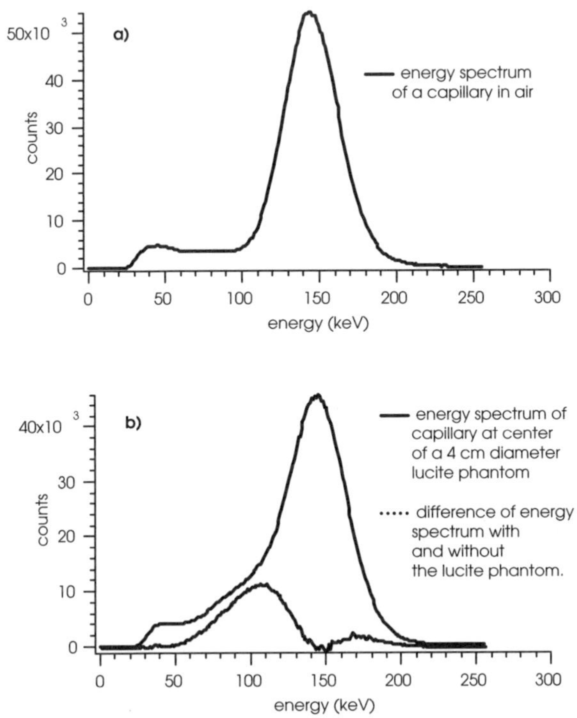

Fig. 10.9. (a) Energy spectrum obtained with a capillary filled with ^{99m}Tc placed in air. (b) Energy spectrum (black line) obtained with the same capillary but placed at the center of a 4 cm diameter lucite cylinder. The dotted curve represents the difference of the spectrum in (a) and (b) showing the contribution of scattered photons.

From these graphs it is clear how efficiency correction and scatter correction become difficult and depend on energy threshold and geometrical setup. A very simple, but successful technique, to eliminate scattered events is to select only those events above 140.5 keV, where the contribution due to scatter becomes negligible.

10.3.1.3 *Rate of acquisition and detector speed*

Typically the activities administered to small animals are about 100MBq. With an efficiency of about $10^{-4} - 10^{-5}$ this leads to a maximum count rate of about $10^{3} - 10^{4}$ events per second. Count rate problems are usually limited and low cost, relatively high Z, slow scintillators like CsI(Tl), which has a high light output for good energy resolution, can be used very successfully.

10.3.2 *Collimator Geometries*

Several kinds of collimator designs can be found: pinhole, parallel-hole, converging and diverging holes, fan beam and more. Here we will only discuss briefly the principle geometries used in small animal imaging: pinhole and parallel-hole collimators. An example of a small animal scanner developed using a converging collimator will be described in Section 10.3.3.3.

10.3.2.1 *Pinhole collimator*

At the moment the more widely used collimator geometry for small animal SPECT imaging is the pinhole design [30–35]. Although this design allows high spatial resolution, basically equal to the diameter of the pinhole, the sensitivity depends strongly on the distance from the collimator and falls off very rapidly with increasing distance as can be seen in Fig. 10.10. In fact the fraction of isotropically emitted gamma rays from a source, which pass through the pinhole, is given by the fraction of the solid angle of the pinhole, as seen from the source.

If d is the diameter of the hole, b is the distance of the object from the pinhole and ϑ is the angle shown in Fig. 10.11, the geometrical sensitivity is given by [44]

$$g_{pinhole} = \left(\frac{d}{4b}\right)^2 \sin^3 \vartheta \tag{10.11}$$

In Fig. 10.10 one can see the sensitivity plotted as a function of the distance from the collimator hole for an on axis point source.

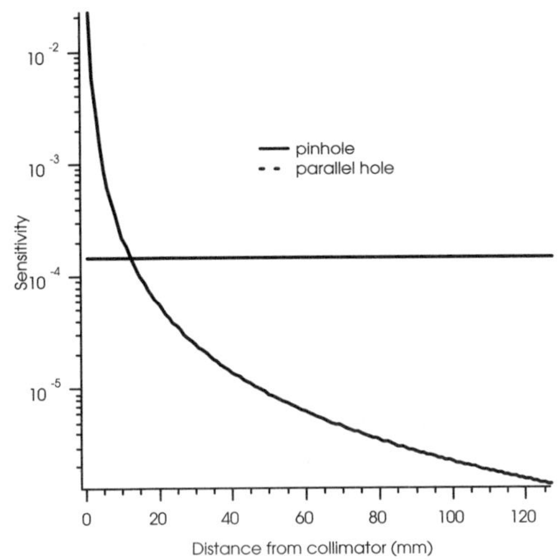

Fig. 10.10. Calculated sensitivity of a pinhole collimator compared to the sensitivity of a parallel-hole collimator as a function of distance. Both collimators have the same intrinsic spatial resolution of 1 mm.

The field of view of a pinhole collimator is defined by the cone projected through the pinhole onto the detector plane. It is clear that where the sensitivity is higher the field of view is smaller. This kind of design is therefore suitable for very small animals such as mice.

The spatial resolution R_p of a pinhole is given by

$$R_p = \frac{d_e(a+b)}{a} \tag{10.12}$$

where a is the detector pinhole distance, b is the object pinhole distance and d_e is the effective pinhole diameter. The effective pinhole diameter is greater than the geometrical diameter because the incident radiation will penetrate through the edge of the pinhole. An estimate of d_e can be obtained by the expression

$$d_e \approx \sqrt{d\left(d + \frac{2}{\mu}\tan\left(\frac{\alpha}{2}\right)\right)} \tag{10.13}$$

where d is the geometrical pinhole diameter, μ is the linear attenuation coefficient of the pinhole material and α is the angle of the pinhole as shown in Fig. 10.11.

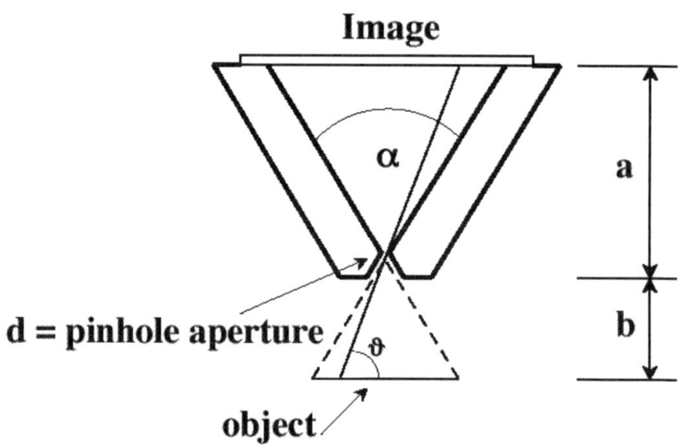

Fig. 10.11. Scheme of a pinhole collimator.

If the detector has a spatial resolution R_d the system will have a spatial resolution R_S

$$R_S = \sqrt{R_p^{\ 2} + \left(R_d \frac{b}{a}\right)^2} = \sqrt{\left(\frac{d_e(a+b)}{a}\right)^2 + \left(R_d \frac{b}{a}\right)^2} \tag{10.14}$$

where the factor $\frac{b}{a}$ multiplying R_d accounts for the demagnification of the detector resolution intrinsic in pinhole systems.

It must be noted that an interesting characteristic of pinhole imaging is that, according to where the detector is placed behind the pinhole, there can be an enlargement factor of the projected image. Given a large enough detector, one can obtain a spatial resolution limited only by the pinhole even if the detector itself has a poor spatial resolution.

Very small pinholes (100μm) have been successfully used to obtain submillimeter spatial resolutions. In this case ^{125}I, which emits fluorescence X-rays in the range between 28–33 keV, was used. At these lower energies the penetration of the X-rays across the pinhole edges is small.

To try to solve the sensitivity problem in the use of pinhole collimators, some research groups have developed multi pinhole systems [39, 44].

10.3.2.2 *Parallel-hole collimator*

The parallel-hole collimator design is the geometry typically used in human scanners where a large field of view is required. In small animal imaging, where one would like a spatial resolution between 1 and 3 mm, one of the difficulties of this design is to make the very small diameter but long collimator holes. Collimators made of lead with holes 2 cm long and diameter 0.6 mm exist.

Each hole of a parallel-hole collimator defines a cone whose total angular aperture is given by

$$\vartheta_{Tot} = \frac{d}{L} \tag{10.15}$$

where d is the diameter of the holes and L is their length. As the distance D of the object from the collimator increases so does the diameter of the subtended cone, and therefore the spatial resolution degrades. The spatial resolution of a parallel-hole collimator can be approximated by the expression [45]

$$R_{Coll} = d + D\frac{d}{L}. \tag{10.16}$$

If the detector behind the collimator has a spatial resolution R_d, then the total response will be of the form

$$R_{Tot} = \sqrt{R_{Coll}^{\ 2} + R_d^{\ 2}} = \sqrt{\left(d + D\frac{d}{L}\right)^2 + R_d^{\ 2}}. \tag{10.17}$$

A graph of the measured spatial resolution as a function of distance from a collimator can be seen in Fig. 10.17 and Fig. 10.14. In this example the collimator has 0.6 mm diameter holes, 20 mm long.

Here too, therefore, it is important to keep the object under examination near to the collimator for optimal spatial resolution.

Unlike the pinhole collimator configuration, which can take advantage of an enlarging factor, to take full advantage of a parallel-hole collimator's spatial resolution, the detector must have a spatial resolution comparable to the spatial resolution of the collimator.

The geometrical sensitivity of a parallel-hole collimator can be approximated with the expression [45]

$$g = K^2 \left(\frac{d}{L}\right)^2 \frac{d^2}{(d+h)^2} \qquad (10.18)$$

where h is the thickness of the septae and K is a factor depending on the geometry of the holes. In the case of hexagonal holes $K = 0.26$. It must be noted that, unlike the pinhole collimator, the sensitivity does not depend on the source collimator distance nor does it depend on the lateral position of the source within the field of view. This kind of collimator is therefore more suitable in the studies of rats and larger animals.

10.3.3 *Small Animal SPECT Scanner Examples*

There are not many dedicated small animal SPECT scanners. This is rather unfortunate due to the extensively studied chemical techniques for pharmaceutical labeling with ^{99m}Tc, and for the ease in obtaining this radioisotope. As mentioned at the beginning of this chapter, PET has the drawback of needing a cyclotron in the same research center as the scanner. From the examples below we will see that spatial resolution can be better than in PET giving a high potential for pharmaceutical studies especially in mice where a small field of view is sufficient.

10.3.3.1 *Pinhole collimator scanners*

Typically, pinhole systems are dedicated to mice imaging, due to the very small field of view. Pinholes as small as 100μm [31] have been successfully tested with ^{125}I (28–33 keV X-rays), obtaining spatial resolutions of 0.4 mm at a distance of 10 mm from the pinhole. This extremely high spatial resolution was obtained despite the detector used, which had a spatial resolution of 3.3 mm with this isotope (30 keV).

The detector used was based on a 2×2×6 mm pixellated array of NaI(Tl) coupled to an array of 5×5, 1″ position sensitive photomultiplier tubes.

When imaging ^{125}I with 0.1 mm and 0.25 mm pinholes, the measured sensitivity was respectively 2.7 and 13.5 cps/MBθ at a source collimator distance of 10 mm.

With ^{99m}Tc and a 1 mm pinhole the spatial resolution obtained varied from 1 mm at 10 mm to 1.2 mm at 30 mm from the source. Again the spatial resolution of the scanner is better than the intrinsic detector spatial resolution of about 2 mm at 140.5 keV.

In Fig. 10.12 one can see a bone scan of a mouse obtained with a 1 mm pinhole and 35 mm radius of rotation.

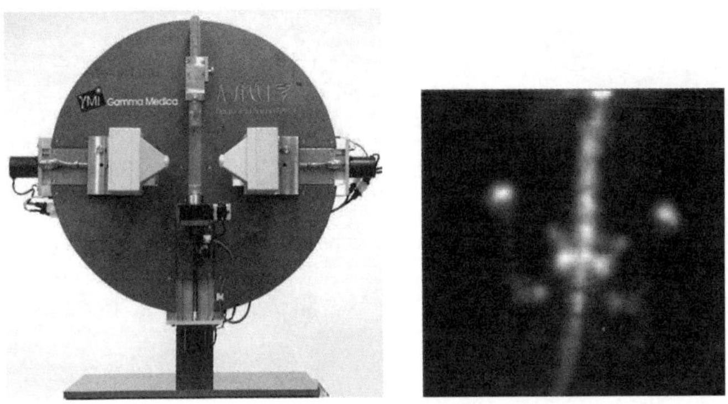

Fig. 10.12. Photograph of the MicroSPECT scanner and a reconstructed image of a mouse bone scan. (From Gamma Medica.) http://www.gammamedica.com/products/a_spect/spect.html

Due to the enlargement factor in pinhole imaging, conventional SPECT cameras can also be implemented with high resolution pinhole collimators. A good example of this is a system developed by B. Tsui and collaborators [33]. In this case a GE 400 AC/T scintillation camera with a 40 cm diameter 3/8″ thick continuous NaI(Tl) crystal was used. The distance between the pinhole and the crystal was 27 cm producing reconstructed images with ~1 mm spatial resolution as can be seen in Fig. 10.13. The phantom is 34 mm in height and 45 mm in diameter with cold rods inserted. The diameters of the rods are 1.2, 1.6, 2.4, 3.2, 4.0, and 4.8 mm in diameter.

Other examples of dedicated pinhole SPECT scanners can be found in literature [30–35].

To increase sensitivity several independent cameras can be mounted together. This is the case of the FASTSPECT [39] scanner developed by Arizona Health Science Center which has implemented 24 cameras. With a field of view of 3.0×3.2×3.2 cm^3 and a spatial resolution of 1.5 mm the sensitivity is 4240 cps/MBq.

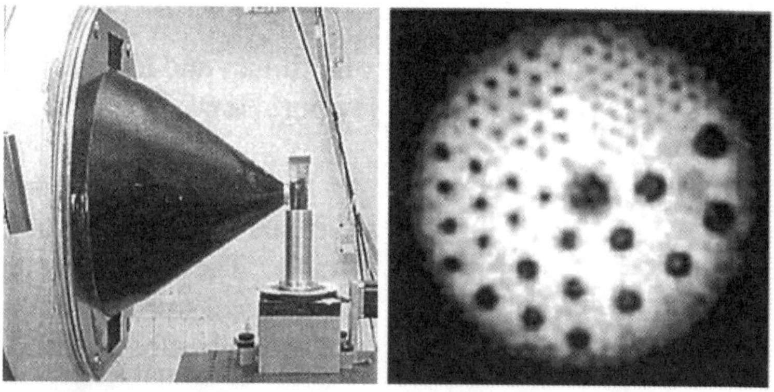

Fig. 10.13. Left: Setup showing the large conventional GE 400 AC/T detector with a pinhole collimator mounted. Right: Transaxial slice of a reconstructed pinhole SPECT image of a cold rod Derenzo phantom. The rod diameters are 1.2, 1.6, 2.4, 3.2, 4.0, and 4.8 mm [33].

10.3.3.2 *Parallel-hole collimator scanners*

Imaging of slightly larger animals requires a larger field of view. Due to the strong dependency of sensitivity when using pinhole scanners, in this case it is preferable to use parallel-hole collimators [36, 36, 40]. High resolution parallel-hole collimators can now be found with holes down to 0.6 mm in diameter, 20 mm long.

The difficulty here is to couple a detector with a spatial resolution comparable to the resolution of the collimator in order to take full advantage of its performances.

A classical scheme of a detector for SPECT has been developed by a group in Juelich, Germany [36]. In this scanner a 37 mm thick collimator with 1 mm holes and 0.2 mm septum was placed in front of a detector. In its best version a 2 mm thick, 9 cm diameter continuous NaI(Tl) scintillator coupled to a 5″ position sensitive photomultiplier was used. The field of view is 90 mm and the intrinsic spatial resolution of the detector is 1.4 mm FWHM with an energy resolution of 8.5% FWHM at 140.5 keV. Close to the collimator the planar spatial resolution of the system is 1.8 mm FWHM and 2.5 mm FWHM at a distance of 30 mm. The sensitivity of the system is 8.5 cps/MBq. Measured curves of the sensitivity and spatial resolution as a function of object—collimator distance are shown in Fig. 10.14. Notice how the sensitivity does not depend on the radial position of the source.

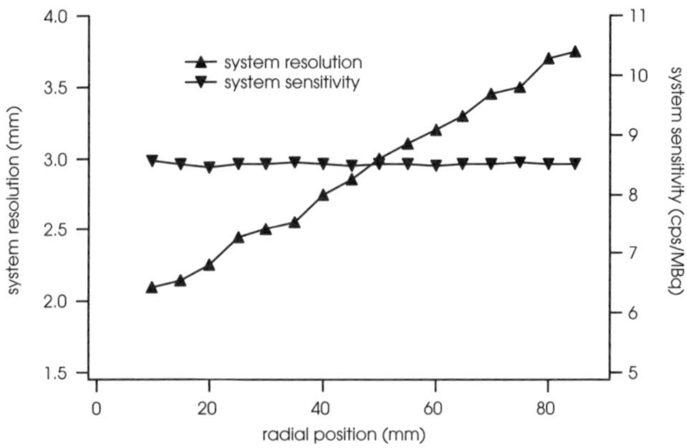

Fig. 10.14. Figure of sensitivity and spatial resolution for the small animal SPECT system [36]. (Reprint authorized by IEEE, Piscataway, NJ, USA.)

When used for tomographic imaging the spatial resolution (in water) is 2.7 mm. An image of the system is shown if Fig. 10.15.

Transaxial sections of tomographic images are also shown for this system in Fig. 10.16.

The phantom is 36 mm in diameter and 32 mm high. The cold rod version (upper image) is a small cylinder containing several rods of different diameters. The smallest is 2 mm and the largest 7 mm. The hot rod phantom consists of two small tubes placed inside the large cylinder.

The concentration of ^{99m}Tc in the different regions was 10 : 5 : 2. Eighty projections were acquired over 360° and the image was reconstructed with an iterative algorithm.

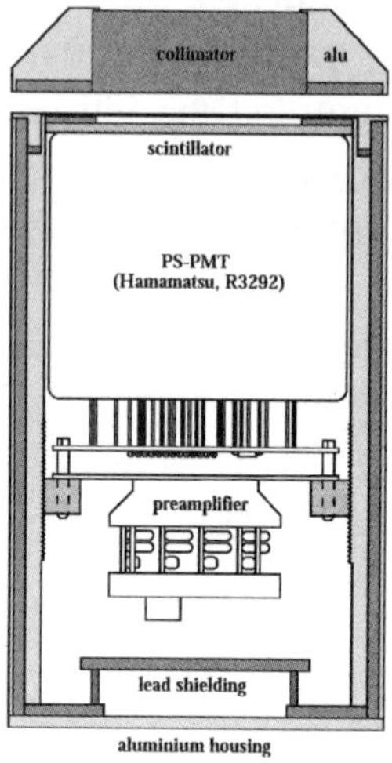

Fig. 10.15. Schematic drawing of the small animal SPECT detector developed by N. Schramm et al. [36]. (Reprint authorized by IEEE Piscataway, NJ, USA.)

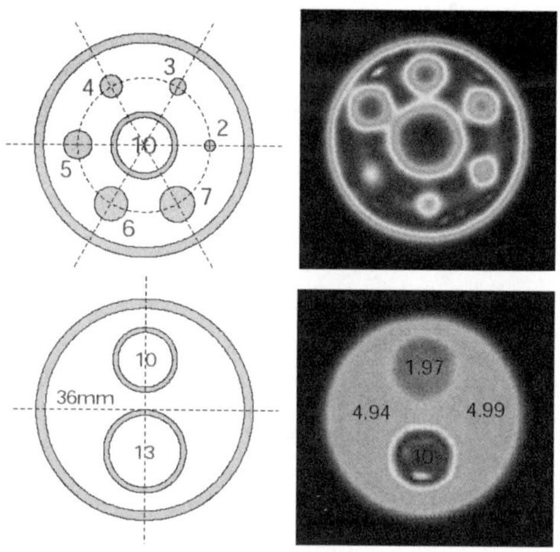

Fig. 10.16. Transaxial sections of two phantoms and the relative reconstructed tomographic images [36]. (Reprint authorized by IEEE Piscataway, NJ, USA.)

A similar system based on pixellated crystals rather that a continuous crystal, is the YAP-(S)PET [22,23] scanner, originally developed for PET and then modified for both PET and SPECT capabilities. Here the collimator is 20 mm thick with 0.6 mm diameter holes and 0.15 mm septae. The scintillator is a matrix of YAP:Ce with 2×2×30 mm^3 pixels optically separated by a very thin (5 μm) reflective layer. Each matrix is directly couple to a 3″ position sensitive photomultiplier. Two heads are mounted on a rotating gantry to slightly increase the sensitivity. The measured sensitivity of each detector head is 20 cps/MBq and the tomographic spatial resolution is 3.5 mm for a field of view of 4 cm in diameter and 4 cm in length. In this configuration the limit in the spatial resolution is the detector pixel size, which was originally chosen for PET.

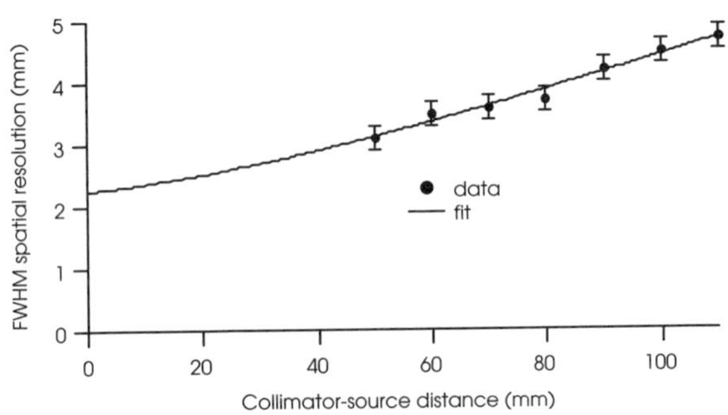

Fig. 10.17. Planar spatial resolution of the YAP-(S)PET scanner as a function of collimator source distance. The data are fit with the expression $R_{ToT} = (R_{Coll}^2 + R_d^2)^{1/2}$ with $d = 0.6$ mm and $L = 20$ mm (see expression (10.17)).

The planar spatial resolution of each detector obtained from the profile of a sinogram taken with a 0.8 mm capillary as a function of distance is plotted in Fig. 10.17. The energy resolution at 140.5 keV is about 30% and is rather poor compared to NaI(Tl) due to the light collection and light output of the scintillator.

In Fig. 10.18a one can see a tomographic image of a modified Derenzo phantom with hot bars 1.5, 2.0, 2.5, 3.0 mm in diameter and with a center to center spacing twice their diameters. In Fig. 10.18b one can also see a transaxial image, reconstructed with an EM algorithm, of an in vivo rat heart after the rat was injected with 100 MBq of ^{99m}Tc(N)(PNP5)(DBODC). Some of the heart's structure can be seen.

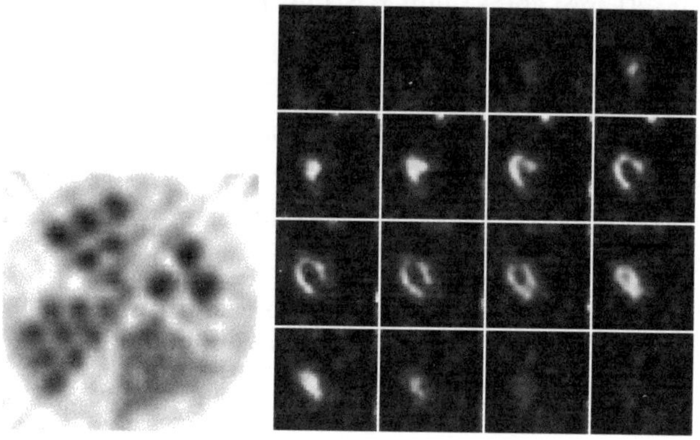

a) b)

Fig. 10.18. (a) Hot Derenzo phantom. The hot bars have diameters 3.0, 2.5, 2.0, and 1.5 mm with a center to center spacing twice their diameter. (b) Coronal sections of the image of an *in-vivo* rat heart taken after injection of ^{99m}Tc(N)(PNP5)(DBODC).

10.3.3.3 *Converging hole collimator scanner*

An interesting alternative to the collimation methods seen above is the converging or focusing collimator. This kind of collimator is used in the TOHR (TOmograph Haute Résolution) scanner [38]. The concept here is to detect only the photons which have been emitted in a small volume: the focus of the collimator. In this way the sensitivity to radiation emitted within the focus can be very high.

A scheme of the principle is shown in Fig. 10.19. In the case of TOHR a solid angle of 3π is covered by placing focusing collimators on 15 of the 20 sides of an icosahedrum. A drawing of the structure is shown in Fig. 10.20 and a photograph of the system is shown in Fig. 10.21. Imaging of a small animal is performed by scanning it along the three spatial directions.

TOHR can also be made to detect two gamma rays or X rays in coincidence. In fact some radioisotopes emit 2 or more angularly uncorrelated photons simultaneously. By detecting two of these in coincidence through the focusing collimators one can further improve the spatial resolution of this scanner and reduce background of radiation emitted from regions outside the focal point. Clearly in this case the efficiency will be lower. Table 10.2 summarizes the spatial resolution and sensitivities for the TOHR scanner in both singles mode and coincidence mode.

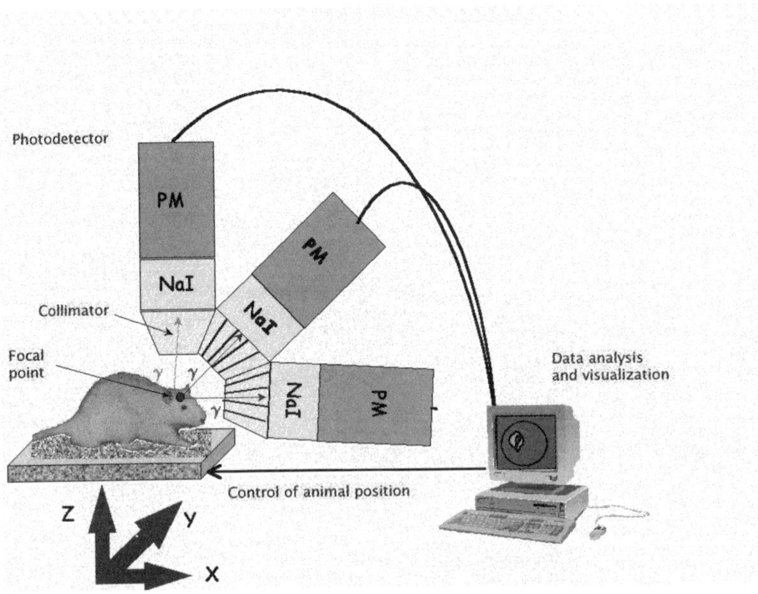

Fig. 10.19. Imaging principle when using converging collimators. http://ipnweb.in2p3.fr/~ipb/home/theme_recherche/theme_animaux/theme_tohr/tohrprinc.htm

Table 10.2. Spatial resolution and detection efficiencies of TOHR for the two modes: single and double photon detection [38].

Detection mode	Spatial resolution	Detection efficiency
Single	2.4 mm	10.2 %
Double	1.4 mm	0.84 %

This dual mode capability can allow fast imaging with a reduced spatial resolution so as to locate a region of interest. Subsequently the coincidence modality can then be taken advantage of for superior image quality. A nice image of a rat thyroid obtained in coincidence mode is shown in Fig. 10.22.

Depending on the size of the region studied, the scan will have a different duration to allow spatial scanning of the animal. In the study of small organs, the high efficiency allows short scans and therefore good dynamic capabilities.

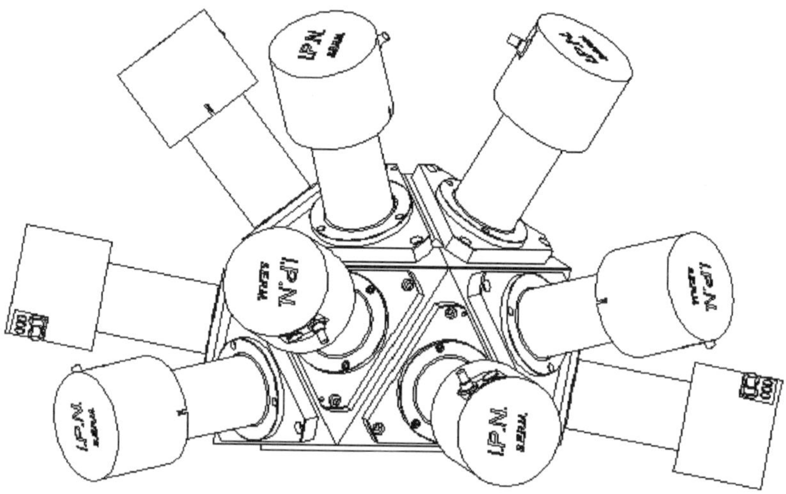

Fig. 10.20. Schematic drawing showing 9 of the 15 detectors placed on the sides of an icosahedrum.
http://ipnweb.in2p3.fr/~ipb/home/theme_recherche/theme_animaux/hometohr.htm#tomo

Fig. 10.21. Photograph of the TOHR scanner.
http://ipnweb.in2p3.fr/~ipb/home/theme_recherche/theme_animaux/hometohr.htm#tomo

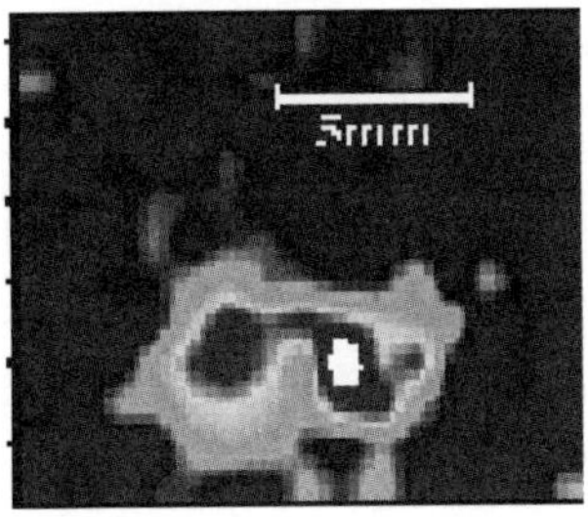

Fig. 10.22. Image of a 15x12 mm^2 area with a 1 mm pixel centered on the thyroid of a rat. The thyroid was labeled with 100 μCi of ^{123}I and acquisition was taken in coincidence mode for increased image quality.
http://ipnweb.in2p3.fr/~ipb/home/theme_recherche/theme_animaux/hometohr.htm#tomo

10.4 Positron Emission Tomography (PET)

The principles of PET imaging have already been treated earlier. Some of the problematics involved in PET have also been treated and these will be revised in the perspective of small animal imaging. In the description of the detector, emphasis will be given to the materials chosen for the detector and geometry of the scanners.

We will discuss the intrinsic limitations to spatial resolution, and also limitations to the sensitivity. As mentioned in the introduction, high resolution scanners must also have high sensitivity in order to acquire high statistics images to fully exploit the scanner's capabilities.

10.4.1 *Physical Limitations to Spatial Resolution*

PET is based on the property that e^+e^- annihilation, at rest, almost always produces two 511 keV gamma rays in opposite directions. The simultaneous detection and determination of the interaction positions of the two gamma rays will naturally define a line of flight. For an accurate reconstruction of the distribution of the β^+ emitter, this line of flight should pass through the position where the β^+ was emitted. Three factors will limit this:

- Deviation from collinearity of the two emitted gamma rays.
- The distance traveled by the β^+ before annihilating.
- The inaccuracy in the determination of the interaction position of the gamma ray.

The first two points originate from the physical processes involved in PET and come into play when determining the ultimate limit in the spatial resolution of a PET scanner. The third point is related to detector performances.

The collinearity of the 511 keV photons is degraded firstly by the so called Fermi motion of the electrons within the source when the positron annihilates and secondly by the scattering of one (or both) of the photons in the surrounding tissue. Their effects in the reconstruction of the lines of flight can be seen in Fig. 10.23 and Fig. 10.32 respectively.

10.4.1.1 *Electron Fermi motion*

Consider the center of mass of a system composed of an electron and a positron which have bonded. After a very brief time interval the pair will annihilate transforming the mass of the system into energy. The annihilation will almost always occur from the singlet *S*-wave ($s = 0$, $L = 0$), in which case the annihilation time is about 125 ps, and to conserve the quantum numbers of the system an even number of photons will be emitted. Two collinear photons is the dominant decay and these will be emitted each carrying away half of the energy of the system: 511 keV. (In about 1 event in 10^4 the positron and electron will annihilate into 3 photons. This situation will not be considered here.)

In the reference frame we are considering, the two photons are generated at exactly 180° from one another so as to conserve energy and momentum.

Since in the laboratory reference frame the center of mass is not at rest, the collinearity of the two photons will be lost due to the angle transformation from one reference frame to the other. This non collinearity results in a dispersion centered at 180° of about 0.4° at FWHM when a positron annihilates in water.

Given the center of mass reference frame $R_{C.M.}$ of the electron-positron system and a laboratory reference frame R_{Lab} with $\mathbf{x}$ and $\mathbf{x}'$

parallel (see Fig. 10.24) and given that $R_{C.M.}$ is moving with velocity **v**´ in the direction **x**´ with respect to R_{Lab}, the transformation for angles with respect to **x** (or **x**´) is given by:

$$\tan\vartheta' = \sin\vartheta/[\gamma(\cos\vartheta+\beta)] \qquad (10.19)$$

where ϑ' is the angle between one of the photons and the **x**´ axis of R_{Lab} and ϑ is the angle between the same photon and the **x** axis of $R_{C.M.}$. γ is the relativistic factor $1/\sqrt{1-\beta^2}$ with $\beta = v'/c$. The minimum value of $\vartheta'_1+\vartheta'_2$ (the greatest deviation from 180° between the two emitted photons) is for $\vartheta_1 = \vartheta_2 = \pi/2$. In this case

$$\tan\vartheta'_{min} = 1/\gamma\beta \qquad (10.20)$$

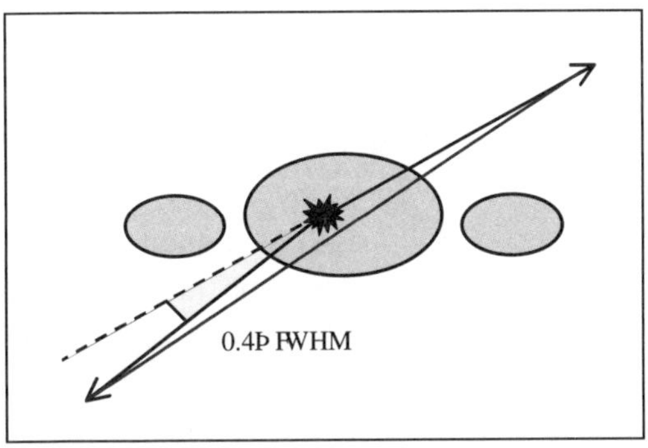

Fig. 10.23. Non collinearity of the emitted photons due to the electron Fermi motion. The dotted line represents the ideal collinearity.

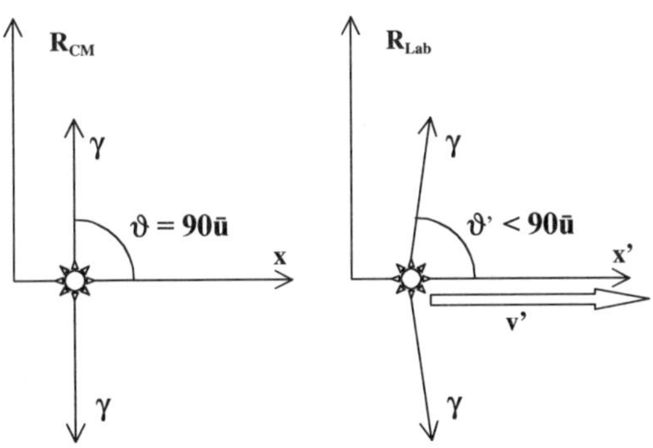

Fig. 10.24. Emission of two back to back gamma rays as seen from the center of mass reference frame, R_{CM}, and from a laboratory reference frame, R_{Lab}. R_{CM} is moving at a velocity **v**´ parallel to the **x**´ direction of R_{Lab}.

In the case of hydrogen, $\beta \sim 1/137$ and $\gamma \sim 1$ resulting in tan $\vartheta' = 137$ and therefore $\vartheta' = 89.6^\circ$ ($\vartheta = 90^\circ$). In this situation both photons undergo the same aberration and the angle between them is no longer 180° but 179.2°.

In the case of large PET scanners where the diameter is about 40 cm the contribution of this noncollinearity to the degradation of the spatial resolution is about 1 mm. In small PET scanners the contribution can be reduced but then must be compared to the improved spatial resolution of small animal scanners.

10.4.1.2 Scattering in the source

Fermi motion is a cause of non-collinearity "at the source". Another important process which will degrade the photon collinearity is Compton scattering of one (or both) of the emitted photons with the electrons in the surrounding tissue.

In human PET scanners the length of tissue traversed by the photons is somewhat greater than their interaction length. In fact in water the mass attenuation length for 511 keV gamma rays is 0.096 cm^2/g. This limitation has been already discussed previously and techniques to reduce the effect of scattering (energy selection of events, etc.) have been considered.

Let us, though, consider small animals and the situation depicted in Fig. 10.32a. One of the emitted photons leaves the animal without an interaction whereas the other scatters. In order to have a coincidence event, both photons must be detected. Therefore the scattered photon must be within the coincidence solid angle of the scanner.

Considering rats (similar to a 4 cm diameter cylinder), in a fraction $S \approx 30\%$ of the annihilations one of the gamma rays will be scattered by the tissue. Given that one of the two gamma rays is detected the second one will be detected only if it is scattered within the scanner's angular coverage, Ω, typically less than 5%. Considering as a rough estimate Compton scattering to be isotropic, the percentage of annihilations generating a coincidence with one of the gamma rays having scattered will be the product $S\ \Omega$. The result is slightly more than one percent of all the detected coincidences. That is why in small animal imaging scattering in the source can be partially overlooked as regards spatial resolution.

The situation depicted in Fig. 10.32c, due to independent decays, may be generated by annihilations outside the field of view. Timing in this case becomes very important and fast detectors with good timing resolution are of great advantage.

Generally both situations described above can be neglected and the acquisition of the energy spectrum is no longer fundamental for background reduction. Other strategies to improve spatial resolution and to reject background events can be taken. If the energy spectrum is discarded, other techniques for detecting the 511 keV photons can also be used where the energy information is lost but the spatial resolution is very good (see for example the HIDAC scanner described later in Section 10.5.3.2).

Moreover dense medium Z scintillators can also be used, some of which have advantages as compared to existing high Z materials. An example, in using medium Z materials is the YAP-(S)PET scanner [22] (see section 10.5.3.1). The Compton interaction in the detector rather than the photoelectric interaction is used to obtain the photon's interaction position. As we will see, in these detectors, most of the events undergo a first Compton interaction and then escape the scintillator allowing the correct determination of the line of flight. This point will be discussed in more detail below.

10.4.1.3 *Positron range*

The second physical limitation to spatial resolution in PET scanners is the distance traveled by the positron before it annihilates. Even if the collinearity of the photons were maintained, this fact leads to a shift of the line of flight from the position of emission of the positron to the position where it annihilates. This is shown schematically in Fig. 10.25. The range a positron will travel in the surrounding tissue is determined by its initial kinetic energy and by the density of the tissue. In the process of β^+ decay the positron is not emitted with a fixed energy but with a continuous spectrum.

The annihilation position will therefore not coincide with the decay position. A point source will generate annihilations in a finite region. This distance can vary, in water, from about 0.6 mm in the case of ^{18}F to several millimeters in the case of ^{82}Rb (see Table 10.3).

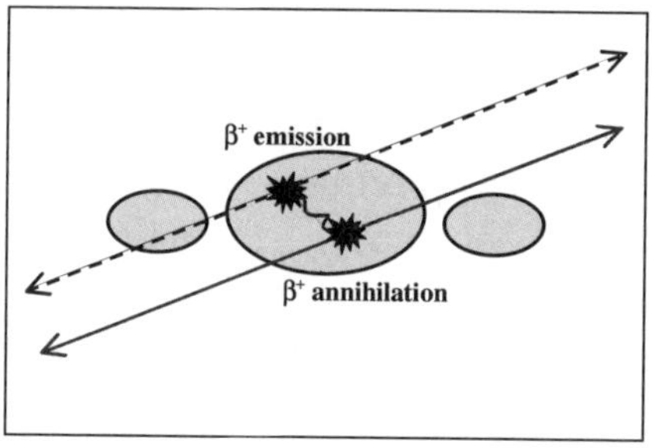

Fig. 10.25. Effect of positron range on the misplacement of a line of flight. The dotted line represents the ideal line of flight.

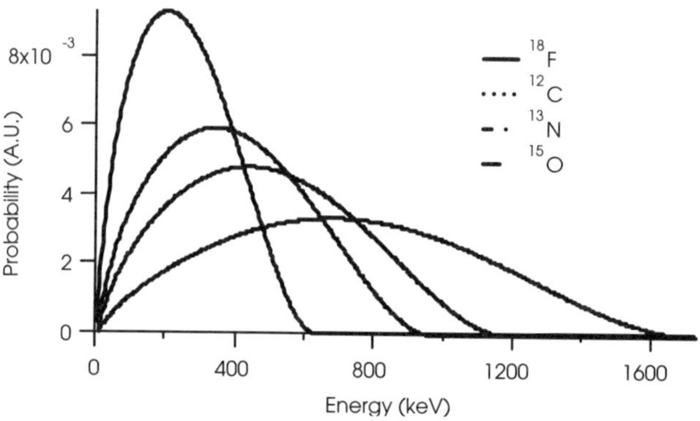

Fig. 10.26. Figure of positron energy spectrum for ^{18}F, ^{12}C, ^{13}N and ^{15}O decay.

When a proton rich nucleus undergoes a β^+ decay, transforming a proton into a neutron, a positron (anti-electron) and an electron neutrino, the emitted positron will have a kinetic energy in a range extending from zero to the endpoint energy corresponding to the Q value of the reaction.

The energy spectrum of the emitted positron in the case of ^{18}F, ^{12}C, ^{13}N and ^{15}O can be seen in Fig. 10.26. Table 10.3 reports the energy endpoints of the spectrum and the mean positron range in water for some β^+ emitting isotopes. The range a positron will travel before thermalizing and annihilating (only a few percent will annihilate in flight at the energies considered), depends on its initial kinetic energy. In Fig. 10.27 one can see a graph of the positron range, expressed in g/cm^2 (linear range X density), as a function of its kinetic energy.

Table 10.3. Characteristics of commonly used β^+ emitters

Isotope	Half life (min)	positron kinetic energy endpoint (MeV)	Mean positron range in water (mm)	Positron range in water (mm) @ endpoint
^{18}F	109.8	0.635	0.6	2.4
^{11}C	20.4	0.961	1.2	3.9
^{13}N	9.97	1.190	1.6	5.1
^{15}O	2.04	1.723	2.8	8.0
^{68}Ga	68.1	1.899	2.9	8.9
^{82}Rb	1.27	3.350	7.0	17.0

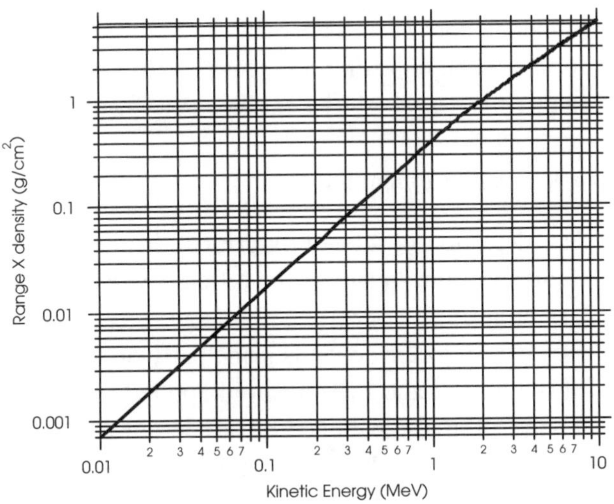

Fig. 10.27. Graph of positron range in water expressed in g/cm^2 (linear range × density).

In the case of ^{15}O the contribution to the source dimensions is of the order 2.5 mm, greater than the spatial resolution of many existing small animal PET scanners.

A detailed evaluation of the effect of positron range on the spatial resolution of PET scanners can be found in Ref. [47]. ^{18}F is the best isotope in this respect and contributes to the spatial resolution with less than 1 mm.

10.4.2 *Efficiency and Coincidence Detection of 511 keV Gamma Rays*

Neglecting for the moment the non-collinearity of the annihilation photons, to correctly determine a line of flight, one needs the three spatial coordinates of any two points along the line of flight. Therefore the two photons must be detected in coincidence in each of two detectors placed along the line of flight. To determine the correct coordinates of the line of flight, each photon must interact only once in each detector. The more accurately one can determine the spatial coordinates of the gamma ray interactions, the better will be the spatial resolution of the tomograph.

Spatial resolution is important but is not the only important parameter of a scanner. Other factors such as sensitivity and time resolution will also come into play when choosing a detector configuration for a small animal PET scanner.

10.4.2.1 *Intrinsic detector efficiency*

In PET scanners, detection efficiency is critical. The reason is that the coincidence efficiency is the square of the efficiency of detection of each single 511 keV photon.

The efficiency of a detector depends principally on the material used, the thickness of the crystal and the threshold applied to the signals. The choice of each one of these parameters, to optimize efficiency, may deteriorate other aspects of the tomograph. For example the spatial resolution of a detector depends on the thickness of the crystals used. However, the thicker the crystal the greater will be the parallax error due to the depth of interaction. This is especially true near the edges of the field of view in ring scanners (see Fig. 10.39). Techniques [28] are under study to try to measure the depth of interaction in ring scanners so as to reduce this error. As we will see when discussing the geometry of scanners, the thickness of a crystal is less critical in the case of a scanner with planar geometry.

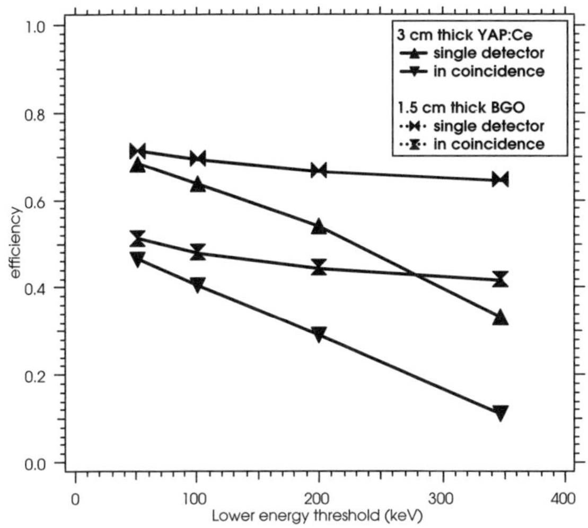

Fig. 10.28. Efficiency as a function of a lower energy threshold for BGO, YAP:Ce. Both singles and coincidence curves are shown.

In Fig. 10.28 one can see the comparison of the detection efficiency for 511 keV photons with BGO and YAP:Ce scintillators as a function of the lower energy threshold. In this example the thickness of both crystals is 1.5 times their gamma mean free path. High *Z* materials will maintain a rather constant efficiency whereas medium *Z* materials have a strong dependence on the lower energy threshold.

10.4.2.2 *Detector Scatter fraction*

Scattering of the emitted gamma rays within the source and surroundings was shown to be a limited cause of loss in spatial resolution. Let us now consider the scattering of 511 keV gamma rays in the detector.

As briefly discussed earlier, 511 keV gamma rays are rather penetrating and the dominant interaction with matter is Compton scattering. Many incident photons will undergo multiple interactions in the detector and the average coordinate of the gamma will usually not coincide with the first interaction, which is the one that lies on the line of flight. Multiple interaction events are very difficult to distinguish from single interaction events and contribute significantly to the scatter fraction of a scanner.

In Fig. 10.29 one can see the profile, through the maximum, of the reconstructed coordinates of events generated by a pencil beam of 511 keV gamma rays after having interacted in a 1.5 cm thick BGO crystal. The contribution due to multiscattering within the detector can be clearly seen. In this example the pencil beam of 511 keV gamma rays interacts at the center of a single pixel within a matrix. Each pixel has a cross section of 2×2 mm^2 and the matrix is 4×4 cm^2. The determination of the coordinates is based on expressions (10.7) with a lower energy threshold of 250 keV.

The Monte Carlo data (dots) are fit with the expression

$$f(x) = Ae^{-\frac{1}{2}\left(\frac{x-x_0}{\sigma}\right)^2} + Be^{-\frac{|x-x_0|}{\lambda}} \tag{10.21}$$

where A and B are the amplitudes of the Gaussian and exponential functions, σ is the sdandard deviation of the Gaussian curve, λ is the decay constant of the exponential function and x_0 is the center of the two curves.

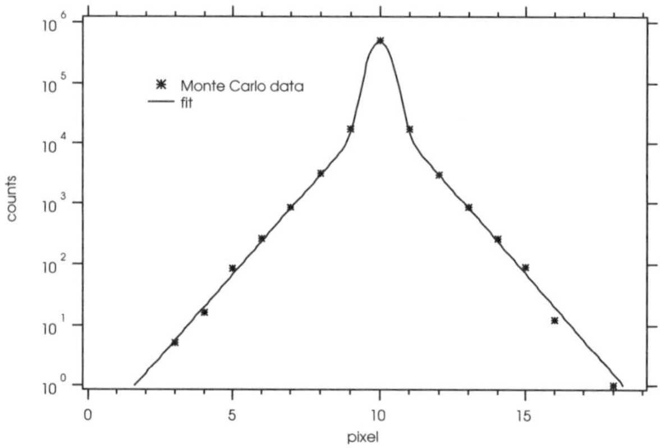

Fig. 10.29. Profile of the reconstructed position for a 511 keV pencil beam incident at the center of the pixel of a matrix. The pixel has a section of 2×2 mm^2. The contribution to multiscattering can be clearly seen even in the favorable case of BGO scintillator, 1.5 cm thick. Superimposed is expression (10.21). The total number of events is 663824 and the events with the correct coordinate are 487435.

There are two contributions to the profile. The first is a narrow Gaussian curve whose standard deviation σ is about 0.7 mm (2 mm bins). These represent events which have originated from interactions within (or very near to) the central pixel, and have therefore generated the correct coordinate for the event. The second, wider curve, represents the events which have given the wrong coordinate and originate from multiscattered events.

It is clear from the Fig. 10.29 that even in the case of BGO, which has the highest photoelectric fraction, multiscattering contributes considerably giving, in this example, a scattered over total ratio $f = 0.27$.

An interesting example for multiple scattering rejection is the YAP-(S)PET scanner based on YAP:Ce scintillator. As reported in Table 10.1 the photofraction is only 4.4%. A typical energy spectrum, shown in Fig. 10.56, of one of the YAP-(S)PET scanner detectors (the thickness of each YAP:Ce matrix is 3 cm) shows a full absorption peak with a population much greater than 4.4%.

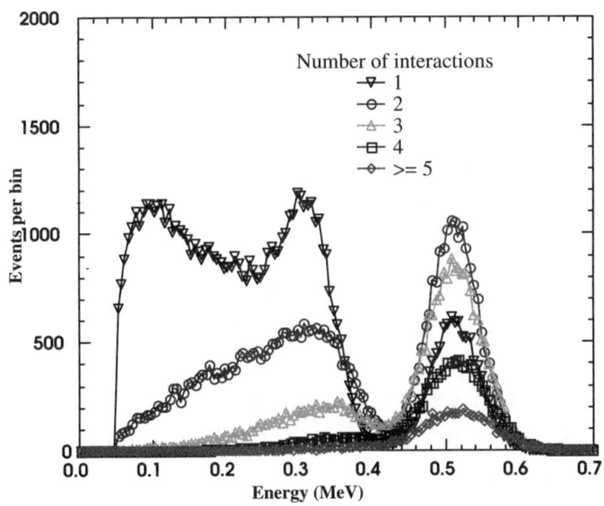

Fig. 10.30. Energy deposit as a function of the number of interactions in the case of a YAP:Ce scintillatore 3 cm thick and 4 x 4 cm^2 surface [40].

In Fig. 10.30 one can see a graph showing the energy deposit as a function of the number of interactions within the crystal of this scanner [40]. It is clear that the events above 400 keV are mainly due to multiple interactions whereas events below 400 keV have a high fraction of single interactions in the detector. By choosing events whose deposited energy is below 400 keV the scatter fraction is reduced from $f = 0.51$ to $f = 0.33$ comparable to BGO. In the example of a 3 cm thick YAP:Ce crystal the price is a reduction in the intrinsic coincidence efficiency from about 46% to about 14%. The advantage of YAP:Ce is in the better timing.

10.4.2.3 *Intrinsic spatial resolution*

Tomographic PET imaging includes a lot of data processing and image reconstruction. These processes will often have an important effect on the reconstructed spatial resolution of a scanner.

When designing a PET scanner it is desirable to define the spatial resolution of a detector or system independently of any reconstruction process. For this reason one distinguishes the intrinsic spatial resolution of a PET detector and the spatial resolution of a PET system.

10.4.2.3.1 Detector intrinsic spatial resolution

The intrinsic spatial resolution in PET [15] will be defined in a similar way to SPECT.

Here any line of flight, in the laboratory reference, which is sampled during a tomographic acquisition can be chosen. The highest spatial resolution will be obtained when the line of flight is perpendicular to the detector crystal. In this way the depth of interaction will not contribute to degrading the spatial resolution. These lines of flight are used to define the intrinsic spatial resolution of a PET detector. In the case of a ring scanner, this situation will correspond to a line in the transaxial plane passing through the center of the scanner whereas for a planar geometry this is not necessarily true.

If a point source is moved perpendicularly to the chosen line of flight, the number of events populating it will be a function of the source position. The FWHM of the distribution will be taken as the intrinsic spatial resolution of the detector.

Ideally the distribution of the number of events for a line of flight passing near the center of the tomograph as a function of the source position should be a triangle whose FWHM is half the pixel size. This definition is schematized in Fig. 10.31.

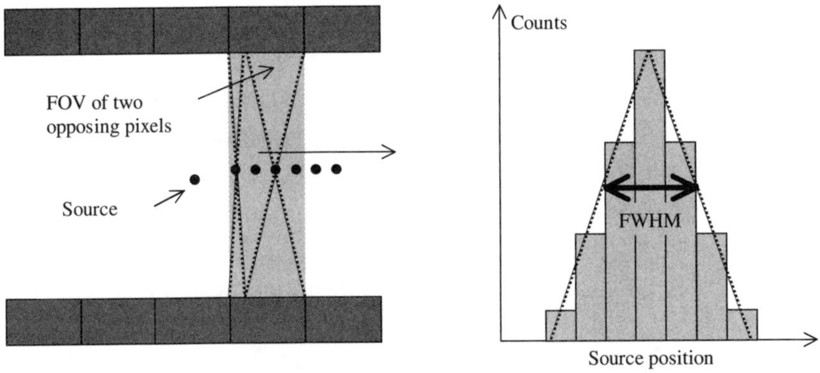

Fig. 10.31. Definition of intrinsic spatial resolution of a PET detector.

An example of the measured intrinsic spatial resolution of the MicroPET scanner is shown in Fig. 10.46. As can be seen, in practice the measured profiles are not triangular due to factors such as multiple scattering in the detector and the FWHM is slightly more than the ideal FWHM which in this case should be about 1.1 mm.

10.4.2.3.2 System intrinsic spatial resolution

In the previous section a very specific choice was made for the line of flight to be chosen when defining the intrinsic detector spatial resolution. The same measurement described in Section 10.4.2.3.1 can be performed for any line of flight acquired during a tomographic acquisition. In the case of a ring geometry (see Section 10.4.3.2) this will correspond to a line joining any two crystal elements. In the case of rotating planar detectors (see Section 10.4.3.1) the same line of flight (in the laboratory reference) will be sampled by different pairs of crystal elements as the detector heads rotate and the contribution of all of these pairs must be included.

Ideally the detector and system intrinsic spatial resolutions should be the same but in practice they are not, due principally to parallax errors from the penetration of the gamma rays in the crystals.

This results in a deterioration of the system spatial resolution compared to the intrinsic detector spatial resolution. There is a significant difference in the depth of interaction effects in the two different geometries considered.

In a ring geometry the depth of interaction effect on spatial resolution is minimum at the center. As one moves away from the center of the scanner, the angle between the crystal surface and the line of flight increases, and depth of interaction becomes important. This effect can be seen in Fig. 10.39.

In a planar geometry the situation is different: near the center of the scanner a line of flight will be sample by many detector elements each having a slightly different angle with the line of flight. The overall effect is a slight degradation of the system spatial resolution compared to the intrinsic spatial resolution. The average effect of depth of interaction, in this geometry though, is less important compared to a ring geometry and as one moves to the edge of the field of view the spatial resolution actually improves slightly due to the orthogonality of the sampled lines of flight with respect to the detector surface. The result is an almost uniform spatial resolution across the whole field of view as can be seen in Fig. 10.35.

10.4.2.4 *Random coincidences and pile up events*

A coincidence detection, in principle, is the detection of two (or more) events simultaneously. To determine whether there is a coincidence between two detectors each detector's signal is sent to a discriminator which generates a digital signal with a width, Δt, which can be set by the user. This signal can be set to be shorter than the scintillator's decay time.

Let us consider two independent detectors A and B whose count rates are R_A and R_B such that $R_A\Delta t \ll 1$ and $R_B\Delta t \ll 1$. The probability that given an event in detector A there will also be at least one event detected in detector B within a time interval $\pm\Delta t$ is

$$P(B \text{ in } [t, t \pm \Delta t] \mid A) = 2 \cdot \left(1 - e^{-\Delta t R_B}\right) \approx 2 \cdot \Delta t \cdot R_B \qquad (10.22)$$

Therefore there is a non zero probability of detecting two independent photons coming from different annihilations as if they were emitted in coincidence. The rate of such events, called random coincidences, will be the rate of events in detector A multiplied by the probability of having an event in detector B, given the event in detector A within the time interval $\pm\Delta t$:

$$R_{random} = R_A \cdot P(B \text{ in } [t, t \pm \Delta t] \mid A) = 2 \cdot R_A \cdot R_B \cdot \Delta t \quad (10.23)$$

All detectors have a finite time resolution. In the case of a scintillator detector, and by triggering on the first detected photoelectron, this limit is given by [46]

$$\sigma_t = \frac{\tau_{scint}}{N_{p.e.}} \quad (10.24)$$

where σ_t is the standard deviation of the time indetermination, τ_{scint} is the scintillation decay time and $N_{p.e.}$ is the total number of photoelectrons generated at the photocathode. For example for a BGO detector with an energy resolution of $\Delta E/E$=15% at 511 keV, the total number of detected photoelectrons can be estimated to be

$$N_{p.e.} = \left(\frac{2.35}{\Delta E / E} \right)^2 = 245 \quad (10.25)$$

The FWHM time resolution of a single BGO detector will therefore be about 2.8 ns ($N_{p.e.}$ = 245, τ_{scint} =300 ns). With two detectors in coincidence this will become $\Delta t = 2.8\sqrt{2}$ ns = 4 ns. Events within Δt = 4 ns will be considered to be in coincidence generating a random coincidence rate given by expression (10.23).

Once a coincidence has been detected, the analog signals must be processed and registered. The time necessary to process the light signals generated by the scintillator to obtain the position information, and if available the energy of the event, is usually a few times the decay time of the scintillator. For BGO this will be about 1–2 μs whereas for a fast scintillators (LSO, YAP:Ce) this time can be 100 ns or less. This imposes that there must not be any other events interacting and generating light in the same readout channel during this processing time otherwise a pile up event will occur mixing information from different interactions. These events must be rejected with a pileup rejection logic.

For both these reasons fast scintillator detectors are desirable so as to have as narrow a coincidence window as possible and a fast processing time.

10.4.2.5 *Energy resolution*

In large clinical PET scanners a good energy resolution is also desirable to partly eliminate those photons which have interacted either in the subject or in the scanner's surroundings. In Fig. 10.32 one can see the different situations which can arise and cause background coincidences.

The energy *E'* of a scattered photon is given by the Compton formula:

$$\frac{1}{E'} - \frac{1}{E} = \frac{1-\cos\vartheta}{m_e c^2} \tag{10.26}$$

where *E* is the energy of the incoming photon, *E'* is the energy of the scattered photon, ϑ is the angle between the scattered photon and the incoming direction, m_e is the rest mass of the electron and *c* is the velocity of light in vacuum.

By applying a lower energy cut E'_{cut} to the detected photons one can discard photons which have scattered outside an angle given by

$$\cos\vartheta_{cut} < 1 - m_e c^2 \left(\frac{1}{E'_{cut}} - \frac{1}{E} \right) \tag{10.27}$$

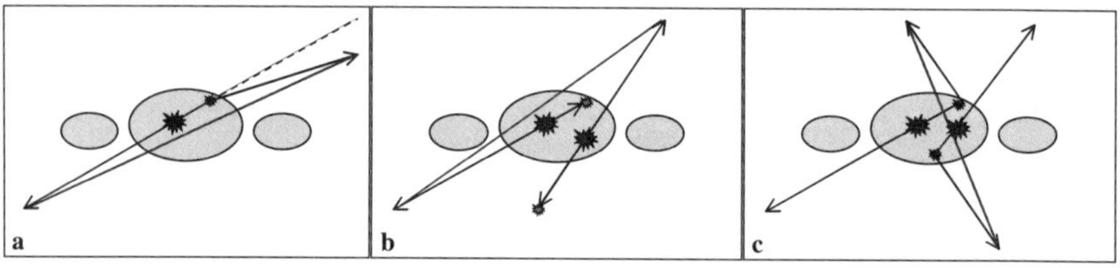

Fig. 10.32. Situations showing how scattered photons (*a*), random coincidences (*b*) and their combination (*c*) may generate incorrect lines of flight.

With an energy resolution of 15% at FWHM one can set a lower energy threshold at about 350 keV. This will discard events scattered with an angle greater than 57°. By combining both energy and timing information, the different background contributions can be reduced. This is especially true for photons emitted outside the field of view which may enter the detectors after having undergone scattering. The generation of false lines of flight are schematically shown in Fig. 10.32.

10.4.3 *Small Animal PET Scanner Geometries*

Small animal scanners can be classified into two geometrical categories: Those with opposite, rather large area, planar detectors put in coincidence (planar geometry) and those with smaller detectors composing a ring (ring geometry). Both have advantages and disadvantages. Typically clinical systems have a ring geometry but in small animal scanners both geometries are found.

10.4.3.1 *Planar geometry*

In Fig. 10.33 is a schematic drawing of a scanner with planar geometry. Two position sensitive detectors are placed opposite to each other and are put in coincidence. Usually there are one or two pairs which rotate around the animal.

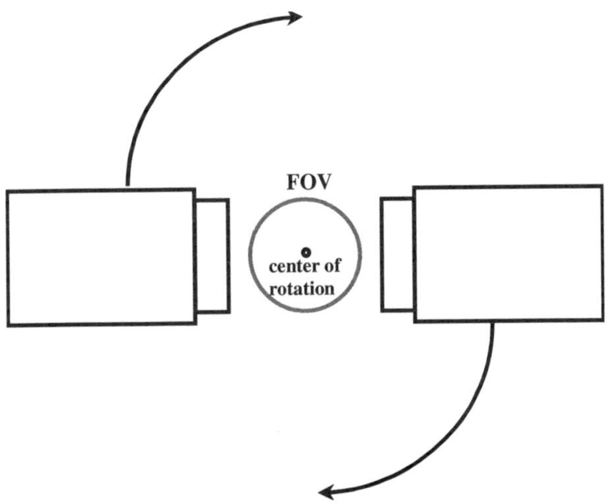

Fig. 10.33. Schematic drawing of a scanner with planar geometry.

To fully sample all the lines of flight one must rotate the detectors around the scanner's axis. By tracing all lines of flight one can see that the resulting FOV is a cylinder whose diameter is equal to the transverse dimensions of the detectors and length is equal to the width of the detectors. The tomographic sensitivity within the FOV has a cylindrical symmetry and is shown in Fig. 10.34 as a function of the transaxial and axial position within the FOV. The central part of the FOV is clearly where the sensitivity is highest and where one will have the best statistics.

The axial, radial and tangential reconstructed spatial resolution of an existing planar small animal PET scanner (TierPET) can be seen in Fig. 10.35. One of the particularities of this geometry is that the spatial resolution is practically constant over the whole FOV with slightly better performance near the edges. The reason of the uniformity of the spatial resolution in the case of a planar geometry is because the parallax contribution due to the depth of interaction becomes less and less relevant as one moves out towards the edge of the FOV. In fact, there are less crystals in lines of coincidence available due to the limited angular covering of the pixels near the edges. In addition the angle of incidence of the available lines of coincidence converge to being perpendicular to the detectors. In the limiting case of perpendicular lines with respect to the detector, the depth of interaction has no influence on the determination of the position of the interaction.

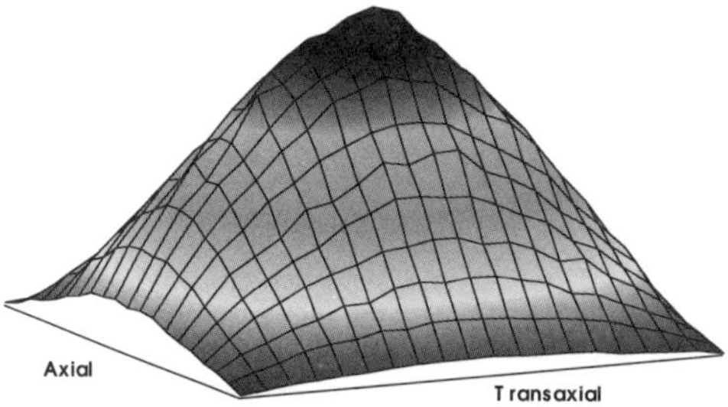

Fig. 10.34. Tomographic sensitivity in function of the transaxial and axial position within the FOV of a planar scanner. The sensitivity is peaked at the center and falls off both along the axial and the transaxial direction.

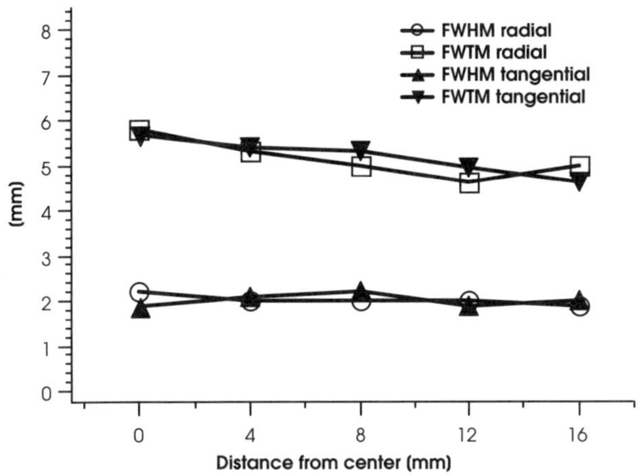

Fig. 10.35. Spatial resolution of the TierPET scanner [27] developed by Juelich. (Reprint authorized by IEEE, Piscataway, NJ, USA.)

In the examples of scanners we will see below, one or two pairs of detectors are used and the matrices are read by a single large area photomultiplier. This is a great advantage to the simplicity of the electronics due to the reduced number of coincidence channels.

10.4.3.2 *Ring geometry*

As one can see in Fig. 10.36 the detectors are placed so as to cover an entire circle around the animal. In this case the detectors are usually smaller than in the case of a planar geometry resulting in a FOV whose axial length is usually smaller. Several rings can be put together to increase the axial FOV.

The transaxial FOV is defined by the number of opposite detectors which are in coincidence with any given detector. In this case by tracing lines of flight one can see that the transaxial sensitivity is almost constant whereas the axial sensitivity is triangular just like in the planar configuration. The sensitivity curve for the UCLA MicroPET scanner is shown in Fig. 10.37.

In this geometry there are no moving parts. The spatial resolution depends on the position within the FOV and deteriorates as one moves radially away from the axis of the scanner. The dominant contribution to the degradation of the spatial resolution is the parallax error introduced by the depth of interaction as can be seen schematically in Fig. 10.38.

The spatial resolution curve can be seen in Fig. 10.39 in the case of the UCLA MicroPET scanner where the radial resolution deteriorates from about 1.5 mm at the center to 3 mm near the edge of the FOV.

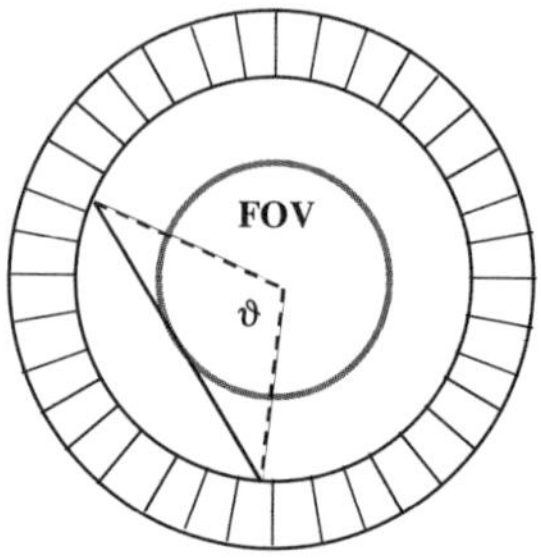

Fig. 10.36. Schematic drawing of a ring PET scanner showing the acceptance angle ϑ which defines the field of view of a ring scanner.

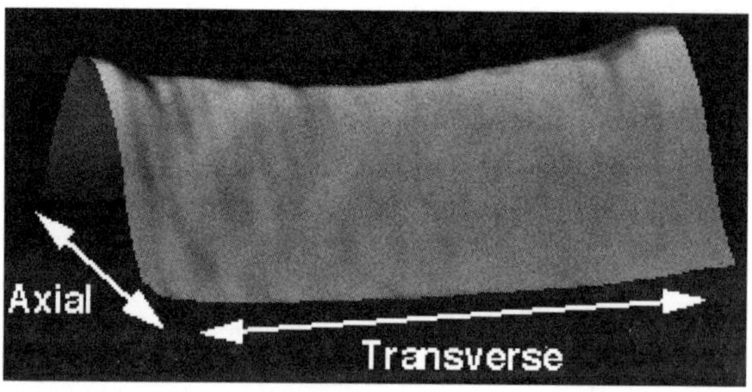

Fig. 10.37. Sensitivity in transaxial plane of a ring scanner [15]. http://www.crump.ucla.edu/user-files/resprojects/microPET/fr_perf.html

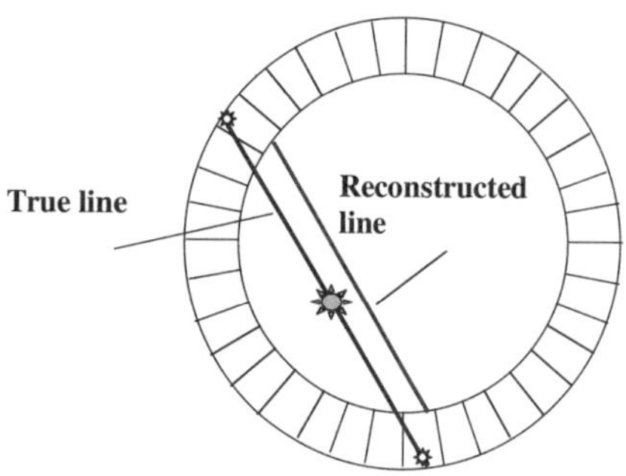

Fig. 10.38. Parallax error introduced by the depth of interaction in ring scanners.

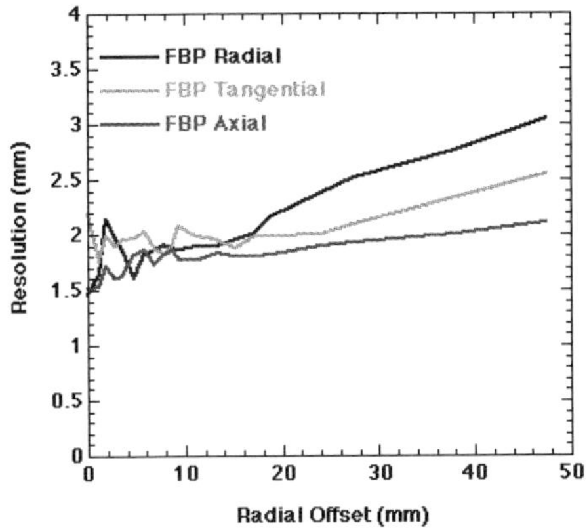

Fig. 10.39. Radial, Tangential and Axial reconstructed spatial resolutions for the UCLA MicroPET small animal scanner [15].
http://www.crump.ucla.edu/user-files/resprojects/microPET/fr_perf.html

10.5 Small Animal PET Scanner Examples

10.5.1 *First Generation Animal Scanners*

In the early 90s two scanners dedicated to the study of primates were designed by Hamamatsu, Japan and CTI PET Systems. These scanners were both based on BGO scintillator and followed the philosophy of clinical ring scanners. The Hamamatsu system was installed at their research center whereas the CTI system was installed at UCLA.

10.5.1.1 *Hamamatsu SHR-2000 and SHR-7700 scanners*

The SHR-2000 scanner [12, 13] used individual 1.7×10×17 mm deep BGO crystals mounted on a position sensitive photomultiplier tube. Four rows of 33 crystals were mounted on each PSPMT to form a block detector. A total of 15 of these blocks were then mounted in a ring to form a 34.8 cm diameter scanner.

The axial FOV is 4.6 cm and the scanner acquires data in 2D mode. The spatial resolution of this scanner is reported to be 3.0 mm in the transaxial direction and 4.4 mm in the axial direction [9]. The sensitivity of this scanner is 0.56 cps/Bq/ml with a lower energy threshold of 300 keV, measured with a 10 cm diameter cylinder.

A more recent version of this scanner, the SHR-7700, uses an 8×4 matrix of 2.8×6.95×30 mm deep BGO crystals coupled to compact PSPMTs. The complete system has 4 rings consisting each of 60 detector blocks. The diameter of the scanner is 50.8 cm and the axial FOV is 11.4 cm. It can acquire data both in 2D and 3D mode. The transaxial spatial resolution is 2.6 mm and the axial resolution (slice thickness) is 3.2 mm. In this version the sensitivity, using a 10 cm diameter cylinder, is 2.27 cps/Bq/ml in 2D mode and 22.8 cps/Bq/ml in 3D mode.

Both the Hamamatsu systems have the unique capability of rotating the ring into a horizontal position so as to acquire data with a trained monkey sitting upright. A picture of this last version can be seen in Fig. 10.40.

Fig. 10.40. Photograph of the SHR-7700 animal PET scanner built by Hamamatsu. From [9]. (Reprint authorized by Springer-Verlag, Germany.)

10.5.1.2 *CTI-PET Systems ECAT-713*

This scanner [9] uses an adaptation of the design of the ECAT HR scanner. The diameter of the scanner is 64 cm and the axial FOV is 5.4

cm. The detector block consists of a 6x8, 3.5×6.25×30 mm deep BGO scintillator coupled to two dual anode photomultipliers.
The reconstructed transaxial spatial resolution is 3.8 mm and the axial one is 4.2 mm. The sensitivity of the UCLA system is 1.57 cps/Bq/ml with a 350 keV energy threshold measured with a 20 cm diameter cylinder. A better measure for sensitivity is the absolute sensitivity at the center which in this case is 0.36%, equivalent to 3.6 cps/kBq.

The first dedicated animal scanners described briefly above did not have quite the enough spatial resolution and sensitivity for performing studies on rodents. In the following sections we will review some examples of small animal scanners dedicated to the study of rats and mice, which have been developed during the past years.

We will divide the scanners into two categories given by their geometry. The peculiarities of each scanner will be pointed out.

10.5.2 *Dedicated Rodent Ring Scanners*

10.5.2.1 *Hammersmith RatPET*

The first scanner specific for rats and mice was the RatPET scanner [14, 15] developed by the Hammersmith Hospital in London in collaboration with CTI PET Systems Inc. It was constructed from commercially available clinical PET detector blocks which were almost identical to the ECAT-713 scanner described above.

The scanner consists of 16 BGO detector blocks placed in a ring with a diameter of 11.5 cm and axial length of 5 cm. Each block consists of 8, 5.95 mm, axial elements by 7, 2.99 mm, transaxial elements with a crystal thickness of 3 cm. Each block is glued directly to two photomultiplier tubes (Hamamatsu R1548), each having two independent channels. The reduced transverse size of these photomultiplier tubes makes this possible. The position determination of the interaction is obtained from the different amount of light arriving on the four channels due to the cuts between each crystal element having different depths. This is shown schematically in Fig. 10.41.

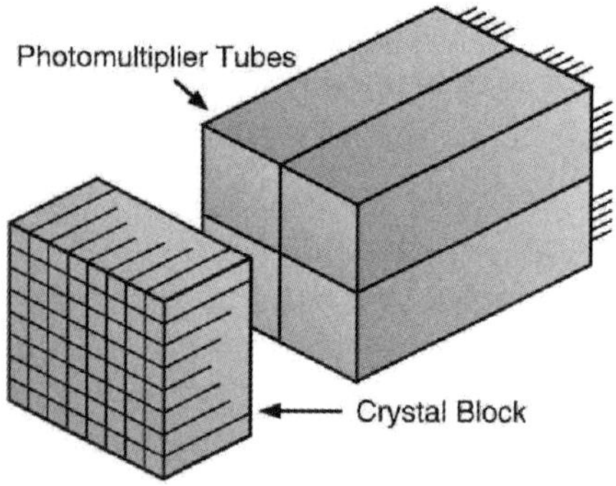

Fig. 10.41. Schematic drawing showing the different depths in the cuts between the pixel elements for a good identification with only 4 conventional photomultipliers.

The scanner therefore consists of 8 rings each with 112 elements. The sinogram generated has 56 (projection bins) by 56 (angular views) bins. Here too the angular views are halved to increase the number of projection bins. The sampling in the sinogram is therefore 1.61 mm and the total field of view is 81.32 mm transaxially and 47.6 mm axially.

The blocks are grouped into 4 'buckets' consisting of 4 blocks each. Each bucket is controlled by its electronics which determines the energy, timing, and position of the event. Coincidences, randoms, and singles are determined by the coincidence processing board. The coincidence window is set to 12 ns and different energy windows can be chosen. Each bucket is put in coincidence with all others.

The spatial resolution of this scanner is 2.3 mm FWHM at the center degrading to 6.6 mm (radial) and 4.4 mm (tangential) at 4 cm from the center. The axial slice width is 4.3 mm FWHM and remains virtually constant along the axial axis.

The solid angle covered by this scanner is very large. The 2D sensitivity (taking only the direct planes and adjacent cross planes) of this scanner is 0.01 cps/Bq at the center for a 250-850 keV energy window whereas the 3D sensitivity for a 380-850 keV energy window is 0.043 cps/Bq which is equivalent to a 4.3% absolute efficiency. A summary of the performances of this scanner are reported in Table 10.4.

Table 10.4. Summary of the performances of the Hammersmith RatPET.

Spatial resolution (mm)	2D	3D
Transaxial		
At centre of FOV:	2.3 FWHM; 5.6 FWTM	2.4 FWHM; 6.7 FWTM
40mm from centre of FOV:		
radial	6.6 FWHM; 10.4 FWTM	6.3 FWHM; 11.2 FWTM
tangential	4.6 FWHM; 14.4 FWTM	4.5 FWHM; 10.56 FWTM
Axial		
Centre of transaxial FOV:	4.3 FWHM; 10.3 FWTM	4.6 FWHM; 15.0 FWTM
20mm from centre of FOV:	4.4 FWHM; 10.6 FWTM	5.0 FWHM; 18.8 FWTM
Sensitivity		
2D	0.01 cps/Bq @ 250-850 keV	
3D	0.043 cps/Bq @ 250-850 keV	

Fig. 10.42. Photograph of the RatPET scanner.
http://www.irsl.org/research/biology/sas_image.html

A photograph of the scanner is shown in Fig. 10.42. The total dimensions of the scanner are about 1×1 m^2.

This is an example of a scanner which has successfully taken advantage of existing commercial technology to design a dedicated small animal scanner with very interesting performances. In this design priority was given to the sensitivity at the price of a significant degradation of the spatial resolution away from the center of the scanner. The primary cause of this, is the parallax error due to the depth of interaction given the thickness of the BGO scintillator. This scanner is practically a rescaled clinical scanner.

The use of this scanner demonstrated the potentiality of dedicated small animal scanners in the study of small animals.

10.5.2.2 *MicroPET*

The MicroPET scanner [15] was developed at the UCLA Crump Institute. It has a ring geometry and is based on LSO (Lutetium Orthosilicate) scintillator matrices with each pixel coupled to a single channel of a multichannel photomultiplier.

Each matrix is composed of 64 (8×8) small 2×2×10 mm^3 crystals packed together and separated by a white diffuser to increase light collection. This results in the pitch of the crystals being 2.25 mm. The readout is performed by a 64 channel multianode photomultiplier (Philips XP1722) coupled to the matrices through a bundle of white fibers as can be seen in Fig. 10.43.

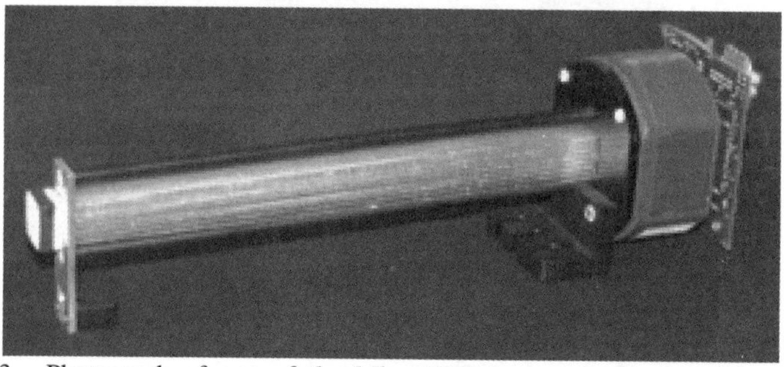

Fig. 10.43. Photograph of one of the MicroPET detector head assembly. The LSO scintillator matrix (far left) is coupled to the multianode photomultiplier through rather long light guides.
http://www.crump.ucla.edu/user-files/resprojects/microPET/fr_perf.html

This design was necessary so as to put the matrices closely packed in a ring. A total of 30 blocks are placed in a ring with a diameter of 17.2 cm. The tomograph therefore has 8 crystal rings with 240 crystals per ring. The field of view of this scanner has a diameter of 11.25 cm and an axial length of 1.8 cm. The principle characteristics of the MicroPET scanner are summarized in Table 10.5 and a photograph of the scanner is shown in Fig. 10.45.

An image of the crystal matrix irradiated with 511 keV gamma rays as seen by the readout system is shown in Fig. 10.44. Each single pixel can be distinguished very nicely.

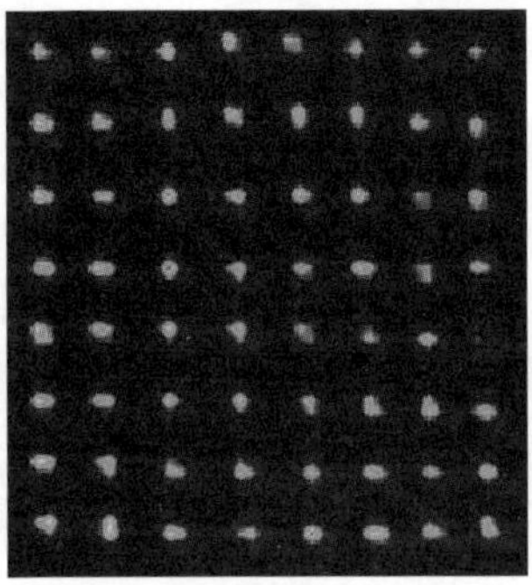

Fig. 10.44. Image of one of the 8x8 crystal matrices and energy spectrum. http://www.crump.ucla.edu/user-files/resprojects/microPET/fr_perf.html

Table 10.5. Principle characteristics of the MicroPET scanner [15]

Number of detector modules:	30
Ring diameter:	172 mm
Animal port:	160 mm
Transaxial field-of-view:	112 mm
Axial field-of-view:	18 mm
Volumetric resolution:	6 μl
Sensitivity:	0.0055 cps/Bq
Coincidence time window:	12 ns
In-plane bed wobble:	0.76 mm radius
Axial bed motion:	0.1 mm accuracy
Number of matrices:	64 (3-D only), 1.54 MBytes
Sinogram size:	100 views x 120 angles
Sampling distance	1.125 mm for non wobbled acquisitions

Typically the scanner is operated with an energy window of 250-650 keV and the sensitivity at the center is 0.0055 cps/Bq. A figure of the sensitivity as a function of the position in the FOV can be seen in Fig. 10.37.

Fig. 10.45. Photograph of the MicroPET small animal scanner.
http://www.crump.ucla.edu/user-files/resprojects/microPET/fr_perf.html

The readout of the electronics was designed so as to be able to use commercially available electronics from clinical manufacturers. In fact the 64 MC-PMT signals are multiplexed so as to emulate the outputs from 4 PMTs of conventional block detectors and these signals are then sent to the analog electronics (CTI PET Systems). This electronics identifies the crystal of interaction and integrates the charge. The coincidence processing is then performed by the same version supplied with the ECAT ART clinical scanner. The system has a coincidence window of 12 ns and allows on-line subtraction of random coincidences. Coincidences are taken between a crystal and the 50 opposing ones generating a transaxial FOV of 11.25 cm. The sinogram sampling is set to 100 bins per projection (linearly spaced by 1.125 mm) and 120 angular projections. Similarly the axial sampling is also 1.125 mm.

In Fig. 10.46 the intrinsic spatial resolution of the scanner is reported, obtained by moving a point like source through the FOV by 0.25 mm steps and by taking the profiles directly from the sinogram.

The bed, which hosts the animal, is controlled electronically and can wobble by 300 μm in the transaxial plane to increase sampling.

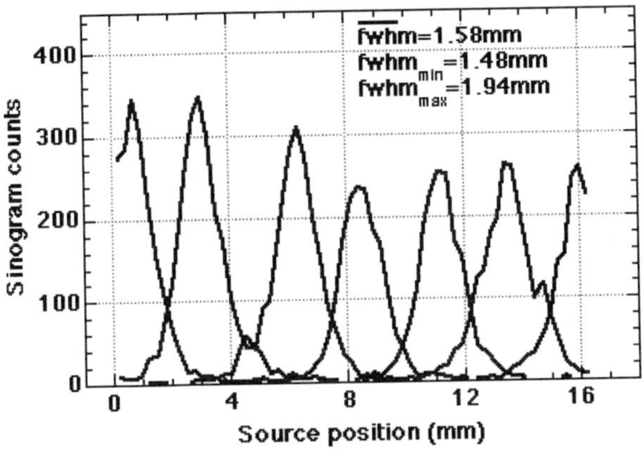

Fig. 10.46. Intrinsic spatial resolution of the MicroPET scanner.
http://www.crump.ucla.edu/user-files/resprojects/microPET/fr_perf.html

The overall scanner measures 120 cm in height by 100 cm in width and 60 cm deep with an imaging aperture of 16 cm.

This small animal PET scanner has a spatial resolution of 1.8 mm at the center and remains below 2.5 mm for the central 5 cm of the FOV. A curve of the spatial resolution as a function of the position within the field of view is shown in Fig. 10.39. The shape is typical of ring scanners. The radial spatial resolution deteriorates whereas the axial one does not. As discussed previously this is due to parallax errors which are present mainly in the transaxial plane.

The image of a hot Derenzo phantom is shown in Fig. 10.47 whereas in Fig. 10.48 one can see the reconstructed image of a whole mouse after having injected it with $^{18}F^{-}$.

The particularity of this scanner is the use of LSO pixellated scintillator coupled to position sensitive multianode photomultiplier tubes. Emphasis in the design and development of this scanner has been put on spatial resolution and use of this new, fast, high Z scintillator.

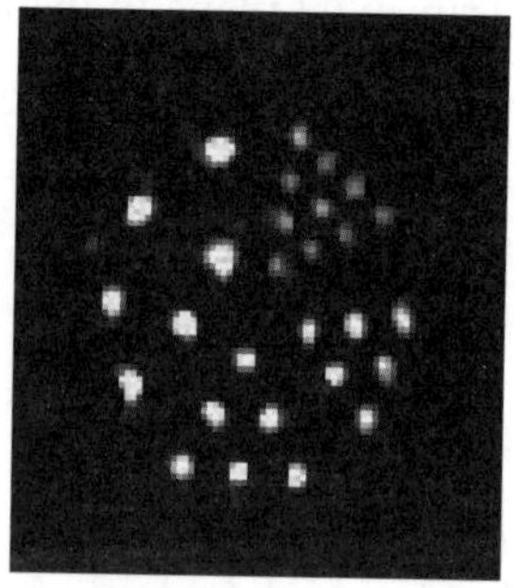

Fig. 10.47. Image of a Derenzo phantom. The hot spots have diameters of 1, 1.25, 1.5, 2, 2.5 mm with a center to center distance between them 4 times their diameter. MAP reconstruction algorithm [40].

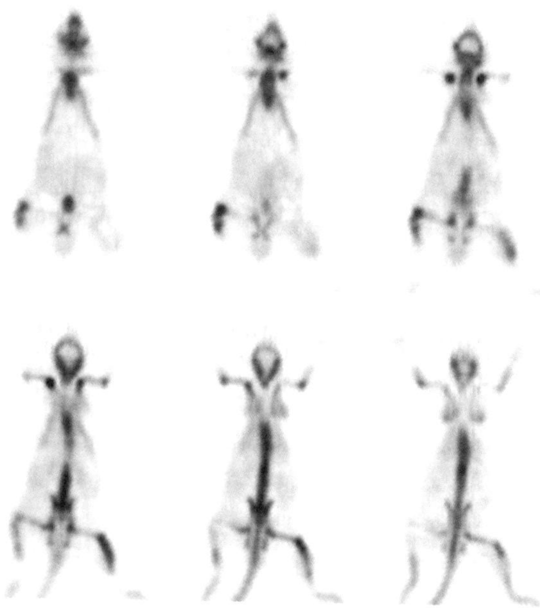

Fig. 10.48. Species: Mouse. Type of Study: Bone Scan (Whole Body). Tracer: $^{18}F^-$: Injected Dose: 1.0 mCi. Imaging Time: 8 min / bed position, 4 positions. http://www.crump.ucla.edu/user-files/resprojects/microPET/fr_perf.html

10.5.2.3 *Sherbrooke PET and the Munich MADPET*

The small scanner developed at Sherbrooke University [16, 18, 19] was the first to replace photomultiplier tubes with solid state photodetectors.

In Table 10.6 one can find the characteristics of this scanner. Each 3×5×20 mm thick BGO scintillator crystal was directly coupled to a single avalanche photodiode. The advantage of APD photodetectors is the high quantum efficiency for detecting light, small physical size and potentially low cost. On the other hand, a major drawback of such light detectors is the high gain dependence on temperature and voltage.

Table 10.6. Characteristics of the Sherbrooke small animal scanner.

Detector type	EG&G C30994 Dual BGO/APD
BGO crystal size	3x5x20 mm (beveled)
Module dimensions	3.8x13.2x33 mm
BGO crystal spacing	3.8 mm tangentially 5.5 mm axially
Number of detectors	256 per ring
Number of detector rings	2 (1 layer of modules)
Ring diameter	310 mm
Port diameter	135 mm
Useful field of view	118 mm
Axial field of view	10.5 mm
Reconstruction planes	3 (2 direct; 1 cross)
Intrinsic spatial resolution (center)	
Transaxial	1.9 mm FWHM; 3.5 mm FWTM
Axial	3.1 mm FWHM; 5.4 mm FWTM
Reconstructed resolution	2.1 mm FWHM; 3.9 mm FWTM
Sensitivity at the center (350 keV)	0.0051 cps/Bq
Energy resolution	<25 % FWHM @ 511 keV
Timing resolution	20 ns FWHM
Timing window	20 to 40 ns

Another difficulty is the necessity to use low noise preamplifiers due to APD gains of 100–1000. The scanner is composed of 2 rings, each with 256 crystal elements, and has a diameter of 31 cm. The system acquires data exclusively in 2D mode due to the reduced number of rings. The axial FOV is 10.5 mm sub-divided into 3 planes. The transaxial spatial resolution is 2.1 mm at the center and the axial slices are 3.1 mm thick.

The reported sensitivity is 3.3 kcps/μCi/ml for a 10.8 cm diameter cylinder and a lower energy threshold of 350 keV. The absolute sensitivity at the center is 0.005 cps/Bq.

Another scanner which also uses an APD readout is the MADPET developed by the Max Planck Institute in Munich [20]. Instead of BGO scintillator, the scanner is equipped with the faster and brighter LSO scintillator. The scanner is composed of two sets of 3 detector blocks which rotate around. The crystal elements have a cross section of 3.7 mm x 3.7 mm and are 12 mm long. The pitch of the elements is 4 mm due to the 0.3 mm reflective layer used. Each single crystal element is coupled directly to a pixel of an APD array. In Fig. 10.49 one can see a schematic drawing of the setup and photographs of the APD array and the LSO elements.

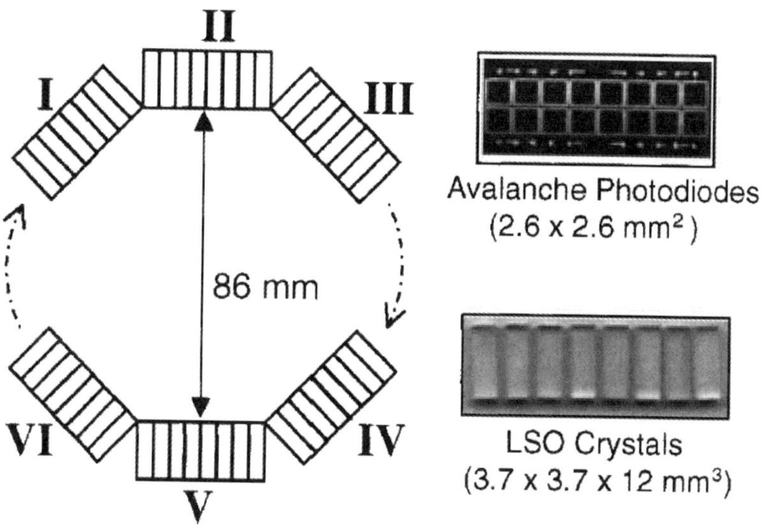

Fig. 10.49. Schematic drawing of the MADPET setup (left) and photographs showing the LSO crystal elements and the array of APDs used to read them out [20]. (Reprint authorized by Springer-Verlag, Germany.)

The measured intrinsic spatial resolution is 2.1±0.1 mm and the sensitivity is 0.00035 cps/Bq measured with a line source placed at the center of the scanner and extending across the whole axial FOV.

10.5.2.4 *The NIH Atlas scanner*

The Advanced Technology Laboratory Animal Scanner (ATLAS) developed by the National Institute of Health [28,29] is the first complete scintillator based system which uses depth of interaction information to reduce the degradation of spatial resolution at the edges of the FOV.

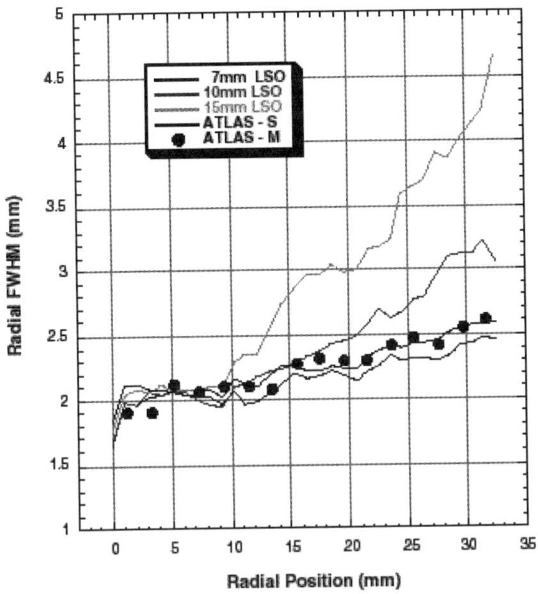

Fig. 10.50. Measured (ATLAS-M) and simulated (ATLAS-S) spatial resolution of the ATLAS system compared to the simulated spatial resolution of a PET scanner with the same diameter but with different crystal thicknesses and without depth of interaction measurement [29]. (Reprint authorized by IEEE, Piscataway, NJ, USA.)

Each pixel element is actually composed of a front scintillator (LGSO) and a back scintillator (LSO) mounted in a phoswich configuration. The LGSO scintillator is 7 mm long and the LSO scintillator is 8 mm long giving a total of 15 mm of high *Z* material. The scintillator in which the interaction has occurred is determined by measuring the decay time of the scintillation time: LGSO decays in 60 ns whereas LSO decays in 40 ns.

Each detector pixel element has a 2 mm $\times$ 2 mm cross section and each module is composed of 9 $\times$ 9 pixels. The pitch of the matrix is

2.25 mm due to the optical reflector. Each matrix is directly coupled to a position sensitive photomultiplier. A total of 18 modules are mounted in a ring generating an 11.8 cm diameter aperture and a 2 cm axial FOV. The sensitivity at the center is 2.7% (0.027 cps/Bq) with a 100 keV energy threshold.

With this technique the spatial resolution is maintained equivalent to that of a scanner with a 7 mm thick crystal as can be seen in Fig. 10.50.

10.5.2.5 *Scanner of the Brussels group: the VUB-PET*

The scanners presented until now all have followed the philosophy adopted by almost all clinical scanners: ring geometry with high *Z* scintillator and read out using photomultipliers. Here the readout of the BaF_2 scintillator light has been replaced by a multiwire proportional chamber.

Fig. 10.51. Photo of BaF2 small animal PET scanner.
http://web.iihe.ac.be/pet/default.htm#new%20PET%20scanner

This scanner [21] was also developed with the idea of reducing costs so as to have a relatively large solid angle coverage and therefore high sensitivity. High spatial resolution position sensitive photomultipliers are difficult to use for reading out large area scintillators. Multiwire proportional chambers are an alternative technique for detecting light emitted from scintillators especially if large area detectors are desired.

The scintillator crystal used in this scanner is BaF_2 which scintillates in the near UV region. It has two components: a fast component which peaks at a wavelength of 220 nm with a decay time of 0.6 ns and a light yield of 1500 photons/MeV; a slow component peaking at 310 nm. The fast component couples nicely to TMAE (tetrakis-dimethylamine-ethylene) gas for photoconversion, thereby generating photoelectrons. This gas has a good photoelectric efficiency to UV light. The photoelectrons are then drifted to an amplification region where the position and timing of the events can be acquired.

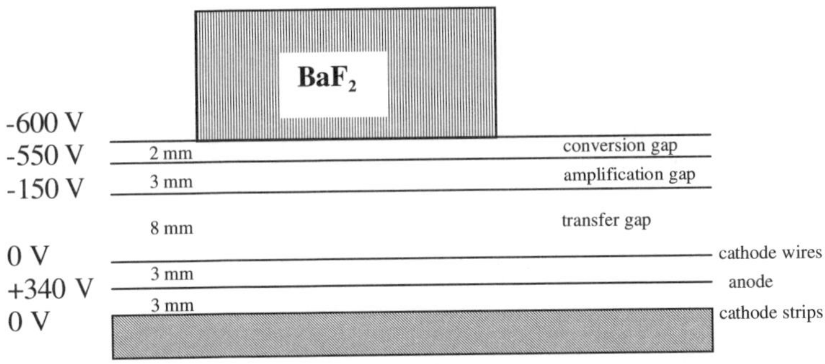

Fig. 10.52. Scheme of the read out system using a two stage multiwire proportional chamber.

The scanner is composed of 2956 BaF_2 crystals each having dimensions of $3\times3\times20$ mm^3. Rows of 17 crystals point towards the axis of the scanner forming a 20 cm diameter cylinder, 5.2 cm long cylinder. The transaxial field of view of the scanner is 11 cm and the axial one is 5.2 cm long.

The light readout is done with a two stage multiwire proportional chamber. In Fig. 10.52 one can see the scheme of the readout.

Due to the efficiency of TMAE the number of photoelectrons generated from 511 keV photons is on average 3.5, not allowing any sort of energy information. The timing of the chambers permits a 29 ns FWHM coincidence. The solid angle coverage and the 3D acquisition give a good sensitivity of this setup: 0.035 cps/Bq at the center.

The spatial resolution of the scanner is shown in Fig. 10.53. At the center the value is 3 mm FWHM both radially and tangentially degrading to 5.5 mm radially and 4 mm tangentially at 4 cm from the center of the FOV.

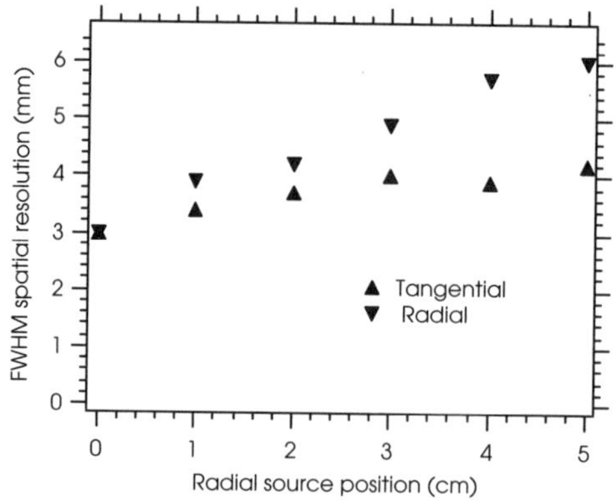

Fig. 10.53. Tangential and Radial FWHM spatial resolution of the Brussels's VUB-PET scanner [21]. (Reprint authorized by IEEE, Piscataway, NJ, USA.)

10.5.3 *Dedicated Rodent Rotating Planar Scanners*

This section will present some PET scanners which have adopted rotating planar geometries for their detectors. Typically the axial and transaxial lengths are very similar giving a relatively large solid angle coverage with a reduced number of detector heads. These are characterized by simple readout schemes and coincidence electronics and good performances. As discussed earlier, these scanner maintain an almost constant spatial resolution within the whole FOV at the cost of a variable radial sensitivity.

10.5.3.1 *YAP-(S)PET and TierPET*

The YAP-(S)PET [22, 23] and TierPET [27] scanners are two very similar scanners. Both have adopted the recently revived scintillator:

YAP:Ce. This scintillator has a good light yield, very fast decay constant and very good machining properties.

The particularity of this scintillator is its high density, but medium Z, resulting in a photoelectric fraction of only 4.4% at 511 keV. Nonetheless the efficiency for detecting 511 keV photons can be made high by increasing the thickness of the scintillator and taking advantage of Compton interactions within the crystal rather than photoelectric interaction. The very good mechanical crystal properties have allowed the manufacturing of crystal matrices with pixel sizes down to 600 μm, 1 cm long separated by a thin 5 μm reflective layer.

The two scanners presented here have both implemented 20 by 20 pixel matrices with a pixel size of 2 mm but with different thicknesses: TierPET has used 1.5 cm thick crystals whereas the YAP-(S)PET has used 3 cm thick crystals. The choice in thickness is dictated by a different philosophy in the event selection as will be discussed below.

The readout of the matrices is done in both scanners with Hamamatsu's position sensitive photomultiplier tubes. The TierPET uses the R2487 3 inch square tube whereas the YAP-(S)PET uses the R2486 round 3 inch tubes. The scanners are equipped with 4 rotating detector heads placed at 90° on a rotating gantry. They have a FOV with a diameter of 4 cm and an axial length of 4 cm. As discussed previously and as will be shown below, the spatial resolution is almost uniform within the FOV. Acquisition is done in 3D mode and the data are stored in list mode.

Both scanners have variable geometries: TierPET varies the radius of the scanner for maintaining a maximum sensitivity compatible with the electronics; YAP-(S)PET varies the tangential position of each pair of opposing detectors so as to increase the transaxial FOV.

The philosophy of event selection in the two scanners is very different: TierPET uses a lower energy threshold of about 250 keV to isolate the full absorption events whereas YAP-(S)PET uses two energy windows: 50 – 650 keV to accept all events, and 50 – 400 keV to discard full absorption events.

The second energy window for YAP-(S)PET is chosen so as to select only those events which have undergone Compton scattering within the detector. The very fast timing of the scanner (2 ns FWHM) and the reduced scattering within the animal make this an interesting alternative for multiple scattering rejection. In fact, as shown in Fig. 10.30, due to the low photofraction, almost all events under the full absorption peak are multiple events which therefore degrade spatial resolution.

TierPET was optimized for spatial resolution when using a classical energy selection scheme, including full absorption events, at the cost of a lower sensitivity. The design of the YAP-(S)PET emphasized both sensitivity and spatial resolution with an unusual event selection which can be successfully applied to YAP:Ce scintillator due to its low Z.

Let us consider the performances of the two scanners separately.

- TierPET

In Fig. 10.54 one can see a photograph of the TiePET scanner. The radial mobility of the detector heads can clearly be seen. The spatial resolution of the TierPET scanner was shown in Fig. 10.35. The value is about 2.1 mm, uniform within the entire FOV.

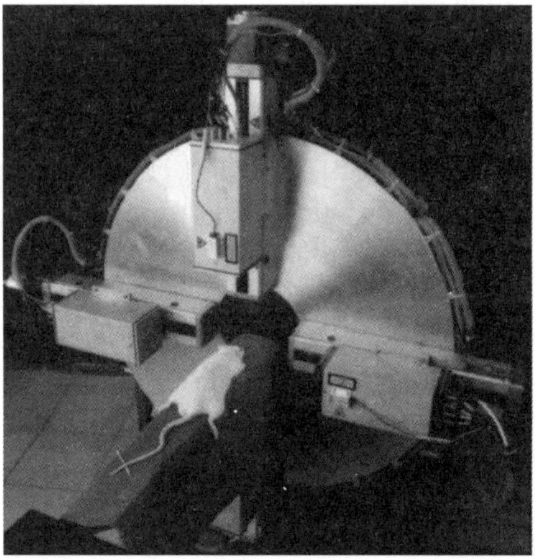

Fig. 10.54. Photograph of the TierPET scanner [27]. (Reprint authorized by IEEE, Piscataway, NJ, USA.)

The maximum sensitivity at the center with the detectors 16 cm apart is 0.0032 cps/Bq equivalent to 0.32% efficiency. This value reflects a little bit the detector configuration and the event selection philosophy chosen. In Fig. 10.55 some images show the very nice capabilities of this scanner.

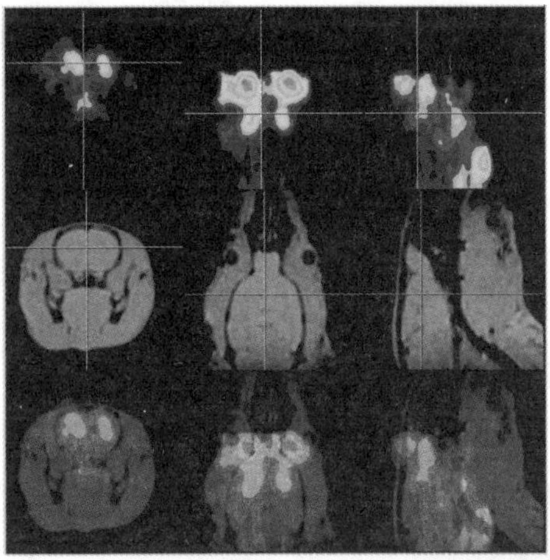

Fig. 10.55. Image of a rat brain: Uptake of the dopamine D2 receptor ligand [^{18}F]methyl-benperidol in the caudate puntamen of a rat brain. Injected activity was 1.4 mCi [27]. (Reprint authorized by IEEE, Piscataway, NJ, USA.)

- YAP-(S)PET

An energy spectrum and a time spectrum are shown in Fig. 10.56. One can see here that the energy spectrum is dominated by Compton events due to the low Z of YAP:Ce scintillator. The full absorption peak accounts for many more events than the 4.4% photofraction indicating that most of these events are multiple interaction events within the crystal. Furthermore a large fraction of the Compton events (c.a. 60%) leave the detector after the first interaction (Fig. 10.30).

The timing properties of YAP:Ce are very interesting as can be seen from the 2 ns FWHM time resolution of a pair of detectors.

The timing information is acquired with the position information allowing the choice of different time windows. The sensitivity of the scanner at the center is 0.017 cps/Bq when choosing all events, decreasing to about 0.007 cps/Bq when choosing only events between 50 and 400 keV.

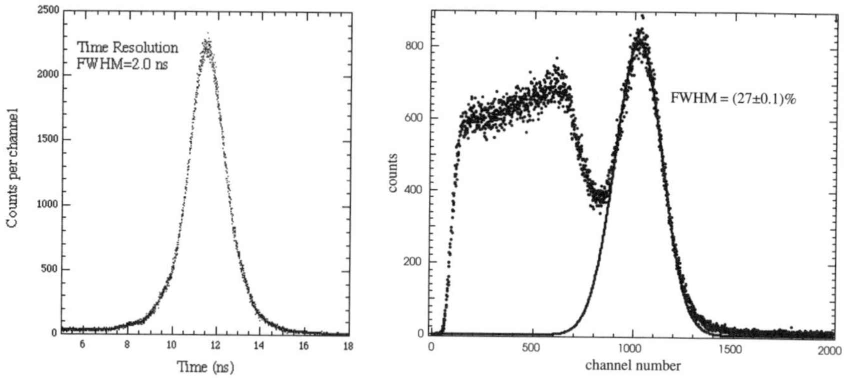

Fig. 10.56. Time distribution (left) between two YAP:Ce detector heads of the YAP–(S)PET scanner and energy spectrum (right).

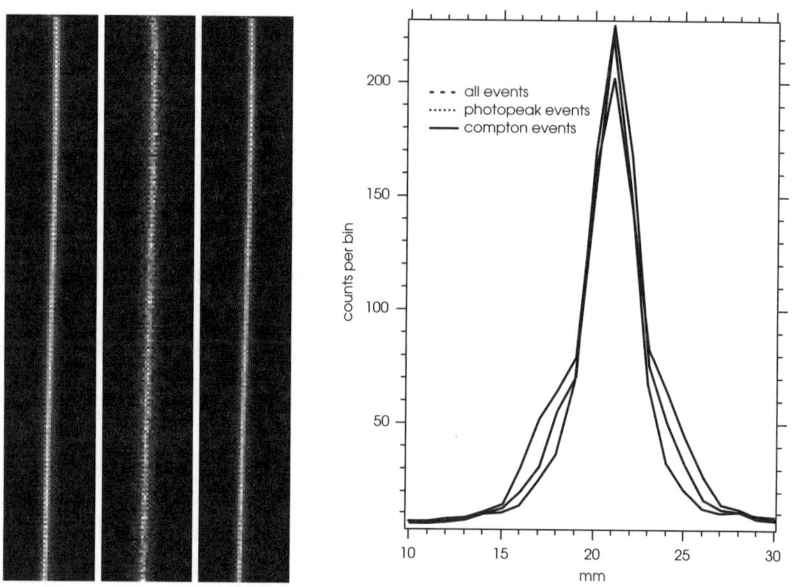

Fig. 10.57. Comparison of sinograms obtained from a 1 mm capillary filled with FDG. Left; all events: Middle; events with E > 400 keV; right; events with E < 400 keV. The graph reports the profiles of the three sinograms.

In Fig. 10.57 one can see the sinograms obtained from a 1 mm diameter capillary placed near the center of the FOV. The difference in the sinograms when applying the different energy windows is clear, the best being the one on the right where only Compton events are selected. A profile of the three sinograms is also shown.

The reconstructed (FBP) FWHM and FWTM profiles of a 0.8 mm ^{22}Na source are shown in Fig. 10.58 with an energy window of 50 – 650 keV.

Figure 10.59 shows the images of a bone scan of the upper part of a rat. These images were obtained after having injected the rat with 500 μCi of ^{18}F^{-}. The data were acquired while moving the bed along the scanner axis at 4 positions, each 2 cm apart and lasted ten minutes each. The images were reconstructed with a limited angle filtered back projection algorithm. Lots of structures are visible but this image suffers from low statistics.

A photograph of the YAP-(S)PET scanner with the cover removed is shown in Fig. 10.60.

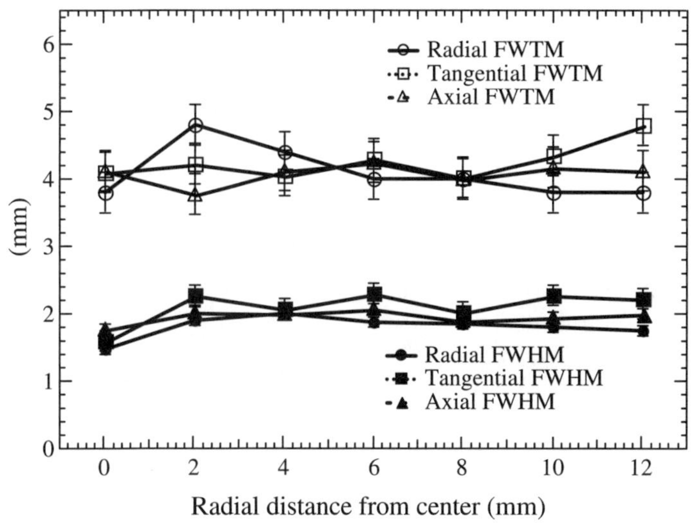

Fig. 10.58. Spatial resolution measured with a 0.8 mm diameter ^{22}Na source [22]. (Reprint authorized by IEEE Piscataway, NJ, USA.)

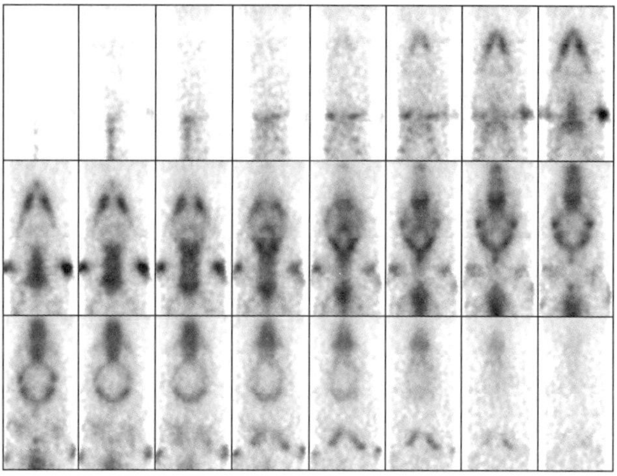

Fig. 10.59. Image of a rate skeleton. The injected activity of $^{18}F^{-}$ was 500 μCi. The data were acquired by placing the bed along the scanner's axis at 4 different positions, each 2 cm apart. Each acquisition lasted ten minutes.

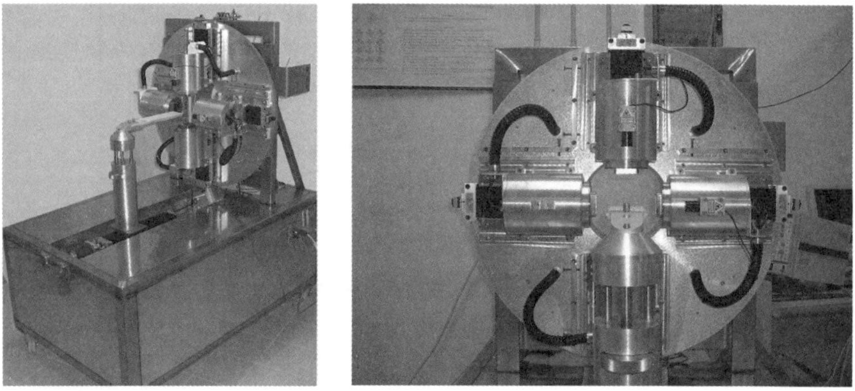

Fig. 10.60. Photograph of the YAP-(S)PET scanner with the cover removed.

10.5.3.2 *HIDAC*

Other techniques, which are not based on scintillator detectors, can also be found to detect 511 keV gamma rays in small animal PET applications.

Using either lead glass capillary tubes packed together [47] or lead foils with small holes drilled in them [25], one can convert 511 keV gamma rays into electrons. A good example is the HIDAC small animal scanner developed by Oxford Positron Systems (Weston-on-the-Green, UK) shown in Fig. 10.61. HIDAC stands for High Density Avalanche Chamber.

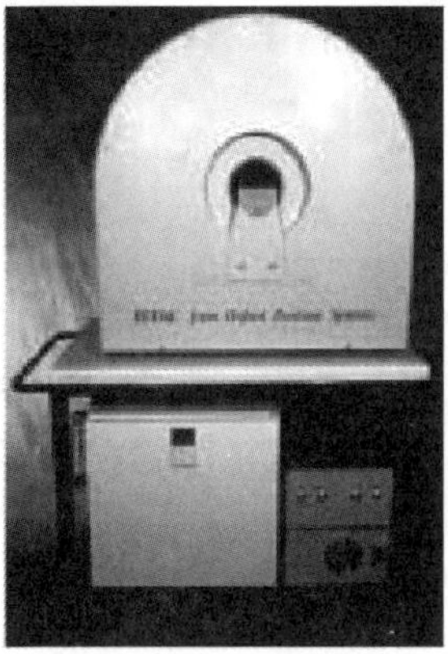

Fig. 10.61. Photograph of the HIDAC scanner (from [9], reprint authorized by Springer-Verlag.)

The detector consists of a multiwire proportional chamber with the addition of high Z conversion plates. These are laminated plates containing interleaved lead and insulating sheets with a mechanically drilled dense matrix of small holes. The interaction of a 511 keV gamma with the lead plates generates electrons which can leave the lead reaching the small holes. These electrons are trapped, amplified, and extracted from the small holes into the MWPC. This process is schematically shown in Fig. 10.62.

Such a technique gives very good spatial resolution. The spatial resolution of the HIDAC-PET detector is plotted in Fig. 10.63 as a function of the gamma ray's incident angle. In this case the holes in the lead foils have a diameter of 0.4 mm and a pitch of 0.5 mm.

Although there is no energy information in the acquired electrical signals, the system has an intrinsic rejection of low energy gamma rays. In fact, the electrons generated in the lead must have a sufficient range, and therefore enough energy to reach the small holes. For low energy gamma rays, most of the electrons remain trapped in the lead and do not generate a signal.

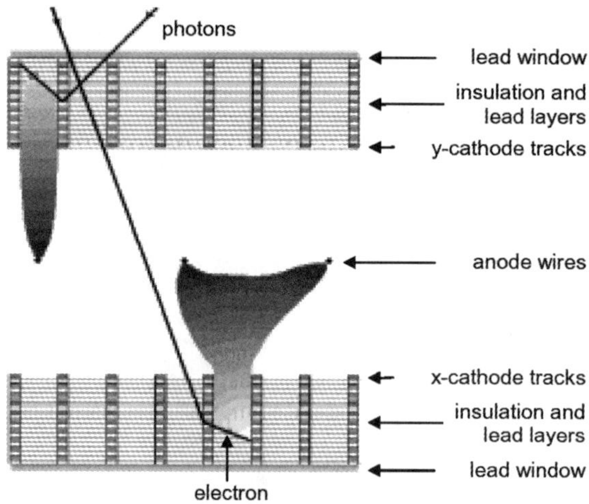

Fig. 10.62. Scheme of the HIDAC MWPC (from [25], reprint authorized by IEEE Piscataway, NJ, USA.)

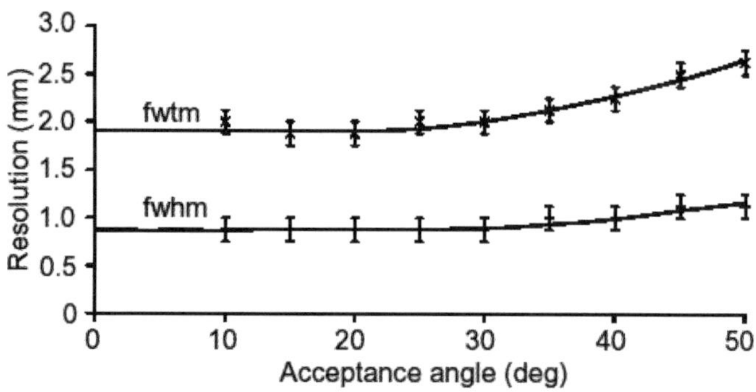

Fig. 10.63. Spatial resolution of a HIDAC detector module in function of the incident angle of a 511 keV gamma ray (from [25], reprint authorized by IEEE, Piscataway, NJ, USA.)

The scanner is composed of 4 planar rotating modules. The axial FOV is 21 cm and the distance between the detectors can vary from 10 to 20 cm. The reconstructed spatial resolution at the center of the FOV, using a filtered back projection algorithm, is 0.95 mm FWHM in the transverse direction and 1.2 mm FWHM in the axial direction.

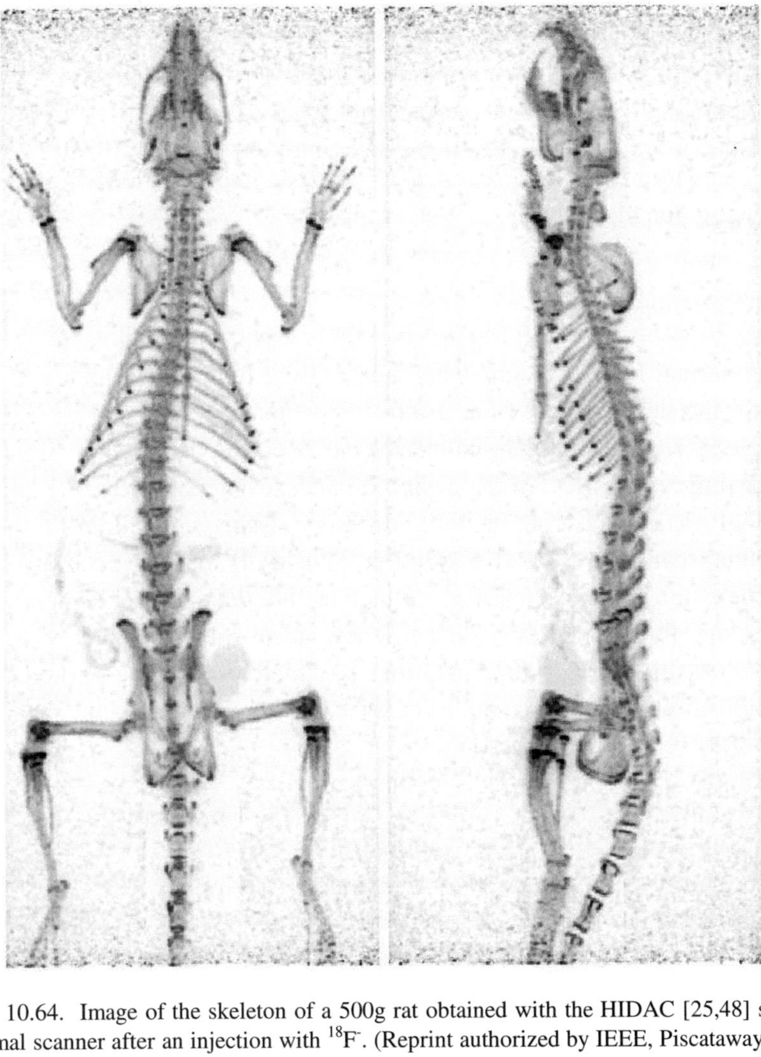

Fig. 10.64. Image of the skeleton of a 500g rat obtained with the HIDAC [25,48] small animal scanner after an injection with $^{18}F^{-}$. (Reprint authorized by IEEE, Piscataway, NJ, USA.)

The detection efficiency of a HIDAC module is not very high and therefore six modules are stack together to make a detector head. This modular structure of each detector head also allows depth of interaction measurement, thereby reducing greatly parallax error which could be important given the large axial FOV.

One drawback of using MWPC is their timing. In fact the FWHM timing resolution of a pair of HIDAC detectors is 29 ns. The timing window is 60 ns and given the large FOV this limits somewhat the count rate capabilities.

In Fig. 10.64 an impressive image of a rat injected with $^{18}F^{-}$ was obtained with the HIDAC scanner [48] with an improved OSEM iterative reconstruction algorithm.

10.6 Conclusions

Small animal *in-vivo* imaging is an expanding field of research and has lead to the development of many different imaging systems. Both PET and SPECT small animal scanners have been described showing very interesting capabilities. In the design of such scanners the various groups have given emphasis to slightly different aspects resulting in rather different scanners. Spatial resolution, sensitivity, cost, versatility and simplicity are some of the aspects which lead to the different designs.

In the future, one will expect new ideas and techniques to be developed to improve the capabilities obtained. Avalanche photodiodes with position sensitive capabilities and very finely pixellated matrices will allow submillimeter resolutions.

Another very interesting development in this field is the design of multi modality systems: PET-CT, PET-MRI, PET-SPECT [49]. A precursor in this sense is the YAP-(S)PET [22,23] scanner described in this chapter, which is capable of both PET and SPECT tomography with interesting results.

References

1. Susan P. Hume, Terry Jones "Positron Emission Tomography (PET) Methodology for Small Animals and Its Application in Radiopharmaceutical Preclinical Investigation". *Nucl Med Biol 25*: 729–732 (1998)
2. Michael E. Phelps "PET: The Merging of Biology and Imaging into Molecular Imaging". *Journal Nucl Med 41*: 661–680 (2000)
3. M. P. Tornai, R. J. Jaszczak, T. G. Turkington, R. E. Coleman. Editorial: "Small-Animal PET: Advent of a New Era of PET Research". *Journal Nucl Med 40*: 1176–1179 (1999)
4. Joanna S. Fowler, Nora D. Volkow, Gene-Jack Wang, Yu-Shin Ding and Stephen L. Dewey. "PET and Drug Research and Development". *Journal Nucl Med 40*: 1154–1163 (1999)
5. M. E. Phelps Inaugural article: "Positron Emission Tomography provides molecular imaging of biological processes". *Proc Natl Acad Sci USA 97*: 9226–9233 (2000)
6. M. E. Phelps. "PET: a biological imaging technique". *Neurochem Res 16*: 929–940 (1991)
7. R. Hichwa "Are Animal Scanners Really Necessary for PET?" *Journal Nucl. Med., 35 (8)*: 1396–1397 (1994)
8. R. Myers. "The Biological Application of Small Animal PET imaging". *Nucl Med Biol 28*: 585–593 (2001)
9. Arion Chatziiouannou. "Molecular imaging of small animals with dedicated PET tomographs". *Europ Jou Nucl Med 29*: 98–114 (2002)
10. P. D. Cutler, S. R. Cherry, E. J. Hoffman, W. M. Digby, M. E. Phelps. "Design features and performance of a PET system for animal research". *J Nucl Med 33*: 595–604 (1992)
11. Simon Cherry. "Watching biology in action". *Physics World* June 29 (2002)
12. M. Watanabe, H. Uchida, K. Shimizu, N. Satoh, E. Yoshikawa, T. Omura, T. Yamashita and E. Tanaka. "A high resolution PET for animal studies". *IEEE Trans Med Imag 11*: 577–580 (1992)
13. M. Watanabe, M. Okada, K. Shimizu, T. Omura, E. Yoshikawa, T. Kosugi, S. Mori, T. Yamashita. "A high resolution animal PET scanner using compact PS-PMT detectors". *IEEE Trans Nucl Sci 44*: 1277–1282 (1997)
14. P. M. Bloomfield, S. Rajeswaran, T. J. Spinks, S. P. Hume, R. Myers, S. Ashworth, K. M. Clifford, W. F. Jones, L. G. Byars, J. Young, M. Andreaco, C. W. Williams, A. A. Lammertsma and T. Jones. "The design and physical characteristics of a small animal positron emission tomograph". *Phys Med Biol 40*: 1105–1126 (1995)
15. M. Bloomfield, R. Myers, S. P. Hume, T. J. Spinks, A. A. Lammertsma and T. Jones. "Three-dimensional performance of a small-diameter positron emission tomograph". *Phys Med Biol 42*: 389–400 (1997)

16. A. F. Chatziiouannou, S. R. Cherry, Y. Shao, R. W. Silverman, K. Meadors, T. H. Farquhar, M. Pedarsani and M. E. Phelps. "Performance Evaluation of microPET: A High-Resolution Lutetium Oxyorthosilicate PET Scanner for Animal Imaging". *J Nucl Med 40*: 1164–1175 (1999)
17. C. J. Marriott, J. Cadorette and R. Lecomte "High-resolution PET Imaging and Quantitation Pharmaceutical Biodistribution in a Small Animal Avalanche Photodiode Detector" *J. Nucl. Med., 35 (8)*: 1390–1395 (1994)
18. R. Lecomte, J. Cadorette, P. Richard, S. Rodrigue and D. Rouleau. "Design and Engineering Aspects of a High Resolution Positron Tomograph for Small Animal Imaging". *IEEE Trans Nucl Sci 41*: 1446–1452 (1994)
19. R. Lecomte, J. Cadorette, S. Rodrigue, D. Lapointe, D. Rouleau, M. Bentoukia, R. Yao and P. Msaki. "Initial Results from the Sherbrooke Avalanche Photodiode Positron Tomograph". *IEEE Trans Nucl Sci 43*: 1952–1957 (1996)
20. S. I. Ziegler, B. J. Pichler, G. Boening, M. Rafecas, W. Pimpl, E. Lorenz, N. Schmitz and M. Schwaiger. "A prototype high resolution animal positron tomograph with avalanche photodiode arrays and LSO crystal". *Eur J Nucl Med 28*: 136–143 (2001)
21. P. Bruyndonckx, L. Xuan, S. Rajeswaran, W. Smolik, S. Tavernier and Z. Shuping. "Design and physical characteristics of a small animal PET using BaF_2 crystals and a photosensitive wire chamber". *Nucl Inst Meth in Phy Res A 382*: 589–600 (1996)
22. A Del Guerra, G. Di Domenico, M. Scandola and G. Zavattini "YAP-PET: first results of a small animal Positron Emission Tomograph based on YAP:Ce finger crystals" *IEEE Trans. Nucl.Sci6, 45 (6)*: 3105–3108 (1998)
23. C. Damiani, A. Del Guerra, G. Di Domenico, M. Gambaccini, A. Motta, N. Sabba and G. Zavattini. "An integrated PET/SPECT imager for small animals". *Nucl Instr Meth Phys Res A 461*: 416–419 (2001)
24. J. A. Correia, C. A. Burnham, D. Kaufman and A. J. Fischman. "Development of a small animal PET imaging device with resolution approaching 1 mm". *IEEE Trans Nucl Sci 46*: 631–635 (1999)
25. S. I. Ziegler, B. J. Pichler, G. Boening, M. Rafecas, W. Pimpl, E. Lorenz, N. Schmitz and M. Schwaiger. "A prototype high-resolution animal positron tomograph with avalanche photodiode arrays and LSO crystal". *Eur J Nucl Med 28*: 136–143 (2001)
26. A. P. Jeavons, R. A. Chandler and C. A. R. Dettmar. "A 3D HIDAC-PET Camera with Sub-millimeter Resolution for Imaging Small Animals". *IEEE Trans Nucl Sci 46*: 468–473 (1999)
27. S. Weber, H. Herzog, M. Cremer, R. Engels, K, Hamacher, F. Kehren, H. Muehlensiepen, L. Ploux, R. Reinartz, P. Reinhart, F. Rongen, F. Sonnenberg, H. H. Coenen and H. Halling. "Evaluation of the TierPET system". *IEEE Trans Nucl Sci 46*: 1177–1183 (1999)
28. J. Seidel, J. J. Vaquero and M. V. Green. "Resolution uniformity and sensitivity of the NIH ATLAS small animal PET scanner: comparison to simulated LSO scanners without depth-of-interaction capability". *IEEE Nucl Sci Symp Conf Rec 3*: 1555–1558 (2001)

29. J. J. Vaquero, J. Seidel, I. Lee, F. J. Barbosa and M. V. Green. "The NIH Alas Small Animal PET Scanner". In *High resolution imaging in small animals*
30. D. P. McElroy, L.R. MacDonald, F.J. Beekman, Yuchuan Wang; B.E. Patt, J.S. Iwanczyk, B.M.W. Tsui and E.J. Hoffman, "Evaluation of A-SPECT: a desktop pinhole SPECT system for small animal imaging" *IEEE Nuclear Science Symposium Conference Record 3*: 1835 –1839. Piscataway, NJ, USA (*2001*)
31. D. P. McElroy, L. R. MacDonald, Y. Wang, B. E. Patt and J. S. Iwanczyk. "Ultra High Resolution In Vivo I-125 and Tc-99m Small Animal Pinhole SPECT Imaging". In High Resolution Imaging in Small Animals: Instrumentation, Applications and Animal Handling, Rockville, Maryland, USA (2001)
32. M. B. Williams, G. Zhang, M. J. Mitali, V. Dabak, S. Krishnamurthi, D. Narayanan, S. S. Berr, D. Pan, M. Robers, S. Majewsk and R. Wojcik "Prototype CT-SPECT Scanner for Small Animal Imaging". In High Resolution Imaging in Small Animals: Instrumentation, Applications and Animal Handling, Rockville, Maryland, USA (2001)
33. B. M. W. Tsui, T. C. Wamg, E. C. Freu and D. E. Wesse. "High Resolution Pinhole SPECT Imaging of Small Animals using Scintillation Camera". In High Resolution Imaging in Small Animals: Instrumentation, Applications and Animal Handling, Rockville, Maryland, USA (2001)
34. M. C. Wu, H. R. Tang, D. W. Gao, A. Ido, J. W. O'Connell, B. H. Hasegawa and M. W. Dae, "ECG-gated pinhole SPECT in mice with millimeter spatial resolution", *IEEE Trans on Nucl Sci 47*: 1218 –1221 (2000)
35. K. Ogawa, T. Kawade, K. Nakamura, A. Kubo and T. Ichihara, "Ultra high resolution pinhole SPECT for small animal study", *IEEE Trans on Nucl Sci 45*: 3122 –3126 (1998)
36. N. Schramm, A. Wirrwar, F. Sonnenberg, H. Halling, "Compact high resolution detector for small animal SPECT". *Nuclear Science Symposium Conference Records, 1999 IEEE,* Part 3, 1479 – 1482 *3*: Piscataway, NJ, USA (1999)
37. A. G. Weisenberger, E. L. Bradley, S. Majewski, M. S. Saha, M. F. Smith, R. E. Welsh and R. Wojcik. "Development of a Miniature SPECT-CT System for Smal Animal Imaging". In High Resolution Imaging in Small Animals: Instrumentation, Applications and Animal Handling, Rockville, Maryland, USA (2001)
38. L. Ploux, A. Valda Ochoa, L. Pinot, A Glatigny, Y. Charon, P. Laniece, F. Lefebvre, P. Hantraye, B. Tavitian, L. Valentin, and R. Mastrippolito. "An Original Approach for Small Animal Emission Tomography: TOHR; Characterisation and First Evaluation". *Nuclear Science Symposium Conference Records, 2000 IEEE 3*: 21/24 – 21/29. Piscataway, NJ, USA (2000)
39. Zhonglin Liu, George A. Kastis, Gail D. Stevenson, Harrison H. Barrett, Lars R. Furenlid, Matthew A. Kupinski, Dennis D. Patton and Donald W. Wilson. "Quantitative Analysis of Acute Myocardial Infarct in Rat Hearts with Ischemia–Reperfusion Using a High-Resolution Stationary SPECT System". *J Nucl Med 43*: 933–939 (2002)

40. Jinyi Qi and R. M. Leahy. "Resolution and Noise Properties of MAP Reconstruction for Fully 3-D PET". *IEEE Trans Med Imag 19*: 493–506 (2000)
41. A. Del Guerra, G. Di domenico, M. Scandola, G. Zavattini D. Bollini M. Galli and A. Bevilacqua "A 3-D Monte Carlo simulation of a small animal Positron Emission Tomograph with millimiter spatial resolution". *IEEE Trans Nucl Sci 49*: 697–671 (1999)
42. Glenn F. Knoll, *Radiation Detection and Measurement*, second edition, Chapter 3. (John Wiley & Sons.)
43. Hamamatsu position sensitive photomultiplier tube R2486 datasheet. HAMAMATSU PHOTONICS K.K., Electoron Tube Center 314-5, Shimokanzo, Toyooka-village, Iwata-gun, Shizuoka-ken, 438–0193, Japan.
44. M. Ivanovic, D.A. Weber and S. Loncaric, "Multi-pinhole collimator optimization for high resolution SPECT imaging" *IEEE Nuclear Science Symposium, 1997. 2*: 1097–1101 (1997)
45. *The Physics of Medical Imaging,* Chapter 6, edited by Steve Webb. Institute of Physics Publishing
46. R. F. Post and L. I. Schiff. "Statistical Limitations on the Resolving Time of a Scintillation Counter". *Phys Rev 80*: 1113 (1950)
47. C. S. Levin and E. J. Hoffman. "Calculation of positron range and its effect on the fundamental limit of positron emission tomography system spatial resolution". *Phys. Med. and Biol. 44*: 781–799 (1999)
48. A. Del Guerra, A. Bandettini, M. Conti, G. De Pascalis, P. Maiano, V. Perez Mendez and C. Rizzo "Test of the HISPET Prototype". *IEEE Trans Nucl Sci 37*: 817–822 (1990)
49. A. J. Reader, S. Ally, F. Bakatselos, R. Manavaki, R. J. Walledge, A. P. Jeavons, P. J. Julyan, Sha Zhao, D. L. Hastings and J. Zweit "Regularized one-pass list-mode EM algorithm for high resolution 3D PET image reconstruction into large arrays" *IEEE J. Nuclear Science Symposium Conference Record 2001. 4*: 1853 –1858 (2002)
50. A. Del Guerra, C. Damiani, G. Di Domenico, M. Gambaccini, A. Motta, N. Sabba and G. Zavattini, "Seeing double: combined modalities in functional imaging". *Nuclear Instruments & Methods in Physics Research Section A-Accelerators Spectrometers Detectors & Associated Equipment. 471*: 145–150 (2001)

CHAPTER 11

DETECTORS FOR RADIOTHERAPY

Mike Partridge

Joint Department of Physics, The Institute of Cancer Research & The Royal Marsden NHS Trust, Sutton, UK

11.1 Introduction

Modern radiotherapy is a highly complex and technically demanding field. The safe, accurate and reproducible delivery of complex three-dimensional radiation distributions must be accomplished in an efficient way. National and international requirements for medical and radiation safety and traceable quality standards must also be fulfilled. Radiation detectors therefore play a crucial supporting role in all aspects of radiotherapy.

A range of different detector systems are used for a variety of different types of measurement. The two main types of measurement are: (i) quality assurance of equipment, and (ii) verification of treatment delivery. The main types of detector fall again into two groups: point detectors, including ionization chambers, semiconductor detectors and thermoluminescent detectors (TLDs), and area detectors, including radiographic film and electronic portal imagers.

In this chapter, the major types of detector system used in radiotherapy will be discussed. The physical principles of operation and typical uses will be presented briefly for each detector. A short introduction to the physics of radiotherapy and an overview of the detection of photons and electrons at the energies commonly used in radiotherapy is also given by way of an introduction. References to more detailed treatments of each topic are included throughout, including reference to some published guidelines and protocols for dosimetric measurements. Radiotherapy dosimetry is a very specialised but well documented and regulated field. Readers wishing to carry out accurate dosimetric measurements are encouraged to read and follow published guidelines.

11.2 Introduction to Radiotherapy

The aim of radiotherapy is to deliver a well-defined high dose of radiation to a particular target volume within the body of the patient. Dose to normal tissues and critical structures surrounding the target volume should be kept as low as possible; and this is indeed often a limiting factor in beam delivery. For superficial targets (i.e. close to the skin surface) electron beams with energies in a typical range from 4 to 25 MeV are used, with the penetration depth varying as a function of energy. For deeper target volumes, bremsstrahlung X-rays from electron linear accelerators (linacs) with energies in the range 6 to 25 MV are used. Gamma ray beams from high activity ^{60}Co sources and medium energy or "orthovoltage" x-ray energies (200–400 keV) are also used, but are being gradually superseded. To deliver a sufficient dose to kill tumour cells, while allowing recovery in normal tissues, radiotherapy treatments are *fractionated*—that is, split into a number of small daily doses, typically 2Gy—over 5 or 6 weeks. Accurate positioning of the patient each day is therefore a crucial part of the radiotherapy delivery process. (Highly accurate single dose treatments can also be performed, but these are usually referred to as radiosurgery). Fractionation allows the normal tissues time to repair and tumour tissue reoxygenation between doses and raises the probability that tumour cells will be irradiated at the most radio-sensitive part of the cell cycle, generally during the cells' G2 and mitosis phases [1].

11.2.1 *External Beam Radiation Delivery*

The simplest form of radiotherapy delivery is to use a series of intersecting radiation beams of rectangular cross section, as shown in

Fig. 11.1. A high-dose volume is formed at the intersection of the beams, with the dose outside this target region kept lower over the rest of the body. The rectangular beams are shaped using two pairs of movable lead or tungsten collimators. More complex shapes, reducing dose to particular organs, can be defined using rectangular or custom-made absorbing metal blocks. The fluence across the beam can also be varied using wedge shaped filters [2].

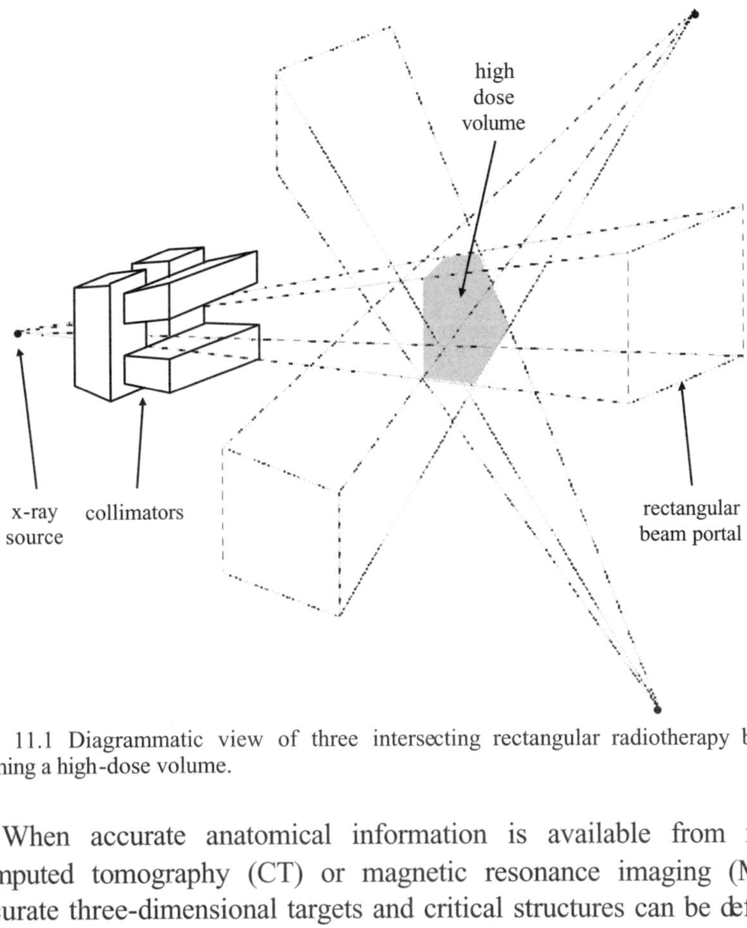

Fig. 11.1 Diagrammatic view of three intersecting rectangular radiotherapy beams forming a high-dose volume.

When accurate anatomical information is available from x-ray computed tomography (CT) or magnetic resonance imaging (MRI), accurate three-dimensional targets and critical structures can be defined. With this knowledge of the geometrical arrangement of the target critical structures, or *organs-at-risk*, more sophisticated beam delivery can sometimes be beneficial. Instead of simply using a series of uniform rectangular beams, each beam can be shaped to match the projection of the target in the beam direction.

This so-called *conformal* radiotherapy can lead to a lower dose to the critical structures while, by including suitable geometric margins to account for organ motion and inaccuracies in patient positioning, also giving a low probability of missing the target. Beam shaping can be carried out using a computer controlled *multileaf collimator* (MLC) integrated into the treatment machine. The edge of the radiation field is in that case defined by two opposing rows of movable tungsten slabs, rather than the rectangular jaws of the simple treatment machines [3].

The most recent advance in radiotherapy has been the clinical introduction of *intensity-modulated* radiotherapy (IMRT), where not only is the beam outline conformed to the projection of the target, but the fluence across the beam is also varied in an attempt to give a three-dimensional dose distribution that matches as closely as possible the shape of the target volume. IMRT offers the greatest flexibility in designing complex high-dose targets volumes while keeping the doses to particular organs-at-risk low [4].

This range of "external-beam" irradiation techniques, using megavoltage X-ray or electron beams, are the main focus of this chapter. Other methods of radiation therapy include the use of radioactive sources inside the body (brachytherapy) or irradiation using heavy, charged particles (protons or carbon ions), these are not discussed further in this chapter. (Although PET detectors—which can be used to image activation produced as a result of irradiation with heavy, charged particles—are discussed in Chapter 8).

11.2.2 *Requirements for Standards and Reporting*

In many countries there are statutory radiation safety requirements for protection of both the general public, workers and patients undergoing medical examination or radiation therapy. In the European Union, Council Directive 97/43/EURATOM covers radiation protection and medical exposure.

The International Commission on Radiation Units and Measurements (ICRU) also issues recommendations aimed at standardizing the measurement, calculation and reporting of radiation dose. ICRU Report 60 [5] suggests a general system of units for reporting dose with ICRU Report 50 [6] suggesting a common language for the specifying and reporting dose specifically within radiotherapy, including a careful definition of the treated volumes. This should enable meaningful comparison of the treatment outcomes of different centres and enable the development of consistent treatment policies. It is recommended that the dose to a well-defined reference point be recorded together with the maximum and minimum doses to a well-defined treatment volume. It is also recommended that the mean dose, standard deviation and dose/volume histograms are recorded. (Note: the placement of the reference point at the centre of the tumour makes direct *in vivo* measurement impossible, so indirect methods such as entrance and exit dose measurement using diodes or thermoluminescent dosimeters (TLDs) are often used, see Section 11.4).

In the UK absolute dosimetric standards are derived from direct measurement of absorbed energy using a graphite calorimeter. This kind of measurement is not practical for routine use, so air ionization chamber measurements are used to disseminate secondary standards. In addition to providing a calibration service for Farmer-type ionization chambers (see Section 11.4.1) standards can also be disseminated, and audit checks on local centres made, using mailed TLDs (Section 11.4.2) or Fricke solutions (Section 11.7.1).

At a local level, regular dosimetric checks can then be made using equipment with calibrations traceable to national standards, either calibrated directly be a standards laboratory, or provided by equipment vendors. Clear codes of practice for making routine dosimetric measurements are published by the IAEA (2000) [7], with useful notes on practical implementation published by ESTRO [8, 9].

11.3 The Physics of Detection for Radiotherapy

11.3.1 *Photon interaction mechanisms*

For the energies commonly used in radiotherapy the major interaction mechanisms of photons with matter are summarised in Fig. 11.2. The fraction of the total collision cross-section arising from each interaction mechanism is plotted as a function of energy. The physics of each interaction is then discussed very briefly below [2, 10].

Photoelectric effect. A photon is absorbed completely by an atom and an electron ejected. The probability of such an interaction is roughly proportional to

$$Z^4/(h\nu)^3, \tag{11.1}$$

where Z is the atomic number and $h\nu$ is the photon energy. The strong inverse energy dependence means that this mechanism dominates at low (diagnostic X-ray) energies. The strong Z dependence gives rise to the excellent contrast seen between bone (high Z) and tissue (low Z) in diagnostic imaging.

Coherent scattering, or Rayleigh scattering. The incident photon is scattered elastically by free electrons, leaving the photon energy unchanged. For radiotherapy coherent scattering effects are small and are also very forward-peaked at high energy.

Compton scattering. When a photon scatters inelastically with an atom, it is deflected from its original course and loses energy, which is transferred to a recoil electron. The collision cross-section for Compton scattering depends on the incident energy and the electron density of the medium, the collision cross-section per free electron is given by the Klein-Nishina equation

$$\sigma_c = 2\pi r_0^2 \left\{ \frac{1+\alpha}{\alpha^2} \left[\frac{2(1+\alpha)}{1+2\alpha} - \frac{\ln(1+2\alpha)}{\alpha} \right] + \frac{\ln(1+2\alpha)}{2\alpha} - \frac{1+3\alpha}{(1+2\alpha)^2} \right\}, \tag{11.2}$$

where α is the photon energy in units of 0.511 MeV and r_0 is the classical electron radius. This is the dominant interaction mechanism for radiotherapy photon beams, as shown in Fig. 11.2.

Pair production. A photon interacting directly with the nucleus can be completely absorbed and its energy converted into an electron/positron pair. In general, for photon energies of a few MeV the pair-production cross section is proportional to Z^2 for low atomic numbers. This becomes weaker with higher Z due to nuclear screening, with the ejected particles being distributed in the forward direction. Pair production becomes steadily more significant with increase in photon energy, see Fig. 11.2.

(γ,n) reactions. There is a small probability that a photon will be absorbed by the nucleus and cause a neutron to be ejected. This may produce an unstable atomic nucleus, thus activating the irradiated material. For radiotherapy, these effects are normally only significant in the bremsstrahlung target inside the linear accelerator. During servicing—when the target is exposed—care should be taken to ensure that they are adequately shielded, or sufficient time left for the activity to decay.

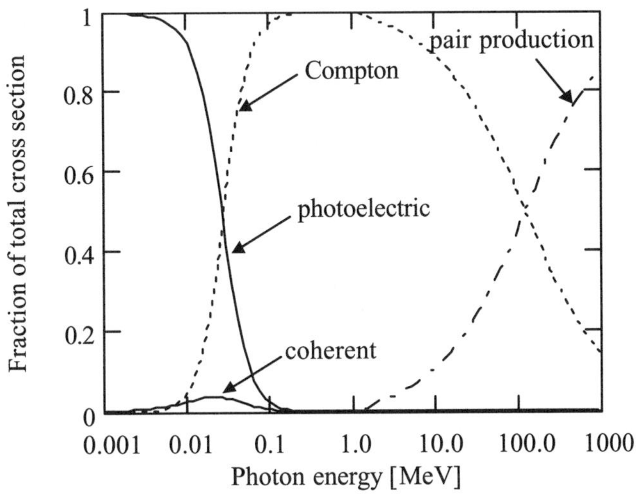

Fig. 11.2 Photon interaction mechanisms as a function of energy for a material with atomic number $Z = 12$.

11.3.2 *Electron interaction mechanisms*

A brief review of the main physical interaction mechanism of electrons passing through a medium is given here [11]. Unlike photons, electrons usually only lose energy a small amount at a time and therefore may undergo tens of thousands of interactions before coming to rest[a]. The two main categories of interaction are collision losses and radiative losses.

Collision losses. Electrons can lose energy during a collision with an atom either by excitation or ionization of the atom. If the distance from the passing electron to the atom is large compared to the size of the atom, then the atom as a whole reacts to the force exerted by the passing electron and is excited. The energy required to promote an atomic electron from an inner to an outer shell is only a few eV, so the energy losses caused by this sort of interaction are correspondingly small.

If the passing electron approaches to a distance similar to the dimensions of the atom, interaction can occur directly with one of atomic electrons rather than with the atom as a whole. If the incident electron has sufficient energy, the atomic electron can be ejected completely and the atom ionized. If the energy of the ejected electron is greater than about 100 eV—as happens occasionally—the ejected electron has enough energy to produce excitations and ionizations of its own and is sometimes called a *delta ray*.

Radiative losses. If the distance of closest approach of the incident electron is smaller than the atomic radius, the incident electron can be influenced by the nuclear Coulomb field and deflected from its original trajectory with a loss of energy. This energy is emitted as a broad spectrum of electromagnetic radiation up to the kinetic energy of the incident electron and is know as bremsstrahlung radiation. Direct interaction with atomic K-shell electrons is also possible—although the probability is low—producing electron–electron bremsstrahlung at the characteristic K-radiation frequencies. Energy loss per atom for electron–nucleus bremsstrahlung is roughly proportional to the square of the atomic number. For low incident energy electrons ($E \ll mc^2$) the maximum bremsstrahlung intensity is perpendicular to the beam, for example in conventional X-ray tubes, whereas for high incident electron energy ($E \gg mc^2$) a narrow forward-peaked photon beam is produced. This high-energy bremsstrahlung radiation is the mechanism used for producing photon beams in medical linear accelerators: a high-energy

[a] i.e. until the energy is sufficiently low as to make no further significant contribution to dose.

electron beam is incident on a thick, high Z target (typically tungsten or gold) giving rise to a narrow forward-peaked photon beam on the far side of the target [12].

The final radiative mechanism is Cerenkov radiation, emitted when a charged particle passes though a dielectric medium at a velocity greater than the phase velocity of light in that medium. This is generally a small effect in radiotherapy, but depends on the dielectric constant of the material.

11.3.3 *Units*

The following definitions of exposure, absorbed dose and air kerma from ICRU60 [5] will be used throughout this chapter.

Exposure X is defined as the total charge of all the ions of one sign produced in air when all of the electrons liberated by an X-ray photon in a volume of air of mass dm are completely stopped in air

$$X = \frac{dQ}{dm}. \qquad \text{[units C / kg]} \qquad (11.3)$$

Absorbed dose D, measured in Gray, is defined as the mean energy $\bar{\varepsilon}$ imparted by ionizing radiation to a mass of matter dm (1 Gy = 1 J / kg)

$$D_m = \frac{d\bar{\varepsilon}}{dm}. \qquad \text{[units Gy]} \qquad (11.4)$$

Kerma. If we consider a collimated beam of radiation with a fluence Φ incident upon an absorber of thickness dx, kinetic energy transfer coefficient μ_k and density ρ, the kerma[b] describes the kinetic energy transferred to electrons in the absorber

$$K_m = \frac{d\Phi}{\rho\, dx} = \Phi\left(\frac{\mu_k}{\rho}\right). \qquad \text{[units Gy]} \qquad (11.5)$$

Not all of the energy absorbed from a radiation beam is converted into kinetic energy of electrons, so kerma is always smaller than the absorbed dose. In addition, not all kerma remains local, some may be radiated away from the local area as bremsstrahlung.

---

[b] Kinetic Energy Released per unit Mass of Absorber.

Finally, we can write down an expression for the connection between these units. Exposure and kerma can be linked simply (i.e. not including bremsstrahlung) using the mean energy to form an electron/hole pair in air ε_{air} and the collision kerma in air $K_{\text{air,c}}$

$$X\varepsilon_{\text{air}} = K_{\text{air,c}}\,. \tag{11.6}$$

11.3.4 *Charged particle equilibrium and cavity theory*

Charged particle equilibrium (CPE) is said to exist in a medium if, for each particle leaving a given volume, an identical particle with identical energy enters. In practice true charged particle equilibrium is almost impossible to achieve because of attenuation. If the number of primary particles varies with depth in a material, then the number of secondaries can never be constant. However, *transient charged particle equilibrium* (TCPE) can be said to exist in regions where the absorbed dose is proportional to kerma. For uniform fields and media, this is approximately true for depths greater than the depth of maximum dose D_{max}. Where transient charged particle equilibrium exists, the relationship between the value given by a particular detector D_d in a medium m and the dose to a uniform medium in the absence of the detector D_m can be calculated

$$D_d \overset{\text{TCPE}}{=} \Phi_d(\mu_{en}/\rho)_d\,, \tag{11.7}$$

$$D_m \overset{\text{TCPE}}{=} \Phi_m(\mu_{en}/\rho)_m\,, \tag{11.8}$$

where Φ_m and Φ_d are the photon fluences at the measurement point. $(\mu_{en}/\rho)_d$ and $(\mu_{en}/\rho)_m$ are the mass/energy attenuation coefficients for the medium and the detector respectively (in the case of an electron beam the stopping power is used). If the assumption is made that the detector does not significantly perturb the particle fluence, then Φ_d and Φ_m are equal and the above equations can be combined to give

$$\frac{D_m}{D_d} \overset{\text{TCPE}}{=} \frac{(\mu_{en}/\rho)_m}{(\mu_{en}/\rho)_d}\,. \tag{11.9}$$

This is the standard result of *cavity theory* for “large detectors” (i.e. where the detector is large enough for CPE to exist within its sensitive volume). For common air ionization chambers, the range of high-energy electrons is often larger than the sensitive volume, so this result would not seem to be true. However, if the cavity is sufficiently small that it does not perturb the electron fluence set up by primary photon interactions in the surrounding medium—which means that the dose to the gas in the cavity comes from electrons arising from interactions in the surrounding medium, rather than the cavity itself—then the result is still valid. This result is the standard *Bragg-Gray* cavity theory and is widely applied to high-energy photon dosimetry. For electron dosimetry, the presence of the detector can have a larger perturbing effect on particle fluence, so perturbations to the standard theory have to be considered.

An extension to Bragg-Gray cavity theory is the *Fano Theorem* which states that, in a medium of a given composition exposed to a uniform primary radiation fluence, the secondary radiation fluence will also be uniform and independent of the density of the medium. Almost all dosimetry measurements rely on these results.

11.3.5 *Effects of measurement depth*

For accurate, practical dosimetry measurements great care is usually taken to ensure that a good approximation to transient charged particle equilibrium exists at the measurement point. This is achieved by making measurements at (at least) the depth of maximum dose in water D_{max} (see Fig. 11.3 for illustrations of depth-dose curves for typical 6 MV photon and electron beams.) A *build-up cap* can also be used, which is a plastic sleeve that is either incorporated into the detector, or fits closely over it, with a water-equivalent thickness of at least D_{max}.

Figure 11.5 shows a photograph of a Farmer ionization chamber with the build-up cap removed. Diode detectors often have build-up material incorporated into the device packaging.

Where measurements have to be made in steep dose gradients—such as the measurement of depth-dose curves—then a detector with a small sensitive volume that perturbs the field as little as possible should be chosen, such as a Markus ionization chamber or a diamond detector.

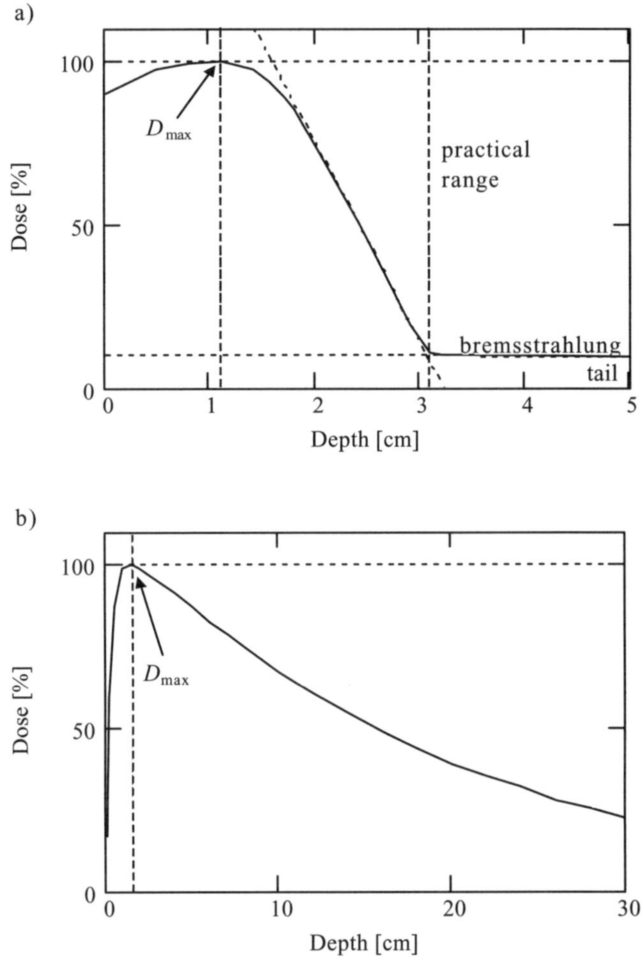

Fig. 11.3 Illustration of the change in measured dose with depth in water for a typical 6 MeV electron beam (a) and 6 MV photon beam (b).

11.3.6 *Quality Assurance and Verification Measurements*

Dosimetric quality assurance (QA) measurements are carried out to ensure that radiation delivery is consistent from day to day, treatment machine to treatment machine and hospital to hospital [13]. Measurement of absorbed dose to water is typically performed using standard Farmer ionization chambers, calibrated with reference to national standards, in water tanks or in solid plastic blocks, or *phantoms*[c]. Typical periodic QA measurements for linear accelerators include:

- beam calibration and output checks; an ionization chamber in a standard PMMA phantom is used to measure the dose rate on the central axis of the beam; these measurements are usually performed daily for each beam energy of each treatment machine using a standard sized radiation field;
- beam flatness, symmetry and penumbra; an ionization chamber is scanned along the principle axes of standard sized square fields at standard depths in a water tank;
- depth-dose; the dose on the central axis of standard-sized radiation fields is measured as a function of water depth using an ionization chamber; typical depth-dose curves for 6 MV photon and electron beams are shown in Fig. 11.3; separate measurements must be made for each available photon and electron energy;
- variation of output with field size; due to a lack of charged particle equilibrium and differences in the scattered photon fluences, the effective linac output changes with field size; these field size factors are determined using ionization chamber measurements in water.

These QA measurements are carried out periodically according to a written schedule, and must be repeated after adjustment or replacement of key linac beam delivery components.

Patient treatment verification measurements include verification of the patient position immediately before or during treatment, using film or electronic portal imaging, and verification of delivered dose, using either thermoluminescent dosimeters (TLDs) or semiconductor diodes. The prescribed dose is usually quoted at a point in the centre of the tumour, hence making direct measurement impossible. The *entrance* dose, or dose at the skin surface of the patient at the centre of one of the beams, is therefore often used. To ensure charged particle equilibrium a build-up cap or encapsulated detector is often used. This can lead to unacceptably high doses being delivered to the patient's skin if measurements are to be

---

[c] Appropriate correction factors must be applied if the measurement medium is not water-equivalent.

carried out for every fraction, so *exit* dose measurements can also be made. Interpretation of exit dose measurements is more difficult, since calculation of the expected dose at the exit surface is less accurate.

11.4 Point Detectors

11.4.1 *Ionization Chambers*

Perhaps the most important type of detector for radiotherapy dosimetry is the air ionization chamber. Radiation entering the chamber causes ionization, these ions then drift apart in an applied electric field (typically 200 to 400 V) and are collected by read-out electrodes.

Air ionization chambers form the basis of the traceable standard dosimetry systems operating in many countries because of their stability, accuracy and relative ease of use in a clinical environment. Air is a useful material because it has a similar atomic number Z to tissue, and therefore similar radiation interaction mechanisms operate. There are two basic chamber designs: cylindrical and parallel-plate. The most widely used design is the cylindrical "thimble" chamber, or *Farmer* chamber, see Figs. 11.4 and 11.5. In this design a cylindrical graphite cup surrounds a central aluminium electrode enclosing a volume of 0.6 cm^3 (again, graphite and aluminium are used because they have similar atomic numbers to air).

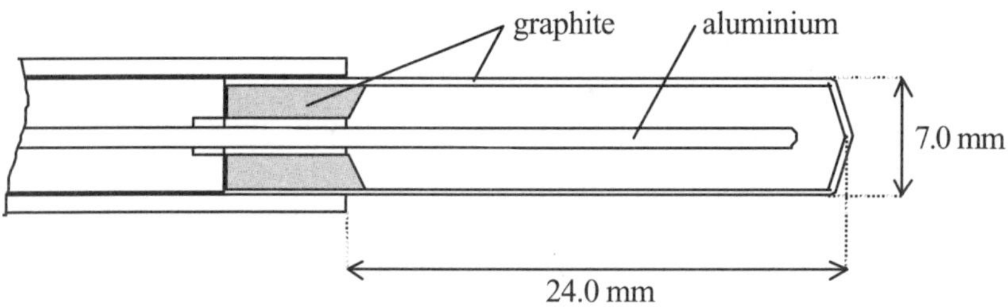

Fig. 11.4 Schematic diagram of a Farmer-type ionisation chamber.

For megavoltage use, the thin graphite chamber walls are significantly smaller than the mean secondary electron range. To ensure that all ionization in the chamber arises from interactions either in the cavity or the chamber wall, a *build-up cap*—close fitting plastic sleeve that covers the chamber—is used to provide approximate charged particle equilibrium.

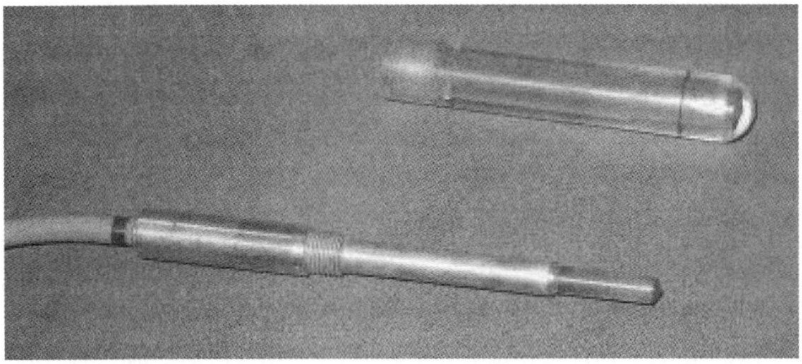

Fig. 11.5 Photograph of a Farmer type air ionisation chamber showing the build-up cap removed.

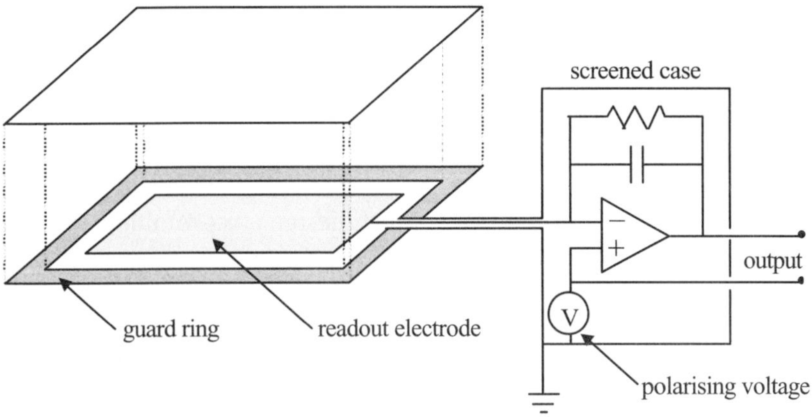

Figure 11.6 Schematic of a parallel-plate chamber showing the guard ring defining the sensitive volume of the chamber and a schematic of a typical electrometer circuit.

It is important to choose a chamber design suited to the desired measurement task. For low radiation field gradients cylindrical Farmer-type chamber measurements in water are common, with routine quality assurance measurements being made in plastic quasi-water-equivalent phantoms. For measurements in strong radiation field gradients—such as in the build-up region—Farmer chambers have the disadvantage that the effective measurement volume is quite large and high spatial resolution measurements are therefore difficult. This is particularly true for electron beams.

In these cases, parallel-plate chambers are preferable and measurements are often performed in plastic slab phantoms, providing greater geometrical accuracy than water tanks. The chamber and entrance window thicknesses in the particle beam direction can be made small to minimize perturbations of the field introduced by the chamber, and to more accurately define the effective point of measurement. The *Markus* chamber, commonly used for electron dosimetry, has an entrance window thickness of 0.03 mm polythene[d]. Figure 11.6 shows a schematic diagram of a parallel-plate ionisation chamber. An important design feature to note is the *guard ring*, which forms a conducting path just outside the edge of the read-out electrode. This has two effects, reducing the potential across the insulator—and thus reducing leakage currents—and also helping to define the sensitive volume of the chamber.

Parallel plate chambers are also important in radiotherapy because they provide the fundamental dosimetry system of the treatment linear accelerators, the *monitor chamber*. The monitor chamber is typically situated after the bremsstrahlung target, flattening filter and scattering foils and before the collimators. Most designs use multiple parallel plates, constructed from 0.1 mm thick plastic sheets coated with very thin aluminium conducting layers. Double-sided chambers are common, where two planes of readout electrodes are mounted, one on each side of a polarizing electrode. One channel provides the main clinical dosimetry system and the second channel acts as a back-up safety system which can terminate beam delivery if the first channel fails. To remove changes in sensitivity due to changes in temperature and pressure, accelerator monitor chambers are often sealed, maintaining a constant mass of air in the chamber [12].

Although in principle the charge measured by an air ionization chamber should be directly proportional to dose, for accurate practical

---

[d] PTW Freiburg, Lörracher Strasse 7, 79115 Freiburg, Germany.

dosimetry a number of corrections must be made. These include corrections for:

- recombination
- scattered radiation
- temperature and pressure
- polarity effects
- leakage

and are discussed below. For megavoltage X-ray beams, when sufficiently high polarizing voltage is applied, recombination effects are small (a few per cent). Scattered radiation, or "stem effects", can add spurious signals where the scatter from the stem of the chamber perturbs the local radiation field. These either need to be estimated from experiment, or measurement geometries designed that minimize them.

Polarity effects are a problem seen if the signal detected by the chamber changes with the direction of the polarising field. One cause of this is direct collection of primary electrons, leading to an additional charge added to the "true" chamber measurement. To minimize this, the collecting electrode should be made as thin as possible (as is the case in many parallel-plate chambers designed for electron dosimetry). Another possible problem is differences in the positive and negative ion mobilities, leading to different collection efficiencies. This is really a problem of recombination and can in part be eliminated by raising the polarizing voltage. Leakage, as discussed earlier, can be minimized by careful chamber design and the incorporation of suitable guard rings. Many air chambers are unsealed, so require corrections for temperature and pressure. Calibration factors are quoted for standard reference conditions and can be converted using the following multiplication factor

$$\frac{P_0}{P}\cdot\frac{(273.2+T)}{(273.2+T_0)}, \tag{11.10}$$

where T and P are the temperature and pressure at the time of measurement and T_0 and P_0 are the calibration values. Sensitivity variations can also occur because of changes in relative humidity, but these effects are small.

11.4.2 *Thermolumiscent Detectors*

When a semiconductor material is subject to ionizing radiation, electron/hole pairs are created. Thermoluminescent dosimetry uses materials that trap the charge formed by ionization in metastable states. Either electrons, holes or both can be trapped. For the purposes of this discussion we will assume that the electron trap is the deepest and therefore controls the material behaviour. These metastable states are deep enough to prevent escape of the majority of electrons at room temperature, but if the material is then deliberately heated, sufficient energy is given to promote electrons back to the conduction band where they will subsequently recombine with the holes. If a suitable material is selected, the energy released by recombination can be sufficient to emit optical photons which can then be detected.

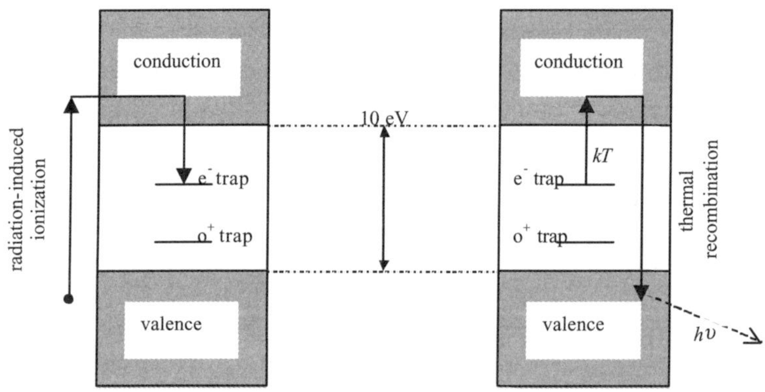

Fig. 11.7 Energy level diagram showing trapping of ionization electrons during irradiation (left panel) and release of a visible photon during subsequent thermal excitation (right panel).

For a given temperature T the probability of escape from a trap of depth E is given by

$$p = \alpha \exp(-E/kT) \tag{11.11}$$

where k is the Boltzman constant and α is a constant of proportionality called the "frequency factor" ($\sim 10^9$ s^{-1}). Clearly if the temperature is gradually increased, then the number of trapped electrons will also increase reaching a maximum as the majority of the trapped electrons are released. This temperature for maximum release T_m is related to the rate of heating q by the following

$$\frac{E}{kT_m^2} = \frac{\alpha}{\exp(-E/kT_m)}. \tag{11.12}$$

T_m is 216 °C for a trap depth E of 1 eV and q of 1 K s^{-1}. The most common material for thermoluminescent dosimeters (TLDs) is LiF doped with mixtures of Mg, Cu, P and Ti, although calcium fluoride, calcium sulphate and lithium borate are also used. The material is often in the form of rectangular "chips" $3.2 \times 3.2 \times 0.9$ mm^3, but can also be used in the form of thin rods or powder in capsules or tubes. Composites of PTFE and LiF have also been produced for easier handling and more tissue-equivalent behaviour. LiF:Mg,Cu,P TLDs have a useful dose range from 0.5 μGy to 12 Gy and LiF:Mg,Ti from 0.05 mGy to 500 Gy[e].

Practical measurements are made using automated TLD reading machines that heat each sample individually in a light-tight enclosure and measure the light emitted as a function of temperature, or *glow curve*, using a photomultiplier tube. The exposure is roughly linearly related to the light output. Heating is usually performed in an atmosphere of dry nitrogen to prevent surface oxidation effects, and great care should be taken to ensure that the TLD surfaces remain clean to prevent spurious signals. Chips are usually heated to 300 or 400 °C during read-out then annealed at lower temperature for up to 24 hours before re-use.

Since the signal fades with time (since the electrons will have a finite probability of escape from the trapped states) the same length of time should be left between irradiation and readout for calibration and clinical measurements. One method of ensuring this is to perform calibration experiments immediately before or after the clinical irradiation and then read out both sets of TLDs together. To compensate for background radiation effects, a control batch of unirradiated chips can also be read out at the same time. To improve statistical accuracy, multiple TLDs are often used for each measurement. In normal clinical use the expected accuracy is between 2 and 3%.

---

[e]PTW Freiburg, Lörracher Strasse 7, 79115 Freiburg, Germany.

TLDs are frequently used for *in vivo* dosimetry, particularly for treatments where accurate dose calculation is difficult, like total body irradiation. They can also be used internally in catheters or easily placed inside phantoms for quality assurance measurements. They offer the advantage that they are easy to use, since they can be simply taped to the skin, but have the disadvantage that they can only be read out after irradiation and careful calibration and handling is required to get reasonably accurate results. Mailed TLDs are used for quality assurance checks or audit checks by standards laboratories and regulatory authorities.

11.4.3 *Diode Detectors*

Semiconductor diodes are widely used for relative dosimetry; they are relatively robust when suitably packaged, easy to use and give instantaneous results. Packages are available for QA measurements in water tanks and for *in vivo* dosimetry, either on the skin surface for entry and exit dose measurements, or internally.

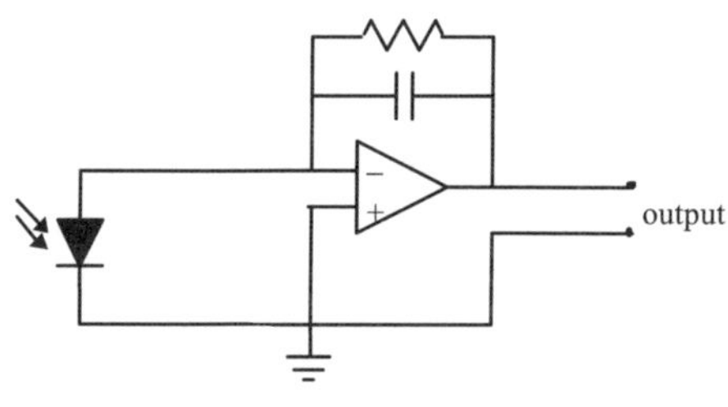

Fig. 11.8 Simplified schematic of an electrometer circuit.

For radiotherapy, both *p*-type and *n*-type silicon diodes are available, with each type having its own advantages and disadvantages. For a typical *p*-type diode, a *p*-type bulk material (Si doped with a group III material) is counter-doped using a thin layer of an *n*-type (group V) material to form the *p–n* junction. A carrier-free region—the *depletion layer*—forms at the junction, typically a few 10's of microns wide, with an intrinsic potential of 0.7 V. Electron-hole pairs created during irradiation drift along the intrinsic potential causing a reverse current to flow. This current is comprised almost entirely of electrons in a *p*-type device and holes in the *n*-type. The current is roughly proportional to the exposure rate and can be measured using a relatively simple electrometer with a low input impedance and low offset voltage.

A proportion of carriers are trapped by defects in the silicon and, since holes are trapped more easily than electrons, this effect is larger in *n*-type devices. Recombination is also affected by dose rate. At high dose rates the recombination centres can become saturated and lead to a proportionally higher signal. Again this effect is more pronounced in *n*-type devices, which have a lower doping level and therefore lower numbers of recombination centres.

Radiation damage to the Si crystal lattice also forms recombination centres, thus reducing the detected signal gradually with use. Typical values for sensitivity loss as a function of radiation damage are about 1% per kGy, although this depends on the beam energy and previous irradiation history. Sensitivity tends to fall off faster during initial exposure, so commercial diode detectors are often available pre-irradiated to minimize this effect. Regular recalibration of any diode detector is, however, necessary if it is to be used for accurate, absolute dosimetry.

Although reverse-biasing the diode would lead to a higher sensitivity and a lower leakage current, diodes are usually used unbiased because the leakage current decreases more rapidly than the radiation-induced current as the bias voltage drops to zero. Leakage currents are primarily caused by thermal motion of carriers and are therefore temperature dependent, with typical values 0.1 to 0.4 % per °C. Temperature dependence again tends to increase with accumulated dose, but again the rate of change of sensitivity slows down with increasing exposure. Calibration of diodes should therefore be carried out at the same temperature used for measurement (skin temperature or internal body temperature for *in vivo* measurements).

The average energy to form an electron/hole pair in silicon is around 3 eV (10 times less than that in air), and the density of Si is 1000 times higher than air giving a relative efficiency per unit volume for silicon diodes 10^4 times greater than an air ionization chamber. Silicon diodes can therefore be made with much smaller sensitive volumes. The shape of the sensitive volume and surrounding package cause a significant angular dependence in the response, with up to 15% variation with angle being common. Care must therefore be taken in checking that the same orientations are used for calibration and measurement.

Fig. 11.9 Photograph of commercially-available diode arrays, courtesy IBA Scandatronix Wellhöfer[f].

Finally, the materials the diode is comprised of are not tissue equivalent, so significant energy dependence is seen, with lower energy radiation giving a disproportionally high signal. *Compensated* diodes are produced where the diode encapsulation (stainless steel and epoxy resin) is designed to preferentially absorb the lower energy radiation to give a more water-equivalent behaviour. These ensure that photon scatter conditions are as similar as possible to those at the D_{max} in water to minimize the correction factors, although large build-up caps can cause significant perturbations to the radiation field and are more difficult to attach to the patient's skin.

[f] IBA Scandatronix Wellhöfer, Bahnhofstrasse 5, 90592 Schwarzenbruck, Germany.

For practical clinical measurements regular calibrations against an air ionization chamber under standard reference conditions are recommended. The diode response should be calibrated as a function of the beam energy with suitable correction factors measured for beam energy, field size, wedges, blocks and accessories, angle of incidence, source to surface distance and temperature. In general it is good practice to make the calibration conditions as close to those used in clinical practice as possible. For example, in total body irradiation (TBI), where large source to surface distances and low dose rates are used, diodes should be calibrated under similar conditions. It is good practice to keep a record of changes in calibration factor with time to establish how often re-calibration is necessary, noting that frequency of calibration will depend on the level of use and age of the diode. An excellent booklet with suggested guidelines for implementation of *in vivo* diode dosimetry is published by ESTRO [9].

Linear arrays of diode and ionisation chamber detectors are also commercially available, potentially speeding up routine QA measurements and enabling measurement of dynamic wedged deliveries. Both 12 and 25 channel diode arrays are shown in Fig. 11.9. To summarise, in spite of the many factors that can influence diode dose measurements, they remain very attractive for routine clinical use because of their ruggedness, ease of use and instantaneous readout.

11.4.4 *Diamond Detectors*

In addition to diode detectors, another type of semiconductor point detector used for radiotherapy is the diamond detector [14]. Both natural diamond and artificial, diamond-like thin films made by chemical vapour deposition (CVD) can be used, although the use of natural diamond is currently more common in radiotherapy. Diamond can be classed as a "large" band gap semiconductor, with a gap of 5.54 eV for natural diamond (5.47 for CVD diamond) compared to 1.12 eV for silicon. This large band gap means that there are very few free charge carriers present at room temperature, leading to a very high resistivity and correspondingly low leakage currents.

To make a useful detector, diamond does not therefore have to be depleted, so no diode structure is necessary. An electric field (typically 100 V) is placed across contacts attached to the detector, electron/hole pairs created by charged particles interacting with the diamond drift towards the contacts. A simple charge integrating amplifier will give a signal proportional to the induced charge, in the same way as for the silicon diode detector. The relatively high energy required to create an electron/hole pair in diamond compared to silicon is partially made up for by the good electron and hole mobilities, low dielectric constant, high saturation velocity and very good radiation hardness.

Table 11.1 Properties of silicon and diamond at 293 K (values are for CVD diamond except where indicated † where values for natural diamond are given in parentheses).

	silicon	diamond
band gap [eV]	1.12	5.47 (5.54)†
resistivity [Ω cm]	2.3×10^5	$>10^{11}$
energy to form e/h pair [eV]	3.6	13
electron mobility [$cm^2 V^{-1} s^{-1}$]	1350	1800
hole mobility [$cm^2 V^{-1} s^{-1}$]	480	1200

The average distance that electron/hole pairs drift apart before they are trapped is called the *collection distance*. In diamond this is typically smaller than the detector thickness; the collection distance is given by

$$d_e = \mu \tau E \, , \tag{11.13}$$

where μ is the mobility of the charge carriers, τ is the carrier mean lifetime and E is the applied electric field. A property of diamond detectors that is very important for radiotherapy dosimetry is the fact that the charge collection distance increases rapidly at low dose, then reaches a plateau as low as 50% of its pre-irradiated value. This effect is caused by charge traps being filled—and therefore neutralized—by the initial irradiation. This saturation state is referred to as the "pumped state" and can be stable for several months if the detector is kept at room temperature and dark in CVD diamond. To avoid the possibility of serious dosimetry errors arising from lack of equilibrium, manufacturers of diamond detectors for radiotherapy recommend exposure of the detector to pre-measurement dose of 5 Gy before each usage.

Commercial diamond detectors have the advantage of a very small measurement volume (a few cubic mm) giving excellent spatial resolution and almost no directional dependence or temperature dependence. Diamond is also quasi-water equivalent, at least in terms of atomic number, so it is particularly attractive for measurements in radiation fields where electron equilibrium cannot be assumed. Diamond detectors have been shown to have a sensitivity that is independent of energy for photon beams from 4 to 25 MV and electron beams from 5 to 20 MeV [15]. One effect that must be taken into account however, which arises because of the very short electron/hole recombination time, is a decreasing response with increasing dose rate. If the detector is to be calibrated at one dose rate and then used at another, a correction should be applied. It has been shown that the following simple empirical formula can be used to relate the detector reading M to the dose per pulse (measured in water) D_w:

$$M = \alpha D_w^{\Delta}, \tag{11.14}$$

where α is a constant and the correction factor Δ has a value close to 1.00. These dose-rate correction factors have been shown to be energy independent.

In contrast to diode detectors, diamond detectors show no strong directional or energy dependence. Both detector types have small sensitive volumes and give instantaneous results, making them attractive for both QA and *in vivo* use. However, a major potential drawback of diamond detectors is the need for pre-irradiation, which limits their usefulness for *in vivo* work. The main use of diamond detectors is therefore in QA, where their small sensitive volume, favourable energy response and directional invariance make them highly attractive.

11.5 Film

X-ray film is usually comprised of a suspension of silver halide crystals (typically 1 μm silver bromide) in a gelatine matrix [16, 17]. This photographic emulsion is bonded to either one or both sides of a base material which is usually a synthetic polyester, although some films use a cellulose acetate base. Polyester is more common for x-ray films because of its greater physical strength, tear resistance and chemical stability. A thin protective layer can then cover the emulsion.

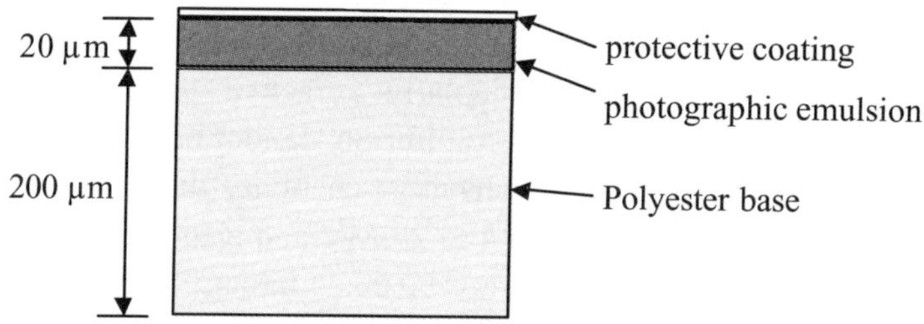

Fig. 11.10 Diagrammatic cross section showing typical x-ray film construction.

Photons interact photoelectrically with the halide and the electrons produced reduce the silver ions forming a "latent image" of metallic silver atoms. When the film is developed, the film grains containing more than a critical number of non-ionized silver atoms are then completely reduced to metallic silver. Grains with less than this critical number are removed by the fixing process. X-ray film images, on a microscopic scale, are therefore binary images formed by a noisy amplification process. On a macroscopic scale however the average optical transmittance varies smoothly over a wide range. The transmittance T of a film illuminated by a source with irradiance I_0 is given by

$$T = \frac{I}{I_0}, \tag{11.15}$$

where I is the transmitted irradiance. It is conventional however to express film appearance in terms of *optical density* (OD) which is simply the logarithm of transmittance

$$OD = \log_{10}\left(\frac{I}{I_0}\right). \tag{11.16}$$

A simple physical model can be derived to relate the optical density of a film to the exposure X, giving the following expression

$$OD = OD_{\max}\left(1 - e^{-kX}\right). \tag{11.17}$$

The maximum optical density of the film is given by the transmittance when every silver halide grain is reduced (i.e. saturation of the film) and the rate constant k describing the sensitivity. In practice, as well as the limited maximum film density, there is also a minimum density caused by the fact that some unexposed silver ions are still reduced during developing, giving rise to a background "fog" or base density level. Figure 11.11 shows a plot of equation (11.17), including the limitations for the non-zero film base density. This standard form of curve is called an H-D curve, after Hurter and Driffield who presented the first film sensitivity curves in 1890. The film characteristic over the region where the optical density varies roughly linearly with the logarithm of exposure is called the *gamma* of the film, and is defined thus

$$OD = \Gamma \log_{10}\left(\frac{X}{X_0}\right) \tag{11.18}$$

where X_0 is the apparent exposure indicated by unexposed processed film (i.e. the base density level).

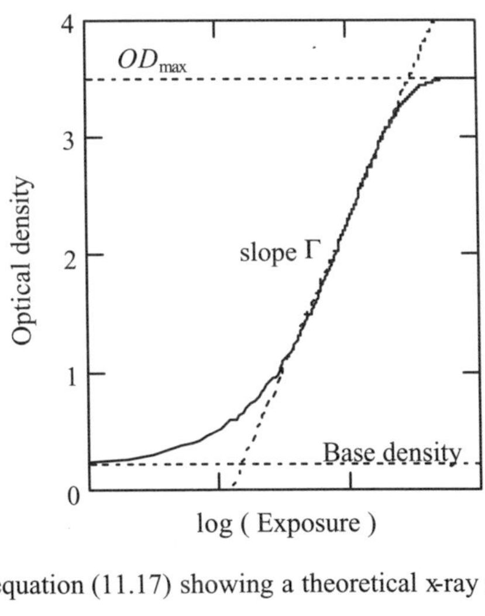

Fig. 11.11 Plot of equation (11.17) showing a theoretical x-ray film response curve.

Practical film usage in radiotherapy can be generalized using two categories: (i) patient localization checks, and (ii) QA of equipment. Localization measurements are normally carried out using a *film/screen system*, where the film is contained within a light-tight cassette. The cassettes consist of either one or two phosphorescent screens either side of the film, and thin lead build-up and/or backscatter plates, depending on the manufacturer. These localization port films are typically taken using a large-area megavoltage beam immediately before treatment to check the position of the patient with reference to previously acquired Simulator images or digitally reconstructed radiographs derived from CT images. Since these imaging beams cover a much larger area than the treatment beam, the film/screen systems are designed to maximize image contrast for the smallest possible dose (typically a few cGy).

In contrast to localization images, dosimetric quality assurance measurements are normally performed using higher doses (50 cGy). These QA films are usually individually packaged in light-tight paper envelopes. This makes them easy to place in water-equivalent phantoms to make measurements that are as close to water-equivalent as possible, and therefore quantitatively more accurate than a film/screen system when measuring absorbed dose to water. For accurate absolute dosimetry careful film handling and processing is required. Sensitivity variations between film batches should be checked and care taken to ensure consistency with film processing and read-out.

Film is very widely used in radiotherapy, both for dosimetric QA and localization imaging—where although it takes time to process, it is very convenient and easy to handle and archive.

11.6 Electronic Portal Imaging

An increasingly important group of detectors used in radiotherapy are the so-called *electronic portal imaging devices* or EPIDs. These area detectors measure the therapy energy beam after it has exited the patient (the *portal* beam). Systems are usually integrated into treatment delivery machines, mounted on retractable or easily removable arms so that they do not restrict the operation of the treatment machine. The main technologies are camera based and liquid-ionization-chamber-based systems, with amorphous-silicon-based flat-panel systems coming into clinical use over the past few years [18, 19].

EPIDs are typically used to check patient set-up and correct field placement. For patient set-up, images using the treatment beam portals are often difficult to interpret since, particularly with small fields, little anatomical information may be present in the image. Because of this, larger area low dose fields are often used for imaging prior to treatment, (in a similar way to portal films). Digitally reconstructed radiographs (DRRs) or radiotherapy simulator images are often generated so that a direct comparison with the portal image can be made using software tools provided by the equipment manufacturers. If the patient position is judged to be correct, treatment can proceed and the portal imager can be used to verify the shape of the treatment beam. It is important to use patient set-up information as part of a well-designed intervention protocol [19].

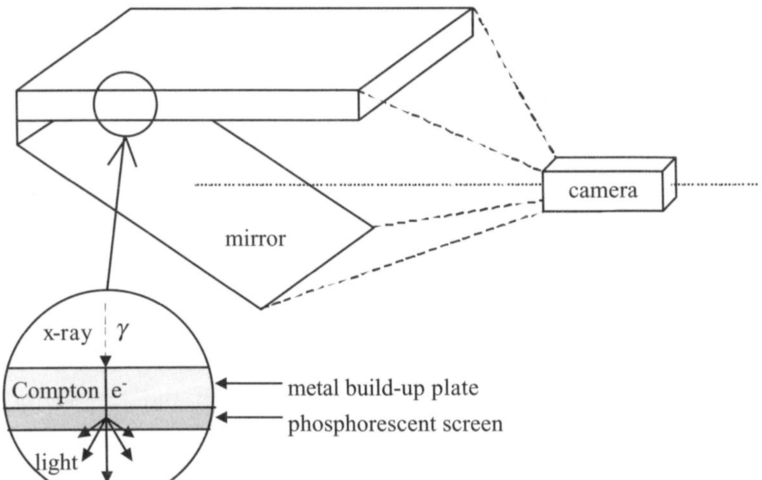

Fig. 11.12 Schematic view of a camera-based electronic portal imaging system. Radiation is incident from the top of the page.

Another use of portal imaging devices is dosimetry, although this is at present largely limited to research. Problems with performing accurate dosimetry include modelling the response of the system and accounting for radiation scattered by the patient, although much research work has been done in this field. Finally there is increasing current interest in the use of EPIDs for IMRT verification.

11.6.1 *Camera-Based Systems*

The radiation detector in the camera-based systems works using a very similar principle to a film cassette: a metal plate (1–2 mm brass, or steel) and phosphorescent screen detect the incoming high energy photons through a mixture of Compton and photoelectric interactions. The energetic electrons produced excite the phosphor screen (usually Gd_2O_2S:Tb) and the scintillation light is then detected using a camera. The camera cannot be placed in the direct beam because of the radiation damage that would occur, so instead is mounted at 90° to the screen, collecting light with the aid of a 45° mirror, see Fig. 11.12. The relatively large phosphor-screen-to-camera distance leads to the largest disadvantage of this type of system; because the light emitted by the phosphor is essentially Lambertian, substantial losses in signal are seen between the optical photon creation in the phosphor and detection by the camera.

To counteract this large optical loss, system designers have concentrated on increasing the photon yield by using a thicker phosphor screen and increasing the light collection efficiency of the camera by using large-aperture lenses. Limitations of this include optical blurring in the thick phosphor screens and problems of spherical aberration, vignetting and geometric distortion in the large aperture lenses.

Both CCD and tube-type cameras (Plumbicon and Newvicon) have been used in commercial systems. The CCD cameras offer the advantage of intrinsically good geometric stability, but can suffer from progressive radiation damage leading to so-called "salt and pepper noise", where defective camera pixels show either zero or maximum readings. Such damage can be removed by median window filtering up to a point, but cameras may need replacing periodically. Tube cameras show good long-term stability, being highly resistant to radiation damage, but can show significant geometric distortion especially at the edges of the field of view and show some "memory effect" with a portion of one frame contributing to the next. For the kind of accurate numerical measurement necessary for dosimetry, another problem is a radiation field-size dependent optical cross-talk effect caused by multiple optical reflections between the mirror and phosphor screen, although solutions to this problem have been demonstrated.

Advantages of the camera-based systems include fast data acquisition with disadvantages including poor light collection efficiency and aberrations introduced by the optical systems that may necessitate some post-processing.

11.6.2 *Liquid Ionization Chamber Based Systems*

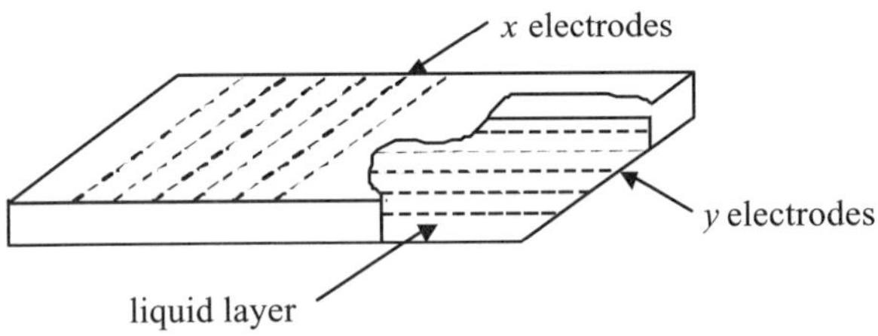

Fig. 11.13 Schematic diagram of a liquid ionization chamber-based electronic portal imaging system.

The liquid ionization chamber design consists of two sets of linear electrodes arranged normal to each other, shown in Fig. 11.13. The 0.8 mm gap between the two sets of electrodes is filled with 2,4,4-trimethyl pentane. Ions generated by irradiation of the liquid migrate to the electrodes under the influence of an applied electric field. The advantage of using 2,4,4-trimethyl pentane is that is has a relatively long ion recombination time, leading to an effective charge integration time in the system of 0.5 s. The signal recorded by the device is proportional to the square root of the dose rate. The electrodes making up the ion chamber matrix are arranged in two rows of 256, with a pitch of 1.27 mm. Two versions of the system are commercially available, one using a polarizing voltage of 300 V with a frame read-out time of 5.5 s, and the other using 500 V with a read-out time of 1.25 s [20].

Advantages of this system include its relatively compact and robust design, with the disadvantage of a relatively slow frame rate, and therefore a relatively high dose required per image. The image quality and spatial resolution of the camera-based and liquid ionization chamber-based systems are very similar.

11.6.3 *Amorphous Silicon Flat-Panel Systems*

A recent advance in electronic portal imaging has been the introduction of detectors comprised of arrays of sensitive elements fabricated using large-area amorphous silicon technology [21].

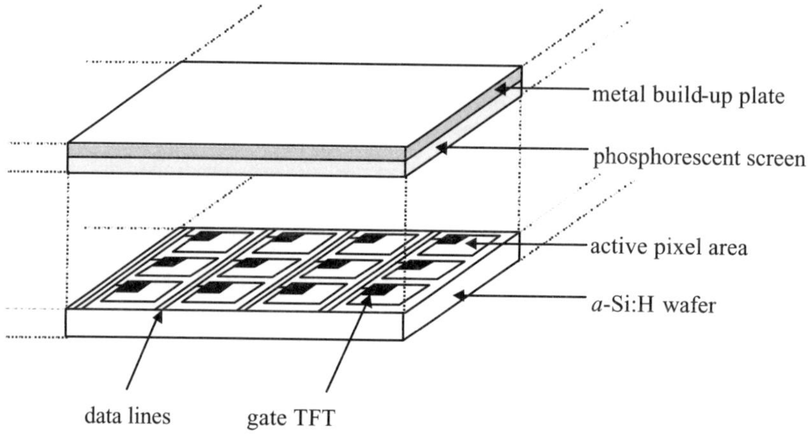

Fig. 11.14 Schematic diagram of an amorphous silicon flat-panel imaging system. Note: the phosphorescent screen is in contact with the active pixel surface, an expanded view of a small region is shown here simply for clarity.

Typical devices are comprised of hydrogenated amorphous silicon (*a*-Si:H) thin films deposited on glass substrates. An *n*-doped layer is generated over the bottom metal contact, followed by an intrinsic and a *p*-doped layer with a transparent conducting layer at the top. Devices are normally reverse biased to fully deplete the intrinsic layer. A thin film transistor (TFT) is fabricated in the corner of each pixel to read out each element. Light from the phosphor screen created electron/hole pairs, which are stored as charge in each pixel. When the device is to be read out, the TFT's are switched, connecting each pixel to the data lines running down each column. By switching each row of TFT's in turn, the detector array can be read out a line at a time.

The majority of the systems currently commercially available are *indirect* detectors based on very similar metal plate/phosphor screen primary detectors to the camera based portal imaging systems described in Section 11.6.1. Their major advantage, however, is that the amorphous silicon photodiode array can be placed in contact with the phosphor screen leading to very much more efficient light collection. The result is X-ray quantum limited noise behaviour, where image quality is related primarily to the quantum efficiency of the metal plate/phosphor screen combination.

Further advantages of the current commercially available systems are high resolution (40 cm × 40 cm, 400 μm pitch) and large dynamic range, producing significantly better image quality than the camera or liquid-ionization chamber based systems. The amorphous silicon devices themselves are comparatively very radiation hard, but a clear disadvantage of the systems is the fact that the read-out electronics surrounding the array *is* radiation sensitive, and can easily be damaged by accidental irradiation.

11.7 Radio-Sensitive Chemical Detectors

11.7.1 *Fricke Dosimetry*

Solutions of ferrous ammonium sulphate in water are useful for radiation dosimetry [22]. The effect of ionizing radiation is to convert the Fe^{2+} ions in the solution to Fe^{3+} causing a visible colour change. By accurately measuring the change in light absorption δA at about 300 nm—where ferric ions absorb, but the ferrous ions do not—the absorbed dose D can be calculated

$$D = \frac{\delta A}{(\varepsilon G \rho l)}, \tag{11.19}$$

where ρ is the density of the solution, l is the optical path length, ε is the molar extinction coefficient and G is the radiation chemical yield. Values of εG are tabulated for various energies, with a value for electrons in the range 1–30 MeV of 352×10^{-6} m^2 J^{-1} [23].

The Fricke system is relatively insensitive, but has a linear range from 20 to 100 Gy and is mostly used for quality assurance check and transfer of standards between National standards laboratories and clinics. Although great care should be taken to ensure cleanliness when preparing the chemical solutions, once prepared and sealed, they can easily be transported or sent by mail between sites. When handled correctly, accuracies of 0.5% are achievable.

An interesting development of the Fricke dosimetry system was to fix the Fe^{3+} ions in a gelatine matrix, and therefore preserve the spatial distribution of the absorbed dose in three dimensions. The T1 and T2 magnetic resonance relaxation times of protons in water are both affected by the local Fe^{3+} ion concentration, so irradiated gels can be read out using magnetic resonance imaging (MRI). A major problem with this system is the relatively fast diffusion rate of Fe ions in the gels. For good spatial resolution results the MR imaging has to be performed within less than an hour of irradiation. This limits Fricke gel dosimetry largely to research applications.

11.7.2 *Polymer Gels*

A gel system using bis and acrylic monomers, that polymerize as a result of the formation of free radicals by the ionizing radiation, has been shown to be relatively stable with time, not exhibiting the diffusion problems seen in the Fricke system [24]. The gels show good sensitivity and can also be analysed using MRI. Known problems with these gels, however, include a marked temperature dependence of the MR properties, which is particularly important when using long MR read-out sequences where significant RF heating can be caused. Some of the constituent chemicals are also toxic, so require careful handling. Care must also be taken to eliminate oxygen when preparing the gels, since oxygen would react with the free radicals and lower the sensitivity. This limits their use to purpose designed phantoms, but with careful calibration accurate three-dimensional results are possible, providing a useful research tool, especially for the verification of intensity-modulated radiotherapy. A recent development that might make the use of polymer gels more widespread is the use of optical read-out. The polymerized regions of the gel become opaque so, with suitably designed phantoms and read-out optics, fast three-dimensional results can be obtained without the use of MRI.

References

1. G. G. Steel, *Basic Clinical Radiology.* London: Arnold (1997)
2. H. E. Johns, and J. R. Cunningham, *The Physics of Radiotherapy.* (4th Edition) Springfield: C. C. Thomas (1983)
3. W. Schlegel, and A. Mahr, *3D Conformal Radiation Therapy, a multimedia introduction to methods and techniques*. Springer Verlag, Heidlberg (2001)
4. S. Webb, *Intensity-Modulated Radiation Therapy.* Bristol: IOP Publishing (2001)
5. *Fundamental Quantities and Units for Ionising Radiation. ICRU Report 60.* Bethsada MD: International Commission on Radiation Units and Measurement (1998)
6. *Prescribing, Recording, and Reporting Photon Beam Therapy. ICRU Report 50.* Bethsada MD: International Commission on Radiation Units and Measurement (1993)
7. IAEA, *Absorbed dose determination in external beam radiotherapy: an international code of practice for dosimetry based on standards of absorbed dose to water*. Vienna: IAEA (2000)
8. D. Thwaites, et al*, Practical guidelines for the implementation of quality system in radiotherapy.* Brussels: ESTRO (1994)
9. D. Huyskens, et al, *Practical guidelines for the implementation of in vivo dosimetry with diodes in external radiotherapy with photon beams (entrance dose).* Brussels: ESTRO (2001)
10. F. H. Attix, *Introduction to Radiological Physics and Radiation Dosimetry.* New York: Wiley (1986)
11. S. C. Klevenhagen, *Physics of Electron Beam Therapy.* Bristol: Adam Hilger (1985)
12. D. Green, and Williams. *Linear Accelerators for Radiation Therapy.* Bristol: IOP Publishing (1997)
13. D. Thwaites, "Quality assurance into the next century". *Radiother. Oncol* **54** vii-ix (2000)
14. M. Krammer, et al., "Status of diamond particle detectors", *Nucl. Instr. Meth. A* **418**, 196–202 (1998)

15. A. Fidanzio, L. Azario, R. Miceli, A. Russo, and A. Piermattei, "PTW-diamond detector: dose rate and particle type dependence". *Med. Phys.* **27** 2589–2593 (2000)
16. H. H. Barrett, and W. Swindell, *Radiological Imaging*. New York: Academic Press (1981)
17. S. Webb, *The Physics of Medical Imaging.* Bristol: IOP Publishing (1988)
18. A. L. Boyer, et al., "A review of electronic portal imaging devices (EPIDs)", *Med. Phys.* **19** 1–16 (1991)
19. M. G. Herman, et al., "Clinical use of electronic portal imaging: Report of AAPM Radiation Therapy Committee Task Group 58". *Med. Phys.* **28** 712–737 (2001)
20. M. Van Herk, "Physical aspects of a liquid-filled ionisation chamber with pulsed polarizing voltage", *Med. Phys.* **18** 692–702 (1991)
21. R. A. Street, *Hydrogenated amorphous silicon*. Cambridge University Press, Cambridge (1991)
22. H. Fricke, and E. J. Hart, *Radiation Dosimetry*, ed. F. J. Attix (1966)
23. ICRU, *Radiation Dosimetry: Electron beams with energies between 1 and 50 MeV. ICRU Report 35*. Bethsada MD: International Commission on Radiation Units and Measurement (1984).
24. M. J. Maryanski,, J. C. Gore, R. P. Kennan, and R. J. Schultz, "NMR relaxation enhancement by ionising radiation: a new approach to 3D dosimetry by MRI". *Mag. Res. Imag.* **11** 253–258 (1993)

ANALYTICAL INDEX